Derivatives

1. $\dfrac{d(au)}{dx} = a\dfrac{du}{dx}$

2. $\dfrac{d(u+v-w)}{dx} = \dfrac{du}{dx} + \dfrac{dv}{dx} - \dfrac{dw}{dx}$

3. $\dfrac{d(uv)}{dx} = u\dfrac{dv}{dx} + v\dfrac{du}{dx}$

4. $\dfrac{d(u/v)}{dx} = \dfrac{v(du/dx) - u(dv/dx)}{v^2}$

5. $\dfrac{d(u^n)}{dx} = nu^{n-1}\dfrac{du}{dx}$

6. $\dfrac{d(u^v)}{dx} = vu^{v-1}\dfrac{du}{dx} + u^v(\ln u)\dfrac{dv}{dx}$

7. $\dfrac{d(e^u)}{dx} = e^u\dfrac{du}{dx}$

8. $\dfrac{d(e^{au})}{dx} = ae^{au}\dfrac{du}{dx}$

9. $\dfrac{da^u}{dx} = a^u(\ln a)\dfrac{du}{dx}$

10. $\dfrac{d(\ln u)}{dx} = \dfrac{1}{u}\dfrac{du}{dx}$

11. $\dfrac{d(\log_a u)}{dx} = \dfrac{1}{u(\ln a)}\dfrac{du}{dx}$

12. $\dfrac{d\sin u}{dx} = \cos u\dfrac{du}{dx}$

13. $\dfrac{d\cos u}{dx} = -\sin u\dfrac{du}{dx}$

14. $\dfrac{d\tan u}{dx} = \sec^2 u\dfrac{du}{dx}$

15. $\dfrac{d\cot u}{dx} = -\csc^2 u\dfrac{du}{dx}$

16. $\dfrac{d\sec u}{dx} = \tan u\sec u\dfrac{du}{dx}$

17. $\dfrac{d\csc u}{dx} = -(\cot u)(\csc u)\dfrac{du}{dx}$

18. $\dfrac{d\sin^{-1}u}{dx} = \dfrac{1}{\sqrt{1-u^2}}\dfrac{du}{dx}$

19. $\dfrac{d\cos^{-1}u}{dx} = \dfrac{-1}{\sqrt{1-u^2}}\dfrac{du}{dx}$

20. $\dfrac{d\tan^{-1}u}{dx} = \dfrac{1}{1+u^2}\dfrac{du}{dx}$

21. $\dfrac{d\cot^{-1}u}{dx} = \dfrac{-1}{1+u^2}\dfrac{du}{dx}$

22. $\dfrac{d\sec^{-1}u}{dx} = \dfrac{1}{u\sqrt{u^2-1}}\dfrac{du}{dx}$

23. $\dfrac{d\csc^{-1}u}{dx} = \dfrac{-1}{u\sqrt{u^2-1}}\dfrac{du}{dx}$

24. $\dfrac{d\sinh u}{dx} = \cosh u\dfrac{du}{dx}$

25. $\dfrac{d\cosh u}{dx} = \sinh u\dfrac{du}{dx}$

26. $\dfrac{d\tanh u}{dx} = \operatorname{sech}^2 u\dfrac{du}{dx}$

27. $\dfrac{d\coth u}{dx} = -(\operatorname{csch}^2 u)\dfrac{du}{dx}$

28. $\dfrac{d\operatorname{sech} u}{dx} = -(\operatorname{sech} u)(\tanh u)\dfrac{du}{dx}$

29. $\dfrac{d\operatorname{csch} u}{dx} = -(\operatorname{csch} u)(\coth u)\dfrac{du}{dx}$

30. $\dfrac{d\sinh^{-1}u}{dx} = \dfrac{1}{\sqrt{1+u^2}}\dfrac{du}{dx}$

31. $\dfrac{d\cosh^{-1}u}{dx} = \dfrac{1}{\sqrt{u^2-1}}\dfrac{du}{dx}$

32. $\dfrac{d\tanh^{-1}u}{dx} = \dfrac{1}{1-u^2}\dfrac{du}{dx}$

33. $\dfrac{d\coth^{-1}u}{dx} = \dfrac{1}{1-u^2}\dfrac{du}{dx}$

34. $\dfrac{d\operatorname{sech}^{-1}u}{dx} = \dfrac{-1}{u\sqrt{1-u^2}}\dfrac{du}{dx}$

35. $\dfrac{d\operatorname{csch}^{-1}u}{dx} = \dfrac{-1}{|u|\sqrt{1+u^2}}\dfrac{du}{dx}$

Continued on overleaf

A Brief Table of Integrals

(An arbitrary constant may be added to each integral.)

1. $\int x^n \, dx = \frac{1}{n+1} x^{n+1} \quad (n \neq -1)$
2. $\int \frac{1}{x} \, dx = \ln|x|$
3. $\int e^x \, dx = e^x$
4. $\int a^x \, dx = \frac{a^x}{\ln a}$
5. $\int \sin x \, dx = -\cos x$
6. $\int \cos x \, dx = \sin x$
7. $\int \tan x \, dx = -\ln|\cos x|$
8. $\int \cot x \, dx = \ln|\sin x|$
9. $\int \sec x \, dx = \ln|\sec x + \tan x| = \ln\left|\tan\left(\frac{1}{2}x + \frac{1}{4}\pi\right)\right|$
10. $\int \csc x \, dx = \ln|\csc x - \cot x| = \ln\left|\tan \frac{1}{2}x\right|$
11. $\int \sin^{-1}\frac{x}{a} \, dx = x \sin^{-1}\frac{x}{a} + \sqrt{a^2 - x^2} \quad (a > 0)$
12. $\int \cos^{-1}\frac{x}{a} \, dx = x \cos^{-1}\frac{x}{a} - \sqrt{a^2 - x^2} \quad (a > 0)$
13. $\int \tan^{-1}\frac{x}{a} \, dx = x \tan^{-1}\frac{x}{a} - \frac{a}{2}\ln(a^2 + x^2) \quad (a > 0)$
14. $\int \sin^2 mx \, dx = \frac{1}{2m}(mx - \sin mx \cos mx)$
15. $\int \cos^2 mx \, dx = \frac{1}{2m}(mx + \sin mx \cos mx)$
16. $\int \sec^2 x \, dx = \tan x$
17. $\int \csc^2 x \, dx = -\cot x$
18. $\int \sin^n x \, dx = -\frac{\sin^{n-1}x \cos x}{n} + \frac{n-1}{n}\int \sin^{n-2}x \, dx$
19. $\int \cos^n x \, dx = \frac{\cos^{n-1}x \sin x}{n} + \frac{n-1}{n}\int \cos^{n-2}x \, dx$
20. $\int \tan^n x \, dx = \frac{\tan^{n-1}x}{n-1} - \int \tan^{n-2}x \, dx \quad (n \neq 1)$
21. $\int \cot^n x \, dx = -\frac{\cot^{n-1}x}{n-1} - \int \cot^{n-2}x \, dx \quad (n \neq 1)$
22. $\int \sec^n x \, dx = \frac{\tan x \sec^{n-2}x}{n-1} + \frac{n-2}{n-1}\int \sec^{n-2}x \, dx \quad (n \neq 1)$
23. $\int \csc^n x \, dx = -\frac{\cot x \csc^{n-2}x}{n-1} + \frac{n-2}{n-1}\int \csc^{n-2}x \, dx \quad (n \neq 1)$
24. $\int \sinh x \, dx = \cosh x$
25. $\int \cosh x \, dx = \sinh x$
26. $\int \tanh x \, dx = \ln|\cosh x|$
27. $\int \coth x \, dx = \ln|\sinh x|$
28. $\int \operatorname{sech} x \, dx = \tan^{-1}(\sinh x)$

This table is continued on the endpapers at the back.

Undergraduate Texts in Mathematics

Jerrold Marsden
Alan Weinstein

CALCULUS II

Second Edition

With 297 Figures

Springer-Verlag New York Berlin Heidelberg Tokyo

Jerrold Marsden
Department of Mathematics
University of California
Berkeley, California 94720
U.S.A.

Alan Weinstein
Department of Mathematics
University of California
Berkeley, California 94720
U.S.A.

AMS Subject Classification: 26-01

Cover photograph by Nancy Williams Marsden.

Library of Congress Cataloging in Publication Data
Marsden, Jerrold E.
Calculus II.
(Undergraduate texts in mathematics)
Includes index.
1. Calculus. II. Weinstein, Alan.
II. Marsden, Jerrold E. Calculus. III. Title.
IV. Title: Calculus two. V. Series.
QA303.M3372 1984b 515 84-5480

Typeset by Computype, Inc., St. Paul, Minnesota.
Printed and bound by Halliday Lithograph, West Hanover, Massachusetts.
Printed in the United States of America.

9 8 7 6 5 4 3 2 1

ISBN 0-387-90975-3 Springer-Verlag New York Berlin Heidelberg Tokyo
ISBN 3-540-90975-3 Springer-Verlag Berlin Heidelberg New York Tokyo

To Nancy and Margo

Preface

The goal of this text is to help students learn to use calculus intelligently for solving a wide variety of mathematical and physical problems.

This book is an outgrowth of our teaching of calculus at Berkeley, and the present edition incorporates many improvements based on our use of the first edition. We list below some of the key features of the book.

Examples and Exercises

The exercise sets have been carefully constructed to be of maximum use to the students. With few exceptions we adhere to the following policies.

- The section *exercises are graded* into three consecutive groups:

(a) The first exercises are routine, modelled almost exactly on the examples; these are intended to give students confidence.
(b) Next come exercises that are still based directly on the examples and text but which may have variations of wording or which combine different ideas; these are intended to train students to think for themselves.
(c) The last exercises in each set are difficult. These are marked with a star (★) and some will challenge even the best students. Difficult does not necessarily mean theoretical; often a starred problem is an interesting application that requires insight into what calculus is really about.

- The *exercises come in groups* of two and often four similar ones.
- *Answers* to odd-numbered exercises are available in the back of the book, and every other odd exercise (that is, Exercise 1, 5, 9, 13, . . .) has a complete solution in the student guide. Answers to even-numbered exercises are not available to the student.

Placement of Topics

Teachers of calculus have their own pet arrangement of topics and teaching devices. After trying various permutations, we have arrived at the present arrangement. Some highlights are the following.

- *Integration* occurs early in Chapter 4; *antidifferentiation* and the $\int$ notation with motivation already appear in Chapter 2.

- *Trigonometric functions* appear in the first semester in Chapter 5.
- The *chain rule* occurs early in Chapter 2. We have chosen to use rate-of-change problems, square roots, and algebraic functions in conjunction with the chain rule. Some instructors prefer to introduce $\sin x$ and $\cos x$ early to use with the chain rule, but this has the penalty of fragmenting the study of the trigonometric functions. We find the present arrangement to be smoother and easier for the students.
- *Limits* are presented in Chapter 1 along with the derivative. However, while we do not try to hide the difficulties, technicalities involving epsilonics are deferred until Chapter 11. (Better or curious students can read this concurrently with Chapter 2.) Our view is that it is very important to teach students to differentiate, integrate, and solve calculus problems as quickly as possible, without getting delayed by the intricacies of limits. After some calculus is learned, the details about limits are best appreciated in the context of l'Hôpital's rule and infinite series.
- *Differential equations* are presented in Chapter 8 and again in Sections 12.7, 12.8, and 18.3. Blending differential equations with calculus allows for more interesting applications early and meets the needs of physics and engineering.

Prerequisites and Preliminaries

A historical introduction to calculus is designed to orient students before the technical material begins.

Prerequisite material from algebra, trigonometry, and analytic geometry appears in Chapters R, 5, and 14. These topics are treated completely: in fact, analytic geometry and trigonometry are treated in enough detail to serve as a first introduction to the subjects. However, high school algebra is only lightly reviewed, and knowledge of some plane geometry, such as the study of similar triangles, is assumed.

Several *orientation quizzes* with answers and a *review section* (Chapter R) contribute to bridging the gap between previous training and this book. Students are advised to assess themselves and to take a pre-calculus course if they lack the necessary background.

Chapter and Section Structure

The book is intended for a three-semester sequence with six chapters covered per semester. (Four semesters are required if pre-calculus material is included.)

The length of chapter sections is guided by the following typical course plan: If six chapters are covered per semester (this typically means four or five student contact hours per week) then approximately two sections must be covered each week. Of course this schedule must be adjusted to students' background and individual course requirements, but it gives an idea of the pace of the text.

Proofs and Rigor

Proofs are given for the most important theorems, with the customary omission of proofs of the intermediate value theorem and other consequences of the completeness axiom. Our treatment of integration enables us to give particularly simple proofs of some of the main results in that area, such as the fundamental theorem of calculus. We de-emphasize the theory of limits, leaving a detailed study to Chapter 11, after students have mastered the fundamentals of calculus—differentiation and integration. Our book *Calculus Unlimited* (Benjamin/Cummings) contains all the proofs omitted in this text and additional ideas suitable for supplementary topics for good students. Other references for the theory are Spivak's *Calculus* (Benjamin/Cummings & Publish or Perish), Ross' *Elementary Analysis: The Theory of Calculus* (Springer) and Marsden's *Elementary Classical Analysis* (Freeman).

Calculators

Calculator applications are used for motivation (such as for functions and composition on pages 40 and 112) and to illustrate the numerical content of calculus (see, for instance, p. 405 and Section 11.5). Special calculator discussions tell how to use a calculator and recognize its advantages and shortcomings.

Applications

Calculus students should not be treated as if they are already the engineers, physicists, biologists, mathematicians, physicians, or business executives they may be preparing to become. Nevertheless calculus is a subject intimately tied to the physical world, and we feel that it is misleading to teach it any other way. Simple examples related to distance and velocity are used throughout the text. Somewhat more special applications occur in examples and exercises, some of which may be skipped at the instructor's discretion. Additional connections between calculus and applications occur in various section supplements throughout the text. For example, the use of calculus in the determination of the length of a day occurs at the end of Chapters 5, 9, and 14.

Visualization

The ability to visualize basic graphs and to interpret them mentally is very important in calculus and in subsequent mathematics courses. We have tried to help students gain facility in forming and using visual images by including plenty of carefully chosen artwork. This facility should also be encouraged in the solving of exercises.

Computer-Generated Graphics

Computer-generated graphics are becoming increasingly important as a tool for the study of calculus. High-resolution plotters were used to plot the graphs of curves and surfaces which arose in the study of Taylor polynomial approximation, maxima and minima for several variables, and three-dimensional surface geometry. Many of the computer drawn figures were kindly supplied by Jerry Kazdan.

Supplements

Student Guide Contains

- Goals and guides for the student
- Solutions to every other odd-numbered exercise
- Sample exams

Instructor's Guide Contains

- Suggestions for the instructor, section by section
- Sample exams
- Supplementary answers

Misprints

Misprints are a plague to authors (and readers) of mathematical textbooks. We have made a special effort to weed them out, and we will be grateful to the readers who help us eliminate any that remain.

Acknowledgments

We thank our students, readers, numerous reviewers and assistants for their help with the first and current edition. For this edition we are especially grateful to Ray Sachs for his aid in matching the text to student needs, to Fred Soon and Fred Daniels for their unfailing support, and to Connie Calica for her accurate typing. Several people who helped us with the first edition deserve our continued thanks. These include Roger Apodaca, Grant Gustafson, Mike Hoffman, Dana Kwong, Teresa Ling, Tudor Ratiu, and Tony Tromba.

Jerry Marsden
Alan Weinstein

Berkeley, California

How to Use this Book: A Note to the Student

Begin by orienting yourself. Get a rough feel for what we are trying to accomplish in calculus by rapidly reading the Introduction and the Preface and by looking at some of the chapter headings.

Next, make a preliminary assessment of your own preparation for calculus by taking the quizzes on pages 13 and 14. If you need to, study Chapter R in detail and begin reviewing trigonometry (Section 5.1) as soon as possible.

You can learn a little bit about calculus by reading this book, but you can learn to use calculus only by practicing it yourself. You should do many more exercises than are assigned to you as homework. The answers at the back of the book and solutions in the student guide will help you monitor your own progress. There are a lot of examples with complete solutions to help you with the exercises. The end of each example is marked with the symbol ▲.

Remember that even an experienced mathematician often cannot "see" the entire solution to a problem at once; in many cases it helps to begin systematically, and then the solution will fall into place.

Instructors vary in their expectations of students as far as the degree to which answers should be simplified and the extent to which the theory should be mastered. In the book we have arranged the theory so that only the proofs of the most important theorems are given in the text; the ends of proofs are marked with the symbol ■. Often, technical points are treated in the starred exercises.

In order to prepare for examinations, try reworking the examples in the text and the sample examinations in the Student Guide without looking at the solutions. Be sure that you can do all of the assigned homework problems.

When writing solutions to homework or exam problems, you should use the English language liberally and correctly. A page of disconnected formulas with no explanatory words is incomprehensible.

We have written the book with your needs in mind. Please inform us of shortcomings you have found so we can correct them for future students. We wish you luck in the course and hope that you find the study of calculus stimulating, enjoyable, and useful.

Jerry Marsden
Alan Weinstein

Contents

Chapter 10
Further Techniques and Applications of Integration

Chapter 11
Limits, L'Hôpital's Rule, and Numerical Methods

Chapter 12
Infinite Series

Contents of Volume I

Contents of Volume III

Chapter 7

Basic Methods of Integration

Learning the art of integration requires practice.

In this chapter, we first collect in a more systematic way some of the integration formulas derived in Chapters 4–6. We then present the two most important general techniques: integration by substitution and integration by parts. As the techniques for evaluating integrals are developed, you will see that integration is a more subtle process than differentiation and that it takes practice to learn which method should be used in a given problem.

7.1 Calculating Integrals

The rules for differentiating the trigonometric and exponential functions lead to new integration formulas.

In this section, we review the basic integration formulas learned in Chapter 4, and we summarize the integration rules for trigonometric and exponential functions developed in Chapters 5 and 6.

Given a function $f(x)$, $\int f(x)\,dx$ denotes the general antiderivative of f, also called the indefinite integral. Thus

$$\int f(x)\,dx = F(x) + C,$$

where $F'(x) = f(x)$ and C is a constant. Therefore,

$$\frac{d}{dx}\int f(x)\,dx = f(x).$$

The definite integral is obtained via the fundamental theorem of calculus by evaluating the indefinite integral at the two limits and subtracting. Thus:

$$\int_a^b f(x)\,dx = F(x)\Big|_a^b = F(b) - F(a).$$

We recall the following general rules for antiderivatives (see Section 2.5), which may be deduced from the corresponding differentiation rules. To check the sum rule, for instance, we must see if

$$\frac{d}{dx}\left[\int f(x)\,dx + \int g(x)\,dx\right] = f(x) + g(x).$$

But this is true by the sum rule for derivatives.

Sum and Constant Multiple Rules for Antiderivatives

$$\int [f(x) + g(x)]\,dx = \int f(x)\,dx + \int g(x)\,dx;$$

$$\int cf(x)\,dx = c\int f(x)\,dx.$$

The antiderivative rule for powers is given as follows:

Power Rule for Antiderivatives

$$\int x^n\,dx = \begin{cases} \dfrac{x^{n+1}}{n+1} + C, & n \neq -1, \\ \ln|x| + C, & n = -1. \end{cases}$$

The power rule for integer n was introduced in Section 2.5, and was extended in Section 6.3 to cover the case $n = -1$ and then to all real numbers n, rational or irrational.

Example 1 Calculate (a) $\int\left(3x^{2/3} + \frac{8}{x}\right)dx$; (b) $\int\left(\frac{x^3 + 8x + 3}{x}\right)dx$; (c) $\int(x^{\pi} + x^3)\,dx$.

Solution (a) By the sum and constant multiple rules,

$$\int\left(3x^{2/3} + \frac{8}{x}\right)dx = 3\int x^{2/3}\,dx + 8\int\frac{1}{x}\,dx.$$

By the power rule, this becomes

$$3\cdot\frac{x^{5/3}}{5/3} + 8\ln|x| + C = \frac{9}{5}x^{5/3} + 8\ln|x| + C.$$

(b) $$\int\frac{x^3 + 8x + 3}{x}\,dx = \int\left(x^2 + 8 + \frac{3}{x}\right)dx = \frac{x^3}{3} + 8x + 3\ln|x| + C.$$

(c) $$\int(x^{\pi} + x^3)\,dx = \frac{x^{\pi+1}}{\pi+1} + \frac{x^4}{4} + C. \ \blacktriangle$$

Applying the fundamental theorem to the power rule, we obtain the rule for definite integrals of powers:

Definite Integral of a Power

$$\int_a^b x^n\,dx = \left.\frac{x^{n+1}}{n+1}\right|_a^b = \frac{b^{n+1} - a^{n+1}}{n+1} \qquad \text{for } n \text{ real}, \quad n \neq -1.$$

If $n = -2, -3, -4, \ldots,$ a and b must have the same sign. If n is not an integer, a and b must be positive (or zero if $n > 0$).

$$\int_a^b \frac{1}{x}\,dx = \ln|x|\Big|_a^b = \ln|b| - \ln|a| = \ln\left(\frac{b}{a}\right).$$

Again a and b must have the same sign.

The extra conditions on a and b are imposed because the integrand must be defined and continuous on the domain of integration; otherwise the fundamental theorem does not apply. (See Exercise 46.)

Example 2 Evaluate (a) $\int_0^1 (x^4 - 3\sqrt{x})\,dx$; (b) $\int_1^2 \left(\sqrt{x} + \frac{2}{x}\right) dx$;

(c) $\int_{1/2}^1 \left(\frac{x^4 + x^6 + 1}{x^2}\right) dx.$

Solution (a)

$$\int_0^1 (x^4 - 3\sqrt{x})\,dx = \int (x^4 - 3\sqrt{x})\,dx\bigg|_0^1 = \frac{x^5}{5} - 3\cdot\frac{x^{3/2}}{3/2}\bigg|_0^1$$

$$= \frac{1}{5} - 2 = -\frac{9}{5}.$$

(b)

$$\int_1^2 \left(\sqrt{x} + \frac{2}{x}\right) dx = \left(\frac{x^{3/2}}{3/2} + 2\ln|x|\right)\bigg|_1^2$$

$$= \frac{2}{3}2^{3/2} + 2\ln 2 - \left(\frac{2}{3} + 0\right) = \frac{4\sqrt{2} - 2}{3} + 2\ln 2.$$

(c)

$$\int_{1/2}^1 \left(\frac{x^4 + x^6 + 1}{x^2}\right) dx = \int_{1/2}^1 \left(x^2 + x^4 + \frac{1}{x^2}\right) dx$$

$$= \left(\frac{x^3}{3} + \frac{x^5}{5} - \frac{1}{x}\right)\bigg|_{1/2}^1$$

$$= \left(\frac{1}{3} + \frac{1}{5} - 1\right) - \left(\frac{1}{3\cdot 8} + \frac{1}{5\cdot 32} - 2\right)$$

$$= \frac{713}{480}. \ \blacktriangle$$

In the following box, we recall some general properties satisfied by the definite integral. These properties were discussed in Chapter 4.

Properties of the Definite Integral

1. *Inequality rule*: If $f(x) \leqslant g(x)$ for all x in $[a,b]$, then

$$\int_a^b f(x)\,dx \leqslant \int_a^b g(x)\,dx.$$

2. *Sum rule*:

$$\int_a^b [f(x) + g(x)]\,dx = \int_a^b f(x)\,dx + \int_a^b g(x)\,dx.$$

3. *Constant multiple rule*:

$$\int_a^b cf(x)\,dx = c\int_a^b f(x)\,dx, \qquad c \text{ a constant.}$$

4. *Endpoint additivity rule*:

$$\int_a^c f(x)\,dx = \int_a^b f(x)\,dx + \int_b^c f(x)\,dx, \qquad a < b < c.$$

5. *Wrong-way integrals*:

$$\int_b^a f(x)\,dx = -\int_a^b f(x)\,dx.$$

If we consider the integral as the area under the graph, then the endpoint additivity rule is just the principle of addition of areas (see Fig. 7.1.1).

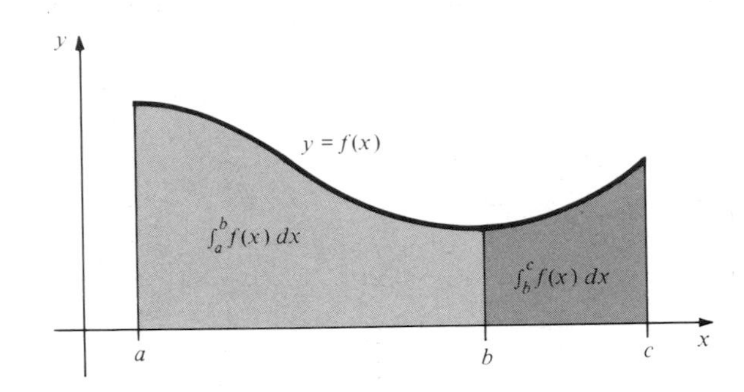

Figure 7.1.1. The area of the entire figure is $\int_a^c f(x)\,dx = \int_a^b f(x)\,dx + \int_b^c f(x)\,dx$, which is the sum of the areas of the two subfigures.

Example 3 Let

$$f(t) = \begin{cases} \frac{1}{2} & 0 \leqslant t < \frac{1}{2}, \\ t, & \frac{1}{2} \leqslant t \leqslant 1. \end{cases}$$

Draw a graph of f and evaluate $\int_0^1 f(t)\,dt$.

Solution The graph of f is drawn in Fig. 7.1.2. To evaluate the integral, we apply the endpoint additivity rule with $a = 0$, $b = \frac{1}{2}$, and $c = 1$:

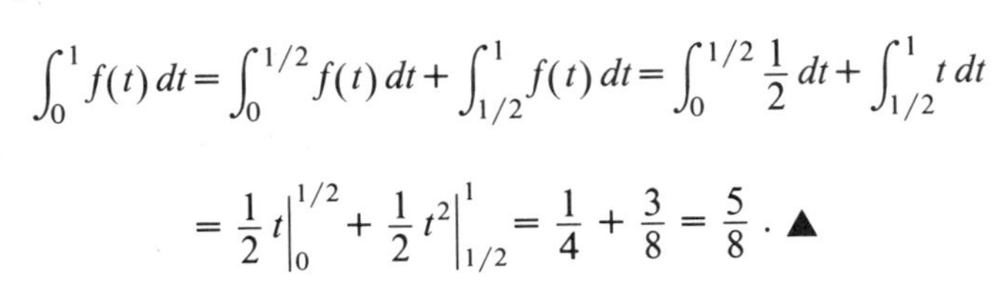

$$\int_0^1 f(t)\,dt = \int_0^{1/2} f(t)\,dt + \int_{1/2}^1 f(t)\,dt = \int_0^{1/2} \frac{1}{2}\,dt + \int_{1/2}^1 t\,dt$$

$$= \frac{1}{2}t\Big|_0^{1/2} + \frac{1}{2}t^2\Big|_{1/2}^1 = \frac{1}{4} + \frac{3}{8} = \frac{5}{8}. \blacktriangle$$

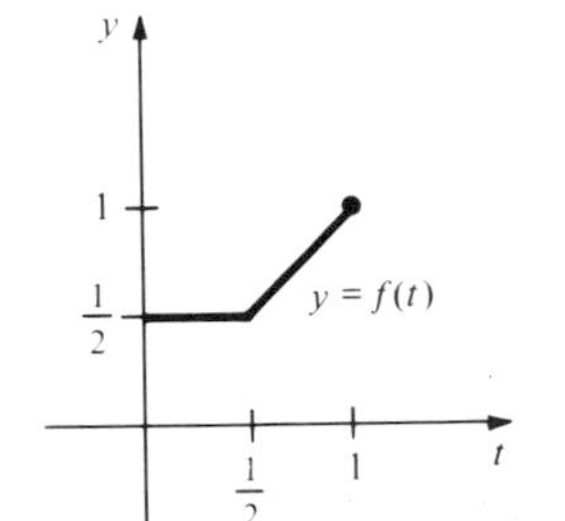

Figure 7.1.2. The integral of f on $[0, 1]$ is the sum of its integrals on $[0, \frac{1}{2}]$ and $[\frac{1}{2}, 1]$.

Let us recall that the alternative form of the fundamental theorem of calculus states that if f is continuous, then

$$\frac{d}{dx}\int_a^x f(t)\,dt = f(x).$$

Example 4 Find $\displaystyle\frac{d}{dt}\int_0^{t^2}\sqrt{1+2s^3}\,ds$.

Solution We write $g(t) = \int_0^{t^2}\sqrt{1+2s^3}\,ds$ as $f(t^2)$, where $f(u) = \int_0^u \sqrt{1+2s^3}\,ds$. By the fundamental theorem (alternative version), $f'(u) = \sqrt{1+2u^3}$; by the chain rule, $g'(t) = f'(t^2)[d(t^2)/dt] = \sqrt{1+2t^6}\cdot 2t$. ▲

As we developed the calculus of the trigonometric and exponential functions, we obtained formulas for the antiderivatives of certain of these functions. For convenience, we summarize those formulas. Here are the formulas from Chapter 5:

Trigonometric Formulas

1. $\int \cos\theta\,d\theta = \sin\theta + C$
2. $\int \sin\theta\,d\theta = -\cos\theta + C$
3. $\int \sec^2\theta\,d\theta = \tan\theta + C$
4. $\int \csc^2\theta\,d\theta = -\cot\theta + C$
5. $\int \tan\theta\sec\theta\,d\theta = \sec\theta + C$
6. $\int \cot\theta\csc\theta\,d\theta = -\csc\theta + C$

Inverse Trigonometric Formulas

1. $\int \frac{dx}{\sqrt{1-x^2}} = \sin^{-1}x + C, \qquad -1 < x < 1.$
2. $\int \frac{-dx}{\sqrt{1-x^2}} = \cos^{-1}x + C, \qquad -1 < x < 1.$
3. $\int \frac{dx}{1+x^2} = \tan^{-1}x + C, \qquad -\infty < x < \infty.$
4. $\int \frac{-dx}{1+x^2} = \cot^{-1}x + C, \qquad -\infty < x < \infty.$
5. $\int \frac{dx}{\sqrt{x^2(x^2-1)}} = \sec^{-1}x + C, \qquad -\infty < x < -1 \text{ or } 1 < x < \infty.$
6. $\int \frac{-dx}{\sqrt{x^2(x^2-1)}} = \csc^{-1}x + C, \qquad -\infty < x < -1 \text{ or } 1 < x < \infty.$

By combining the fundamental theorem of calculus with these formulas and the ones in the tables on the endpapers of this book, we can compute many definite integrals.

Example 5 Evaluate (a) $\int_0^{\pi}(x^4 + 2x + \sin x)\,dx$; (b) $\int_0^{\pi/6}\cos 3x\,dx$; (c) $\int_{-1/2}^{1/2}\frac{dy}{\sqrt{1-y^2}}$.

Solution (a) We begin by calculating the indefinite integral, using the sum and constant multiple rules, the power rule, and the fact that the antiderivative of $\sin x$ is $-\cos x + C$:

$$\int(x^4 + 2x + \sin x)\,dx = \int x^4\,dx + 2\int x\,dx + \int \sin x\,dx$$
$$= x^5/5 + x^2 - \cos x + C.$$

The fundamental theorem then gives

$$\int_0^{\pi}(x^4 + 2x + \sin x)\,dx$$
$$= \left(\frac{x^5}{5} + x^2 - \cos x\right)\Bigg|_0^{\pi} = \frac{\pi^5}{5} + \pi^2 - \cos\pi - (0 + 0 - \cos 0)$$
$$= \frac{\pi^5}{5} + \pi^2 + 1 + 1 = 2 + \pi^2 + \frac{\pi^5}{5} \approx 73.07.$$

(b) An antiderivative of $\cos 3x$ is, by guesswork, $\frac{1}{3}\sin 3x$. Thus

$$\int_0^{\pi/6}\cos 3x\,dx = \frac{1}{3}\sin 3x\Bigg|_0^{\pi/6} = \frac{1}{3}\sin\frac{\pi}{2} = \frac{1}{3}.$$

(c) From the preceding box, we have

$$\int \frac{1}{\sqrt{1-y^2}}\,dy = \sin^{-1}y + C,$$

and so by the fundamental theorem,

$$\int_{-1/2}^{1/2} \frac{1}{\sqrt{1-y^2}}\,dy = \sin^{-1}y\big|_{-1/2}^{1/2} = \sin^{-1}\left(\frac{1}{2}\right) - \sin^{-1}\left(-\frac{1}{2}\right)$$

$$= \frac{\pi}{6} - \left(-\frac{\pi}{6}\right) = \frac{\pi}{3}. \blacktriangle$$

The following box summarizes the antidifferentiation formulas obtained in Chapter 6.

Exponential and Logarithm

$$\int e^x\,dx = e^x + C,$$

$$\int b^x\,dx = \frac{b^x}{\ln b} + C,$$

$$\int \frac{1}{x}\,dx = \ln|x| + C.$$

Example 6 Find (a) $\int_{-1}^{1} 2^x\,dx$; (b) $\int_0^1 (3e^x + 2\sqrt{x})\,dx$; (c) $\int_0^1 2^{2y}\,dy$.

Solution (a) $\displaystyle \int_{-1}^{1} 2^x\,dx = \frac{2^x}{\ln 2}\bigg|_{-1}^{1} = \frac{2}{\ln 2} - \frac{2^{-1}}{\ln 2} = \frac{3}{2\ln 2} \approx 2.164.$

(b)
$$\int_0^1 (3e^x + 2\sqrt{x})\,dx = 3\int_0^1 e^x\,dx + 2\int_0^1 x^{1/2}\,dx$$

$$= 3e^x\big|_0^1 + 2\left(\frac{x^{3/2}}{3/2}\right)\bigg|_0^1$$

$$= 3(e^1 - e^0) + \frac{4}{3}(1^{3/2} - 0^{3/2})$$

$$= 3e - 3 + \frac{4}{3} = 3e - \frac{5}{3} \approx 6.488.$$

(c) By a law of exponents, $2^{2y} = (2^2)^y = 4^y$. Thus,

$$\int_0^1 2^{2y}\,dy = \int_0^1 4^y\,dy = \frac{4^y}{\ln 4}\bigg|_0^1 = \frac{1}{\ln 4}(4-1) = \frac{3}{2\ln 2}.$$

Example 7 (a) Differentiate $x\ln x$. (b) Find $\int \ln x\,dx$. (c) Find $\int_2^5 \ln x\,dx$.

Solution (a) By the product rule for derivatives,

$$\frac{d}{dx}(x\ln x) = \ln x + x\cdot\frac{1}{x} = \ln x + 1.$$

(b) From (a), $\int(\ln x + 1)\,dx = x\ln x + C$. Hence,

$$\int \ln x\,dx = x\ln x - x + C.$$

(c)
$$\int_2^5 \ln x\,dx = (x\ln x - x)\big|_2^5 = (5\ln 5 - 5) - (2\ln 2 - 2)$$

$$= 5\ln 5 - 2\ln 2 - 3. \blacktriangle$$

Finally we recall by means of a few examples how integrals can be used to solve area and rate problems.

Example 8 (a) Find the area between the x axis, the curve $y = 1/x$, and the lines $x = -e^3$ and $x = -e$.
(b) Find the area between the graphs of $\cos x$ and $\sin x$ on $[0, \pi/4]$.

Solution (a) For $-e^3 \leqslant x \leqslant -e$, we notice that $1/x$ is negative. Therefore the graph of $1/x$ lies below the x axis (the graph of $y = 0$), and the area is

$$\int_{-e^3}^{-e}\left(0 - \frac{1}{x}\right)dx = -\ln|x|\Big|_{-e^3}^{-e} = -(\ln e - \ln e^3) = -(1-3) = 2.$$

See Fig. 7.1.3.

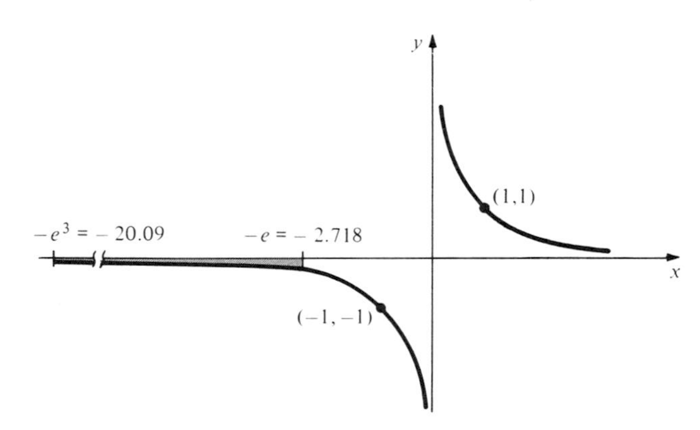

Figure 7.1.3. Find the shaded area.

(b) Since $0 \leqslant \sin x \leqslant \cos x$ for x in $[0, \pi/4]$ (see Fig. 7.1.4), the formula

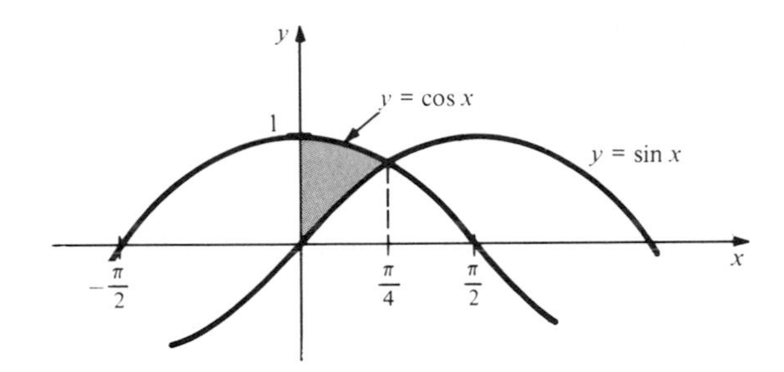

Figure 7.1.4. Find the area of the shaded region.

for the area between two graphs (see Section 4.6) gives

$$\int_0^{\pi/4}(\cos x - \sin x)\,dx = (\sin x + \cos x)\Big|_0^{\pi/4} = \frac{1}{\sqrt{2}} + \frac{1}{\sqrt{2}} - 1 = \sqrt{2} - 1. \blacktriangle$$

Example 9 Water flows into a tank at the rate of $2t + 3$ liters per minute, where t is the time measured in *hours* after noon. If the tank is empty at noon and has a capacity of 1000 liters, when will it be full?

Solution First we should express everything in terms of the same unit of time. Choosing hours, we convert the rate of $2t + 3$ liters per minute to $60(2t + 3) = 120t + 180$ liters per hour. The total amount of water in the tank at time T hours past noon is the integral

$$\int_0^T (120t + 180)\,dt = \frac{120}{2}(T^2 - 0^2) + 180(T - 0) = 60T^2 + 180T.$$

The tank is full when $60T^2 + 180T = 1000$. Solving for T by the quadratic formula, we find $T \approx 2.849$ hours past noon, so the tank is full at 2:51 P.M. $\blacktriangle$

Example 10 Let $P(t)$ denote the population of bacteria in a certain colony at time t. Suppose that $P(0) = 100$ and that P is increasing at a rate of $20e^{3t}$ bacteria per day at time t. How many bacteria are there after 50 days?

Solution We are given $P'(t) = 20e^{3t}$ and $P(0) = 100$. Taking the antiderivative of $P'(t)$ gives $P(t) = \frac{20}{3}e^{3t} + C$. Substituting $P(0) = 100$ gives $C = 100 - \frac{20}{3}$. Hence $P(t) = 100 + \frac{20}{3}(e^{3t} - 1)$, and $P(50) = 100 + \frac{20}{3}(e^{150} - 1) \approx 9.2 \times 10^{65}$ bacteria. (This exceeds the number of atoms in the universe, so growth cannot go on at such a rate and our model for bacterial growth must become invalid.) ▲

Exercises for Section 7.1

Evaluate the indefinite integrals in Exercises 1–8.

1. $\int (3x^2 + 2x + x^{-3})\,dx$
2. $\int (8x^2 + 3x^{-4} + x^{-8})\,dx$
3. $\int (e^x + 2x)\,dx$
4. $\int (e^{-2x} - 8x^2)\,dx$
5. $\int (\sin 2x + 3x)\,dx$
6. $\int (\cos 3x - 2x + 1)\,dx$
7. $\int (e^{-x} + 2\cos x + 5x^2)\,dx$
8. $\int (e^{3x} - 8\sin 2x + x^{-4})\,dx$

Evaluate the definite integrals in Exercises 9–34.

9. $\int_{-2}^{2} (x^8 + 2x^2 - 1)\,dx$
10. $\int_{-2}^{2} (x^{16} + x^9)\,dx$
11. $\int_3^6 (1 - y + y^2)\,dy$
12. $\int_0^a (6x^2 + 3x + 2)\,dx$
13. $\int_{16}^{81} \sqrt[4]{s}\,ds$
14. $\int_1^{81} s\sqrt[4]{s}\,ds$
15. $\int_{-4}^{-3} \frac{1}{r^2}\,dr$
16. $\int_1^3 \left(\frac{1}{x^2} + \frac{1}{x^3}\right)dx$
17. $\int_{-\pi}^{\pi} \cos x\,dx$
18. $\int_0^{\pi/2} \sin 5x\,dx$
19. $\int_0^{\pi} (3\sin\theta + 4\cos\theta)\,d\theta$
20. $\int_0^{\pi/2} (3\sin 4x + 4\cos 3x)\,dx$
21. $\int_0^1 \frac{3}{x^2 + 1}\,dx$
22. $\int_0^1 \frac{ds}{1 + s^2}$
23. $\int_{\sqrt{2}}^{2} \frac{du}{u\sqrt{u^2 - 1}}$
24. $\int_0^{\sqrt{2}/2} (4 - 4s^2)^{-1/2}\,ds$
25. $\int_0^{\pi/4} \sec^2 x\,dx$
26. $\int_0^{\pi/4} \left(e^x - \frac{3}{\cos^2 x}\right)dx$
27. $\int_1^2 (e^{3x} + x^{2/3})\,dx$
28. $\int_{-1}^{1} e^{-4x}\,dx$
29. $\int_1^5 \frac{1}{t}\,dt$
30. $\int_{3/2}^{2} \frac{dx}{20x}$
31. $\int_1^2 \frac{1 + 2x + 3x^2 + 4x^3}{x^4}\,dx$
32. $\int_1^2 \left[x + \left(\frac{1}{x}\right)\right]^2 dx$
33. $\int_{-200}^{200} (90x^{21} - 80x^{33} + 5580x^{97} + 1)\,dx$
34. $\int_{-243.8}^{243.8} (65x^{73} + 48x^{29} - 3x^{13} + 15x^5 - 2x)\,dx$.
35. Check the formula
$$\int x\sqrt{1 + x}\,dx = \frac{2}{15}(3x - 2)(1 + x)^{3/2} + C$$
and evaluate $\int_0^3 x\sqrt{1 + x}\,dx$.
36. (a) Check the integral
$$\int \frac{1}{x\sqrt{x - 1}}\,dx = 2\tan^{-1}\sqrt{x - 1} + C.$$
(b) Evaluate $\int_2^4 (1/x\sqrt{x - 1})\,dx$.
37. (a) Verify that $\int xe^{x^2}\,dx = \frac{1}{2}e^{x^2} + C$.
(b) Evaluate $\int_1^e (2xe^{x^2} + 3\ln x)\,dx$ (see Example 7).

38. (a) Verify the formula

$$\int\left[\frac{\sqrt{x^2-1}}{x}\right]dx=\sqrt{x^2-1}-\sec^{-1}x+C.$$

(b) Evaluate $\int_1^{3/2}[\sqrt{x^2-1}/x]\,dx$.

39. Suppose that $\int_0^2 f(t)\,dt=5$, $\int_2^5 f(t)\,dt=6$, and $\int_0^7 f(t)\,dt=3$. (a) Find $\int_0^5 f(t)\,dt$. (b) Find $\int_5^7 f(t)\,dt$. (c) Show that $f(t)<0$ for some t in $(5,7)$.

40. Find $\int_1^3[4f(s)+3/\sqrt[3]{s}\,]\,ds$, where $\int_1^3 f(s)\,ds=6$.

41. Find $\dfrac{d}{dt}\int_{t^2}^4\sqrt{e^x+\sin 5x^2}\,dx$.

42. Compute $\dfrac{d}{d\alpha}\int_0^{\alpha^2}(\sin^2 t+e^{\cos t})^{3/2}\,dt$.

43. Let

$$f(t)=\begin{cases}2 & -1\leqslant t<0,\\ t & 0\leqslant t\leqslant 2,\\ -1 & 2<t\leqslant 3.\end{cases}$$

Compute $\int_{-1}^3 f(t)\,dt$.

44. Let

$$h(x)=\begin{cases}x & 0\leqslant x<\frac{3}{4},\\ \frac{3}{4} & \frac{3}{4}\leqslant x<1.\end{cases}$$

Compute $\int_0^1 h(x)\,dx$.

45. Let $f(x)=\sin x$,

$$g(x)=\begin{cases}1 & -\pi\leqslant x\leqslant 2,\\ 2 & 2<x\leqslant\pi,\end{cases}$$

and $h(x)=1/x^2$. Find:

(a) $\int_{-\pi/2}^{\pi/2}f(x)g(x)\,dx$; (b) $\int_1^3 g(x)h(x)\,dx$;

(c) $\int_{\pi/2}^x f(t)g(t)\,dt$, for x in $(0,\pi]$. Draw a graph of this function of x.

46. We have $1/x^4>0$ for all x. On the other hand, $\int(dx/x^4)=\int x^{-4}\,dx=(x^{-3}/-3)+C$, so

$$\int_{-1}^1\frac{dx}{x^4}=\frac{1^{-3}-(-1)^{-3}}{-3}=\frac{1+1}{-3}=-\frac{2}{3}.$$

How can a positive function have a negative integral?

Find the area under the graph of each of the functions in Exercises 47–50 on the stated interval.

47. $\dfrac{x^3+1}{x^2+1}$ on $[0,2]$. [*Hint:* Divide.]

48. $\dfrac{1}{x^2+1}$ on $[0,2]$.

49. $\dfrac{x^2+2}{\sqrt{x}}$ on $[1,4]$.

50. $\sin x-\cos 2x$ on $\left[\dfrac{\pi}{4},\dfrac{\pi}{2}\right]$.

51. Find the area under the graph of $y=e^{2x}$ between $x=0$ and $x=1$.

52. A region containing the origin is cut out by the curves $y=1/\sqrt{x}$, $y=-1/\sqrt{x}$, $y=1/\sqrt{-x}$, and $y=-1/\sqrt{-x}$ and the lines $x=\pm 4$, $y=\pm 4$; see Fig. 7.1.5. Find the area of this region.

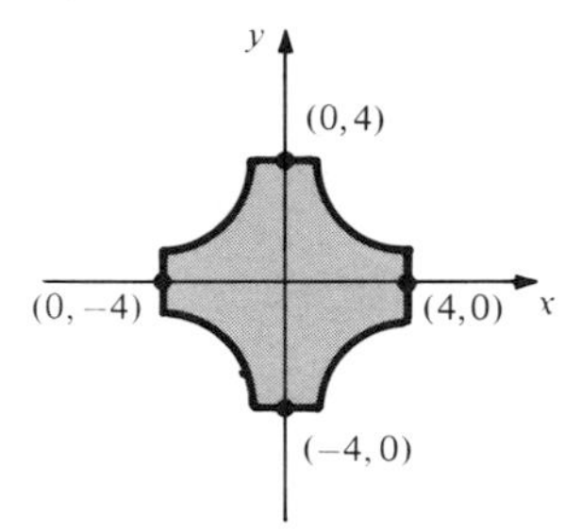

Figure 7.1.5. Find the area of the shaded region.

53. Find the area of the shaded region in Fig. 7.1.6.

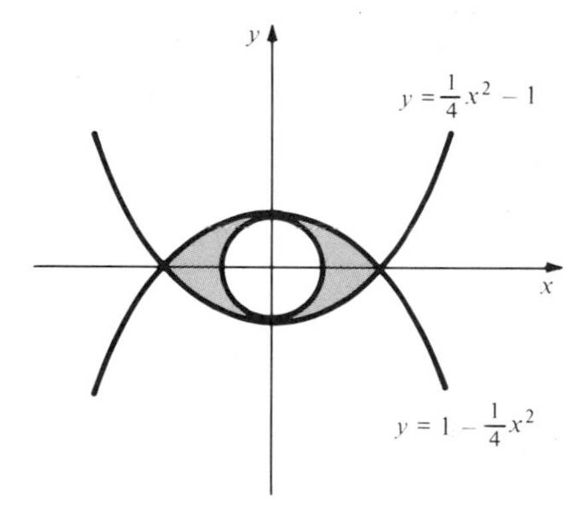

Figure 7.1.6. Find the area of the "retina."

54. Find the area of the shaded "flower" in Fig. 7.1.7.

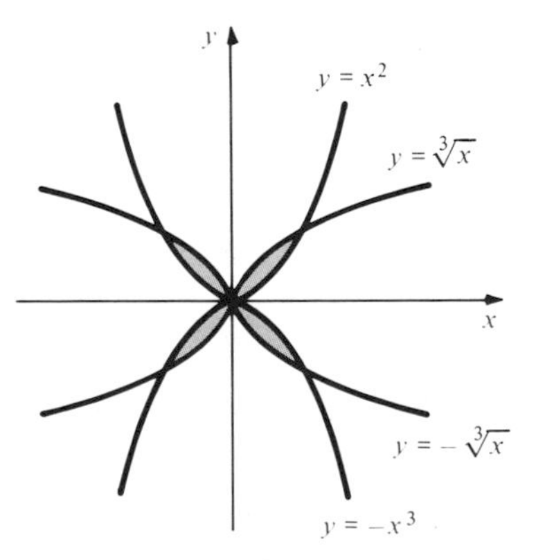

Figure 7.1.7. Find the shaded area.

55. Illustrate in terms of areas the fact that

$$\int_0^{n\pi}\sin x\,dx=\begin{cases}2, & \text{if } n \text{ is an odd positive integer;}\\ 0, & \text{if } n \text{ is an even positive integer.}\end{cases}$$

56. Find the area of the shaded region in Fig. 7.1.8.

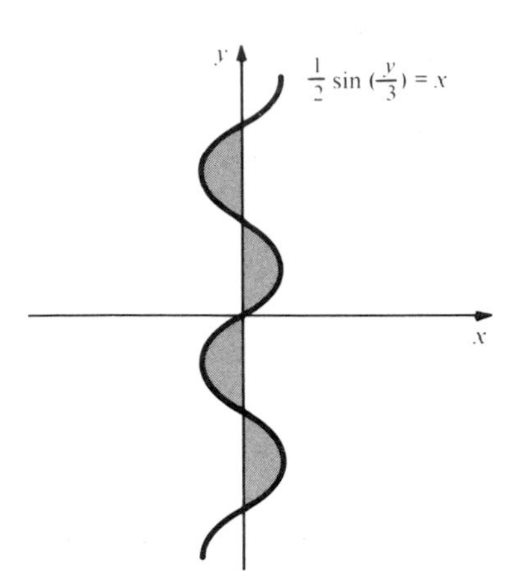

Figure 7.1.8. Find the area of the shaded region.

57. Assuming without proof that

$$\int_0^{\pi/2}\sin^2 x\,dx = \int_0^{\pi/2}\cos^2 x\,dx \quad \text{(see Fig. 7.1.9)},$$

find $\int_0^{\pi/2}\sin^2 x\,dx$. (*Hint*: $\sin^2 x + \cos^2 x = 1$.)

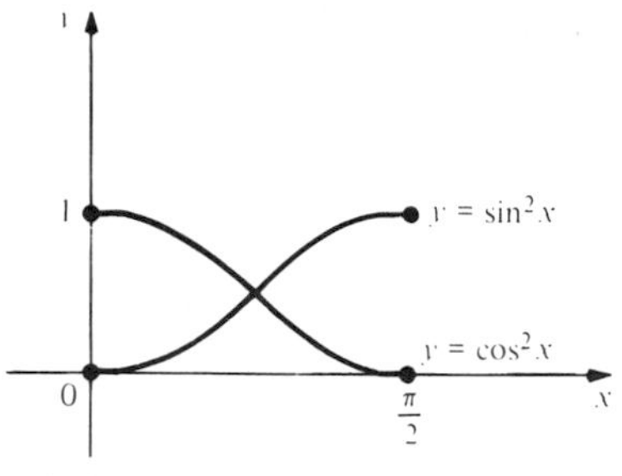

Figure 7.1.9. The areas under the graphs of $\sin^2 x$ and $\cos^2 x$ on $[0, \pi/2]$ are equal.

58. Find:
   (a) $\int \cos 2x\,dx$;
   (b) $\int (\cos^2 x - \sin^2 x)\,dx$;
   (c) $\int (\cos^2 x + \sin^2 x)\,dx$;
   (d) $\int \cos^2 x\,dx$ (use parts (b) and (c));
   (e) $\int_0^{\pi/2}\cos^2 x\,dx$ and $\int_0^{\pi/2}\sin^2 x\,dx$ (compare with Exercise 57).

59. (a) Show that $\int \sin t\cos t\,dt = \frac{1}{2}\sin^2 t + C$.
   (b) Using the identity $\sin 2t = 2\sin t\cos t$, show that $\int \sin t\cos t\,dt = -\frac{1}{4}\cos 2t + C$.
   (c) Use each of parts (a) and (b) to compute $\int_{\pi/6}^{\pi/4}\sin t\cos t\,dt$. Compare your answers.

60. Find the area of the shaded region in Fig. 7.1.10.

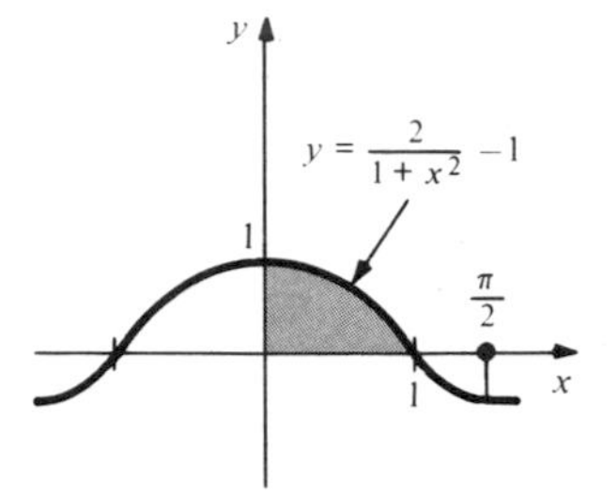

Figure 7.1.10. Find the shaded area.

61. Show that the area under the graph of $f(x) = 1/(1 + x^2)$ on $[a, b]$ is less than π, no matter what the values of a and b may be.
62. Show: the area under the graph of $1/(x^2 + x^6)$ between $x = 2$ and $x = 3$ is smaller than $\frac{1}{68}$.
63. A particle starts at the origin and has velocity $v(t) = 7 + 4t^3 + 6\sin(\pi t)$ centimeters per second after t seconds. Find the distance travelled in 200 seconds.
64. The sales of a clothing company t days after January 1 are given by $S(t) = 260e^{(0.1)t}$ dollars per day.
   (a) Set up a definite integral which gives the accumulated sales on $0 \leqslant t \leqslant 10$.
   (b) Find the accumulated sales for the first 10 days.
   (c) How many days must pass before sales exceed \$900 per day?
65. Each unit in a four-plex rents for \$230/month. The owner will trade the property in five years. He wants to know the *capital value* of the property over a five-year period for continuous interest of 8.25%, that is, the amount he could borrow *now* at 8.25% continuous interest, to be paid back by the rents over the next five years. This amount A is given by $A = \int_0^T Re^{-kt}\,dt$, where R = annual rents, k = annual continuous interest rate, T = period in years.
   (a) Verify that $A = (R/k)(1 - e^{-kT})$.
   (b) Find A for the four-plex problem.
66. The *strain energy* V_e for a simply supported uniform beam with a load P at its center is

$$V_e = \frac{1}{EI}\int_0^{l/2}\left(\frac{Px}{2}\right)^2 dx.$$

The *flexural rigidity* EI and the *bar length* l are constants, $EI \neq 0$ and $l > 0$. Find V_e.
67. A manufacturer determines by curve-fitting methods that its marginal revenue is given by $R'(t) = 1000e^{t/2}$ and its marginal cost by $C'(t) = 1000 - 2t$, t days after January 1. The revenue and cost are in dollars.
   (a) Suppose $R(0) = 0$, $C(0) = 0$. Find, by means of integration, formulas for $R(t)$ and $C(t)$.
   (b) The total profit is $P = R - C$. Find the total profit for the first seven days.

68. The probability P that a capacitor manufactured by an electronics company will last between three and five years with normal use is given approximately by $P = \int_3^5 (22.05)t^{-3}\,dt$.
    (a) Find the probability P.
    (b) Verify that $\int_3^7 (22.05)t^{-3}\,dt = 1$, which says that all capacitors have expected life between three and seven years.

69. Using the identity $\frac{1}{t} - \frac{1}{t+1} = \frac{1}{t(t+1)}$, find $\int_1^e \frac{dt}{t(t+1)}$.

★70. Compute $\int \frac{dt}{t^2(t+1)}$ by writing
$$\frac{1}{t^2(t+1)} = \frac{A}{t} + \frac{B}{t^2} + \frac{C}{(t+1)}$$
for suitable constants A, B, C.

7.2 Integration by Substitution

Integrating the chain rule leads to the method of substitution.

The method of integration by substitution is based on the chain rule for differentiation. If F and g are differentiable functions, the chain rule tells us that $(F \circ g)'(x) = F'(g(x))g'(x)$; that is, $F(g(x))$ is an antiderivative of $F'(g(x))g'(x)$. In indefinite integral notation, we have
$$\int F'(g(x))g'(x)\,dx = F(g(x)) + C.$$
As in differentiation, it is convenient to introduce an intermediate variable $u = g(x)$; then the preceding formula becomes
$$\int F'(u)\frac{du}{dx}\,dx = F(u) + C.$$
If we write $f(u)$ for $F'(u)$, so that $\int f(u)\,du = F(u) + C$, we obtain the formula
$$\int f(u)\frac{du}{dx}\,dx = \int f(u)\,du. \tag{1}$$
This formula is easy to remember, since one may "cancel the dx's."

To apply the method of substitution one must find in a given integrand an expression $u = g(x)$ whose derivative $du/dx = g'(x)$ also occurs in the integrand.

Example 1 Find $\int 2x\sqrt{x^2+1}\,dx$ and check the answer by differentiation.

Solution None of the rules in Section 7.1 apply to this integral, so we try integration by substitution. Noticing that $2x$, the derivative of $x^2 + 1$, occurs in the integrand, we are led to write $u = x^2 + 1$; then we have
$$\int 2x\sqrt{x^2+1}\,dx = \int \sqrt{x^2+1}\cdot 2x\,dx = \int \sqrt{u}\left(\frac{du}{dx}\right)dx.$$
By formula (1), the last integral equals $\int \sqrt{u}\,du = \int u^{1/2}du = \frac{2}{3}u^{3/2} + C$. At this point we substitute $x^2 + 1$ for u, which gives
$$\int 2x\sqrt{x^2+1}\,dx = \tfrac{2}{3}(x^2+1)^{3/2} + C.$$
Checking our answer by differentiating has educational as well as insur-

ance value, since it will show how the chain rule produces the integrand we started with:

$$\frac{d}{dx}\left[\frac{2}{3}(x^2+1)^{3/2}+C\right]=\frac{2}{3}\cdot\frac{3}{2}(x^2+1)^{1/2}\frac{d}{dx}(x^2+1)=\left[\sqrt{x^2+1}\,\right]2x,$$

as it should be. ▲

Sometimes the derivative of the intermediate variable is "hidden" in the integrand. If we are clever, however, we can still use the method of substitution, as the next example shows.

Example 2 Find $\int \cos^2 x \sin x\, dx$.

Solution We are tempted to make the substitution $u = \cos x$, but du/dx is then $-\sin x$ rather than $\sin x$. No matter—we can rewrite the integral as

$$\int(-\cos^2 x)(-\sin x)\,dx.$$

Setting $u = \cos x$, we have

$$\int -u^2\frac{du}{dx}\,dx=\int -u^2\,du=-\frac{u^3}{3}+C,$$

so

$$\int \cos^2 x \sin x\,dx=-\tfrac{1}{3}\cos^3 x+C.$$

You may check this by differentiating. ▲

Example 3 Find $\int \frac{e^x}{1+e^{2x}}\,dx$.

Solution We cannot just let $u = 1 + e^{2x}$, because $du/dx = 2e^{2x} \neq e^x$; but we may recognize that $e^{2x} = (e^x)^2$ and remember that the derivative of e^x is e^x. Making the substitution $u = e^x$ and $du/dx = e^x$, we have

$$\begin{aligned}\int\frac{e^x}{1+e^{2x}}\,dx&=\int\frac{1}{1+(e^x)^2}\cdot e^x\,dx\\&=\int\frac{1}{1+u^2}\cdot\frac{du}{dx}\cdot dx=\int\frac{1}{1+u^2}\,du\\&=\tan^{-1}u+C=\tan^{-1}(e^x)+C.\end{aligned}$$

Again you should check this by differentiation. ▲

We may summarize the method of substitution as developed so far (see Fig. 7.2.1).

Integration by Substitution

To integrate a function which involves an intermediate variable u and its derivative du/dx, write the integrand in the form $f(u)(du/dx)$, incorporating constant factors as required in $f(u)$. Then apply the formula

$$\int f(u)\frac{du}{dx}\,dx=\int f(u)\,du.$$

Finally, evaluate $\int f(u)\,du$ if you can; then substitute for u its expression in terms of x.

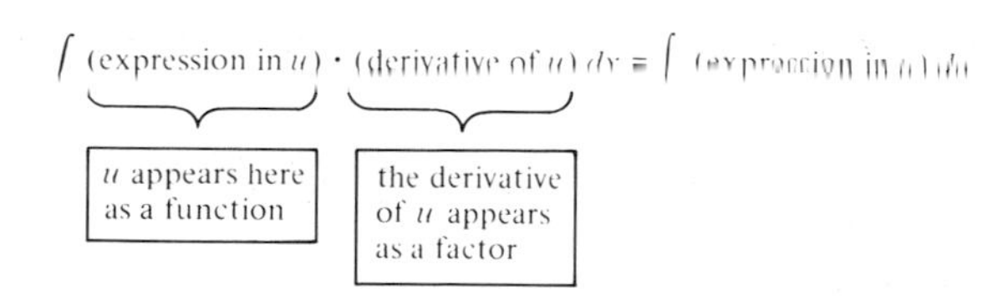

Figure 7.2.1. How to spot u in a substitution problem.

Example 4 Find (a) $\int x^2\sin(x^3)\,dx$, (b) $\int \sin 2x\,dx$.

Solution (a) We observe that the factor x^2 is, apart from a factor of 3, the derivative of x^3. Substitute $u = x^3$, so $du/dx = 3x^2$ and $x^2 = \frac{1}{3}\,du/dx$. Thus

$$\int x^2\sin(x^3)\,dx = \int \frac{1}{3}\frac{du}{dx}\sin u\,dx = \frac{1}{3}\int(\sin u)\frac{du}{dx}\,dx$$
$$= \frac{1}{3}\int \sin u\,du = -\frac{1}{3}\cos u + C.$$

Hence $\int x^2\sin(x^3)\,dx = -\frac{1}{3}\cos(x^3) + C$.
(b) Substitute $u = 2x$, so $du/dx = 2$. Then

$$\int \sin 2x\,dx = \int \frac{1}{2}(\sin 2x)2\,dx = \frac{1}{2}\int \sin u\frac{du}{dx}\,dx$$
$$= \frac{1}{2}\int \sin u\,du = -\frac{1}{2}\cos u + C.$$

Thus

$$\int \sin 2x\,dx = -\frac{1}{2}\cos 2x + C. \blacktriangle$$

Example 5 Evaluate: (a) $\int \frac{x^2}{x^3+5}\,dx$, (b) $\int \frac{dt}{t^2-6t+10}$ [*Hint*: Complete the square in the denominator], and (c) $\int \sin^2 2x\cos 2x\,dx$.

Solution (a) Set $u = x^3 + 5$; $du/dx = 3x^2$. Then

$$\int \frac{x^2}{x^3+5}\,dx = \int \frac{1}{3(x^3+5)}3x^2\,dx = \frac{1}{3}\int \frac{1}{u}\frac{du}{dx}\,dx$$
$$= \frac{1}{3}\int \frac{du}{u} = \frac{1}{3}\ln|u| + C = \frac{1}{3}\ln|x^3+5| + C.$$

(b) Completing the square (see Section R.1), we find

$$t^2 - 6t + 10 = (t^2 - 6t + 9) - 9 + 10$$
$$= (t-3)^2 + 1$$

We set $u = t - 3$; $du/dt = 1$. Then

$$\int \frac{dt}{t^2-6t+10} = \int \frac{dt}{1+(t-3)^2} = \int \frac{1}{1+u^2}\frac{du}{dt}\,dt$$
$$= \int \frac{1}{1+u^2}\,du = \tan^{-1}u + C,$$

so

$$\int \frac{dt}{t^2-6t+10} = \tan^{-1}(t-3) + C.$$

(c) We first substitute $u = 2x$, as in Example 4(b). Since $du/dx = 2$,

$$\int \sin^2 2x\cos 2x\,dx = \int \sin^2 u\cos u\frac{1}{2}\frac{du}{dx}\,dx = \frac{1}{2}\int \sin^2 u\cos u\,du.$$

At this point, we notice that another substitution is appropriate: we set $s = \sin u$ and $ds/du = \cos u$. Then

$$\frac{1}{2}\int \sin^2 u \cos u \, du = \frac{1}{2}\int s^2 \frac{ds}{du}\, du = \frac{1}{2}\int s^2\, ds$$
$$= \frac{1}{2}\cdot\frac{1}{3}s^3 + C = \frac{s^3}{6} + C.$$

Now we must put our answer in terms of x. Since $s = \sin u$ and $u = 2x$, we have

$$\int \sin^2 2x \cos 2x \, dx = \frac{s^3}{6} + C = \frac{\sin^3 u}{6} + C = \frac{\sin^3 2x}{6} + C.$$

You should check this formula by differentiating.

You may have noticed that we could have done this problem in one step by substituting $u = \sin 2x$ in the beginning. We did the problem the long way to show that you can solve an integration problem even if you do not see everything at once. ▲

Two simple substitutions are so useful that they are worth noting explicitly. We have already used them in the preceding examples. The first is the *shifting rule*, obtained by the substitution $u = x + a$, where a is a constant. Here $du/dx = 1$.

Shifting Rule

To evaluate $\int f(x+a)\,dx$, first evaluate $\int f(u)\,du$, then substitute $x + a$ for u:

$$\int f(x+a)\,dx = F(x+a) + C, \qquad \text{where } F(u) = \int f(u)\,du.$$

The second rule is the *scaling rule*, obtained by substituting $u = bx$, where b is a constant. Here $du/dx = b$. The substitution corresponds to a change of scale on the x axis.

Scaling Rule

To evaluate $\int f(bx)\,dx$, evaluate $\int f(u)\,du$, divide by b and substitute bx for u:

$$\int f(bx)\,dx = \frac{1}{b}F(bx) + C, \qquad \text{where } F(u) = \int f(u)\,du.$$

Example 6 Find (a) $\int \sec^2(x+7)\,dx$ and (b) $\int \cos 10x\,dx$.

Solution (a) Since $\int \sec^2 u\,du = \tan u + C$, the shifting rule gives

$$\int \sec^2(x+7)\,dx = \tan(x+7) + C.$$

(b) Since $\int \cos u\,du = \sin u + C$, the scaling rule gives

$$\int \cos 10x\,dx = \tfrac{1}{10}\sin(10x) + C. \text{ ▲}$$

You do not need to memorize the shifting and scaling rules as such; however, the underlying substitutions are so common that you should learn to use them rapidly and accurately.

To conclude this section, we shall introduce a useful device called *differential notation*, which makes the substitution process more mechanical. In particular, this notation helps keep track of the constant factors which must be distributed between the $f(u)$ and du/dx parts of the integrand. We illustrate the device with an example before explaining why it works.

Example 7 Find $\int \frac{x^4+2}{(x^5+10x)^5}\,dx$.

Solution We wish to substitute $u = x^5 + 10x$; note that $du/dx = 5x^4 + 10$. Pretending that du/dx is a fraction, we may "solve for dx," writing $dx = du/(5x^4+10)$. Now we substitute u for $x^5 + 10x$ and $du/(5x^4+10)$ for dx in our integral to obtain

$$\int \frac{x^4+2}{(x^5+10x)^5}\,dx = \int \frac{x^4+2}{u^5}\,\frac{du}{5x^4+10} = \int \frac{x^4+2}{5(x^4+2)}\,\frac{du}{u^5} = \int \frac{1}{5}\,\frac{du}{u^5}\,.$$

Notice that the (x^4+2)'s cancelled, leaving us an integral in u which we can evaluate:

$$\frac{1}{5}\int \frac{du}{u^5} = \frac{1}{5}\left(-\frac{1}{4}u^{-4}\right) + C = -\frac{1}{20u^4} + C.$$

Substituting $x^5 + 10x$ for u gives

$$\int \frac{x^4+2}{(x^5+10x)^5}\,dx = -\frac{1}{20(x^5+10x)^4} + C. \blacktriangle$$

Although du/dx is not really a fraction, we can still justify "solving for dx" when we integrate by substitution. Suppose that we are trying to integrate $\int h(x)\,dx$ by substituting $u = g(x)$. Solving $du/dx = g'(x)$ for dx amounts to replacing dx by $du/g'(x)$ and hence writing

$$\int h(x)\,dx = \int \frac{h(x)}{g'(x)}\,du. \tag{2}$$

Now suppose that we can express $h(x)/g'(x)$ in terms of u, i.e., $h(x)/g'(x) = f(u)$ for some function f. Then we are saying that $h(x) = f(u)g'(x) = f(g(x))g'(x)$, and equation (2) just says

$$\int f(g(x))g'(x)\,dx = \int f(u)\,du,$$

which is the form of integration by substitution we have been using all along.

Example 8 Find $\int \left(\frac{e^{1/x}}{x^2}\right) dx$.

Solution Let $u = 1/x$; $du/dx = -1/x^2$ and $dx = -x^2\,du$, so

$$\int \left(\frac{1}{x^2}\right) e^{1/x}\,dx = \int \left(\frac{1}{x^2}\right) e^u(-x^2\,du) = -\int e^u\,du = -e^u + C$$

and therefore

$$\int \left(\frac{1}{x^2}\right) e^{1/x}\,dx = -e^{1/x} + C. \blacktriangle$$

Integration by Substitution (Differential Notation)

To integrate $\int h(x)\,dx$ by substitution:

1. Choose a new variable $u = g(x)$.
2. Differentiate to get $du/dx = g'(x)$ and then solve for dx.
3. Replace dx in the integral by the expression found in step 2.
4. Try to express the new integrand completely in terms of u, eliminating x. (If you cannot, try another substitution or another method.)
5. Evaluate the new integral $\int f(u)\,du$ (if you can).
6. Express the result in terms of x.
7. Check by differentiating.

Example 9 (a) Calculate the following integrals: (a) $\displaystyle\int \frac{x^2+2x}{\sqrt[3]{x^3+3x^2+1}}\,dx$, (b) $\int \cos x[\cos(\sin x)]\,dx$, and (c) $\displaystyle\int\left(\frac{\sqrt{1+\ln x}}{x}\right)dx$.

Solution (a) Let $u = x^3+3x^2+1$; $du/dx = 3x^2+6x$, so $dx = du/(3x^2+6x)$ and

$$\int \frac{x^2+2x}{\sqrt[3]{x^3+3x^2+1}}\,dx = \int \frac{1}{\sqrt[3]{u}}\,\frac{x^2+2x}{3x^2+6x}\,du$$

$$= \frac{1}{3}\int \frac{1}{\sqrt[3]{u}}\,du = \frac{1}{3}\cdot\frac{3}{2}u^{2/3} + C.$$

Thus

$$\int \frac{x^2+2x}{\sqrt[3]{x^3+3x^2+1}}\,dx = \frac{1}{2}(x^3+3x^2+1)^{2/2} + C.$$

(b) Let $u = \sin x$; $du/dx = \cos x$, $dx = du/\cos x$, so

$$\int \cos x\big[\cos(\sin x)\big]\,dx = \int \cos x\big[\cos(\sin x)\big]\frac{du}{\cos x}$$

$$= \int \cos u\,du = \sin u + C,$$

and therefore

$$\int \cos x\big[\cos(\sin x)\big]\,dx = \sin(\sin x) + C.$$

(c) Let $u = 1+\ln x$; $du/dx = 1/x$, $dx = x\,du$, so

$$\int \frac{\sqrt{1+\ln x}}{x}\,dx = \int \frac{\sqrt{1+\ln x}}{x}(x\,du) = \int u^{1/2}\,du = \frac{2}{3}u^{3/2} + C,$$

and therefore

$$\int \frac{\sqrt{1+\ln x}}{x}\,dx = \frac{2}{3}(1+\ln x)^{3/2} + C. \blacktriangle$$

Exercises for Section 7.2

Evaluate each of the integrals in Exercises 1–6 by making the indicated substitution, and check your answers by differentiating.

1. $\int 2x(x^2+4)^{3/2}\,dx$; $u = x^2+4$.
2. $\int (x+1)(x^2+2x-4)^{-4}\,dx$; $u = x^2+2x-4$.
3. $\int \frac{2y^7+1}{(y^8+4y-1)^2}\,dy$; $x = y^8+4y-1$.
4. $\int \frac{x}{1+x^4}\,dx$; $u = x^2$.
5. $\int \frac{\sec^2\theta}{\tan^3\theta}\,d\theta$; $u = \tan\theta$.
6. $\int \tan x\,dx$; $u = \cos x$.

Evaluate each of the integrals in Exercises 7–22 by the method of substitution, and check your answer by differentiating.

7. $\int (x+1)\cos(x^2+2x)\,dx$
8. $\int u\sin(u^2)\,du$
9. $\int \frac{x^3}{\sqrt{x^4+2}}\,dx$
10. $\int \frac{x}{(x^2+3)^2}\,dx$
11. $\int \frac{t^{1/3}}{(t^{4/3}+1)^{3/2}}\,dt$
12. $\int \frac{x^{1/2}}{(x^{3/2}+2)^2}\,dx$
13. $\int 2r\sin(r^2)\cos^3(r^2)\,dr$
14. $\int e^{\sin x}\cos x\,dx$
15. $\int \frac{x^3}{1+x^8}\,dx$
16. $\int \frac{dx}{\sqrt{1-4x^2}}$
17. $\int \sin(\theta+4)\,d\theta$
18. $\int \frac{1}{x^2}\sin\frac{1}{x}\,dx$
19. $\int (5x^4+1)(x^5+x)^{100}\,dx$
20. $\int (1+\cos s)\sqrt{s+\sin s}\,ds$
21. $\int \left(\frac{t+1}{\sqrt{t^2+2t+3}}\right)dt$
22. $\int \frac{dx}{x^2+4}$

Evaluate the indefinite integrals in Exercises 23–36.

23. $\int t\sqrt{t^2+1}\,dt$.
24. $\int t\sqrt{t+1}\,dt$.
25. $\int \cos^3\theta\,d\theta$. [*Hint*: Use $\cos^2\theta+\sin^2\theta=1$.]
26. $\int \cot x\,dx$.
27. $\int \frac{dx}{x\ln x}$.
28. $\int \frac{dx}{\ln(x^x)}$.
29. $\int \sqrt{4-x^2}\,dx$. [*Hint*: Let $x = 2\sin u$.]
30. $\int \sin^2 x\,dx$. (Use $\cos 2x = 1-2\sin^2 x$.)
31. $\int \frac{\cos\theta}{1+\sin\theta}\,d\theta$.
32. $\int \sec^2 x(e^{\tan x}+1)\,dx$.
33. $\int \frac{\sin(\ln t)}{t}\,dt$.
34. $\int \frac{e^{2s}}{1+e^{2s}}\,ds$.
35. $\int \frac{\sqrt[3]{3+1/x}}{x^2}\,dx$.
36. $\int \frac{1}{x^3}\left(1-\frac{1}{x^2}\right)^{1/3}dx$.
37. Compute $\int \sin x\cos x\,dx$ by each of the following three methods: (a) Substitute $u = \sin x$, (b) substitute $u = \cos x$, (c) use the identity $\sin 2x = 2\sin x\cos x$. Show that the three answers you get are really the same.
38. Compute $\int e^{ax}\,dx$, where a is constant, by each of the following substitutions: (a) $u = ax$; (b) $u = e^x$. Show that you get the same answer either way.

★39. For which values of m and n can $\int \sin^m x\cos^n x\,dx$ be evaluated by using a substitution $u = \sin x$ or $u = \cos x$ and the identity $\cos^2 x+\sin^2 x = 1$?

★40. For which values of r can $\int \tan^r x\,dx$ be evaluated by the substitution suggested in Exercise 39?

7.3 Changing Variables in the Definite Integral

When you change variables in a definite integral, you must keep track of the endpoints.

We have just learned how to evaluate many indefinite integrals by the method of substitution. Using the fundamental theorem of calculus, we can use this knowledge to evaluate definite integrals as well.

Example 1 Find $\int_0^2 \sqrt{x+3}\,dx$.

Solution Substitute $u = x + 3$, $du = dx$. Then

$$\int \sqrt{x+3}\,dx = \int \sqrt{u}\,du = \frac{2}{3}u^{3/2} + C = \frac{2}{3}(x+3)^{3/2} + C.$$

By the fundamental theorem of calculus,

$$\int_0^2 \sqrt{x+3}\,dx = \frac{2}{3}(x+3)^{3/2}\Big|_0^2 = \frac{2}{3}(5^{3/2} - 3^{3/2}) \approx 3.99.$$

To check this result we observe that, on the interval $[0,2]$, $\sqrt{x+3}$ lies between $\sqrt{3}$ (≈ 1.73) and $\sqrt{5}$ (≈ 2.24), so the integral must lie between $2\sqrt{3}$ (≈ 3.46) and $2\sqrt{5}$ (≈ 4.47). (This check actually enabled the authors to spot an error in their first attempted solution of this problem.) ▲

Notice that we must express the indefinite integral in terms of x before plugging in the endpoints 0 and 2, since they refer to values of x. It is possible, however, to evaluate the definite integral directly in the u variable—*provided that we change the endpoints*. We offer an example before stating the general procedure.

Example 2 Find $\int_1^4 \frac{x}{1+x^4}\,dx$.

Solution Substitute $u = x^2$, $du = 2x\,dx$, that is, $x\,dx = du/2$. As x runs from 1 to 4, $u = x^2$ runs from 1 to 16, so we have

$$\int_1^4 \frac{x}{1+x^4}\,dx = \int_1^{16} \frac{x}{1+x^4}\frac{du}{2x} = \frac{1}{2}\int_1^{16}\frac{du}{1+u^2}$$

$$= \frac{1}{2}\tan^{-1}u\Big|_1^{16} = \frac{1}{2}(\tan^{-1}16 - \tan^{-1}1) \approx 0.361. ▲$$

In general, suppose that we have an integral of the form $\int_a^b f(g(x))g'(x)\,dx$. If $F'(u) = f(u)$, then $F(g(x))$ is an antiderivative of $f(g(x))g'(x)$; by the fundamental theorem of calculus, we have

$$\int_a^b f(g(x))g'(x)\,dx = F(g(b)) - F(g(a)).$$

However, the right-hand side is equal to $\int_{g(a)}^{g(b)} f(u)\,du$, so we have the formula

$$\int_a^b f(g(x))g'(x)\,dx = \int_{g(a)}^{g(b)} f(u)\,du.$$

Notice that $g(a)$ and $g(b)$ are the values of $u = g(x)$ when $x = a$ and b, respectively. Thus we can evaluate an integral $\int_a^b h(x)\,dx$ by writing $h(x)$ as

$f(g(x))g'(x)$ and using the formula

$$\int_a^b h(x)\,dx = \int_{g(a)}^{g(b)} f(u)\,du.$$

Example 3 Evaluate $\int_0^{\pi/4} \cos 2\theta\, d\theta$.

Solution Let $u = 2\theta$; $d\theta = \frac{1}{2}\,du$; $u = 0$ when $\theta = 0$, $u = \pi/2$ when $\theta = \pi/4$. Thus

$$\int_0^{\pi/4} \cos 2\theta\, d\theta = \frac{1}{2}\int_0^{\pi/2} \cos u\, du = \frac{1}{2}\sin u\Big|_0^{\pi/2} = \frac{1}{2}\left(\sin\frac{\pi}{2} - \sin 0\right) = \frac{1}{2}\,. \;\blacktriangle$$

Definite Integral by Substitution

Given an integral $\int_a^b h(x)\,dx$ and a new variable $u = g(x)$:

1. Substitute $du/g'(x)$ for dx and then try to express the integrand $h(x)/g'(x)$ in terms of u.
2. Change the endpoints a and b to $g(a)$ and $g(b)$, the corresponding values of u.

Then

$$\int_a^b h(x)\,dx = \int_{g(a)}^{g(b)} f(u)\,du,$$

where $f(u) = h(x)/(du/dx)$. Since $h(x) = f(g(x))g'(x)$, this can be written as

$$\int_a^b f(g(x))g'(x)\,dx = \int_{g(a)}^{g(b)} f(u)\,du.$$

Example 4 Evaluate $\int_1^5 \frac{x}{x^4 + 10x^2 + 25}\,dx$.

Solution Seeing that the denominator can be written in terms of x^2, we try $u = x^2$, $dx = du/(2x)$; $u = 1$ when $x = 1$ and $u = 25$ when $x = 5$. Thus

$$\int_1^5 \frac{x}{x^4 + 10x^2 + 25}\,dx = \frac{1}{2}\int_1^{25} \frac{du}{u^2 + 10u + 25}\,.$$

Now we notice that the denominator is $(u + 5)^2$, so we set $v = u + 5$, $du = dv$; $v = 6$ when $u = 1$, $v = 30$ when $u = 25$. Therefore

$$\frac{1}{2}\int_1^{25} \frac{du}{u^2 + 10u + 25} = \frac{1}{2}\int_6^{30} \frac{dv}{v^2} = \frac{1}{2}\left(-\frac{1}{v}\right)\Big|_6^{30}$$

$$= -\frac{1}{60} + \frac{1}{12} = \frac{1}{15}\,.$$

If you see the substitution $v = x^2 + 5$ right away, you can do the problem in one step instead of two. ▲

Example 5 Find $\int_0^{\pi/4} (\cos^2\theta - \sin^2\theta)\,d\theta$.

Solution It is not obvious what substitution is appropriate here, so a little trial and error is called for. If we remember the trigonometric identity $\cos 2\theta = \cos^2\theta - \sin^2\theta$, we can proceed easily:

$$\int_0^{\pi/4}(\cos^2\theta - \sin^2\theta)\,d\theta = \int_0^{\pi/4}\cos 2\theta\,d\theta = \int_0^{\pi/2}\cos u\,\frac{du}{2} \qquad (u = 2\theta)$$

$$= \frac{\sin u}{2}\bigg|_0^{\pi/2} = \frac{1-0}{2} = \frac{1}{2}\,.$$

(See Exercise 32 for another method.) ▲

Example 6 Evaluate $\int_0^1 \frac{e^x}{1+e^x}\,dx$.

Solution Let $u = 1 + e^x$; $du = e^x\,dx$, $dx = du/e^x$; $u = 1 + e^0 = 2$ when $x = 0$ and $u = 1 + e$ when $x = 1$. Thus

$$\int_0^1 \frac{e^x}{1+e^x}\,dx = \int_2^{1+e}\frac{1}{u}\,du = \ln u\bigg|_2^{1+e} = \ln(1+e) - \ln 2 = \ln\left(\frac{1+e}{2}\right).\ ▲$$

Substitution does not always work. We can always make a substitution, but sometimes it leads nowhere.

Example 7 What does the integral $\int_0^2 \frac{dx}{1+x^4}$ become if you substitute $u = x^2$?

Solution If $u = x^2$, $du/dx = 2x$ and $dx = du/2x$, so

$$\int_0^2 \frac{dx}{1+x^4} = \int_0^4 \frac{1}{1+u^2}\,\frac{du}{2x}\,.$$

We must solve $u = x^2$ for x; since $x \geqslant 0$, we get $x = \sqrt{u}$, so

$$\int_0^2 \frac{dx}{1+x^4} = \int_0^4 \frac{du}{2\sqrt{u}\,(1+u^2)}\,.$$

Unfortunately, we do not know how to evaluate the integral in u, so all we have done is to equate two unknown quantities. ▲

As in Example 7, after a substitution, the integral $\int f(u)\,du$ might still be something we do not know how to evaluate. In that case it may be necessary to make another substitution or use a completely different method. There is an infinite choice of substitutions available in any given situation. It takes practice to learn to choose one that works.

In general, integration is a trial-and-error process that involves a certain amount of educated guessing. What is more, the antiderivatives of such innocent-looking functions as

$$\frac{1}{\sqrt{(1-x^2)(1-2x^2)}} \quad \text{and} \quad \frac{1}{\sqrt{3-\sin^2 x}}$$

cannot be expressed in any way as algebraic combinations and compositions of polynomials, trigonometric functions, or exponential functions. (The proof of a statement like this is not elementary; it belongs to a subject known as "differential algebra".) Despite these difficulties, you can learn to integrate many functions, but the learning process is slower than for differentiation, and practice is more important than ever.

Since integration is harder than differentiation, one often uses tables of integrals. A short table is available on the endpapers of this book, and extensive books of tables are on the market. (Two of the most popular are Burington's and the CRC tables, both of which contain a great deal of mathematical data in addition to the integrals.) Using these tables requires a knowledge of the basic integration techniques, though, and that is why you still need to learn them.

Example 8 Evaluate $\int_1^3 \frac{dx}{x\sqrt{1+x}}$ using the tables of integrals.

Solution We search the tables for a form similar to this and find number 49 with $a = 1$, $b = 1$. Thus

$$\int \frac{dx}{x\sqrt{1+x}} = \ln\left|\frac{\sqrt{1+x}-1}{\sqrt{1+x}+1}\right| + C.$$

Hence

$$\int_1^3 \frac{dx}{x\sqrt{1+x}} = \ln\left|\frac{\sqrt{4}-1}{\sqrt{4}+1}\right| - \ln\left|\frac{\sqrt{2}-1}{\sqrt{2}+1}\right| = \ln\frac{1}{3} - \ln\left|\frac{\sqrt{2}-1}{\sqrt{2}+1}\right|$$

$$= \ln\left[\frac{\sqrt{2}+1}{3(\sqrt{2}-1)}\right] = \ln\left(1 + \frac{2}{3}\sqrt{2}\right). \blacktriangle$$

Exercises for Section 7.3

Evaluate the definite integrals in Exercises 1–22.

1. $\int_{-1}^{1} \sqrt{x+2}\, dx$
2. $\int_2^3 \frac{dt}{t-1}$
3. $\int_0^2 x\sqrt{x^2+1}\, dx$
4. $\int_0^1 t\sqrt{t^2+1}\, dt$
5. $\int_2^4 (x+1)(x^2+2x+1)^{5/4}\, dx$
6. $\int_1^2 \frac{\sqrt{1+\ln x}}{x}\, dx$
7. $\int_1^3 \frac{3x}{(x^2+5)^2}\, dx$
8. $\int_1^2 \frac{t^2+1}{\sqrt{t^3+3t+3}}\, dt$
9. $\int_0^1 xe^{(x^2)}\, dx$
10. $\int_0^1 \frac{e^x}{1+e^{2x}}\, dx$
11. $\int_0^{\pi/6} \sin(3\theta+\pi)\, d\theta$
12. $\int_0^{\pi} \sin(\theta/2+\pi/4)\, d\theta$
13. $\int_{-\pi/2}^{\pi/2} 5\cos^2 x \sin x\, dx.$
14. $\int_{\pi/4}^{\pi/2} \frac{\csc^2 y}{\cot^2 y + 2\cot y + 1}\, dy.$
15. $\int_0^{\sqrt{\pi}} x\sin(x^2)\, dx.$
16. $\int_0^1 \frac{x^2}{x^3+1}\, dx.$
17. $\int_{\pi/8}^{\pi/4} \tan\theta\, d\theta.$
18. $\int_{\pi/4}^{\pi/2} \cot\theta\, d\theta.$
19. $\int_0^{\pi/2} \sin x \cos x\, dx.$
20. $\int_1^{\pi/2} [\ln(\sin x) + (x\cot x)](\sin x)^x\, dx.$
21. $\int_1^3 \frac{x^3+x-1}{x^2+1}\, dx$ (simplify first).
22. $\int_1^e \frac{2\ln(x^x)+1}{x^2}\, dx.$
23. Using the result $\int_0^{\pi/2} \sin^2 x\, dx = \pi/4$ (See Exercise 57, Section 7.1), compute each of the following integrals: (a) $\int_0^{\pi} \sin^2(x/2)\, dx$; (b) $\int_{\pi/2}^{\pi} \sin^2(x-\pi/2)\, dx$; (c) $\int_0^{\pi/4} \cos^2(2x)\, dx$.
24. (a) By combining the shifting and scaling rules, find a formula for $\int f(ax+b)\, dx$.
    (b) Find $\int_2^3 \frac{dx}{4x^2+12x+9}$ [*Hint*: Factor the denominator.]
25. What happens in the integral
$$\int_0^1 \frac{(x^2+3x)}{\sqrt[3]{x^3+3x^2+1}}\, dx$$
if you make the substitution $u = x^3+3x^2+1$?
26. What becomes of the integral $\int_0^{\pi/2} \cos^4 x\, dx$ if you make the substitution $u = \cos x$?

Evaluate the integrals in Exercises 27–30 using the tables.

27. $\int_0^1 \frac{dx}{3x^2+2x+1}$
28. $\int_1^2 \frac{\sqrt{x^2-1}}{x}\, dx$
29. $\int_0^1 \frac{dx}{\sqrt{3x^2+2x+1}}$
30. $\int_2^3 \frac{\sqrt{x^2-2}}{x^4}\, dx$

31. Given two functions f and g, define a function h by
$$h(x) = \int_0^1 f(x-t)g(t)\,dt.$$
Show that
$$h(x) = \int_{x-1}^{x} g(x-t)f(t)\,dt.$$
32. Give another solution to Example 5 by writing $\cos^2\theta - \sin^2\theta = (\cos\theta - \sin\theta)(\cos\theta + \sin\theta)$ and using the substitution $u = \cos\theta + \sin\theta$.
33. Find the area under the graph of the function $y = (x+1)/(x^2+2x+2)^{3/2}$ from $x = 0$ to $x = 1$.
34. The curve $x^2/a^2 + y^2/b^2 = 1$, where a and b are positive, describes an *ellipse* (Fig. 7.3.1). Find the area of the region inside this ellipse. [*Hint*: Write half the area as an integral and then change variables in the integral so that it becomes the integral for the area inside a semicircle.]

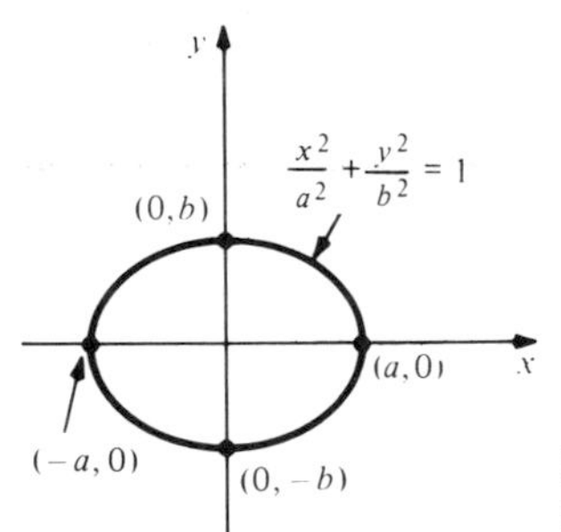

Figure 7.3.1. Find the area inside the ellipse.

35. The curve $y = x^{1/3}$, $1 \leqslant x \leqslant 8$, is revolved about the y axis to generate a surface of revolution of area s. In Chapter 10 we will prove that the area is given by $s = \int_1^2 2\pi y^3\sqrt{1+9y^4}\,dy$. Evaluate this integral.

★36. Let $f(x) = \int_1^x (dt/t)$. Show, using substitution, and without using logarithms, that $f(a) + f(b) = f(ab)$ if $a, b > 0$. [*Hint*: Transform $\int_a^{ab} \frac{dt}{t}$ by a change of variables.]

37. (a) Find $\int_0^{\pi/2} \cos^2 x \sin x\,dx$ by substituting $u = \cos x$ and changing the endpoints.
★(b) Is the formula
$$\int_a^b f(g(x))g'(x)\,dx = \int_{g(a)}^{g(b)} f(u)\,du$$
valid if $a < b$, yet $g(a) > g(b)$? Discuss.

7.4 Integration by Parts

Integrating the product rule leads to the method of integration by parts.

The second of the two important new methods of integration is developed in this section. The method parallels that of substitution, with the chain rule replaced by the product rule.

The product rule for derivatives asserts that
$$(FG)'(x) = F'(x)G(x) + F(x)G'(x).$$
Since $F(x)G(x)$ is an antiderivative for $F'(x)G(x) + F(x)G'(x)$, we can write
$$\int [F'(x)G(x) + F(x)G'(x)]\,dx = F(x)G(x) + C.$$
Applying the sum rule and transposing one term leads to the formula
$$\int F(x)G'(x)\,dx = F(x)G(x) - \int F'(x)G(x)\,dx + C.$$
If the integral on the right-hand side can be evaluated, it will have its own constant C, so it need not be repeated. We thus write
$$\int F(x)G'(x)\,dx = F(x)G(x) - \int F'(x)G(x)\,dx, \tag{1}$$
which is the formula for integration by parts. To apply formula (1) we need to break up a given integrand as a product $F(x)G'(x)$, write down the right-hand side of formula (1), and hope that we can integrate $F'(x)G(x)$. Integrands involving trigonometric, logarithmic, and exponential functions are often good candidates for integration by parts, but practice is necessary to learn the best way to break up an integrand as a product.

Example 1 Evaluate $\int x \cos x \, dx$.

Solution If we remember that $\cos x$ is the derivative of $\sin x$, we can write $x \cos x$ as $F(x)G'(x)$, where $F(x) = x$ and $G(x) = \sin x$. Applying formula (1), we have

$$\int x \cos x \, dx = x \cdot \sin x - \int 1 \cdot \sin x \, dx = x \sin x - \int \sin x \, dx$$
$$= x \sin x + \cos x + C.$$

Checking by differentiation, we have

$$\frac{d}{dx}(x \sin x + \cos x) = x \cos x + \sin x - \sin x = x \cos x,$$

as required. ▲

It is often convenient to write formula (1) using differential notation. Here we write $u = F(x)$ and $v = G(x)$. Then $du/dx = F'(x)$ and $dv/dx = G'(x)$. Treating the derivatives as if they were quotients of "differentials" du, dv, and dx, we have $du = F'(x)\,dx$ and $dv = G'(x)\,dx$. Substituting these into formula (1) gives

$$\int u \, dv = uv - \int v \, du \tag{2}$$

(see Fig. 7.4.1).

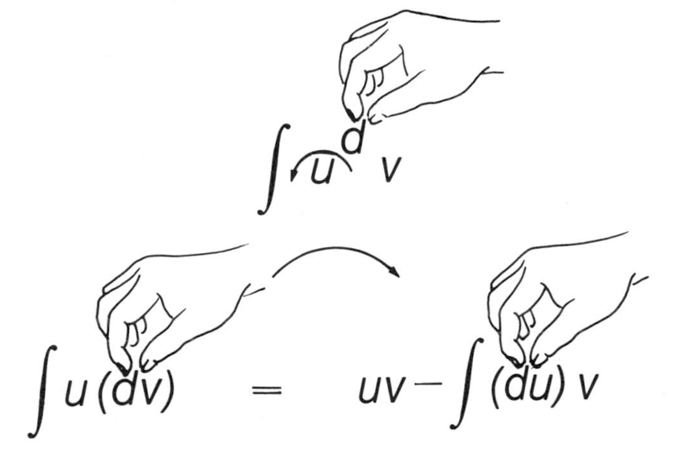

Figure 7.4.1. You may move "d" from v to u if you switch the sign and add uv.

Integration by Parts

To evaluate $\int h(x)\,dx$ by parts:

1. Write $h(x)$ as a product $F(x)G'(x)$, where the antiderivative $G(x)$ of $G'(x)$ is known.
2. Take the derivative $F'(x)$ of $F(x)$.
3. Use the formula

$$\int F(x)G'(x)\,dx = F(x)G(x) - \int F'(x)G(x)\,dx,$$

i.e., with $u = F(x)$ and $v = G(x)$,

$$\int u \, dv = uv - \int v \, du.$$

When you use integration by parts, to integrate a function h write $h(x)$ as a product $F(x)G'(x) = u\,dv/dx$; the factor $G'(x)$ is a function whose antiderivative $v = G(x)$ can be found. With a good choice of $u = F(x)$ and $v = G(x)$, the integral $\int F'(x)G(x)\,dx = \int v\,du$ becomes simpler than the original problem $\int u\,dv$. The ability to make good choices of u and v comes with practice. A last reminder—don't forget the minus sign.

Example 2 Find (a) $\int x \sin x\,dx$ and (b) $\int x^2 \sin x\,dx$.

Solution (a) (*Using formula* (1)) Let $F(x) = x$ and $G'(x) = \sin x$. Integrating $G'(x)$ gives $G(x) = -\cos x$; also, $F'(x) = 1$, so

$$\begin{aligned}\int x \sin x\,dx &= -x\cos x - \int -\cos x\,dx \\ &= -x\cos x - (-\sin x) + C \\ &= -x\cos x + \sin x + C.\end{aligned}$$

(b) (*Using formula* (2)) Let $u = x^2$, $dv = \sin x\,dx$. To apply formula (2) for integration by parts, we need to know v. But $v = \int dv = \int \sin x\,dx = -\cos x$. (We leave out the arbitrary constant here and will put it in at the end of the problem.)

Now

$$\begin{aligned}\int x^2 \sin x\,dx &= uv - \int v\,du \\ &= -x^2\cos x - \int -\cos x \cdot 2x\,dx \\ &= -x^2\cos x + 2\int x\cos x\,dx.\end{aligned}$$

Using the result of Example 1, we obtain

$$-x^2\cos x + 2(x\sin x + \cos x) + C = -x^2\cos x + 2x\sin x + 2\cos x + C.$$

Check this result by differentiating—it is nice to see all the cancellation. ▲

Integration by parts is also commonly used in integrals involving e^x and $\ln x$.

Example 3 (a) Find $\int \ln x\,dx$ using integration by parts. (b) Find $\int xe^x\,dx$.

Solution (a) Here, let $u = \ln x$, $dv = 1\,dx$. Then $du = dx/x$ and $v = \int 1\,dx = x$. Applying the formula for integration by parts, we have

$$\begin{aligned}\int \ln x\,dx &= uv - \int v\,du = (\ln x)x - \int x\,\frac{dx}{x} \\ &= x\ln x - \int 1\,dx = x\ln x - x + C.\end{aligned}$$

(Compare Example 7, Section 7.1.)

(b) Let $u = x$ and $v = e^x$, so $dv = e^x\,dx$. Thus, using integration by parts,

$$\begin{aligned}\int xe^x\,dx &= \int u\,dv = uv - \int v\,du \\ &= xe^x - \int e^x\,dx = xe^x - e^x + C. \ \blacktriangle\end{aligned}$$

Next we consider an example involving both e^x and $\sin x$.

Example 4 Apply integration by parts twice to find $\int e^x \sin x\,dx$.

Solution Let $u = \sin x$ and $v = e^x$, so $dv = e^x\,dx$ and

$$\int e^x \sin x\,dx = e^x \sin x - \int e^x \cos x\,dx. \tag{3}$$

Repeating the integration by parts,

$$\int e^x \cos x\,dx = e^x \cos x + \int e^x \sin x\,dx, \tag{4}$$

where, this time, $u = \cos x$ and $v = e^x$. Substituting formula (4) into (3), we get

$$\int e^x \sin x\,dx = e^x \sin x - e^x \cos x - \int e^x \sin x\,dx.$$

The unknown integral $\int e^x \sin x\,dx$ appears twice in this equation. Writing "I" for this integral, we have

$$I = e^x \sin x - e^x \cos x - I,$$

and solving for I gives

$$I = \tfrac{1}{2} e^x (\sin x - \cos x),$$

i.e.,

$$\int e^x \sin x\,dx = \tfrac{1}{2} e^x (\sin x - \cos x) + C.$$

Some students like to remember this as "the I method." ▲

Some special purely algebraic expressions can also be handled by a clever use of integration by parts, as in the next example.

Example 5 Find $\int x^7 (x^4 + 1)^{2/3}\,dx$.

Solution By taking x^3 out of x^7 and grouping it with $(x^4 + 1)^{2/3}$, we get an expression which we can integrate. Specifically, we set $dv = 4x^3(x^4 + 1)^{2/3}\,dx$, leaving $u = x^4/4$. Using integration by substitution, we get $v = \frac{3}{5}(x^4 + 1)^{5/3}$, and differentiating, we get $du = x^3\,dx$. Hence

$$\int x^7 (x^4 + 1)^{2/3}\,dx = \frac{3x^4}{20}(x^4 + 1)^{5/3} - \frac{3}{5}\int x^3 (x^4 + 1)^{5/3}\,dx.$$

Substituting $w = (x^4 + 1)$ gives

$$\int x^3 (x^4 + 1)^{5/3}\,dx = \tfrac{3}{32}(x^4 + 1)^{8/3} + C;$$

hence

$$\begin{aligned}\int x^7 (x^4 + 1)^{2/3}\,dx &= \tfrac{3}{20} x^4 (x^4 + 1)^{5/3} - \tfrac{9}{160}(x^4 + 1)^{8/3} + C \\ &= \tfrac{3}{160}(x^4 + 1)^{5/3}(5x^4 - 3) + C. \ ▲\end{aligned}$$

Using integration by parts and then the fundamental theorem of calculus, we can calculate definite integrals.

Example 6 Find $\int_{-\pi/2}^{\pi/2} x \sin x\,dx$.

Solution From Example 2 (a) we have $\int x \sin x = -x \cos x + \sin x + C$, so

$$\begin{aligned}\int_{-\pi/2}^{\pi/2} x \sin x\,dx &= (-x \cos x + \sin x)\Big|_{-\pi/2}^{\pi/2} \\ &= \left(-\frac{\pi}{2}\cos\frac{\pi}{2} + \sin\frac{\pi}{2}\right) - \left[\frac{\pi}{2}\cos\left(-\frac{\pi}{2}\right) + \sin\left(-\frac{\pi}{2}\right)\right] \\ &= (0 + 1) - [0 + (-1)] = 2. \ ▲\end{aligned}$$

Example 7 Find (a) $\int_0^{\ln 2} e^x \ln(e^x+1)\,dx$ and (b) $\int_1^e \sin(\ln x)\,dx$.

Solution (a) Notice that e^x is the derivative of (e^x+1), so we first make the substitution $t = e^x + 1$. Then

$$\int_0^{\ln 2} e^x \ln(e^x+1)\,dx = \int_2^3 \ln t\,dt,$$

and, from Example 3, $\int \ln t\,dt = t\ln t - t + C$. Therefore

$$\int_0^{\ln 2} e^x \ln(e^x+1)\,dx = (t\ln t - t)\Big|_2^3 = (3\ln 3 - 3) - (2\ln 2 - 2)$$
$$= 3\ln 3 - 2\ln 2 - 1 \approx 0.9095.$$

(b) Again we begin with a substitution. Let $u = \ln x$, so that $x = e^u$ and $du = (1/x)\,dx$. Then $\int \sin(\ln x)\,dx = \int (\sin u)e^u\,du$, which was evaluated in Example 4. Hence

$$\int_1^e \sin(\ln x)\,dx = \int_0^1 e^u \sin u\,du = \frac{1}{2}e^u(\sin u - \cos u)\Big|_0^1$$
$$= \left[\frac{1}{2}e^1(\sin 1 - \cos 1)\right] - \left[\frac{1}{2}e^0(\sin 0 - \cos 0)\right]$$
$$= \frac{e}{2}\left(\sin 1 - \cos 1 + \frac{1}{e}\right). \blacktriangle$$

Example 8 Find the area under the nth bend of $y = x\sin x$ in the first quadrant (see Fig. 7.4.2).

Solution The nth bend occurs between $x = (2n-2)\pi$ and $(2n-1)\pi$. (Check $n = 1$ and $n = 2$ with the figure.) The area under this bend can be evaluated using integration by parts [Example 2(a)]:

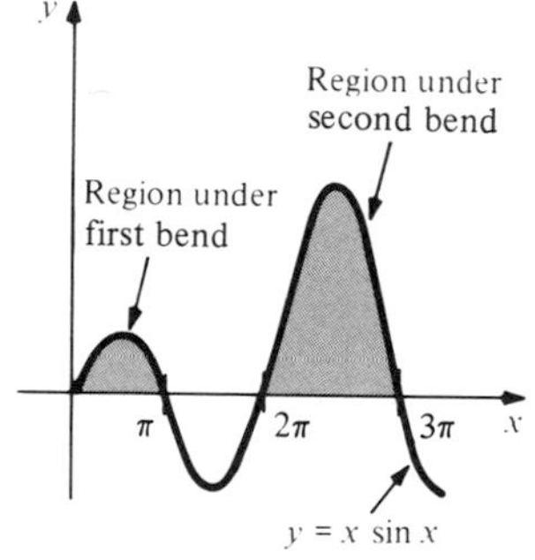

Figure 7.4.2. What is the area under the nth bend?

$$\int_{(2n-2)\pi}^{(2n-1)\pi} x\sin x\,dx = -x\cos x + \sin x\Big|_{(2n-2)\pi}^{(2n-1)\pi}$$
$$= -(2n-1)\pi\cos[(2n-1)\pi] + \sin[(2n-1)\pi]$$
$$+ (2n-2)\pi\cos(2n-2)\pi - \sin(2n-2)\pi$$
$$= -(2n-1)\pi(-1) + 0 + (2n-2)\pi(1) - 0$$
$$= (2n-1)\pi + (2n-2)\pi = (4n-3)\pi.$$

Thus the areas under successive bends are π, 5π, 9π, 13π, and so forth. ▲

We shall now use integration by parts to obtain a formula for the integral of the inverse of a function.

If f is a differentiable function, we write $f(x) = 1 \cdot f(x)$; then

$$\int f(x)\,dx = \int 1 \cdot f(x)\,dx = xf(x) - \int xf'(x)\,dx. \tag{5}$$

Introducing $y = f(x)$ as a new variable, with $dx = dy/f'(x)$, we get

$$\int f(x)\,dx = xy - \int x\,dy. \tag{6}$$

Assuming that f has an inverse function g, we have $x = g(y)$, and equation (6) becomes

$$\int f(x)\,dx = xf(x) - \int g(y)\,dy. \tag{7}$$

Thus we can integrate f if we know how to integrate its inverse. In the notation $y = f(x)$, equation (7) becomes

$$\int y\,dx = xy - \int x\,dy. \tag{8}$$

Notice that equation (8) looks just like the formula for integration by parts, but we are now considering x and y as functions of one another rather than as two functions of a third variable.

Example 9 Use equation (8) to compute $\int \ln x\,dx$.

Solution Viewing $y = \ln x$ as the inverse function of $x = e^y$, equation (8) reads

$$\int \ln x\,dx = xy - \int e^y\,dy = x\ln x - e^y + C = x\ln x - x + C,$$

which is the same result (and essentially the same method) as in Example 3. ▲

We can also state our result in terms of antiderivatives. If $G(y)$ is an antiderivative for $g(y)$, then

$$F(x) = xf(x) - G(f(x)) \tag{9}$$

is an antiderivative for f. (This can be checked by differentiation.)

Example 10 (a) Find an antiderivative for $\cos^{-1}x$. (b) Find $\int \csc^{-1}\sqrt{x}\,dx$.

Solution (a) If $f(x) = \cos^{-1}x$, then $g(y) = \cos y$ and $G(y) = \sin y$. By formula (9),

$$F(x) = x\cos^{-1}x - \sin(\cos^{-1}x);$$

Figure 7.4.3. $\sin(\cos^{-1}x) = \sqrt{1-x^2}$

But $\sin(\cos^{-1}x) = \sqrt{1-x^2}$ (Fig. 7.4.3), so

$$F(x) = x\cos^{-1}x - \sqrt{1-x^2}$$

is an antiderivative for $\cos^{-1}x$. This may be checked by differentiation.

(b) If $y = \csc^{-1}\sqrt{x}$, we have $\csc y = \sqrt{x}$ and $x = \csc^2 y$. Then

$$\begin{aligned}
\int \csc^{-1}\sqrt{x}\,dx = \int y\,dx = xy \quad \int x\,dy \\
&= x\csc^{-1}\sqrt{x} - \int \csc^2 y\,dy \\
&= x\csc^{-1}\sqrt{x} + \cot y + C \\
&= x\csc^{-1}\sqrt{x} + \cot(\csc^{-1}\sqrt{x}) + C \\
&= x\csc^{-1}\sqrt{x} + \sqrt{x-1} + C \qquad \text{(see Fig. 7.4.4).}
\end{aligned}$$ ▲

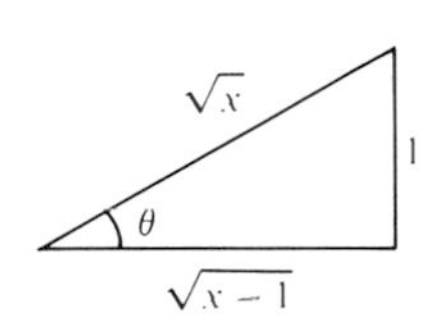

Figure 7.4.4. $\theta = \csc^{-1}\sqrt{x}$.

Example 11 (a) Find $\int \sqrt{\sqrt{x}+1}\,dx$. (b) Find $\int x\cos^{-1}x\,dx$, $0 < x < 1$.

Solution (a) If $y = \sqrt{\sqrt{x}+1}$, then $y^2 = \sqrt{x} + 1$, $\sqrt{x} = y^2 - 1$, and $x = (y^2-1)^2$. Thus we have

$$\begin{aligned}
\int \sqrt{\sqrt{x}+1}\,dx &= xy - \int x\,dy = x\sqrt{\sqrt{x}+1} - \int (y^4 - 2y^2 + 1)\,dy \\
&= x\sqrt{\sqrt{x}+1} - \frac{1}{5}y^5 + \frac{2}{3}y^3 - y + C \\
&= x\sqrt{\sqrt{x}+1} - \frac{1}{5}(\sqrt{x}+1)^{5/2} + \frac{2}{3}(\sqrt{x}+1)^{3/2} \\
&\quad - (\sqrt{x}+1)^{1/2} + C.
\end{aligned}$$

(b) Integrating by parts,

$$\int x\cos^{-1}x\,dx = \frac{x^2}{2}\cos^{-1}x + \int\frac{x^2}{2}\cdot\frac{1}{\sqrt{1-x^2}}\,dx.$$

The last integral may be evaluated by letting $x=\cos u$:

$$\int\frac{x^2}{\sqrt{1-x^2}}\,dx = -\int\frac{\cos^2 u}{\sin u}\sin u\,du = -\int\cos^2 u\,du$$

But $\cos^2 u = \dfrac{\cos 2u + 1}{2}$, so

$$\int\cos^2 u\,du = \frac{1}{4}\sin 2u + \frac{u}{2} + C = \frac{1}{2}\sin u\cos u + \frac{u}{2} + C.$$

Thus,

$$\begin{aligned}\int x\cos^{-1}x\,dx &= \frac{x^2}{2}\cos^{-1}x - \frac{1}{4}\sin(\cos^{-1}x)x - \frac{\cos^{-1}x}{4} + C\\ &= \frac{x^2}{2}\cos^{-1}x - \frac{x}{4}\sqrt{1-x^2} - \frac{1}{4}\cos^{-1}x + C.\ \blacktriangle\end{aligned}$$

Exercises for Section 7.4

Evaluate the indefinite integrals in Exercises 1–26 using integration by parts.

1. $\int(x+1)\cos x\,dx$
2. $\int(x-2)\sin x\,dx$
3. $\int x\cos(5x)\,dx$
4. $\int x\sin(10x)\,dx$
5. $\int x^2\cos x\,dx$
6. $\int x^2\sin x\,dx$
7. $\int(x+2)e^x\,dx$
8. $\int(x^2-1)e^{2x}\,dx$
9. $\int\ln(10x)\,dx$
10. $\int x\ln x\,dx$
11. $\int x^2\ln x\,dx$
12. $\int\ln(9+x^2)\,dx$
13. $\int s^2e^{3s}\,ds$
14. $\int(s+1)^2e^s\,ds.$
15. $\int\dfrac{x^5}{(x^3-4)^{2/3}}\,dx$
16. $\int\dfrac{x^2}{(x^2+1)^2}\,dx$
17. $\int 2t^3\cos t^2\,dt$
18. $\int\dfrac{x^3}{\sqrt{x^2+1}}\,dx$
19. $\int\dfrac{1}{x^3}\cos\dfrac{1}{x}\,dx$
20. $\int x\sin(\ln x)\,dx$
21. $\int\tan x\ln(\cos x)\,dx$
22. $\int e^{2x}e^{(e^x)}\,dx$
23. $\int\cos^{-1}(2x)\,dx$
24. $\int\sin^{-1}x\,dx$
25. $\int\sqrt{\dfrac{1}{y}-1}\,dy$
26. $\int(\sqrt{x}-2)^{1/5}\,dx$
27. Find $\int\sin x\cos x\,dx$ by using integration by parts with $u=\sin x$ and $dv=\cos x\,dx$. Compare the result with substituting $u=\sin x$.
28. Compute $\int\sqrt{x}\,dx$ by the rule for inverse functions. Compare with the result given by the power rule.
29. What happens in Example 2(a) if you choose $F'(x)=x$ and $G(x)=\sin x$?
30. What would have happened in Example 5, if in the integral $\int e^x\cos x\,dx$ obtained in the first integral by parts, you had taken $u=e^x$ and $v=\sin x$ and integrated by parts a second time?

Evaluate the definite integrals in Exercises 31–46.

31. $\int_0^{\pi/5}(8+5\theta)(\sin 5\theta)\,d\theta$
32. $\int_1^2 x\ln x\,dx$
33. $\int_1^3\ln x^3\,dx$
34. $\int_0^1 xe^x\,dx$
35. $\int_0^{\pi/4}(x^2+x-1)\cos x\,dx$
36. $\int_0^{\pi/2}\sin 3x\cos 2x\,dx$
37. $\int_{1/8}^{1/4}\cos^{-1}(4x)\,dx$
38. $\int_0^1 x\tan^{-1}x\,dx$
39. $\int_1^e(\ln x)^2\,dx$
40. $\int_0^{\pi/2}\sin 2x\cos x\,dx.$
41. $\int_{-\pi}^{\pi}e^{2x}\sin(2x)\,dx.$
42. $\int_0^{\pi^2}\sin\sqrt{x}\,dx$. [*Hint*: Change variables first.]
43. $\int_1^2 x^{1/3}(x^{2/3}+1)^{3/2}\,dx.$
44. $\int_0^1\dfrac{x^3}{(x^2+1)^{1/2}}\,dx.$
45. $\int_0^{1/2\sqrt{2}}\sin^{-1}2x\,dx.$
46. $\int_0^1\cos^{-1}(\sqrt{y})\,dy.$

47. Show that

$$\int_0^1 \sqrt{2-x^2}\,dx - \int_0^{\sqrt{2}} \sqrt{2-x^2}\,dx = (1-\pi/2)/2.$$

48. Find $\int_2^{34} f(x)\,dx$, where f is the inverse function of $g(y) = y^5 + y$.

49. Find $\int_0^{2\pi} x \sin ax\,dx$ as a function of a. What happens to this integral as a becomes larger and larger? Can you explain why?

50. (a) Integrating by parts twice (see Example 4), find $\int \sin ax \cos bx\,dx$, where $a^2 \neq b^2$.
(b) Using the formula $\sin 2x = 2 \sin x \cos x$, find $\int \sin ax \cos bx\,dx$ when $a = \pm b$.
(c) Let $g(a) = (4/\pi)\int_0^{\pi/2} \sin x \sin ax\,dx$. Find a formula for $g(a)$. (The formula will have to distinguish the cases $a^2 \neq 1$ and $a^2 = 1$.)
(d) Evaluate $g(a)$ for $a = 0.9$, 0.99, 0.999, 0.9999, and so on. Compare the results with $g(1)$. Also try $a = 1.1$, 1.01, 1.001, and so on. What do you guess is true about the function g at $a = 1$?

51. (a) Integrating by parts twice, show that

$$\int e^{ax} \cos bx\,dx = e^{ax}\left(\frac{b \sin bx + a \cos bx}{a^2 + b^2}\right) + C.$$

(b) Evaluate $\int_0^{\pi/10} e^{3x} \cos 5x\,dx$.

52. (a) Prove the following *reduction formula*:

$$\int x^n e^x\,dx = x^n e^x - n\int x^{n-1} e^x\,dx.$$

(b) Evaluate $\int_0^3 x^3 e^x\,dx$

53. (a) Prove the following *reduction formula*:

$$\int \cos^n x\,dx = \frac{\cos^{n-1} x \sin x}{n} + \frac{n-1}{n}\int \cos^{n-2} x\,dx.$$

(b) Use part (a) to show that

$$\int \cos^2 x\,dx = \frac{1}{2}(\cos x \sin x + x) + C$$

and

$$\int \cos^4 x\,dx = \frac{1}{4}\left(\cos^3 x \sin x + \frac{3}{2}\cos x \sin x + \frac{3x}{2}\right) + C.$$

54. The mass density of a beam is $\rho = x^2 e^{-x}$ kilograms per centimeter. The beam is 200 centimeters long, so its mass is $M = \int_0^{200} \rho\,dx$ kilograms. Find the value of M.

55. The volume of the solid formed by rotation of the plane region enclosed by $y = 0$, $y = \sin x$, $x = 0$, $x = \pi$, around the y axis, will be shown in Chapter 9 to be given by $V = \int_0^{\pi} 2\pi x \sin x\,dx$. Find V.

56. The *Fourier series* analysis of the sawtooth wave requires the computation of the integral

$$b_m = \frac{\omega^2 A}{2\pi^2}\int_{-\pi/\omega}^{\pi/\omega} t \sin(m\omega t)\,dt,$$

where m is an integer and ω and A are nonzero constants. Compute it.

57. The current i in an underdamped RLC circuit is given by

$$i = EC\left(\frac{\alpha^2}{\omega} + \omega\right)e^{-\alpha t}\sin(\omega t).$$

The constants are E = constant emf, switched on at $t = 0$, C = capacitance in farads, R = resistance in ohms, L = inductance in henrys, $\alpha = R/2L$, $\omega = (1/2L)(4L/C - R^2)^{1/2}$.
(a) The charge Q in coulombs is given by $dQ/dt = i$, and $Q(0) = 0$. Find an integral formula for Q, using the fundamental theorem of calculus.
(b) Determine Q by integration.

58. A critically damped RLC circuit with a steady emf of E volts has current $i = EC\alpha^2 t e^{-\alpha t}$, where $\alpha = R/2L$. The constants R, L, C are in ohms, henrys, and farads, respectively. The charge Q in coulombs is given by $Q(T) = \int_0^T i\,dt$. Find it explicitly, using integration by parts.

★59. Draw a figure to illustrate the formula for integration of inverse functions:

$$\int_a^b f(x)\,dx = bf(b) - af(a) - \int_{f(a)}^{f(b)} g(y)\,dy,$$

where $0 < a < b$, $0 < f(a) < f(b)$, f is increasing on $[a, b]$, and g is the inverse function of f.

★60. (a) Suppose that $\phi'(x) > 0$ for all x in $[0, \infty)$ and $\phi(0) = 0$. Show that if $a \geqslant 0$, $b \geqslant 0$, and b is in the domain of ϕ^{-1}, then *Young's inequality* holds:

$$ab \leqslant \int_0^a \phi(x)\,dx + \int_0^b \phi^{-1}(y)\,dy,$$

where ϕ^{-1} is the inverse function to ϕ. [*Hint*: Express $\int_0^b \phi^{-1}(y)\,dy$ in terms of an integral of ϕ by using the formula for integrating an inverse function. Consider separately the cases $\phi(a) \leqslant b$ and $\phi(a) \geqslant b$. For the latter, prove the inequality $\int_{\phi^{-1}(b)}^a \phi(x)\,dx \geqslant \int_{\phi^{-1}(b)}^a b\,dx = b[a - \phi^{-1}(b)]$.]
(b) Prove (a) by a geometric argument based on Exercise 59.
(c) Using the result of part (a), show that if $a, b \geqslant 0$ and $p, q > 1$, with $1/p + 1/q = 1$, then *Minkowski's inequality* holds:

$$ab \leqslant \frac{a^p}{p} + \frac{b^q}{q}.$$

★61. If f is a function on $[0, 2\pi]$, the numbers

$$a_n = (1/\pi)\int_0^{2\pi} f(x) \cos nx\,dx,$$

$$b_n = (1/\pi)\int_0^{2\pi} f(x) \sin nx\,dx$$

are called the *Fourier coefficients* of f ($n = 0, \pm 1, \pm 2, \ldots$). Find the Fourier coefficients of: (a) $f(x) = 1$; (b) $f(x) = x$; (c) $f(x) = x^2$; (d) $f(x) = \sin 2x + \sin 3x + \cos 4x$.

★62. Following Example 5, find a general formula for $\int x^{2n-1}(x^n + 1)^m\,dx$, where n and m are rational numbers with $n \neq 0$, $m \neq -1, -2$.

Review Exercises for Chapter 7

Evaluate the integrals in Exercises 1–46.

1. $\int (x + \sin x)\, dx$
2. $\int \left(x + \frac{1}{\sqrt{1-x^2}}\right) dx$
3. $\int (x^3 + \cos x)\, dx$
4. $\int (8t^4 - 5\cos t)\, dt$
5. $\int \left(e^x - x^2 - \frac{1}{x} + \cos x\right) dx$
6. $\int \left(3^x - \frac{3}{x} + \cos x\right) dx$
7. $\int (e^\theta + \theta^2)\, d\theta$
8. $\int \frac{\sqrt[3]{x^2} - x^{5/2}}{\sqrt{x}}\, dx$
9. $\int x^2 \sin x^3\, dx$
10. $\int \tan x \sec^2 x\, dx$
11. $\int x^2 e^{(x^3)}\, dx$
12. $\int x e^{(x^2)}\, dx$
13. $\int (x+2)^5\, dx$
14. $\int \frac{dx}{3x+4}$
15. $\int x^2 e^{(4x^3)}\, dx$
16. $\int (1 + 3x^2)\exp(x + x^3)\, dx$
17. $\int 2\cos^2 2x \sin 2x\, dx$
18. $\int 3\sin 3x \cos 3x\, dx$
19. $\int x \tan^{-1} x\, dx$
20. $\int x\sqrt{5 - x^2}\, dx$
21. $\int \left[\frac{1}{\sqrt{4-t^2}} + t^2\right] dt$
22. $\int \frac{e^{2x}}{1 + e^{4x}}\, dx$
23. $\int x e^{4x}\, dx$
24. $\int x e^{6x}\, dx$
25. $\int x^2 \cos x\, dx$
26. $\int x^2 e^{2x}\, dx$
27. $\int e^{-x}\cos x\, dx$
28. $\int e^{2x}\tan e^{2x}\, dx$
29. $\int x^2 \ln 3x\, dx$
30. $\int x^3 \ln x\, dx$
31. $\int x\sqrt{x+3}\, dx$
32. $\int x^2\sqrt{x+1}\, dx$
33. $\int x\cos 3x\, dx$
34. $\int t\cos 2t\, dt$
35. $\int 3x\cos 2x\, dx$
36. $\int \sin 2x \cos x\, dx$
37. $\int x^3 e^{(x^2)}\, dx$
38. $\int x^5 e^{(x^3)}\, dx$
39. $\int x(\ln x)^2\, dx$
40. $\int (\ln x)^2\, dx$
41. $\int e^{\sqrt{x}}\, dx$
42. $\int \frac{dx}{x^2 + 2x + 3}$ (Complete the square.)
43. $\int [\cos x]\ln(\sin x)\, dx$
44. $\int \frac{\ln\sqrt{x}}{\sqrt{x}}\, dx$
45. $\int \tan^{-1} x\, dx$
46. $\int \cos^{-1}(12x)\, dx$

Evaluate the definite integrals in Exercises 47–58.

47. $\int_{-1}^{0} x e^{-x}\, dx$
48. $\int_1^e x\ln(5x)\, dx$
49. $\int_0^{\pi/5} x\sin 5x\, dx$
50. $\int_0^{\pi/4} x\cos 2x\, dx$
51. $\int_1^2 x^{-2}\cos(1/x)\, dx$
52. $\int_0^{\pi/2} x^2\cos(x^3)\sin(x^3)\, dx$
53. $\int_0^{\pi/4} x\tan^{-1} x\, dx$
54. $\int_1^{\ln(\pi/4)} e^x \tan e^x\, dx$
55. $\int_{a+1}^{a+2} \frac{t}{\sqrt{t-a}}\, dt$ (substitute $x = \sqrt{t-a}$)
56. $\int_0^1 \frac{\sqrt{x}}{x+1}\, dx$
57. $\int_0^1 x\sqrt{2x+3}\, dx$
58. $\int_0^{\sqrt{3}} \frac{3}{3+u^2}\, du$

In Exercises 59–66, sketch the region under the graph the given function on the given interval and find its area.

59. $40 - x^3$ on $[0, 3]$
60. $\sin x + 2x$ on$[0, 4\pi]$

61. $3x/\sqrt{x^2+9}$ on $[0,4]$
62. $x\sin^{-1}x+2$ on $[0,1]$
63. $\sin x$ on $[0,\pi/4]$
64. $\sin 2x$ on $[0,\pi/2]$
65. $1/x$ on $[2,4]$
66. xe^{-2x} on $[0,1]$

67. Let R_n be the region bounded by the x axis, the line $x=1$, and the curve $y=x^n$. The area of R_n is what fraction of the area of the triangle R_1?
68. Find the area under the graph of $f(x)=x/\sqrt{x^2+2}$ from $x=0$ to $x=2$.
69. Find the area between the graphs of $y=-x^3-2x-6$ and $y=e^x+\cos x$ from $x=0$ to $x=\pi/2$.
70. Find the area above the bends of $y=x\sin x$ which lie below the x axis. (See Fig. 7.4.2).
71. Water is flowing into a tank with a rate of $10(t^2+\sin t)$ liters per minute after time t. Calculate: (a) the number of liters stored after 30 minutes, starting at $t=0$; (b) the average flow rate in liters per minute over this 30-minute interval.
72. The velocity of a train fluctuates according to the formula $v=(100+e^{-3t}\sin 2\pi t)$ kilometers per hour. How far does the train travel: (a) between $t=0$ and $t=1$?; (b) between $t=100$ and $t=101$?
73. Evaluate $\int\sin(\pi x/2)\cos(\pi x)\,dx$ by integrating by parts two different ways and comparing the results.
74. Do Exercise 73 using the product formulas for sine and cosine.
75. Evaluate $\int\sqrt{(1+x)/(1-x)}\,dx$. [*Hint*: Multiply numerator and denominator by $\sqrt{1+x}$.]
76. Substitute $x=\sin u$ to evaluate

$$\int\frac{x\,dx}{\sqrt{1-x^2}}$$

and

$$\int\frac{x^2\,dx}{\sqrt{1-x^2}}; \qquad 0<x<1.$$

77. Evaluate:

(a) $\displaystyle\int\frac{\ln x}{x}\,dx$,

(b) $\displaystyle\int_{\sqrt{3}}^{3\sqrt{3}}\frac{dx}{x^2\sqrt{x^2+9}}$, (use $x=3\tan u$).

78. (a) Prove the following reduction formula:

$$\int\sin^n x\,dx=-\frac{\sin^{n-1}x\cos x}{n}+\frac{n-1}{n}\int\sin^{n-2}x\,dx$$

if $n\geqslant 2$, by integration by parts, with $u=\sin^{n-1}x$, $v=-\cos x$.

(b) Evaluate $\int\sin^2 x\,dx$ by using this formula.

(c) Evaluate $\int\sin^4 x\,dx$.

79. Find $\int x^n\ln x\,dx$ using $\ln x=(1/(n+1))\ln x^{n+1}$ and the substitution $u=x^{n+1}$.
80. (a) Show that:

$$\int x^m(\ln x)^n\,dx=\frac{x^{m+1}(\ln x)^n}{m+1}-\frac{n}{m+1}\int x^m(\ln x)^{n-1}\,dx.$$

(b) Evaluate $\int_1^2 x^2(\ln x)^2\,dx$.

81. The charge Q in coulombs for an RC circuit with sinusoidal switching satisfies the equation

$$\frac{dQ}{dt}+\frac{1}{0.04}Q=100\sin\left(\frac{\pi}{2-5t}\right),\quad Q(0)=0.$$

The solution is

$$Q(t)=100e^{-25t}\int_0^t e^{25x}\cos 5x\,dx.$$

(a) Find Q explicitly by means of integration by parts.

(b) Verify that $Q(1.01)=0.548$ coulomb. [*Hint*: Be sure to use radians throughout the calculation.]

82. What happens if $\int f(x)\,dx$ is integrated by parts with $u=f(x)$, $v=x$?
★83. Arthur Perverse believes that the product rule for integrals ought to be that $\int f(x)g(x)\,dx$ equals $f(x)\int g(x)\,dx+g(x)\int f(x)\,dx$. We wish to show him that this is not a good rule.

(a) Show that if the functions $f(x)=x^m$ and $g(x)=x^n$ satisfy Perverse's rule, then for fixed n the number m must satisfy a certain quadratic equation (assume $n,m\geqslant 0$).

(b) Show that the quadratic equation of part (a) does not have any real roots for any $n\geqslant 0$.

(c) Are there *any* pairs of functions, f and g, which satisfy Perverse's rule? (Don't count the case where one function is zero.)

★84. Derive an integration formula obtained by reading the quotient rule for derivatives backwards.
★85. Find $\int xe^{ax}\cos(bx)\,dx$.

Chapter 8

Differential Equations

A function may be determined by a differential equation together with initial conditions.

In the first two sections of this chapter, we study two of the simplest and most important differential equations, which describe oscillations, growth, and decay. A variation of these equations leads to the hyperbolic functions, which are important for integration and other applications. To end the chapter, we study two general classes of differential equations whose solutions can be expressed in terms of integrals. These equations, called separable and linear equations, occur in a number of interesting geometrical and physical examples. We shall continue our study of differential equations in Chapter 12 after we have learned more calculus.

8.1 Oscillations

The solution of the equation for simple harmonic oscillations may be expressed in terms of trigonometric functions.

A common problem in physics is to determine the motion of a particle in a given force field. For a particle moving on a line, the force field is given by specifying the force F as a function of the position x and time t. The problem is to write x as a function of the time t so that the equation

$$F = m\frac{d^2x}{dt^2} \qquad \text{(Force = Mass} \times \text{Acceleration)} \tag{1}$$

is satisfied, where m is the mass of the particle. Equation (1) is called *Newton's second law of motion.*[1]

If the dependence of F on x and t is given, equation (1) becomes a *differential equation* in x—that is, an equation involving x and its derivatives with respect to t. It is called *second-order* since the second derivative of x appears. (If the second derivative of x were replaced by the first derivative, we would obtain a *first-order differential equation*—these are studied in the following sections). A *solution* of equation (1) is a function $x = f(t)$ which satisfies equation (1) for all t when $f(t)$ is substituted for x.

[1] Newton always expressed his laws of motion in words. The first one to formulate Newton's laws carefully as differential equations was L. Euler around 1750. (See C. Truesdell, *Essays on the History of Mechanics*, Springer-Verlag, 1968.) In what follows we shall not be concerned with specific units of measure for force—often it is measured in *newtons* (1 newton = 1 kilogram-meter per second^2). Later, in Section 9.5, we shall pay a little more attention to units.

For example, if the force is a *constant* F_0 and we rewrite equation (1) as

$$\frac{d^2x}{dt^2} = \frac{F_0}{m},$$

we can use our knowledge of antiderivatives to conclude that

$$\frac{dx}{dt} = \frac{F_0}{m}t + C_1$$

and

$$x = \frac{1}{2}\frac{F_0}{m}t^2 + C_1t + C_2,$$

where C_1 and C_2 are constants. We see that the position of a particle moving in a constant force field is a quadratic function of time (or a linear function, if the force is zero). Such a situation occurs for vertical motion under the force of gravity near the earth's surface. More generally, if the force is a given function of t, independent of x, we can find the position as a function of time by integrating twice and using the initial position and velocity to determine the constants of integration.

In many problems of physical interest, though, the force is given as a function of *position* rather than time. One says that there is a (time-independent) force *field*, and that the particle feels the force given by the value of the field at the point where the particle happens to be.[2] For instance, if x is the downward displacement from equilibrium of a weight on a spring, then *Hooke's law* asserts that

$$F = -kx, \tag{2}$$

where k is a positive constant called the *spring constant*. (See Fig. 8.1.1.) This law, discovered experimentally, is quite accurate if x is not too large. There is

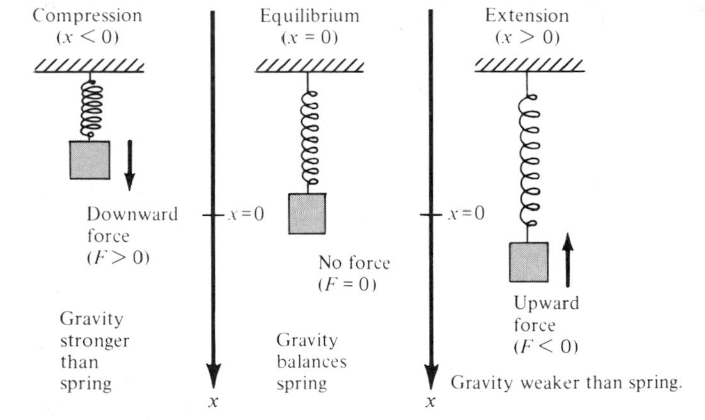

Figure 8.1.1. The force on a weight on a spring is proportional to the displacement from equilibrium.

a minus sign in the formula for F because the force, being directed toward the equilibrium, has the opposite sign to x. Substituting formula (2) into Newton's law (1) gives

$$-kx = m\frac{d^2x}{dt^2} \quad \text{or} \quad \frac{d^2x}{dt^2} = -\left(\frac{k}{m}\right)x.$$

It is convenient to write the ratio k/m as ω^2, where $\omega = \sqrt{k/m}$ is a new constant. This substitution gives us the *spring equation*:

[2] An example of a physical problem in which F depends on *both* x and t is the motion of a charged particle in a time-varying electric or magnetic field—see Exercise 13, Section 14.7.

$$\frac{d^2x}{dt^2} = -\omega^2 x. \tag{3}$$

Since x is an unknown function of t, we cannot find dx/dt by integrating the right-hand side. (In particular, dx/dt is *not* $-\frac{1}{2}\omega^2x^2 + C$, since it is t rather than x which is the independent variable.) Instead, we shall begin by using trial and error.

A good first guess, guided by the observation that weights on springs bob up and down, is

$$x = \sin t.$$

Differentiating twice with respect to t, we get

$$\frac{d^2x}{dt^2} = -\sin t = -x.$$

The factor ω^2 is missing, so we may be tempted to try $x = \omega^2 \sin t$. In this case, we get

$$\frac{d^2x}{dt^2} = -\omega^2 \sin t,$$

which is again $-x$. To bring out a *new factor* when we differentiate, we must take advantage of the chain rule. If we set $x = \sin \omega t$, then

$$\frac{dx}{dt} = \cos \omega t \frac{d(\omega t)}{dt} = \omega \cos \omega t$$

and

$$\frac{d^2x}{dt^2} = -\omega^2 \sin \omega t = -\omega^2 x,$$

which is just what we wanted. Looking back at our wrong guesses suggests that it would not hurt to put a constant factor in front, so that

$$x = B \sin \omega t$$

is also a solution for any B. Finally, we note that $\cos \omega t$ is another solution. In fact, if A and B are any two constants, then

$$x = A \cos \omega t + B \sin \omega t \tag{4}$$

is a solution of the spring equation (3), as you may verify by differentiating (4) twice. We say that the solution (4) is a *superposition* of the two solutions $A \sin \omega t$ and $B \cos \omega t$.

Example 1 Let $x = f(t) = A \cos \omega t + B \sin \omega t$. Show that x is periodic with period $2\pi/\omega$; that is, $f(t + 2\pi/\omega) = f(t)$.

Solution Substitute $t + 2\pi/\omega$ for t:

$$\begin{aligned} f\left(t + \frac{2\pi}{\omega}\right) &= A \cos\left[\omega\left(t + \frac{2\pi}{\omega}\right)\right] + B \sin\left[\omega\left(t + \frac{2\pi}{\omega}\right)\right] \\ &= A \cos[\omega t + 2\pi] + B \sin[\omega t + 2\pi] \\ &= A \cos \omega t + B \sin \omega t = f(t). \end{aligned}$$

Here we used the fact that the sine and cosine functions are themselves periodic with period 2π. ▲

The constants A and B are similar to the constants which arise when antiderivatives are taken. For any value of A and B, we have a solution. If we assign particular values to A and B, we get a *particular solution*. The choice of particular values of A and B is often determined by specifying *initial conditions*.

Example 2 Find a solution of the spring equation $d^2x/dt^2 = -\omega^2 x$ for which $x = 1$ and $dx/dt = 1$ when $t = 0$.

Solution In the solution $x = A\cos\omega t + B\sin\omega t$, we have to find A and B. Now $x = A$ when $t = 0$, so $A = 1$. Also $dx/dt = \omega B\cos\omega t - \omega A\sin\omega t = \omega B$ when $t = 0$. To make $dx/dt = 1$, we choose $B = 1/\omega$, and so the required solution is $x = \cos\omega t + (1/\omega)\sin\omega t$. ▲

In general, if we are given the initial conditions that $x = x_0$ and $dx/dt = v_0$ when $t = 0$, then

$$x = x_0\cos\omega t + \frac{v_0}{\omega}\sin\omega t \tag{5}$$

is the unique function of the form (4) which satisfies these conditions.

Example 3 Solve for x: $d^2x/dt^2 = -x$, $x = 0$ and $dx/dt = 1$ when $t = 0$.

Solution Here $x_0 = 0$, $v_0 = 1$, and $\omega = 1$, so $x = x_0\cos\omega t + (v_0/\omega)\sin\omega t = \sin t$. ▲

Physicists expect that the motion of a particle in a force field is completely determined once the initial values of position and velocity are specified. Our solution of the spring equation will meet the physicists' requirements if we can show that *every* solution of the spring equation (3) is of the form (4). We turn to this task next.

In deriving formula (5) we saw that there are enough solutions of the form (4) so that x and dx/dt can be specified arbitrarily at $t = 0$. Thus if $x = f(t)$ is any solution of equation (3), then the function $g(t) = f(0)\cos\omega t + [f'(0)/\omega]\sin\omega t$ is a solution of the special form (4) with the same initial conditions as f: $g(0) = f(0)$ and $g'(0) = f'(0)$. We will now show that $f(t) = g(t)$ for all t by using the following fact: if $h(t)$ is any solution of equation (3), then the quantity $E = \frac{1}{2}\{[h'(t)]^2 + [\omega h(t)]^2\}$ is constant over time. This expression is called the *energy* of the solution h. To see that E is constant over time, we differentiate using the chain rule:

$$\frac{dE}{dt} = h'(t)h''(t) + \omega^2 h(t)h'(t) = h'(t)\{h''(t) + \omega^2 h(t)\}. \tag{6}$$

This equals zero since $h'' + \omega^2 h = 0$; thus E is constant over time. Now if f and g are solutions of equation (3) with $f(0) = g(0)$ and $f'(0) = g'(0)$, then $h(t) = f(t) - g(t)$ is also a solution with $h(0) = 0$ and $h'(0) = 0$. Thus the energy $E = \frac{1}{2}\{[h'(t)]^2 + [\omega h(t)]^2\}$ is constant; but it vanishes at $t = 0$, so it is identically zero. Thus, since two non-negative numbers which add to zero must both be zero: $h'(t) = 0$ and $\omega h(t) = 0$. In particular, $h(t) = 0$, and so $f(t) = g(t)$ as required.

The solution (4) of the spring equation can also be expressed in the form

$$x = \alpha\cos(\omega t - \theta),$$

where α and θ are constants. In fact, the addition formula for cosine gives

$$\alpha\cos(\omega t - \theta) = \alpha\cos\omega t\cos\theta + \alpha\sin\omega t\sin\theta. \tag{7}$$

This will be equal to $A\cos\omega t + B\sin\omega t$ if

$$\alpha\cos\theta = A \quad \text{and} \quad \alpha\sin\theta = B.$$

Thus α and θ must be the polar coordinates of the point whose cartesian coordinates are (A, B), and so we can always find such an α and θ with $\alpha \geqslant 0$. The form (7) is convenient for plotting, as shown in Fig. 8.1.2.

In Fig. 8.1.2 notice that the solution is a cosine curve with *amplitude* α which is shifted by the *phase shift* θ/ω. The number ω is called the *angular*

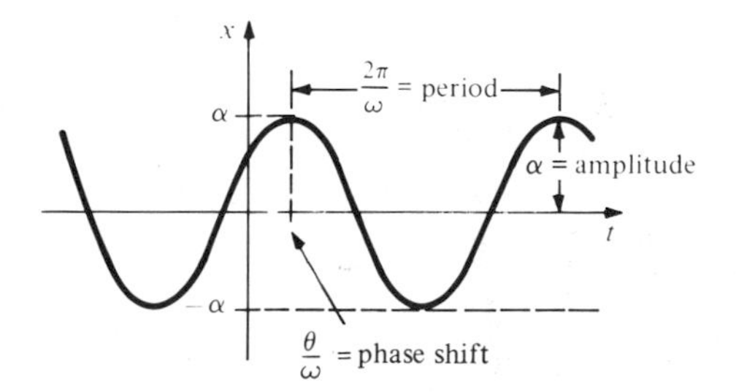

Figure 8.1.2. The graph of $x = \alpha\cos(\omega t - \theta)$.

frequency, since it is the time rate of change of the "angle" $\omega t - \theta$ at which the cosine is evaluated. The number of oscillations per unit time is the *frequency* $\omega/2\pi$ (= 1/period).

The motion described by the solutions of the spring equation is called *simple harmonic motion*. It arises whenever a system is subject to a restoring force proportional to its displacement from equilibrium. Such oscillatory systems occur in physics, biology, electronics, and chemistry.

Simple Harmonic Motion

Every solution of the *spring equation*

$$\frac{d^2x}{dt^2} = -\omega^2 x \qquad \text{has the form} \qquad x = A\cos\omega t + B\sin\omega t,$$

where A and B are constants.

The solution can also be written

$$x = \alpha\cos(\omega t - \theta),$$

where (α, θ) are the polar coordinates of (A, B). [This function is graphed in Fig. 8.1.2.]

If the values of x and dx/dt are specified to be x_0 and v_0 at $t = 0$, then the unique solution is

$$x - x_0\cos\omega t + (v_0/\omega)\sin\omega t.$$

Example 4 Sketch the graph of the solution of $d^2x/dt^2 + 9x = 0$ satisfying $x = 1$ and $dx/dt = 6$ when $t = 0$.

Solution Using (5) with $\omega = 3$, $x_0 = 1$, and $v_0 = 6$, we have

$$x = \cos(3t) + 2\sin(3t) = \alpha\cos(3t - \theta).$$

Since $(A, B) = (1, 2)$, and (α, θ) are its polar coordinates,

$$\alpha = \sqrt{1^2 + 2^2} = \sqrt{5} \approx 2.2$$

and

$$\theta = \tan^{-1}2 \approx 1.1 \text{ radians (or } 63°),$$

so $\theta/\omega \approx 0.37$. The period is $\tau = 2\pi/\omega \approx 2.1$. Thus we can plot the graph as shown in Fig. 8.1.3. ▲

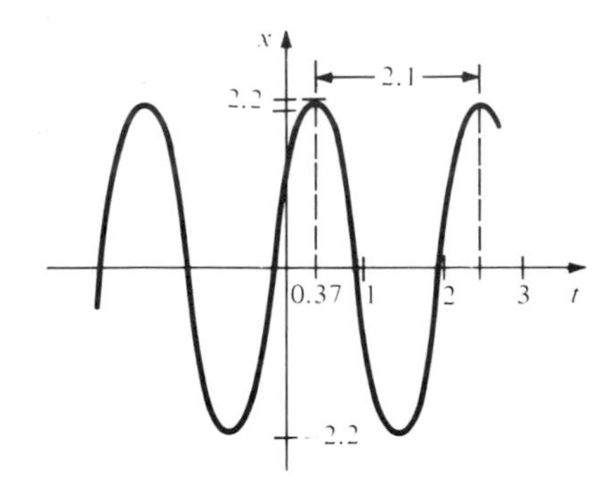

Figure 8.1.3. The graph of $x = 2.2\cos(3t - 1.1)$.

As usual, the independent variable need not always be called t, nor does the dependent variable need to be called x.

Example 5 (a) Solve for y: $d^2y/dx^2 + 9y = 0$, $y = 1$ and $dy/dx = -1$ when $x = 0$.
(b) Sketch the graph of y as a function of x.

Solution (a) Here $y_0 = 1$, $v_0 = -1$, and $\omega = \sqrt{9} = 3$ (using x in place of t and y in place of x), so

$$y = y_0 \cos \omega x + \frac{v_0}{\omega} \sin \omega x = \cos 3x - \tfrac{1}{3} \sin 3x.$$

(b) The polar coordinates of $(1, -\frac{1}{3})$ are given by $\alpha = \sqrt{1 + 1/9} = \sqrt{10/3} \approx 1.05$ and $\theta = \tan^{-1}(-\frac{1}{3}) \approx -0.32$ (or $-18°$). Hence $y = \alpha\cos(\omega x - \theta)$ becomes $y = 1.05\cos(3x + 0.32)$, which is sketched in Fig. 8.1.4. Here $\theta/\omega = -0.1$ and $2\pi/\omega = 2.1$. ▲

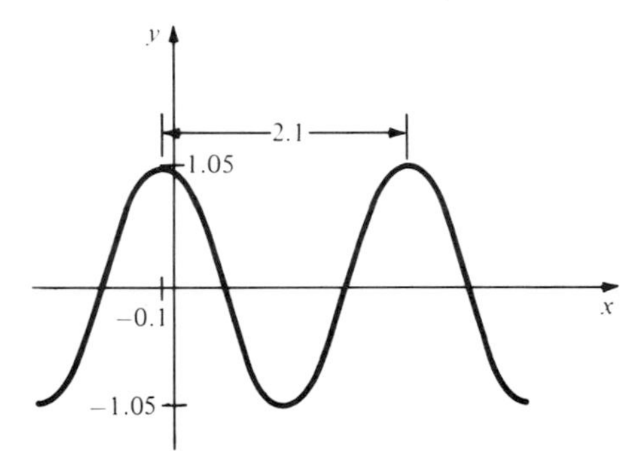

Figure 8.1.4. The graph of $y = (1.05)\cos(3x + 0.32)$.

A Remark on Notation. Up until now we have distinguished *variables*, which are mathematical objects that represent "quantities," and *functions*, which represent relations between quantities. Thus, when $y = f(x)$, we have written $f'(x)$ and dy/dx but *not* y', df/dx, or $y(x)$. It is common in mathematical writing to use the same symbol to denote a function and its dependent variable; thus one sometimes writes $y = y(x)$ to indicate that y is a function of x and then writes "$y' = dy/dx$," "$y(3)$ is the value of y when $x = 3$," and so on. Beginning with the next example, we will occasionally drop our scruples in distinguishing functions from variables and will use this abbreviated notation.

Example 6 Let M be a weight with mass 1 gram on a spring with spring constant $\frac{3}{2}$. Let the weight be initially extended by a distance of 1 centimeter moving at a velocity of 2 centimeters per second.

(a) How fast is M moving at $t = 3$?
(b) What is M's acceleration at $t = 4$?
(c) What is M's maximum displacement from the rest position? When does it occur?
(d) Sketch a graph of the solution.

Solution Let $x = x(t)$ denote the position of M at time t. We use the spring equation (3) with $\omega = \sqrt{k/m}$, where k is the spring constant and m is the mass of M. Since $k = \frac{3}{2}$ and $m = 1$, ω is $\sqrt{3/2}$. At $t = 0$, M is extended by a distance of 1 centimeter and moving at a velocity of 2 centimeters per second, so $x_0 = 1$ and $v_0 = 2$.

Now we have all the information we need to solve the spring equation. Applying formula (5) gives

$$x(t) = \cos\sqrt{3/2}\,t + \frac{2}{\sqrt{3/2}}\sin\sqrt{3/2}\,t.$$

(a) $x'(t) = -\sqrt{3/2}\sin\sqrt{3/2}\,t + 2\cos\sqrt{3/2}\,t$. Substituting $t = 3$ gives

$$x'(3) = -\sqrt{3/2}\sin 3\sqrt{3/2} + 2\cos 3\sqrt{3/2} \approx -1.1 \text{ centimeters per second.}$$

(Negative velocity represents upward motion.)

(b) $$x''(t) = \frac{d}{dt}\left(-\sqrt{\frac{3}{2}}\sin\sqrt{\frac{3}{2}}\,t + 2\cos\sqrt{\frac{3}{2}}\,t\right) = -\frac{3}{2}\cos\sqrt{\frac{3}{2}}\,t - \sqrt{6}\sin\sqrt{\frac{3}{2}}\,t$$

and thus the acceleration at $t = 4$ is

$$x''(4) = -\tfrac{3}{2}\cos 2\sqrt{6} - \sqrt{6}\sin 2\sqrt{6} \approx 2.13 \text{ centimeters per second}^2$$

(c) The simplest way to find the maximum displacement is to use the "phase-amplitude" form (7). The maximum displacement is the amplitude $\alpha = \sqrt{A^2 + B^2}$, where A and B are the coefficients of sin and cos in the solution. Here, $A = 1$ and $B = 2/\sqrt{3/2}$, so $\alpha^2 = 1 + 4/(3/2) = 1 + 8/3 = 11/3$, so $\alpha = \sqrt{11/3} \approx 1.91$ centimeters, which is a little less than twice the initial displacement.

(d) To sketch a graph we also need the phase shift. Now $\theta = \tan^{-1}(B/A)$, which is in the first quadrant since A and B are positive. Thus $\theta = \tan^{-1}(2/\sqrt{3/2}) \approx 1.02$, and so the maximum point on the graph (see Fig. 8.1.2) occurs at $\theta/\omega = 1.02/\sqrt{3/2} \approx 0.83$. The period is $2\pi/\omega \approx 5.13$. The graph is shown in Fig. 8.1.5. ▲

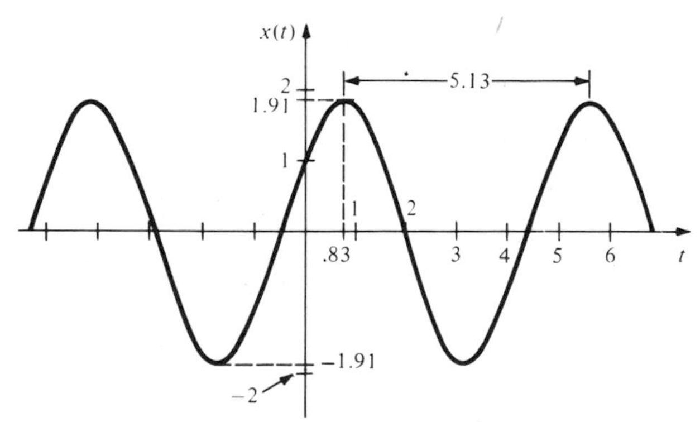

Figure 8.1.5. The graph of $x(t) = \cos\sqrt{\frac{3}{2}}\,t + \sqrt{\frac{8}{3}}\sin\sqrt{\frac{3}{2}}\,t \approx 1.91\cos(\sqrt{\frac{3}{2}}\,t - 1.02)$.

Supplement to Section 8.1: Linearized Oscillations

The spring equation can be applied to determine the *approximate* motion of *any* system subject to a restoring force, even if the force is not linear in the displacement. Such forces occur in more realistic models for springs and in equations for electric circuits. Suppose that we wish to solve the equation of motion

$$m\frac{d^2x}{dt^2} = f(x), \tag{8}$$

where the force function $f(x)$ satisfies the conditions: (i) $f(x_0) = 0$; and (ii) $f'(x_0) < 0$, for some position x_0. The point x_0 is an *equilibrium* position since the constant function $x(t) = x_0$ satisfies the equation of motion, by condition (i). By condition (ii), the force is positive when x is near x_0 and $x < x_0$, while the force is negative when x is near x_0 and $x > x_0$. Thus the particle is being pushed back toward x_0 whenever it is near that point, just as with the spring in Figure 8.1.1.

Rather than trying to solve equation (8) directly, we shall replace $f(x)$ by its linear approximation $f(x_0) + f'(x_0)(x - x_0)$ at x_0. Since $f(x_0) = 0$, the equation (8) becomes

$$m\frac{d^2x}{dt^2} = f'(x_0)(x - x_0), \tag{8'}$$

which is called the *linearization* of equation (8) at x_0. If we write k for the positive number $-f'(x_0)$ and $y = x - x_0$ for the displacement from equilibrium, then we get

$$m\frac{d^2y}{dt^2} = -ky,$$

which is precisely the spring equation.

We thus conclude that, to the degree that the linear approximation of the force is valid, the particle oscillates around the equilibrium point x_0 with period $2\pi/\sqrt{-f'(x_0)/m}$. It can be shown that the particle subject to the exact force law (8) also oscillates around x_0, but with a period which depends upon the amplitude of the oscillations. As the amplitude approaches zero, the period approaches $2\pi/\sqrt{-f'(x_0)/m}$, which is the period for the linearized equation.

Here is an application of these ideas:

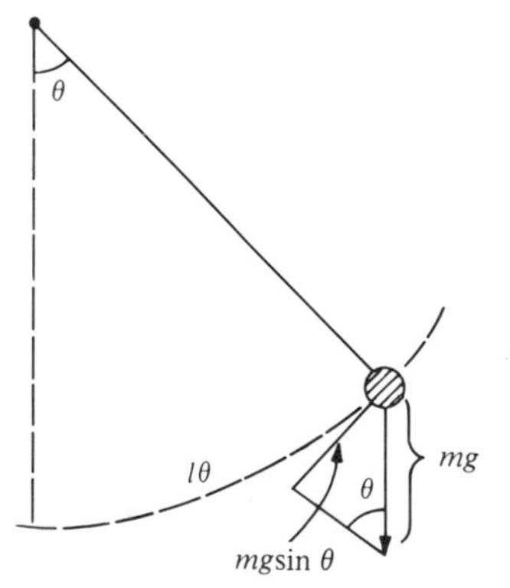

Figure 8.1.6. The forces acting on a pendulum.

By decomposing the gravitational force on a pendulum of mass m and length l into components parallel and perpendicular to the pendulum's axis, it can be shown that the displacement angle θ of the pendulum from its equilibrium (vertical) position satisfies the differential equation $m(d^2\theta/dt^2) = -m(g/l)\sin\theta$, where $g = 9.8$ meters per second^2 is the gravitational constant. (See Fig. 8.1.6). The force function is $f(\theta) = -(mg/l)\sin\theta$. Since $f(0) = 0$, $\theta = 0$ is an equilibrium point. Since $f'(0) = -(mg/l)\sin'(0) = -mg/l$, the linearized equation is $m\,d^2\theta/dt^2 = -(mg/l)\theta$. The period of oscillations for the linearized equation is thus $2\pi/\sqrt{(mg/l)m} = 2\pi\sqrt{l/g}$. (See Review Exercise 83 for information on the solution of the nonlinear equation.)

A point x_0 satisfying the conditions above is called a *stable equilibrium point*. The word *stable* refers to the fact that motions which start near x_0 with small initial velocity stay near x_0.[3] If $f(x_0) = 0$ but $f'(x_0) > 0$, we have an *unstable* equilibrium point (see Section 8.3).

[3] We have only proved stability for the linearized equations. Using conservation of energy, one can show that the motion for the exact equations stays near x_0 as well. (See Exercise 33.)

Exercises for Section 8.1

1. Show that $f(t) = \cos(3t)$ is periodic with period $2\pi/3$.
2. Show that $f(t) = 8\sin(\pi t)$ is periodic with period 2.
3. Show that $f(t) = \cos(6t) + \sin(3t)$ is periodic with period $2\pi/3$.
4. Show that $f(t) = 3\sin(\pi t/2) + 8\cos(\pi t)$ is periodic with period 4.

In Exercises 5–8, find the solution of the given equation with the prescribed values of x and $x' = dx/dt$ at $t = 0$.

5. $x'' + 9x = 0$, $x(0) = 1$, $x'(0) = -2$.
6. $x'' + 16x = 0$, $x(0) = -1$, $x'(0) = -1$.
7. $x'' + 12x = 0$, $x(0) = 0$, $x'(0) = -1$.
8. $x'' + 25x = 0$, $x(0) = 1$, $x'(0) = 0$.

In Exercises 9–12, sketch the graph of the given function and find the period, amplitude, and phase shift.

9. $x = 3\cos(3t - 1)$.
10. $x = 2\cos(5t - 2)$.
11. $x = 4\cos(t + 1)$.
12. $x = 6\cos(3t + 4)$.

In Exercises 13–16, solve the given equation for x and sketch the graph.

13. $\dfrac{d^2x}{dt^2} + 4x = 0$; $x = -1$ and $\dfrac{dx}{dt} = 0$ when $t = 0$.
14. $\dfrac{d^2x}{dt^2} + 16x = 0$, $x = 1$ and $\dfrac{dx}{dt} = 0$ when $t = 0$.
15. $\dfrac{d^2x}{dt^2} + 25x = 0$, $x = 5$ and $\dfrac{dx}{dt} = 5$ when $t = 0$.
16. $\dfrac{d^2x}{dt^2} + 25x = 0$, $x = 5$ at $t = 0$, and $\dfrac{dx}{dt} = 5$ when $t = \pi/4$.
17. Find the solution of $\dfrac{d^2y}{dt^2} = -4y$ for which $y = 1$ and $\dfrac{dy}{dt} = 3$ when $t = 0$.
18. Find $y = f(x)$ if $f'' + 4f = 0$ and $f(0) = 0$, $f'(0) = -1$.
19. Suppose that $f(x)$ satisfies $f'' + 16f = 0$ and $f(0) = 2$, $f'(0) = 0$. Sketch the graph $y = f(x)$.
20. Suppose that $z = g(r)$ satisfies $9z'' + z = 0$ and $z(0) = -1$, $z'(0) = 0$. Sketch the graph $z = g(r)$.
21. A mass of 1 kilogram is hanging from a spring. If $x = 0$ is the equilibrium position, it is given that $x = 1$ and $dx/dt = 1$ when $t = 0$. The weight is observed to oscillate with a frequency of twice a second.
    (a) What is the spring constant?
    (b) Sketch the graph of x as a function of t, indicating the amplitude of the motion on your drawing.
22. An observer sees a weight of 5 grams on a spring undergoing the motion $x(t) = 6.1\cos(2t - \pi/6)$.
    (a) What is the spring constant?
    (b) What is the force acting on the weight at $t = 0$? At $t = 2$?
23. What happens to the frequency of oscillations if three equal masses are hung from a spring where there was one mass before?
24. Find a differential equation of the "spring" type satisfied by the function $y(t) = 3\cos(t/4) - \sin(t/4)$.

★25. A "flabby" spring exerts a force $f(x) = -3x + 2x^3$ when it is displaced a distance x from its equilibrium state, $x = 0$.
    (a) Write the equation of motion for an object of mass 27, vibrating on this spring.
    (b) Write the linearized equation of motion at $x_0 = 0$.
    (c) Find the period of linearized oscillations.

★26. (a) Find the equilibrium position of an object which satisfies the equation of motion

$$4\frac{d^2x}{dt^2} = -x^3 + x^2 - x + 1.$$

(b) What is the frequency of linearized oscillations?

★27. An atom of mass m in a linear molecule is subjected to forces of attraction by its neighbors given by

$$f(x) = k_1(x - x_1)^3 + k_2(x - x_2)^3,$$
$$k_1, k_2 > 0, \quad 0 < x_1 < x < x_2.$$

(a) Compute the equilibrium position.
(b) Show that motion near this equilibrium is unstable.

★28. The equation for a spring with friction is

$$m\frac{d^2x}{dt^2} = -kx - \delta\frac{dx}{dt}$$

(spring equation with damping).
(a) If $\delta^2 < 4km$, check that a solution is

$$x(t) = e^{-\delta t/2m}(A\cos\omega t + B\sin\omega t),$$

where $\omega^2 = k/m - \delta^2/4m^2 > 0$.
(b) Sketch the general appearance of the graph of the solution in (a) and define the "period" of oscillation.
(c) If the force $-kx$ is replaced by a function $f(x)$ satisfying $f(0) = 0$, $f'(0) < 0$, find a formula for the frequency of damped linearized oscillations.

★29. Suppose that $x = f(t)$ satisfies the spring equation. Let $g(t) = at + b$, where a and b are constants. Show that if the composite function $f \circ g$ satisfies the spring equation (with the same ω), then $a = \pm 1$. What about b?

★30. (a) Suppose that $f(t)$ is given and that $y = g(t)$ satisfies $d^2y/dt^2 + \omega^2 y = f(t)$. Show that

$x = y + A \sin \omega t + B \cos \omega t$ represents the general solution of $d^2x/dt^2 + \omega^2 x = f$; that is, x is a solution and any solution has this form. One calls y a *particular* solution and x the *general* solution.

(b) Solve $d^2x/dt^2 + \omega^2 x = k$ if $x = 1$ and $dx/dt = -1$ when $t = 0$; k is a nonzero constant.

(c) Solve $d^2x/dt^2 + \omega^2 x = \omega^2 t$ if $x = -1$ and $dx/dt = 3$ when $t = 0$.

Exercises 31 and 32 outline the complete proof of the following theorem using the "method of variation of constants": *Let $x = f(t)$ be a twice-differentiable function of t such that $(d^2x/dt^2) + \omega^2 x = 0$. Then $x = A \cos \omega t + B \sin \omega t$ for constants A and B.*

★31. Some preliminary calculations are done first. Write

$$x = A(t)\cos \omega t + B(t)\sin \omega t. \tag{9}$$

It is possible to choose $A(t)$ and $B(t)$ in many ways, since for each t either $\sin \omega t$ or $\cos \omega t$ is nonzero. To determine $A(t)$ and $B(t)$ we add a second equation:

$$\frac{dx}{dt} = -\omega A(t)\sin \omega t + \omega B(t)\cos \omega t. \tag{10}$$

This equation is obtained by differentiating (9) *pretending* that $A(t)$ and $B(t)$ are constants. Since this is what we are trying to prove, we should be very suspicious here of circular reasoning. But push on and see what happens. Show that

$$B(t) = x \sin \omega t + \frac{dx/dt}{\omega} \cos \omega t. \tag{11}$$

Similarly, show that

$$A(t) = x \cos \omega t - \frac{dx/dt}{\omega} \sin \omega t. \tag{12}$$

★32. Use the calculations in Exercise 31 to give the proof of the theorem, making sure to avoid circular reasoning. We are given $x = f(t)$ and ω such that $(d^2x/dt^2) + \omega^2 x = 0$. Define $A(t)$ and $B(t)$ by equations (11) and (12). Show that $A(t)$ and $B(t)$ are in fact constants by differentiating (11) and (12) to show that $A'(t)$ and $B'(t)$ are identically zero. Then rewrite formulas (11) and (12) as

$$B = x(t)\sin \omega t + \frac{dx/dt}{\omega} \cos \omega t, \tag{13}$$

and

$$A = x(t)\cos \omega t - \frac{dx/dt}{\omega} \sin \omega t. \tag{14}$$

Use these formulas to show $A \cos \omega t + B \sin \omega t = x$, which proves the theorem.

★33. Suppose that $m(d^2x/dt^2) = f(x)$, where $f(x_0) = 0$ and $f'(x_0) < 0$. Let $V(x)$ be an antiderivative of $-f$.

(a) Show that x_0 is a local minimum of V.

(b) Show that $dE/dt = 0$, where the energy E is given by $E = \frac{1}{2}m(dx/dt)^2 + V(x)$.

(c) Use conservation of energy from (b) to show that if dx/dt and $x - x_0$ are sufficiently small at $t = 0$, then they both remain small.

8.2 Growth and Decay

The solution of the equation for population growth may be expressed in terms of exponential functions.

Many quantities, such as bank balances, populations, the radioactivity of ores, and the temperatures of hot objects change at a rate which is proportional to the current value of the quantity. In other words, if $f(t)$ is the quantity at time t, then f satisfies the differential equation

$$f'(t) = \gamma f(t), \tag{1}$$

where γ is a constant. For example, in the specific case of temperature, it is an experimental fact that the temperature of a hot object decreases at a rate proportional to the difference between the temperature of the object and that of its surroundings. This is called *Newton's law of cooling*.

Example 1 The temperature of a hot bowl of porridge decreases at a rate 0.0837 times the difference between its present temperature and room temperature (fixed at 20°C). Write down a differential equation for the temperature of the porridge. (Time is measured in minutes and temperature in °C.)

Solution Let T be the temperature (°C) of the porridge and let $f(t) = T - 20$ be its temperature above 20°C. Then $f'(t) = dT/dt$ and so

$$f'(t) = -(0.0837)f(t)$$

i.e.,

$$\frac{dT}{dt} = -(0.0837)(T - 20).$$

The minus sign is used because the temperature is *decreasing* when T is greater than 20; $\gamma = -0.0837$. ▲

We solve equation (1) by guesswork, just as we did the spring equation. The answer must be a function which produces itself times a constant when differentiated once. It is reasonable that such a function should be related to the exponential since e^t has the reproductive property $(d/dt)e^t = e^t$. To get a factor γ, we replace t by γt. Then $(d/dt)e^{\gamma t} = \gamma e^{\gamma t}$, by the chain rule. We can also insert a constant factor A to get

$$\frac{d}{dt}(Ae^{\gamma t}) = \gamma(Ae^{\gamma t}).$$

Thus $f(t) = Ae^{\gamma t}$ solves equation (1). If we pick $t = 0$, we see that $A = f(0)$. This gives us a solution of equation (1); we shall show below that it is the only solution.

The Solution of $f' = \gamma f$

Given $f(0)$, there is one and only one solution to the differential equation $f'(t) = \gamma f(t)$, namely

$$f(t) = f(0)e^{\gamma t} \tag{2}$$

To show that formula (2) gives the *only* solution, let us suppose that $g(t)$ also satisfies $g'(t) = \gamma g(t)$ and $g(0) = f(0)$. We will show that $g(t) = f(0)e^{\gamma t}$. To do this, consider the quotient

$$h(t) = \frac{g(t)}{e^{\gamma t}} = e^{-\gamma t}g(t)$$

and differentiate:

$$h'(t) = -\gamma e^{-\gamma t}g(t) + e^{-\gamma t}g'(t) = -\gamma e^{-\gamma t}g(t) + \gamma e^{-\gamma t}g(t) = 0.$$

Since $h'(t) = 0$, we may conclude that h is constant; but $h(0) = e^{-0}g(0) = f(0)$, so $e^{-\gamma t}g(t) = h(t) = f(0)$, and thus

$$g(t) = f(0)e^{\gamma t} = f(t),$$

as required.

Example 2 If $dx/dt = 3x$, and $x = 2$ at $t = 0$, find x for all t.

Solution If $x = f(t)$, then $f(0) = 2$ and $f' = 3f$, so $\gamma = 3$ in the box above. Hence, by formula (2), $x = f(t) = 2e^{3t}$. ▲

Example 3 Find a formula for the temperature of the bowl of porridge in Example 1 if it starts at 80°C. Jane Cool refuses to eat the porridge when it is too cold—namely, if it falls below 50°C. How long does she have to come to the table?

Solution Let $f(t) = T - 20$ as in Example 1. Then $f'(t) = -0.0837f(t)$ and the initial condition is $f(0) = 80 - 20 = 60$. Therefore

$$f(t) = 60e^{-0.0837t}.$$

Hence $T = f(t) + 20 = 60e^{-0.0837t} + 20$. When $T = 50$, we have

$$50 = 60e^{-0.0837t} + 20$$
$$1 = 2e^{-0.0837t}$$
$$e^{0.0837t} = 2$$
$$0.0837t = \ln 2 = 0.693.$$

Thus $t = 8.28$ minutes. Jane has a little more than 8 minutes before the temperature drops to 50°C. ▲

Note how the behavior of solution (2) depends on the sign of γ. If $\gamma > 0$, then $e^{\gamma t} \to \infty$ as $t \to \infty$ (growth); if $\gamma < 0$, then $e^{\gamma t} \to 0$ as $t \to \infty$ (decay). See Fig. 8.2.1.

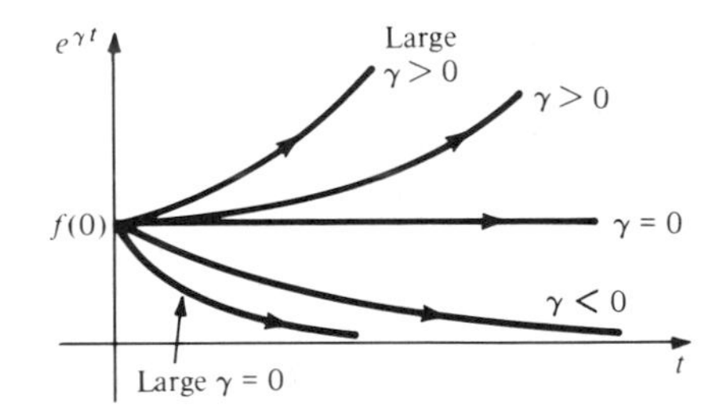

Figure 8.2.1. Growth occurs if $\gamma > 0$, decay if $\gamma < 0$.

A quantity which depends on time according to equation (1) (or, equivalently, (2)) is said to undergo *natural growth* or *decay*.

Natural Growth or Decay

The solution of $f' = \gamma f$ is $f(t) = f(0)e^{\gamma t}$ which grows as t increases if $\gamma > 0$ and which decays as t increases if $\gamma < 0$.

Example 4 Suppose that $y = f(x)$ satisfies $dy/dx + 3y = 0$ and $y = 2$ at $x = 0$. Sketch the graph $y = f(x)$.

Solution The equation may be written $dy/dx = -3y$ which has the form of equation (1) with $\gamma = -3$ and the independent variable t replaced by x. By formula (2) the solution is $y = 2e^{-3x}$. The graph is sketched in Fig. 8.2.2. ▲

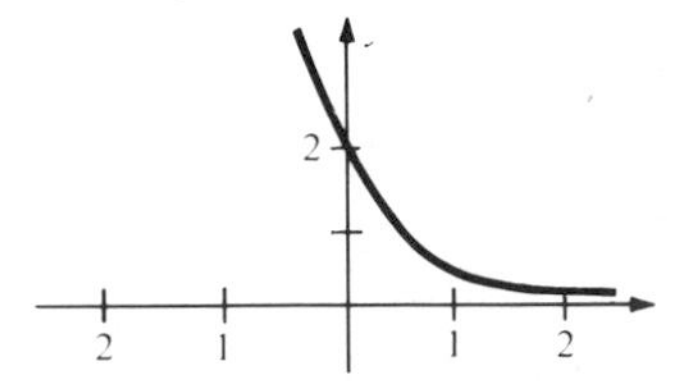

Figure 8.2.2. The graph of $y = 2e^{-3x}$.

If a quantity $f(t)$ is undergoing natural growth or decay, i.e., $f(t) = f(0)e^{\gamma t}$, we notice that

$$\frac{f(t+s)}{f(t)} = \frac{f(0)e^{\gamma(t+s)}}{f(0)e^{\gamma t}} = e^{\gamma s} = \frac{f(s)}{f(0)},$$

so

$$\frac{f(t+s)}{f(t)} = \frac{f(s)}{f(0)}. \tag{3}$$

Thus the percentage increase or decrease in f over a time interval of length s is fixed, independent of when we start. This property, characteristic of natural growth or decay is called *uniform growth* or *decay*. It states, for example, that if you leave money in a bank with a fixed interest rate, then the percentage increase in your balance over each period of a given length (say 3 years) is the same.

We can show that if f undergoes uniform growth (or decay), then f undergoes natural growth (or decay). Indeed, write equation (3) as

$$f(t+s) = \frac{f(t)f(s)}{f(0)}$$

and differentiate with respect to s:

$$f'(t+s) = \frac{f(t)f'(s)}{f(0)}.$$

Now set $s = 0$ and let $\gamma = f'(0)/f(0)$:

$$f'(t) = \gamma f(t),$$

which is the law of natural growth. Thus natural growth and uniform growth are equivalent notions.

We shall now discuss *half-life problems*. It is a physical law that radioactive substances decay at a rate proportional to the amount of the substance present. If $f(t)$ denotes the amount of the substance at time t, then the physical law states that $f'(t) = -\kappa f(t)$ for a positive constant κ. (The *minus* sign is inserted since the substance is *decaying*.) Thus, formula (2) with $\gamma = -\kappa$ gives $f(t) = f(0)e^{-\kappa t}$. The *half-life* $t_{1/2}$ is the time required for half the substance to remain. Therefore $f(t_{1/2}) = \frac{1}{2}f(0)$, so $f(0)e^{-\kappa t_{1/2}} = \frac{1}{2}f(0)$. Hence $2 = e^{\kappa t_{1/2}}$, so

$$t_{1/2} = (1/\kappa)\ln 2. \tag{4}$$

Half-Life

If a quantity decays according to the law $f'(t) = -\kappa f(t)$, it will be half gone after the elapse of time $t_{1/2} = (1/\kappa)\ln 2$; $t_{1/2}$ is called the *half-life*.

Example 5 Radium decreases at a rate of 0.0428% per year. What is its half-life?

Solution *Method 1*. We use the preceding box. Here $\kappa = -0.000428$, so the half-life is $t_{1/2} = (\ln 2)/0.000428 \approx 1620$ years.

Method 2. It is efficient in many cases to rederive the formula for half-life rather than memorizing it. With this approach, the solution looks like this: Let $f(t)$ denote the amount of radium at time t. We have $f'(t) = -0.000428f(t)$, so $f(t) = f(0)e^{-0.000428t}$. If $f(t) = \frac{1}{2}f(0)$, then $\frac{1}{2} = e^{-0.000428t}$; that is, $e^{0.000428t} = 2$, or $0.000428t = \ln 2$. Hence, $t = (\ln 2)/0.000428 \approx 1620$ years. ▲

Example 6 A certain radioactive substance has a half-life of 5085 years. What percentage will remain after an elapse of 10,000 years?

Solution If $f(t)$ is the amount of the substance after an elapse of time t, then $f(t) = f(0)e^{-\kappa t}$ for a constant κ. Since the half-life is 5085, $\frac{1}{2} = e^{-5085\kappa}$, i.e., $\kappa = (1/5085)\ln 2$. The amount after time $t = 10{,}000$ is

$$f(t) = f(0)e^{-10{,}000\kappa} = f(0)e^{-10{,}000\ln 2/5085} = 0.256\,f(0).$$

Thus 25.6% remains. ▲

Another quantity which often changes at a rate proportional to the amount present is a population.

Example 7 The population of the planet $\delta oo\mu$ is increasing at an instantaneous rate of 5% per year. How long will it take for the population to double?

Solution Let $P(t)$ denote the population. Since the rate of increase is 5%, $P'(t) = 0.05P(t)$, so $P(t) = P(0)e^{0.05t}$. In order for $P(t)$ to be $2P(0)$, we should have $2 = e^{0.05t}$; that is, $0.05t = \ln 2$ or $t = 20\ln 2$. Using $\ln 2 = 0.6931$, we get $t \approx 13.862$ years. ▲

In this and similar examples, there is the possibility of confusion over the meaning of phrases like "increases at a rate of 5% per year." When the word "instantaneous" is used, it means that the *rate* is 5%, i.e., $P' = 0.05P$. This does *not* mean that after one year the population has increased by 5%—it will in fact be greater than that.

In Section 6.4 we saw the distinction between annual rates and instantaneous rates in connection with problems of finance. If an initial principal P_0 is left in an account earning r% compounded continuously, this means that the amount of money P in the account at time t changes according to

$$\frac{dP}{dt} = \frac{r}{100}P.$$

Thus, by formula (2),

$$P(t) = e^{rt/100}P_0. \tag{5}$$

The *annual percentage rate* is the percentage increase after one year, namely

$$100\left(\frac{P(1) - P_0}{P_0}\right) = 100(e^{r/100} - 1). \tag{6}$$

This agrees with formula (8) derived in Section 6.4 by a different method.

Example 8 How long does it take for a quantity of money to triple if it is left in an account earning 8.32% interest compounded continuously?

Solution Let P_0 be the amount deposited. By formula (5),

$$P(t) = e^{0.0832t}P_0.$$

If $P(t) = 3P_0$, then

$$3 = e^{0.0832t},$$

$$0.0832t = \ln 3,$$

$$t = \frac{\ln 3}{0.0832} \approx 13.2 \text{ years. } ▲$$

Exercises for Section 8.2

1. The temperature T of a hot iron decreases at a rate 0.11 times the difference between its present temperature and room temperature (20°C). If time is measured in minutes, write a differential equation for the temperature of the iron.
2. A population P of monkeys increases at a rate 0.051 per year times the current population. Write down a differential equation for P.
3. The amount Q in grams of a radioactive substance decays at a rate 0.00028 per year times the current amount present. Write a differential equation for Q.
4. The amount M of money in a bank increases at an instantaneous rate of 13.51% per year times the present amount. Write a differential equation for M.

Solve the differential equations in Exercises 5–12 using the given data.

5. $f' = -3f, f(0) = 2$.
6. $\frac{dx}{dt} = x$, $x = 3$ when $t = 0$.
7. $\frac{dx}{dt} - 3x = 0$, $x = 1$ when $t = 0$.
8. $\frac{du}{dr} - 13u = 0$, $u = 1$ when $r = 0$.
9. $\frac{dy}{dt} = 8y$, $y = 2$ when $t = 1$.
10. $\frac{dy}{dx} = -10y$, $y = 1$ when $x = 1$.
11. $\frac{dv}{ds} + 2v = 0$, $v = 2$ when $s = 3$.
12. $\frac{dw}{dx} + aw = 0$, $w = b$ when $x = c$ (a, b, c constants).
13. If the iron in Exercise 1 starts out at 210°C, how long (in minutes) will it take for it to cool to 100°C?
14. If the population P in Exercise 2 starts out at $P(0) = 800$, how long will it take to reach 1500?
15. If $Q(0) = 1$ gram in Exercise 3, how long will it take until $Q = \frac{1}{2}$ gram?
16. How long does it take the money in Exercise 4 to double?

Solve each equation in Exercises 17–20 for $f(t)$ and sketch its graph.

17. $f' - 3f = 0, f(0) = 1$
18. $f' + 3f = 0, f(0) = 1$
19. $f' = 8f, f(0) = e$
20. $f' = 8f, f(1) = e$

Without solving, tell whether or not the solutions of the equations in Exercises 21–24 are increasing or decreasing.

21. $\frac{dx}{dt} = 3x$, $x = 1$ when $t = 0$.
22. $\frac{dx}{dt} = 3x$, $x = -1$ when $t = 0$.
23. $f' = -3f, f(0) = 1$.
24. $f' = -3f, f(0) = -1$.
25. A certain radioactive substance decreases at a rate of 0.0021% per year. What is its half-life?
26. Carbon-14 decreases at a rate of 0.01238% per year. What is its half-life?
27. It takes 300,000 years for a certain radioactive substance to decay to 30% of its original amount. What is its half-life?
28. It takes 80,000 years for a certain radioactive substance to decrease to 75% of its original amount. Find the half-life.
29. The half-life of uranium is about 0.45 billion years. If 1 gram of uranium is left undisturbed, how long will it take for 90% of it to have decayed?
30. The half-life of substance X is 3,050 years. What percentage of substance X remains after 12,200 years?
31. Carbon-14 is known to satisfy the decay law $Q = Q_0 e^{-0.0001238t}$ for the amount Q present after t years. Find the age of a bone sample in which the carbon-14 present is 70% of the original amount Q_0.
32. Consider two decay laws for radioactive carbon-14: $Q = Q_0 e^{-\alpha t}$, $Q = Q_0 e^{-\beta t}$, where $\alpha = 0.0001238$ and $\beta = 0.0001236$. Find the percentage error between the two exponential laws for predicting the age of a skull sample with 50% of the carbon-14 decayed. (See Exercise 31.)
33. A certain bacterial culture undergoing natural growth doubles in size after 10 minutes. If the culture contains 100 specimens at time $t = 0$, when will the number have increased to 3000 specimens?
34. A rabbit population doubles in size every 18 months. If there are 10,000 rabbits at $t = 0$, when will the population reach 100,000?
35. A bathtub is full of hot water at 110°F. After 10 minutes it will be 90°F. The bathroom is at 65°F. George College refuses to enter water below 100°F. How long can he wait to get in the tub?
36. A blacksmith's hot iron is at 830°C in a room at 32°C. After 1 minute it is 600°C. The blacksmith has to wait until it reaches 450°C. How long does it take after the 600°C temperature is reached?
37. How long does it take for money left in an account earning $7\frac{1}{2}$% interest compounded continuously to quadruple?
38. In a certain bank account, money doubles in 10 years. What is the annual interest rate compounded continuously?

39. A credit card company advertises: "Your interest rate on the unpaid balance is 17% compounded continuously, but federal law requires us to state that your annual interest rate is 18.53%." Explain.
40. If a credit card charges an interest rate of 21% compounded continuously, what is the actual annual percentage rate?
41. A certain calculus textbook sells according to this formula: $S(t) = 2000 - 1000e^{-0.3t}$, where t is the time in years and $S(t)$ is the number of books sold.
    (a) Find $S'(t)$.
    (b) Find $\lim_{t\to\infty} S(t)$ and discuss.
    (c) Graph S.
42. A foolish king, on losing a famous bet, agrees to pay a wizard 1 cent on the first day of the month, 2 cents on the second day, 4 cents on the third, and so on, each day doubling the sum. How much is paid on the thirtieth day?
43. The author of a certain calculus textbook is awake writing in the stillness of 2 A.M. A sound disturbs him. He discovers that the toilet tank fills up fast at first, then slows down as the water is being shut off. Examining the insides of the tank and contemplating for a moment, he thinks that maybe during shutoff the rate of flow of water into the tank is proportional to the height left to go; that is, $dx/dt = c(h - x)$, where x = height of water, h = desired height of water, and c = a constant (depending on the mechanism). Show that $x = h - Ke^{-ct}$. What is K?
    Looking at this formula for x, he says "That explains why my tank is always filling!" and goes to bed.
44. (a) Verify that the solution of $dy/dt = p(t)y$ is $y = y_0 \exp P(t)$, where $P(t)$ is the antiderivative of $p(t)$ with $P(0) = 0$.
    (b) Solve $dy/dt = ty$; $y = 1$ when $t = 0$.
★45. (a) Show that a solution of $t(da/dt) = a + h$ is
$$a(t) = t\int_1^t \frac{h(s)}{s^2}\,ds + tC.$$
    (b) Solve $t(da/dt) = a + e^{-1/t}$, $a(1) = 1$.
★46. Develop a general formula for the *doubling time* of a population in terms of its growth rate.
★47. Develop a general formula for the half-life of the amplitude of a damped spring (Exercise 28, Section 8.1).

8.3 The Hyperbolic Functions

The points $(\cos t, \sin t)$ *lie on a circle, and* $(\cosh t, \sinh t)$ *lie on a hyperbola.*

The hyperbolic functions are certain combinations of exponential functions which satisfy identities very similar to those for the trigonometric functions. We shall see in the next section that the inverse hyperbolic functions are important in integration.

A good way to introduce the hyperbolic functions is through a differential equation which they solve. Recall that $\sin t$ and $\cos t$ are solutions of the equation $d^2x/dt^2 + x = 0$. Now we switch the sign and consider $d^2x/dt^2 - x = 0$. (This corresponds to a negative spring constant!)

We already know one solution to this equation: $x = e^t$. Another is e^{-t}, because when we differentiate e^{-t} twice we bring down, via the chain rule, two minus signs and so recover e^{-t} again. The combination

$$x = Ae^t + Be^{-t}$$

is also a solution, as is readily verified.

If we wish to find a solution analogous to the sine function, with $x = 0$ and $dx/dt = 1$ when $t = 0$, we must pick A and B so that

$$0 = A + B,$$
$$1 = A - B,$$

so $A = \frac{1}{2}$ and $B = \frac{1}{2}$, giving $x = (e^t - e^{-t})/2$.

Similarly, if we wish to find a solution analogous to the cosine function, we should pick A and B such that $x = 1$ and $dx/dt = 0$ when $t = 0$; that is,

$$1 = A + B,$$
$$0 = A - B,$$

so $A = B = \frac{1}{2}$, giving $x = (e^t + e^{-t})/2$.

This reasoning leads to the following definitions.

Hyperbolic Sine and Cosine

The *hyperbolic sine function*, written $\sinh t$, is defined by

$$\sinh t = \frac{e^{t} - e^{-t}}{2}. \tag{1}$$

The *hyperbolic cosine function*, written $\cosh t$, is defined by

$$\cosh t = \frac{e^{t} + e^{-t}}{2}. \tag{2}$$

(See Fig. 8.3.1.)

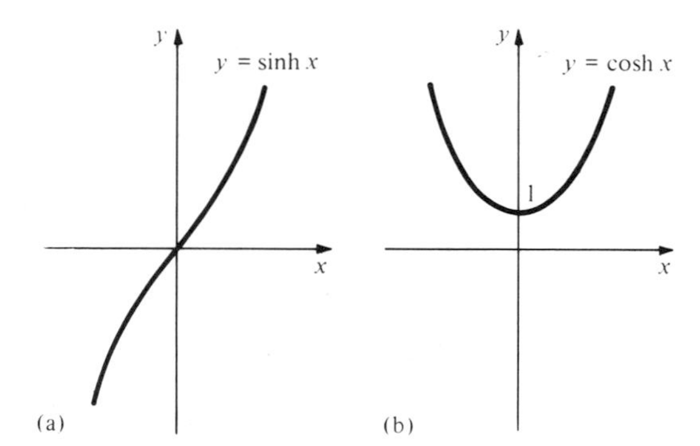

Figure 8.3.1. The graphs of $y = \sinh x$ and $y = \cosh x$.

The usual trigonometric functions $\sin t$ and $\cos t$ are called the *circular functions* because $(x, y) = (\cos t, \sin t)$ parametrizes the circle $x^2 + y^2 = 1$. The functions $\sinh t$ and $\cosh t$ are called *hyperbolic functions* because $(x, y) = (\cosh t, \sinh t)$ parametrizes one branch of the hyperbola $x^2 - y^2 = 1$; that is, for any t, we have the identity

$$\cosh^2 t - \sinh^2 t = 1. \tag{3}$$

(See Fig. 8.3.2.)

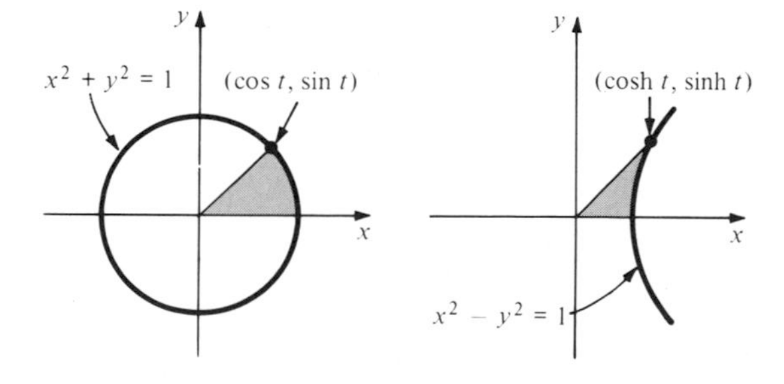

Figure 8.3.2. The points $(\cos t, \sin t)$ lie on a circle, while $(\cosh t, \sinh t)$ lie on a hyperbola.

To prove formula (3), we square formulas (1) and (2):

$$\cosh^2 t = \tfrac{1}{4}(e^{t} + e^{-t})^2 = \tfrac{1}{4}(e^{2t} + 2 + e^{-2t})$$

and

$$\sinh^2 t = \tfrac{1}{4}(e^{t} - e^{-t})^2 = \tfrac{1}{4}(e^{2t} - 2 + e^{-2t}).$$

Subtracting gives formula (3).

Example 1 Show that $e^x = \cosh x + \sinh x$.

Solution By definition,

$$\cosh x = \frac{e^x + e^{-x}}{2} \quad \text{and} \quad \sinh x = \frac{e^x - e^{-x}}{2}.$$

Adding, $\cosh x + \sinh x = e^x/2 + e^{-x}/2 + e^x/2 - e^{-x}/2 = e^x$. ▲

Similarly, $e^{-x} = \cosh x - \sinh x$.

The other hyperbolic functions can be introduced by analogy with the trigonometric functions:

$$\tanh x = \frac{\sinh x}{\cosh x}, \qquad \coth x = \frac{\cosh x}{\sinh x}, \tag{4}$$

$$\operatorname{sech} x = \frac{1}{\cosh x}, \qquad \operatorname{csch} x = \frac{1}{\sinh x}.$$

Various general identities can be proved exactly as we proved formula (3); for instance, the addition formulas are:

$$\sinh(x + y) = \sinh x \cosh y + \cosh x \sinh y, \tag{5a}$$

$$\cosh(x + y) = \cosh x \cosh y + \sinh x \sinh y. \tag{5b}$$

Example 2 Prove identity (5a).

Solution By definition,

$$\sinh(x + y) = \frac{e^{x+y} - e^{-x-y}}{2} = \frac{e^x e^y - e^{-x} e^{-y}}{2}.$$

Now we plug in $e^x = \cosh x + \sinh x$ and $e^{-x} = \cosh x - \sinh x$ to get

$$\sinh(x + y) = \tfrac{1}{2}\big[(\cosh x + \sinh x)(\cosh y + \sinh y) - (\cosh x - \sinh x)(\cosh y - \sinh y)\big].$$

Expanding,

$$\begin{aligned}\sinh(x + y) &= \tfrac{1}{2}(\cosh x \cosh y + \cosh x \sinh y + \sinh x \cosh y + \sinh x \sinh y \\ &\quad - \cosh x \cosh y + \cosh x \sinh y + \sinh x \cosh y - \sinh x \sinh y) \\ &= \cosh x \sinh y + \sinh x \cosh y. \ \blacktriangle\end{aligned}$$

Notice that in formula (5b) for $\cosh(x + y)$ there is no minus sign. This is one of several differences in signs between rules for the hyperbolic and circular functions. Another is in the following:

$$\frac{d}{dx} \sinh x = \cosh x, \tag{6a}$$

$$\frac{d}{dx} \cosh x = \sinh x. \tag{6b}$$

Example 3 Prove formula (6a).

Solution By definition, $\sinh x = (e^x - e^{-x})/2$, so

$$\frac{d}{dx} \sinh x = \frac{e^x - (-1)e^{-x}}{2} = \frac{e^x + e^{-x}}{2} = \cosh x. \ \blacktriangle$$

We also note that

$$\sinh(-x) = -\sinh x \qquad (\sinh \text{ is } odd)$$

and (7)

$$\cosh(-x) = \cosh x \qquad (\cosh \text{ is } even).$$

From formulas (5a) and (5b) we get the half-angle formulas:

$$\sinh^2 x = \frac{\cosh 2x - 1}{2} \quad \text{and} \quad \cosh^2 x = \frac{\cosh 2x + 1}{2}. \tag{8}$$

Example 4 Prove that $\frac{d}{dx}\tanh x = \operatorname{sech}^2 x$.

Solution Since $\tanh x = \sinh x/\cosh x$, the quotient rule gives

$$(d/dx)\tanh x = (\cosh x \cdot \cosh x - \sinh x \cdot \sinh x)/\cosh^2 x = 1 - \tanh^2 x.$$

From $\cosh^2 x - \sinh^2 x = 1$ we get $1 - \sinh^2 x/\cosh^2 x = 1/\cosh^2 x$; that is, $1 - \tanh^2 x = \operatorname{sech}^2 x$. ▲

Hyperbolic Functions and Their Derivatives

$$\sinh x = \frac{e^x - e^{-x}}{2}, \qquad \frac{d}{dx}\sinh x = \cosh x,$$

$$\cosh x = \frac{e^x + e^{-x}}{2}, \qquad \frac{d}{dx}\cosh x = \sinh x,$$

$$\tanh x = \frac{e^x - e^{-x}}{e^x + e^{-x}} = \frac{\sinh x}{\cosh x}, \qquad \frac{d}{dx}\tanh x = \operatorname{sech}^2 x,$$

$$\coth x = \frac{e^x + e^{-x}}{e^x - e^{-x}} = \frac{1}{\tanh x}, \qquad \frac{d}{dx}\coth x = -\operatorname{csch}^2 x,$$

$$\operatorname{sech} x = \frac{2}{e^x + e^{-x}} = \frac{1}{\cosh x}, \qquad \frac{d}{dx}\operatorname{sech} x = -\operatorname{sech} x \tanh x,$$

$$\operatorname{csch} x = \frac{2}{e^x - e^{-x}} = \frac{1}{\sinh x}, \qquad \frac{d}{dx}\operatorname{csch} x = -\operatorname{csch} x \coth x.$$

Example 5 Differentiate: (a) $\sinh(3x + x^3)$; (b) $\cos^{-1}(\tanh x)$; (c) $3x/(\cosh x + \sinh 3x)$.

Solution (a)

$$\frac{d}{dx}\sinh(3x + x^3) = \frac{d}{du}\sinh u \cdot \frac{du}{dx},$$

where $u = 3x + x^3$. We compute

$$\frac{d}{du}\sinh u = \cosh u \quad \text{and} \quad \frac{du}{dx} = 3 + 3x^2.$$

Expressing everything in terms of x, we have

$$\frac{d}{dx}\sinh(3x + x^3) = \cosh(3x + x^3) \cdot 3(1 + x^2).$$

(b) $\frac{d}{dx}\cos^{-1}(\tanh x) = \frac{d}{du}\cos^{-1}u \cdot (du/dx), \qquad \text{where} \quad u = \tanh x.$

We find

$$\frac{d}{du}\cos^{-1}u = \frac{-1}{\sqrt{1-u^2}} = \frac{-1}{\sqrt{1-\tanh^2 x}} = \frac{-1}{\operatorname{sech} x}$$

(from the identity $1-\tanh^2 x = \operatorname{sech}^2 x$ proved in Example 4), and

$$\frac{du}{dx} = \operatorname{sech}^2 x$$

(also from Example 4). Hence

$$\frac{d}{dx}\cos^{-1}(\tanh x) = -\operatorname{sech} x.$$

(c) $$\frac{d}{dx}\left(\frac{3x}{\cosh x + \sinh 3x}\right) = \frac{3(\cosh x + \sinh 3x) - 3x(\sinh x + 3\cosh 3x)}{(\cosh x + \sinh 3x)^2}$$

(by the quotient rule). ▲

Let us return now to the equation $d^2x/dt^2 - \omega^2 x = 0$. Its solution can be summarized as follows.

The Equation $d^2x/dt^2 - \omega^2 x = 0$

The solution of

$$\frac{d^2x}{dt^2} - \omega^2 x = 0 \tag{9}$$

is

$$x = x_0\cosh\omega t + \frac{v_0}{\omega}\sinh\omega t, \tag{10}$$

where $x = x_0$ and $dx/dt = v_0$ when $t = 0$.

That formula (10) gives a solution of equation (9) is easy to see:

$$\frac{dx}{dt} = \omega x_0 \sinh\omega t + v_0\cosh\omega t,$$

using formula (6) and the chain rule. Differentiating again, we get

$$\frac{d^2x}{dt^2} = \omega^2 x_0\cosh\omega t + \omega v_0\sinh\omega t = \omega^2 x,$$

so equation (9) is verified.

One may prove that (10) gives the only solution of equation (9) just as in the case of the spring equation (Exercise 54).

Example 7 Solve for $f(t)$: $f'' - 3f = 0$, $f(0) = 1$, $f'(0) = -2$.

Solution We use formula (10) with $\omega^2 = 3$ (so $\omega = \sqrt{3}$), $x_0 = f(0) = 1$, $v_0 = f'(0) = -2$, and with $f(t)$ in place of x. Thus our solution is

$$f(t) = \cosh\sqrt{3}\,t - \frac{2}{\sqrt{3}}\sinh\sqrt{3}\,t. \blacktriangle$$

Example 8 Prove that $\cosh x$ has a minimum value of 1 at $x = 0$.

Solution $(d/dx)\cosh x = \sinh x$ vanishes only if $x = 0$ since $e^{x} = e^{-x}$ exactly when $e^{2x} = 1$; that is, $x = 0$. Also, $(d^2/dx^2)\cosh x = \cosh x$, which is 1 at $x = 0$, so $\cosh x$ has a minimum at $x = 0$, by the second derivative test. ▲

The kind of reasoning in the preceding example, together with the techniques of graphing, enable us to graph all the hyperbolic functions. These are shown in Fig. 8.3.3.

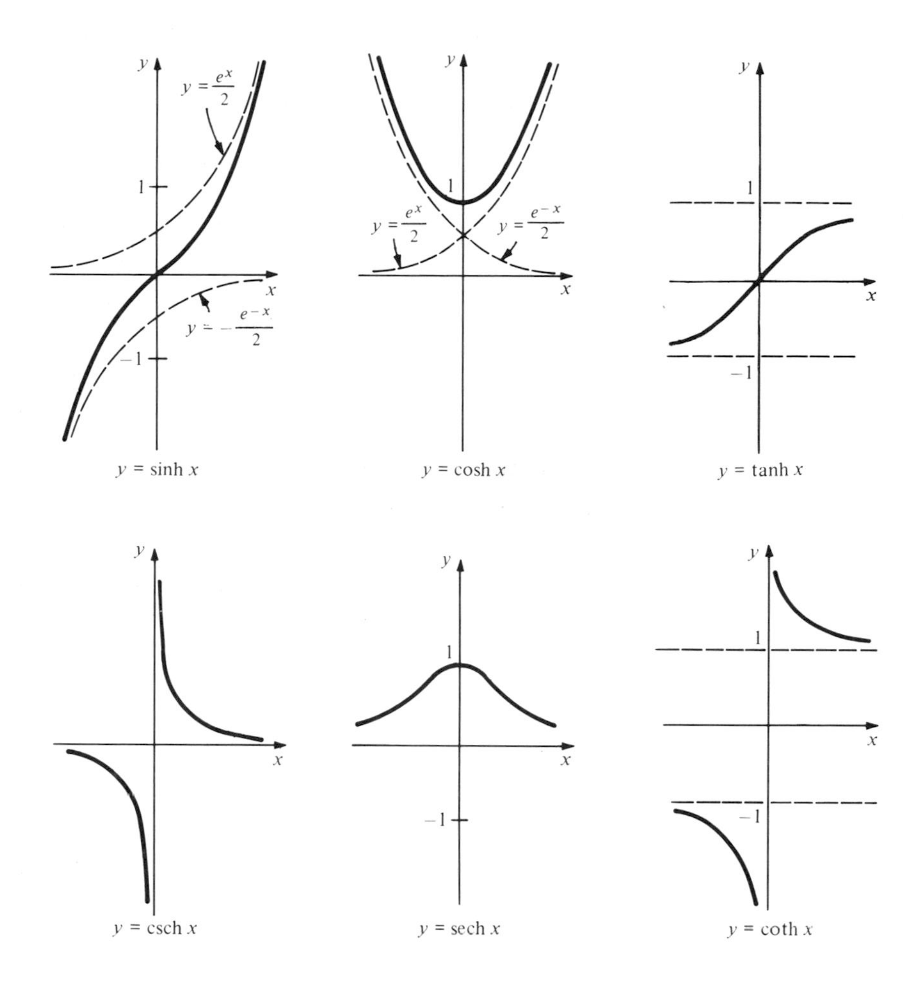

Figure 8.3.3. Graphs of the hyperbolic functions.

Finally, the differentiation formulas for the hyperbolic functions lead to integration formulas.

Antidifferentiation Formulas for Hyperbolic Functions

$$\int \cosh x\,dx = \sinh x + C, \qquad \int \operatorname{csch}^2 x\,dx = -\coth x + C,$$

$$\int \sinh x\,dx = \cosh x + C, \qquad \int \operatorname{sech} x \tanh x\,dx = -\operatorname{sech} x + C,$$

$$\int \operatorname{sech}^2 x\,dx = \tanh x + C, \qquad \int \operatorname{csch} x \coth x\,dx = -\operatorname{csch} x + C.$$

Example 9 Compute the integrals (a) $\int(\sinh 3x + x^3)\,dx$, (b) $\int \tanh x\,dx$, (c) $\int \cosh^2 x\,dx$, (d) $\int \frac{\sinh x}{1+\cosh^2 x}\,dx$.

Solution (a) $\int(\sinh 3x + x^3)\,dx = \int \sinh 3x\,dx + \int x^3\,dx = \frac{1}{3}\cosh 3x + \frac{x^4}{4} + C.$

(b) $\int \tanh x\,dx = \int \frac{\sinh x}{\cosh x}\,dx = \ln|\cosh x| + C = \ln(\cosh x) + C$

(since $\cosh x \geqslant 1$, $|\cosh x| = \cosh x$).

(c) Here we use the half-angle formula (8):

$$\int \cosh^2 x\,dx = \int \frac{\cosh 2x + 1}{2}\,dx = \frac{1}{4}\sinh 2x + \frac{x}{2} + C.$$

(d)
$$\int \frac{\sinh x}{1+\cosh^2 x}\,dx = \int \frac{du}{1+u^2} \qquad (u = \cosh x)$$
$$= \tan^{-1} u + c = \tan^{-1}(\cosh x) + C. \; \blacktriangle$$

Supplement to Section 8.3: Unstable Equilibrium Points

In Section 8.1, we studied approximations to the differential equation of motion $m(d^2x/dt^2) = f(x)$, where the force function $f(x)$ satisfied the equilibrium condition $f(x_0) = 0$ at some position x_0. The linearized equation was

$$m\frac{d^2x}{dt^2} = f'(x_0)(x - x_0)$$

or, setting $y = x - x_0$, the displacement from equilibrium,

$$m\frac{d^2y}{dt^2} = f'(x_0)y.$$

If $f'(x_0) < 0$, this is the spring equation which has oscillating solutions given by trigonometric functions.

Now we can use the hyperbolic functions to study the case $f'(x_0) > 0$. The general solution is $y = A\cosh(\sqrt{f'(x_0)}\;t) + B\sinh(\sqrt{f'(x_0)}\;t)$, with A and B depending on the initial values of y and dy/dt, as in formula (10). We can also write this solution as $y = (A+B)e^{\sqrt{f'(x_0)}\,t} + (A-B)e^{-\sqrt{f'(x_0)}\,t}$. No matter how small the initial values, unless they are chosen so carefully that $A + B = 0$, the solution will approach $+\infty$ or $-\infty$ as $t \to \infty$. We say that the point x_0 is an *unstable equilibrium* position. In contrast to the stable case treated in Section 8.1, the linearization is not useful for all t, since most solutions eventually leave the region where the linear approximation is valid. Still, we can correctly conclude that solutions starting arbitrarily close to x_0 do not usually stay close, and that there are special solutions which approach x_0 as $t \to \infty$.[4]

[4] This analysis is useful in more advanced studies of differential equations. See, for instance, *Elementary Differential Equations and Boundary Value Problems*, by W. Boyce and R. DiPrima, Chapter 9, Third Edition, Wiley (1977) and *Differential Equations and Their Applications*, by M. Braun, Chapter 4, Third Edition, Springer-Verlag (1983).

For instance, let us find the unstable equilibrium point for the pendulum equation

$$m\frac{d^2\theta}{dt^2} = -m\frac{g}{l}\sin\theta = f(\theta).$$

First note that $f(\theta) = 0$ at $\theta = 0$ and π, corresponding to the bottom and top of pendulum swing (see Fig. 8.1.6). At $\theta = \pi$, $f'(\theta) = -(mg/l)\cos\theta = -(mg/l)\cdot(-1) = mg/l > 0$, so the top position is an unstable equilibrium point. At that point, the linearized equation is given by $(d^2/dt^2)(\theta - \theta_0) = (mg/l)(\theta - \theta_0)$, with solutions $\theta - \theta_0 = Ce^{\sqrt{mg/l}}t + De^{-\sqrt{mg/l}}t$. For most initial conditions, C will not equal zero, and so the pendulum will move away from the equilibrium point. If the initial conditions are chosen just right, we will have $C = 0$, and the pendulum will gradually approach the top position as $t \to \infty$, but it will never arrive.

A very different application of hyperbolic functions, to the shape of a hanging cable, is given in Example 6, Section 8.5.

Exercises for Section 8.3

Prove the identities in Exercises 1–8.

1. $\tanh^2 x + \operatorname{sech}^2 x = 1$.
2. $\coth^2 x = 1 + \operatorname{csch}^2 x$.
3. $\cosh(x + y) = \cosh x \cosh y + \sinh x \sinh y$.
4. $\sinh^2 x = (\cosh 2x - 1)/2$.
5. $\dfrac{d}{dx}\cosh x = \sinh x$.
6. $\dfrac{d}{dx}\coth x = -\operatorname{csch}^2 x$.
7. $\dfrac{d}{dx}\operatorname{sech} x = -\operatorname{sech} x \tanh x$.
8. $\dfrac{d}{dx}\operatorname{csch} x = -\operatorname{csch} x \coth x$.

Differentiate the functions in Exercises 9–24.

9. $\sinh(x^3 + x^2 + 2)$
10. $\tan^{-1}(\cosh x)$
11. $\sinh x \sinh 5x$
12. $\dfrac{\sinh x}{1 + \cosh x}$
13. $\sinh(\cos(8x))$
14. $\cos(\sinh(x^2))$
15. $\sinh^2 x + \cosh^2 x$
16. $\sinh^4 x + \cosh^4 x$
17. $\coth 3x$
18. $(\tanh x)(\operatorname{sech} x)$
19. $\exp(\tanh 2x)$
20. $\sin^{-1}(\tanh x)$
21. $\dfrac{\cosh x}{1 + \tanh x}$
22. $(\operatorname{csch} 2x)(1 + \tan x)$
23. $(\cosh x)\left(\int \dfrac{dx}{1 + \tanh^2 x}\right)$
24. $(\sinh x)\left(\int \dfrac{dx}{1 + \operatorname{sech}^2 x}\right)$

Solve the differential equations in Exercises 25–28.

25. $y'' - 9y = 0$, $y(0) = 0$, $y'(0) = 1$.
26. $f'' - 81f = 0$, $f(0) = 1$, $f'(0) = -1$.
27. $g'' - 3g = 0$, $g(0) = 2$, $g'(0) = 0$.
28. $h'' - 9h = 0$, $h(0) = 2$, $h'(0) = 4$.
29. Find the solution of the equation $d^2x/dt^2 - 9x = 0$, for which $x = 1$ and $dx/dt = 1$ at $t = 0$.
30. Solve $d^2y/dt^2 - 25y = 0$, where $y = 1$ and $dy/dt = -1$ when $t = 0$.
31. Find $f(t)$ if $f'' = 36f$, and $f(0) = 2$, $f'(0) = 0$.
32. Find $g(t)$ if $g''(t) = 25g(t)$, and $g(0) = 0$, $g'(0) = 1$.

Sketch the graphs of the functions in Exercises 33–36.

33. $y = 3 + \sinh x$
34. $y = (\cosh x) - 1$
35. $y = \tanh 3x$
36. $y = 3\cosh 2x$

Compute the integrals in Exercises 37–46.

37. $\int \cosh 3x\, dx$
38. $\int [\operatorname{csch}^2(2x) + (3/x)]\, dx$
39. $\int \coth x\, dx$
40. $\int x \tanh(x^2)\, dx$
41. $\int \sinh^2 x\, dx$
42. $\int \cosh^2 9y\, dy$
43. $\int e^x \sinh x\, dx$
44. $\int e^{2t}\cosh 2t\, dt$
45. $\int \cosh^2 x \sinh x\, dx$
46. $\int \dfrac{\sinh x}{\cosh^3 x}\, dx$

Compute dy/dx in Exercises 47–50.

47. $\dfrac{\sinh(x+y)}{xy} = 1$
48. $x + \cosh(xy) = 3$
49. $\tanh(3xy) + \sinh y = 1$
50. $\coth(x - y) - 3y = 6$

★51. (a) Find the *unstable* equilibrium point for the equation of motion $d^2x/dt^2 = x^2 - 1$.
(b) Write down the linearized equation of motion at this point.

★52. An atom of mass m in a linear molecule is subjected to forces of attraction by its neighbors given by

$$f(x) = -\frac{k_1}{(x-x_1)^2} + \frac{k_2}{(x-x_2)^2},$$

where $k_2 > k_1 > 0$, $0 < x_1 < x < x_2$. Show that $x_0 = (x_1 + \alpha x_2)/(1+\alpha)$, where $\alpha = \sqrt{k_1/k_2}$, is an unstable equilibrium. Write the linearized equations at this point.

★53. Prove the identity $(\cosh x + \sinh x)^n = \cosh nx + \sinh nx$.

★54. Prove that the equation $d^2x/dt^2 - \omega^2 x = 0$, with $x = x_0$ and $dx/dt = v_0$ when $t = 0$, has a unique solution given by $x = x_0 \cosh \omega t + (v_0/\omega)\sinh \omega t$. [*Hint*: Study Exercises 31 and 32 of Section 8.1.] Why doesn't the energy method of p. 372 work in this case?

8.4 The Inverse Hyperbolic Functions

The inverse hyperbolic functions occur in several basic integration formulas.

We now study the inverses of the hyperbolic functions using the methods of Section 5.3. As with the inverse trigonometric functions, this yields some interesting integration formulas for algebraic functions.

We turn first to the inverse sinh function. Since $(d/dx)\sinh x = \cosh x$ is positive (Example 8, Section 8.3), $\sinh x$ is increasing. The range of $\sinh x$ is in fact $(-\infty, \infty)$ since $\sinh x \to \pm\infty$ as $x \to \pm\infty$. Thus, from the inverse function test in Section 5.3, we know that $y = \sinh x$ has an inverse function defined on the whole real line, denoted $\sinh^{-1} y$ by analogy with the notation for the inverse trigonometric functions. From the general formula

$$\frac{d}{dy} f^{-1}(y) = \frac{1}{f'(x)} \qquad (\text{where } y = f(x))$$

for the derivative of an inverse function, we get

$$\frac{d}{dy}\sinh^{-1} y = \frac{1}{(d/dx)(\sinh x)} = \frac{1}{\cosh x}.$$

From $\cosh^2 x - \sinh^2 x = 1$, we get

$$\frac{d}{dy}\sinh^{-1} y = \frac{1}{\cosh x} = \frac{1}{\sqrt{1+\sinh^2 x}} = \frac{1}{\sqrt{1+y^2}}. \tag{1}$$

The positive square root is taken because $\cosh x$ is always positive.

Example 1 Calculate (a) $\dfrac{d}{dx}\sinh^{-1}(3x)$ and (b) $\dfrac{d}{dx}[\sinh^{-1}(3\tanh 3x)]$.

Solution (a) Let $u = 3x$, so $\sinh^{-1} 3x = \sinh^{-1} u$. By the chain rule,

$$\frac{d}{dx}\sinh^{-1}(3x) = \left(\frac{d}{du}\sinh^{-1} u\right)\frac{du}{dx}.$$

By formula (1) with y replaced by u, we get

$$\frac{d}{dx}\sinh^{-1}(3x) = \frac{1}{\sqrt{1+u^2}}\,3 = \frac{3}{\sqrt{1+9x^2}}.$$

(b) By the formula $(d/dx)\sinh^{-1}x = 1/\sqrt{1+x^2}$, and the chain rule,

$$\begin{aligned}\frac{d}{dx}\sinh^{-1}(3\tanh 3x) &= \frac{1}{\sqrt{1+9\tanh^2 3x}}\cdot 3\frac{d}{dx}\tanh 3x\\ &= \frac{1}{\sqrt{1+9\tanh^2 3x}}\cdot 3\cdot 3\cdot \operatorname{sech}^2 3x\\ &= \frac{9\operatorname{sech}^2 3x}{\sqrt{1+9\tanh^2 3x}}.\ \blacktriangle\end{aligned}$$

There is an explicit formula for $\sinh^{-1}y$ obtained by solving the equation

$$y = \sinh x = \frac{e^x - e^{-x}}{2}$$

for x. Multiplying through by $2e^x$ and gathering terms on the left-hand side of the equation gives $2e^x y - e^{2x} + 1 = 0$. Hence

$$(e^x)^2 - 2e^x y - 1 = 0,$$

and so, by the quadratic formula,

$$e^x = \frac{2y \pm \sqrt{4y^2+4}}{2} = y \pm \sqrt{y^2+1}\ .$$

Since e^x is positive, we must select the positive square root. Thus, $e^x = y + \sqrt{y^2+1}$, and so $x = \sinh^{-1}y$ is given by

$$\sinh^{-1}y = \ln\left(y + \sqrt{y^2+1}\right). \tag{2}$$

The basic properties of $\sinh^{-1}$ are summarized in the following display.

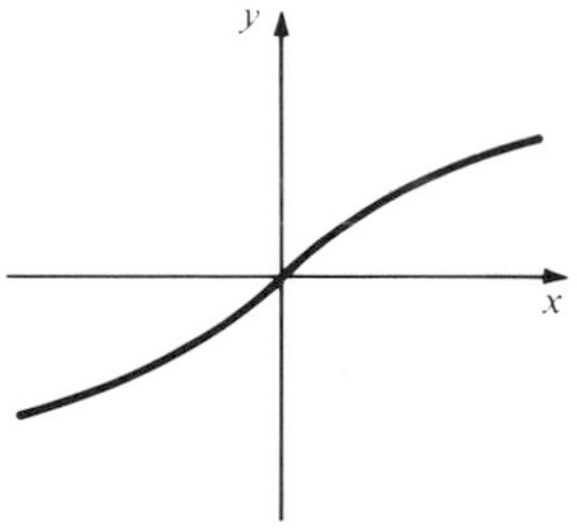

Figure 8.4.1. The graph of the $y = \sinh^{-1}x$.

Inverse Hyperbolic Sine Function

1. $\sinh^{-1}x$ is the inverse function of $\sinh x$; $\sinh^{-1}x$ is defined and is increasing for all x (Fig. 8.4.1); by definition: $\sinh^{-1}x = y$ is that number such that $\sinh y = x$.
2. $\dfrac{d}{dx}\sinh^{-1}x = \dfrac{1}{\sqrt{1+x^2}}$.
3. $\displaystyle\int \frac{dx}{\sqrt{1+x^2}} = \sinh^{-1}x + C = \ln(x + \sqrt{1+x^2}\,) + C.$

Example 2 Find $\sinh^{-1}5$ numerically by using logarithms.

Solution By (2), $\sinh^{-1}5 = \ln(5 + \sqrt{5^2+1}\,) = \ln(5+\sqrt{26}\,) = \ln(10.100) \approx 2.31$. ▲

Example 3 Verify the formula $\displaystyle\int \frac{dx}{\sqrt{1+x^2}} = \ln(x + \sqrt{1+x^2}\,) + C.$

Solution

$$\begin{aligned}&\frac{d}{dx}\ln\left(x + \sqrt{1+x^2}\right)\\ &= \frac{1}{\left(x+\sqrt{1+x^2}\right)}\left(1 + \frac{x}{\sqrt{1+x^2}}\right) \qquad \text{(by the chain rule)}\\ &= \left(\frac{1}{x+\sqrt{1+x^2}}\right)\left[\frac{\sqrt{1+x^2}+x}{\sqrt{1+x^2}}\right] = \frac{1}{\sqrt{1+x^2}}.\end{aligned}$$

Thus the antiderivative for $1/\sqrt{1+x^2}$ is $\ln(x + \sqrt{1+x^2}\,) + C$. ▲

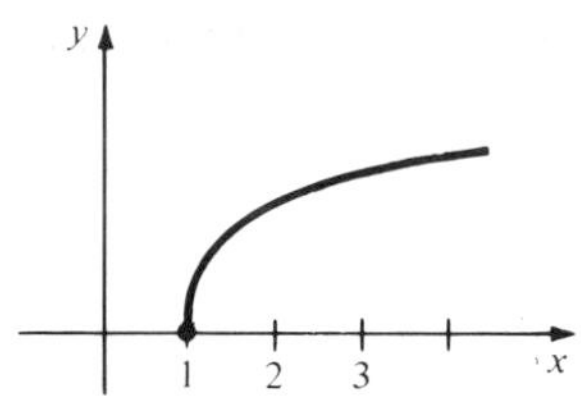

Figure 8.4.2. The graph of $y = \cosh^{-1}x$.

In a similar fashion we can investigate $\cosh^{-1}x$. Since $\cosh x$ is increasing on $[0, \infty)$ and has range $[1, \infty)$, $\cosh^{-1}x$ will be increasing, will be defined on $[1, \infty)$, and will have range $[0, \infty)$. Its graph can be obtained from that of $\cosh x$ by the usual method of looking through the page from the other side (Fig. 8.4.2).

By the same method that we obtained formula (1), we find

$$\frac{d}{dx}\cosh^{-1}x = \frac{1}{\sqrt{x^2-1}} \qquad (x > 1). \tag{3}$$

Example 4 Find $\dfrac{d}{dx}\cosh^{-1}(\sqrt{x^2+1}\,)$, $x \neq 0$.

Solution Let $u = \sqrt{x^2+1}$. Then, by the chain rule,

$$\begin{aligned}\frac{d}{dx}\cosh^{-1}\left(\sqrt{x^2+1}\,\right) &= \left(\frac{d}{du}\cosh^{-1}u\right)\cdot\left(\frac{du}{dx}\right)\\ &= \frac{1}{\sqrt{u^2-1}}\cdot\frac{x}{\sqrt{x^2+1}}\\ &= \frac{1}{\sqrt{x^2}}\cdot\frac{x}{\sqrt{x^2+1}} = \frac{x}{|x|}\cdot\frac{1}{\sqrt{x^2+1}}.\end{aligned}$$

Therefore,

$$\frac{d}{dx}\cosh^{-1}\left(\sqrt{x^2+1}\,\right) = \begin{cases} \dfrac{1}{\sqrt{x^2+1}} & \text{if } x > 0, \\[2ex] \dfrac{-1}{\sqrt{x^2+1}} & \text{if } x < 0. \ \blacktriangle \end{cases}$$

Similarly, we can consider $\tanh^{-1}x$ (see Fig. 8.4.3) and get, for $-1 < x < 1$,

$$\frac{d}{dx}\tanh^{-1}x = \frac{1}{1-x^2}. \tag{4}$$

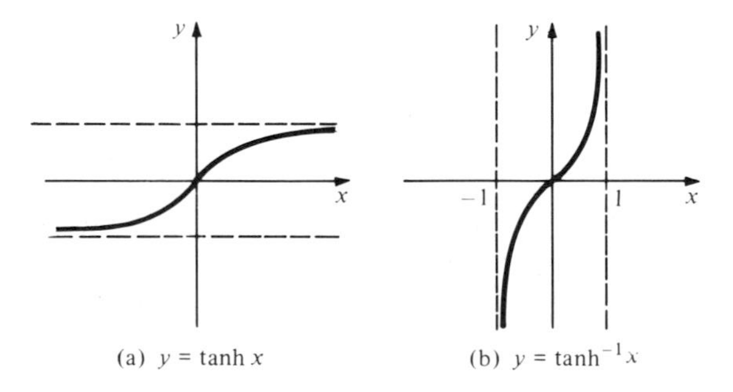

Figure 8.4.3. The graphs of $y = \tanh x$ and $y = \tanh^{-1}x$.

Example 5 Prove that $\tanh^{-1}x = \frac{1}{2}\ln[(1+x)/(1-x)]$, $-1 < x < 1$.

Solution Let $y = \tanh^{-1}x$, so

$$x = \tanh y = \frac{\sinh y}{\cosh y} = \frac{(e^y - e^{-y})}{(e^y + e^{-y})}.$$

Thus $x(e^y + e^{-y}) = e^y - e^{-y}$. Multiplying through by e^y and gathering terms on the left:

$$(x-1)e^{2y} + x + 1 = 0$$

$$e^{2y} = \frac{1+x}{1-x},$$

$$2y = \ln\left(\frac{1+x}{1-x}\right),$$

$$y = \frac{1}{2}\ln\left(\frac{1+x}{1-x}\right),$$

as required. ▲

Example 6 Show directly that the antiderivative of $\frac{1}{1-x^2}$ for $|x| < 1$ is $\frac{1}{2}\ln\frac{1+x}{1-x} + C$ by noticing that

$$\frac{1}{1-x^2} = \frac{1}{2}\left(\frac{1}{1-x} + \frac{1}{1+x}\right).$$

Solution Since

$$\frac{1}{1-x^2} = \frac{1}{2}\left(\frac{1}{1-x} + \frac{1}{1+x}\right),$$

an antiderivative is

$$\frac{1}{2}\ln|1-x| + \frac{1}{2}\ln|1+x| = \frac{1}{2}\ln\left|\frac{1+x}{1-x}\right|, \qquad |x| \neq 1.$$

If $|x| < 1$, $(1+x)/(1-x) > 0$, so the absolute value signs can be removed. ▲

The remaining inverse functions are investigated in a similar way; the results are summarized in Fig. 8.4.4 and the box on the next page.

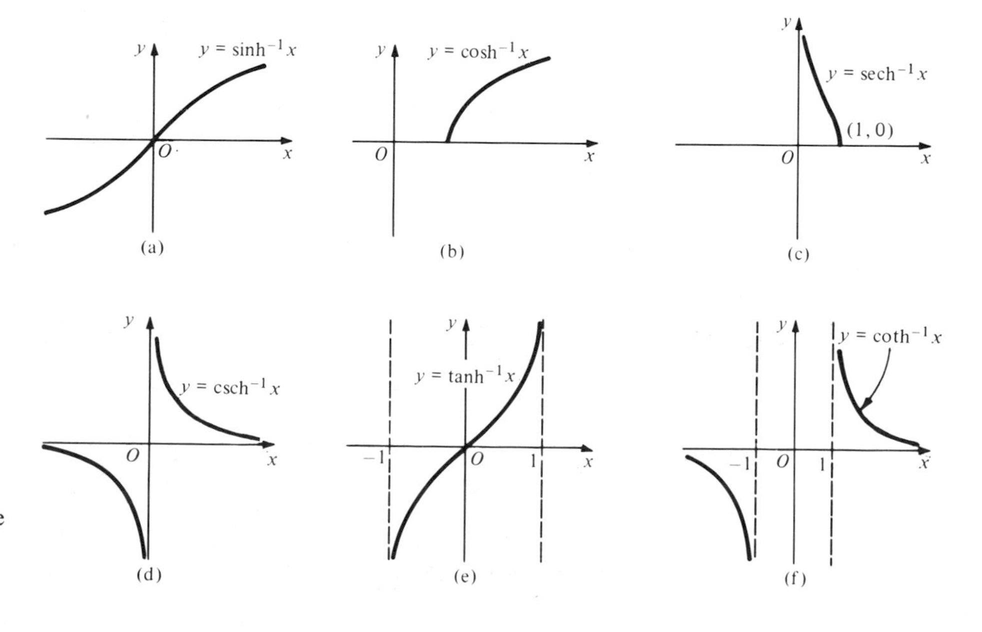

Figure 8.4.4. Graphs of the inverse hyperbolic functions.

The Inverse Hyperbolic Functions

$$\sinh^{-1}x = \ln\left(x + \sqrt{x^2+1}\,\right); \qquad \frac{d}{dx}\sinh^{-1}x = \frac{1}{\sqrt{x^2+1}};$$

$$\cosh^{-1}x = \ln\left(x + \sqrt{x^2-1}\,\right); \qquad \frac{d}{dx}\cosh^{-1}x = \frac{1}{\sqrt{x^2-1}}, \quad |x|>1;$$

$$\tanh^{-1}x = \frac{1}{2}\ln\left(\frac{1+x}{1-x}\right); \qquad \frac{d}{dx}\tanh^{-1}x = \frac{1}{1-x^2}, \quad |x|<1;$$

$$\coth^{-1}x = \frac{1}{2}\ln\left(\frac{x+1}{x-1}\right); \qquad \frac{d}{dx}\coth^{-1}x = \frac{1}{1-x^2}, \quad |x|>1;$$

$$\operatorname{sech}^{-1}x = \ln\left(\frac{1+\sqrt{1-x^2}}{x}\right); \qquad \frac{d}{dx}\operatorname{sech}^{-1}x = \frac{-1}{x\sqrt{1-x^2}}, \quad 0<x<1;$$

$$\operatorname{csch}^{-1}x = \begin{cases} \ln\left(\dfrac{1+\sqrt{1+x^2}}{x}\right), & x>0; \\ -\ln\left(\dfrac{1+\sqrt{1+x^2}}{-x}\right), & x<0; \end{cases} \qquad \frac{d}{dx}\operatorname{csch}^{-1}x = \begin{cases} \dfrac{-1}{x\sqrt{1+x^2}}, & x>0, \\ \dfrac{1}{x\sqrt{1+x^2}}, & x<0; \end{cases}$$

$$\int \frac{dx}{\sqrt{x^2+1}} = \sinh^{-1}x + C = \ln\left(x+\sqrt{x^2+1}\,\right) + C;$$

$$\int \frac{dx}{\sqrt{x^2-1}} = \cosh^{-1}x + C = \ln\left(x+\sqrt{x^2-1}\,\right) + C, \qquad |x|>1;$$

$$\int \frac{dx}{1-x^2} = \frac{1}{2}\ln\left|\frac{1+x}{1-x}\right| + C, \qquad (|x| \neq 1),$$

$$= \begin{cases} \tanh^{-1}x + C = \dfrac{1}{2}\ln\left(\dfrac{1+x}{1-x}\right) + C, & |x|<1, \\ \coth^{-1}x + C = \dfrac{1}{2}\ln\left(\dfrac{x+1}{x-1}\right) + C, & |x|>1; \end{cases}$$

$$\int \frac{dx}{x\sqrt{1-x^2}} = -\operatorname{sech}^{-1}x + C = -\ln\left(\frac{1+\sqrt{1-x^2}}{x}\right) + C, \qquad 0<x<1;$$

$$\int \frac{dx}{x\sqrt{1+x^2}} = -\operatorname{csch}^{-1}x + C = -\ln\left(\frac{1+\sqrt{1+x^2}}{x}\right) + C, \qquad x>0.$$

Example 7 Find $\displaystyle\int \frac{dx}{\sqrt{3x^2-1}}$.

Solution Going through the list of integration formulas for hyperbolic functions, we find $\displaystyle\int \frac{dx}{\sqrt{x^2-1}} = \cosh^{-1}x + C$. Note that $\sqrt{3x^2-1} = \sqrt{(\sqrt{3}\,x)^2-1}$. Using the technique of integration by substitution with $u=\sqrt{3}\,x$ and $du=\sqrt{3}\,dx$, we have

$$\int \frac{dx}{\sqrt{3x^2-1}} \frac{1}{\sqrt{3}} \int \frac{du}{\sqrt{u^2-1}} = \frac{1}{\sqrt{3}} \cosh^{-1}u + C$$
$$= \frac{1}{\sqrt{3}} \cosh^{-1}(\sqrt{3}\,x) + C. \blacktriangle$$

Example 8 Find $\int \frac{dx}{\sqrt{4+x^2}}$.

Solution

$$\int \frac{dx}{\sqrt{x^2+4}} = \frac{1}{2}\int \frac{dx}{\left[\sqrt{1+(x/2)^2}\right]} = \frac{1}{2}\int \frac{2\,du}{\sqrt{1+u^2}} \qquad \left(u = \frac{x}{2}\right)$$
$$= \sinh^{-1}u + C = \sinh^{-1}\left(\frac{x}{2}\right) + C. \blacktriangle$$

Exercises for Section 8.4

Calculate the derivatives of the functions in Exercises 1–12.

1. $\cosh^{-1}(x^2+2)$
2. $\sinh^{-1}(x^3-2)$
3. $\sinh^{-1}(3x+\cos x)$
4. $\cosh^{-1}(x^2-\tan x)$
5. $x\tanh^{-1}(x^2-1)$
6. $x^2\coth^{-1}(x+1)$
7. $\dfrac{x+\cosh^{-1}x}{\sinh^{-1}x+x}$
8. $\dfrac{1+\sinh^{-1}x}{1-\cosh^{-1}x}$
9. $\exp(1+\sinh^{-1}x)$
10. $\exp(3+\cosh^{-1}x)$
11. $\sinh^{-1}[\cos(3x)]$
12. $\cosh^{-1}[2+\sin(x^2)]$

In Exercises 13–16 calculate the indicated values numerically using logarithms.

13. $\tanh^{-1}(0.5)$
14. $\coth^{-1}(1.3)$
15. $\operatorname{sech}^{-1}(0.3)$
16. $\operatorname{csch}^{-1}(1.2)$

Derive the identities in Exercises 17–20.

17. $\cosh^{-1}x = \ln(x+\sqrt{x^2-1}), \quad |x|>1$
18. $\coth^{-1}x = \frac{1}{2}\ln\left(\frac{x+1}{x-1}\right), \quad |x|>1$
19. $\operatorname{sech}^{-1}x = \ln\dfrac{1+\sqrt{1-x^2}}{x}, \quad 0<x<1$
20. $\operatorname{csch}^{-1}x = \ln\dfrac{1+\sqrt{1+x^2}}{x}, \quad x>0.$

Derive the identities in Exercises 21–24.

21. $\dfrac{d}{dx}\tanh^{-1}x = \dfrac{1}{1-x^2}.$
22. $\dfrac{d}{dx}\cosh^{-1}x = \dfrac{1}{\sqrt{x^2-1}}.$
23. $\dfrac{d}{dx}\operatorname{sech}^{-1}x = \dfrac{-1}{x\sqrt{1-x^2}}.$
24. $\dfrac{d}{dx}\coth^{-1}x = \dfrac{1}{1-x^2}.$

In Exercises 25–28 verify the given integration formulas by differentiation.

25. $\displaystyle\int \frac{dx}{\sqrt{x^2-1}} = \ln(x+\sqrt{x^2-1}) + C, \ |x|>1.$
26. $\displaystyle\int \frac{dx}{1-x^2} = \frac{1}{2}\ln\left(\frac{x+1}{x-1}\right) + C, \ |x|>1.$
27. $\displaystyle\int \frac{dx}{x\sqrt{1-x^2}} = -\ln\left(\frac{1+\sqrt{1-x^2}}{x}\right) + C,$ $0<x<1.$
28. $\displaystyle\int \frac{dx}{x\sqrt{1+x^2}} = -\ln\left(\frac{1+\sqrt{1+x^2}}{x}\right) + C, \ x>0.$

Calculate the integrals in Exercises 29–36.

29. $\displaystyle\int \frac{dx}{1-4x^2}$
30. $\displaystyle\int \frac{dx}{4x^2+1}$
31. $\displaystyle\int \frac{dx}{\sqrt{4x^2+1}}$
32. $\displaystyle\int \frac{dx}{x\sqrt{1-4x^2}}$
33. $\displaystyle\int \frac{\cos x}{\sqrt{\sin^2 x+1}}\,dx$
34. $\displaystyle\int \frac{e^x}{e^x\sqrt{1-e^{2x}}}\,dx$
35. $\displaystyle\int \frac{e^x}{1-e^{2x}}\,dx$
36. $\displaystyle\int \frac{\tan x\,dx}{\sqrt{1+\cos^2 x}}$

★37. Is the function $\cosh^{-1}(\sqrt{x^2+1})$ differentiable at all x?

8.5 Separable Differential Equations

Separable equations can be solved by separating the variables and integrating.

The previous sections dealt with detailed methods for solving particular types of differential equations, such as the spring equation and the equation of growth or decay. In this section and the next, we treat a few other classes of differential equations that can be solved explicitly, and we discuss a few general properties of differential equations.

A differential equation of the form

$$\frac{dy}{dx} = g(x)h(y)$$

in which the right-hand side factors into a product of a function of x and a function of y is called *separable*. Note that we use the term separable only for *first-order* equations; that is, equations involving only the first derivative of y with respect to x.

We may solve the above separable equation by rewriting it in differential notation[5] as

$$\frac{dy}{h(y)} = g(x)\,dx \qquad (\text{assuming } h(y) \neq 0)$$

and integrating:

$$\int \frac{dy}{h(y)} = \int g(x)\,dx.$$

If the integrations can be carried out, we obtain an expression relating x and y. If this expression can be solved for y, the problem is solved; otherwise, one has an equation that implicitly defines y in terms of x. The constant of integration may be determined by giving a value y_0 to y for a given value x_0 of x; that is, by specifying *initial conditions*.

Example 1 Solve $dy/dx = -3xy$, $y = 1$ when $x = 0$.

Solution We have

$$\frac{dy}{y} = -3x\,dx.$$

Integrating both sides gives

$$\ln|y| = -\frac{3x^2}{2} + C, \quad \text{and so} \quad y = \pm \exp C \exp\left(-\frac{3x^2}{2}\right).$$

Since $y = 1$ when $x = 0$, we choose the positive solution and $C = 0$, to give

$$y = \exp\left(-\frac{3x^2}{2}\right).$$

The reader may check by using the chain rule that this function satisfies the given differential equation. ▲

[5] For those worried about manipulations with differentials, answers obtained this way can always be checked by implicit differentiation.

The equation of growth (or decay) $y' = \gamma y$ studied in Section 8.2 is clearly separable, and the technique outlined above reproduces our solution $y = Ce^{\gamma x}$. The spring equation is *not* separable since it is of *second-order*; that is, it involves the second derivative of y with respect to x.

Separable Differential Equations

To solve the equation $y' = g(x)h(y)$:

1. Write
$$\frac{dy}{h(y)} = g(x)\,dx.$$
2. Integrate both sides:
$$\int \frac{dy}{h(y)} = \int g(x)\,dx + C.$$
3. Solve for y if possible.
4. The constant of integration C is determined by a given value of y at a given value of x, that is, by given *initial conditions*.

Example 2 Solve $dy/dx = y^2$, with $y = 1$ when $x = 1$, and sketch the solution.

Solution Separating variables and integrating, we get

$$\frac{dy}{y^2} = dx,$$

$$\frac{-1}{y} = x + C,$$

and so the general solution is

$$y = \frac{1}{-C - x}.$$

Substituting the initial conditions $y = 1$ and $x = 1$, we find that C must be -2, so the specific solution we seek in this case is

$$y = \frac{1}{2 - x}.$$

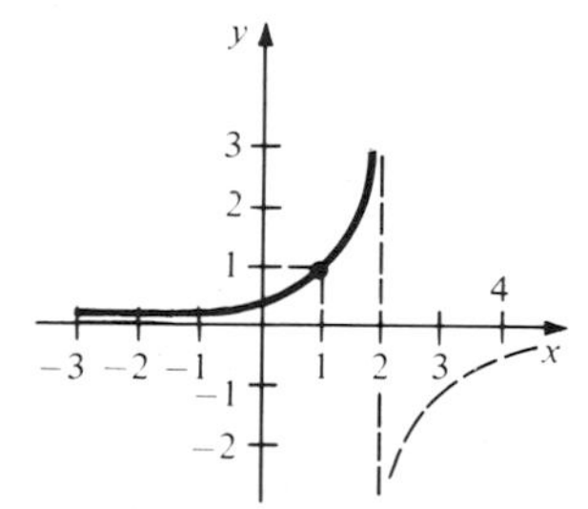

Figure 8.5.1. The solution of $y' = y^2$, $y(1) = 1$.

This function is sketched in Fig. 8.5.1. Notice that the graph has a vertical asymptote, and the function is undefined at $x = 2$. From the point of view of the differential equation, there is really nothing to justify using the portion of the function for $x > 2$, since the equation is not satisfied at $x = 2$. (One could imagine changing the value of C for $x > 2$ and obtaining a new function that still satisfies the differential equation and initial conditions.)

Thus we state that our solution is given by $y = 1/(2 - x)$ for x in $(-\infty, 2)$, and that the solution "blows up" at $x = 2$. ▲

Example 3 Solve

(a) $yy' = \cos 2x$, $y(0) = 1$;
(b) $dy/dx = x/(y + yx^2)$, $y(0) = -1$;
(c) $y' = x^2y^2 + x^2 - y^2 - 1$, $y(0) = 0$.

Solution (a) $y\,dy = \cos 2x\,dx$, so $y^2/2 = \frac{1}{2}\sin 2x + C$. Since $y = 1$ when $x = 0$, $C = \frac{1}{2}$. Thus $y^2 = \sin 2x + 1$ or $y = \sqrt{\sin 2x + 1}$. (We take the + square root since $y = +1$ when $x = 0$.)
(b) $y\,dy = x\,dx/(1 + x^2)$ so $y^2/2 = \frac{1}{2}\ln(1 + x^2) + C$, where the integration was done by substitution. Thus $y^2 = \ln(1 + x^2) + 2C$. Since $y = -1$ when $x = 0$, $C = \frac{1}{2}$. Since y is negative near $x = 0$, we choose the negative root: $y = -\sqrt{1 + \ln(1 + x^2)}$.
(c) The trick is to notice that the right-hand side factors: $y' = (x^2 - 1)(y^2 + 1)$. Thus $dy/(1 + y^2) = (x^2 - 1)\,dx$; integrating gives $\tan^{-1}y = (x^3/3) - x + C$. Since $y(0) = 0$, $C = 0$. Hence $y = \tan[(x^3/3) - x]$. ▲

Many interesting physical problems involve separable equations.

Example 4 **(Electric circuits)** We are told that the equation governing the electric circuit shown in Fig. 8.5.2 is

$$L\frac{dI}{dt} + RI = E$$

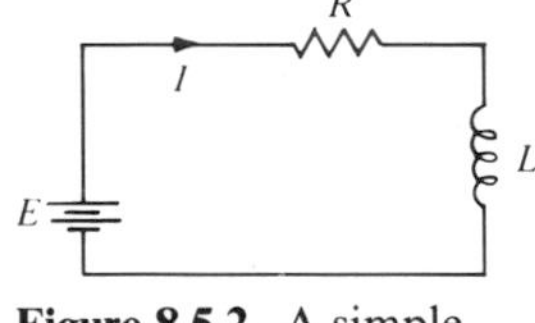

Figure 8.5.2. A simple electric circuit.

and that, in this case,

E (voltage) is a constant;
R (resistance) is a constant > 0;
L (inductance) is a constant > 0; and
I (current) is a function of time.

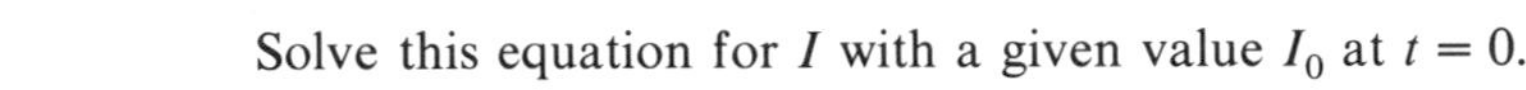

Solve this equation for I with a given value I_0 at $t = 0$.

Solution We separate variables:

$$L\frac{dI}{dt} = E - RI,$$

$$\frac{L}{E - RI}\,dI = dt$$

and then integrate:

$$-\frac{L}{R}\ln|E - RI| = t + C.$$

Thus

$$|E - RI| = \exp\left[-(t + C)\frac{R}{L}\right],$$

and so

$$E - RI = \pm\exp\left(-\frac{R}{L}t\right)\exp\left(-\frac{R}{L}C\right)$$

$$= A\exp\left(-\frac{Rt}{L}\right), \qquad \text{where} \quad A = \pm\exp\left(-\frac{R}{L}C\right).$$

At $t = 0$, $I = I_0$, so $E - RI_0 = A$. Substituting this in the previous equation and simplifying gives

$$I = \frac{E}{R} + \left(I_0 - \frac{E}{R}\right)e^{-Rt/L}.$$

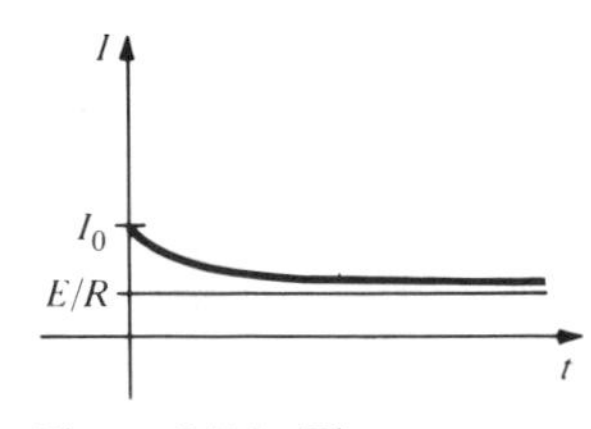

Figure 8.5.3. The current tends to the value E/R as $t \to \infty$.

As $t \to \infty$, I approaches the *steady state part* E/R, while $(I_0 - E/R)e^{-Rt/L}$, which approaches zero as $t \to \infty$, is called the *transient part of* I. (See Fig. 8.5.3.) ▲

Example 5 **(Predator-prey equations)** Consider x predators that feed on y prey. The numbers x and y change as t changes. Imagine the following model (called the *Lotka–Volterra model*).

(i) The prey increase by normal population growth (studied in Section 8.2), at a rate by (b is a positive birth rate constant), but decrease at a rate proportional to the number of predators and the number of prey, that is, $-rxy$ (r is a positive death rate constant). Thus

$$\frac{dy}{dt} = by - rxy.$$

(ii) The predators' population decreases at a rate proportional to their number due to natural decay (starvation) and increases at a rate proportional to the number of predators and the number of prey, that is,

$$\frac{dx}{dt} = -sx + cxy$$

for constants of starvation and consumption s and c.

If we eliminate t by writing

$$\frac{dy}{dx} = \frac{dy/dt}{dx/dt}, \quad \text{we get} \quad \frac{dy}{dx} = \frac{by - rxy}{-sx + cxy}.$$

Solve this equation.

Solution The variables separate:

$$\frac{dy}{dx} = \frac{(b-rx)y}{x(-s+cy)},$$

$$\left(\frac{cy-s}{y}\right)dy = \left(\frac{b-rx}{x}\right)dx.$$

Integrating, we get

$$cy - s\ln y = b\ln x - rx + C$$

for a constant C. This is an implicit form for the parametric curves followed by the predator–prey population. One can show that these curves are closed curves which surround the equilibrium point $(b/r, s/c)$ (the point at which $dx/dt = 0$ and $dy/dt = 0$), as shown in Fig. 8.5.4.[6]

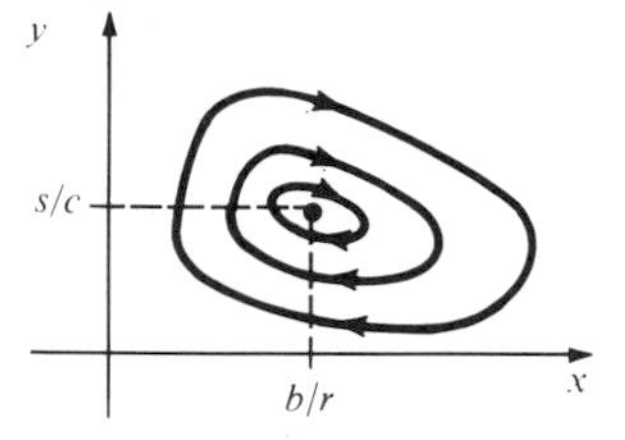

Figure 8.5.4. Solutions of the predator-prey equation.

Variants of this model are important in ecology for predicting and studying cyclic variations in populations. For example, this simple model already shows that if an insect prey and its predator are in equilibrium, killing both predators and prey with an insecticide can lead to a dramatic increase in the population of the prey, followed by an increase in the predators and so on, in cyclic fashion. Similar remarks hold for foxes and rabbits, etc. ▲

Example 6 **(The hanging cable)** Consider a freely hanging cable which weighs m kilograms per meter and is subject to a tension T_0. (See Fig. 8.5.5.) It can be shown[7] that the shape of the cable satisfies

$$\frac{d^2y}{dx^2} = \frac{mg}{T_0}\sqrt{1 + \left(\frac{dy}{dx}\right)^2}$$

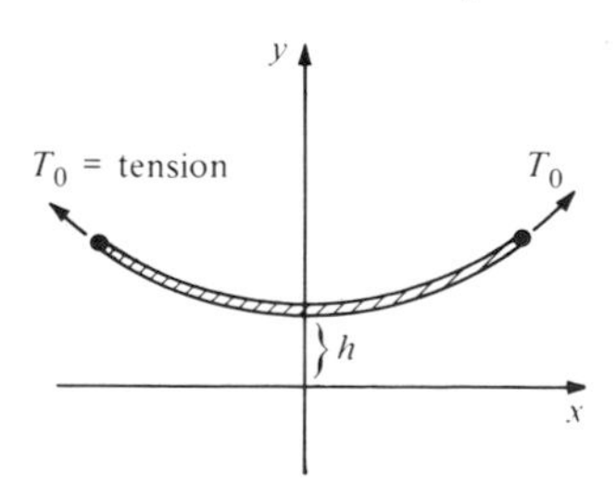

Figure 8.5.5. A cable hanging under its own weight.

Introduce the new variable $w = dy/dx$ and show that w satisfies a separable equation. Solve for w and then y. You may assume the graph to be symmetric about the y axis.

[6] For a proof due to Volterra, see G. F. Simmons, *Differential Equations*, McGraw-Hill (1972) p. 286. There is also a good deal of information, including many references, in Chapter 9 of *Elementary Differential Equations and Boundary Value Problems* by W. Boyce and R. DiPrima, Third Edition, Wiley (1977), and in Section 1.5 of *Differential Equations and their Applications*, by M. Braun, Third Edition, Springer, (1983).

[7] See, for instance, T. M. Creese and R. M. Haralick, *Differential Equations for Engineers*, McGraw-Hill (1978) pp. 71–75.

Solution In terms of w, the equation for the cable is

$$\frac{dw}{dx} = \frac{mg}{T_0}\sqrt{1 + w^2}\,.$$

Separating variables and integrating gives

$$\int \frac{dw}{\sqrt{1 + w^2}} = \frac{mg}{T_0}\int dx,$$

$$\sinh^{-1} w = \frac{mg}{T_0}(x + C).$$

Since the cable is symmetric about the y axis, the slope $w = dy/dx$ is zero when $x = 0$, so the integration constant is zero.

Now $w = \sinh[(mg/T_0)x]$, and so

$$y = \int \frac{dy}{dx}\,dx = \int w\,dx = \int \sinh\left(\frac{mg}{T_0}x\right)dx = \frac{T_0}{mg}\cosh\left(\frac{mg}{T_0}x\right) + C_1\,.$$

The integration constant C_1 is found by setting $x = 0$. Since $\cosh(0) = 1$ and $y = h$ when $x = 0$ (Fig. 8.5.5), we get

$$h = T_0/mg + C_1, \quad \text{so} \quad C_1 = h - T_0/mg = (mgh - T_0)/mg.$$

Thus the equation for the shape of the cable is

$$y = \frac{T_0}{mg}\left[\cosh\left(\frac{mg}{T_0}x\right) - 1\right] + h.$$

The graph of $\cosh x$ takes its name *catenary* from this example and the Latin word *catena*, meaning "chain." ▲

We remark that cables which bear weight, such as the ones on suspension bridges, hang in a *parabolic* form (see Exercise 22).

Example 7 **(Orthogonal trajectories)** Consider the family of parabolas $y = kx^2$ for various constants k. (a) Find a differential equation satisfied by this family that does not involve k by differentiating and eliminating k. (b) Write a differential equation for a family of curves orthogonal (= perpendicular) to each of the parabolas $y = kx^2$ and solve it. Sketch.

Solution (a) If we differentiate, we have $y' = 2kx$; but $y = kx^2$ so $k = y/x^2$, and thus

$$y' = 2kx = 2\left(\frac{y}{x^2}\right)x = \frac{2y}{x}\,.$$

Thus any parabola $y = kx^2$ satisfies the equation $y' = 2y/x$.

(b) The slope of a line orthogonal to a line of slope m is $-1/m$, so the equation satisfied by the orthogonal trajectories is $y' = -x/2y$. This equation is separable:

$$2y\,dy = -x\,dx,$$

$$y^2 = -\frac{x^2}{2} + C,$$

$$y^2 + \frac{x^2}{2} = C.$$

If we write this as $y^2 + (x/\sqrt{2})^2 = C$, we see that these curves are obtained from the family of concentric circles with radii $\sqrt{C}$ centered at the origin by stretching the x axis by a factor of $\sqrt{2}$. (See Fig. 8.5.6.) They are *ellipses*. (See Section 14.1 for a further discussion of ellipses.) ▲

origin by stretching the x axis by a factor of $\sqrt{2}$. (See Fig. 8.5.6.) They are *ellipses*. (See Section 14.1 for a further discussion of ellipses.) ▲

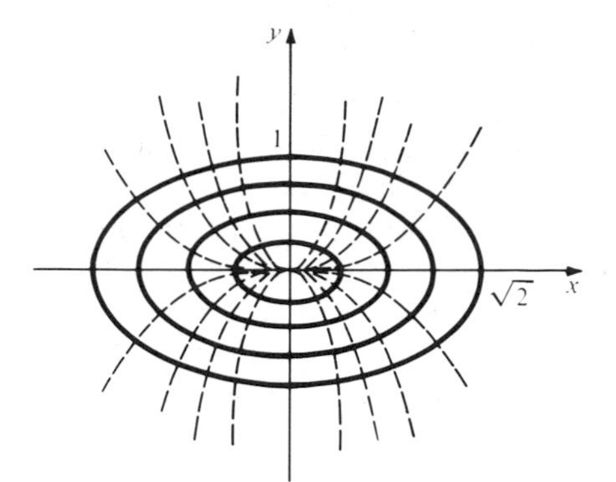

Figure 8.5.6. The orthogonal trajectories of the family of parabolas are ellipses.

Separable differential equations are a special case of the equation

$$\frac{dy}{dx} = F(x, y),$$

where F is a function depending on both x and y.[8] For example,

$$\frac{dy}{dx} = -x^2y + y^3 + 3\sin y + 1$$

is a differential equation that is not separable. There is little hope of solving such equations explicitly, except in rather special cases, such as the separable case. In general, one has to resort to numerical or other approximate methods. To do so, it is useful to have a geometric picture of what is going on.

The given data $dy/dx = F(x, y)$ tell us the slope at each point of the solution $y = f(x)$ that we seek. We can therefore imagine drawing small lines in the xy plane, with slope $F(x, y)$ at the point (x, y), as in Fig. 8.5.7.

The problem of finding a solution to the differential equation is precisely the problem of threading our way through this *direction field* with a curve which is tangent to the given direction at each point. (See Fig. 8.5.8. In this figure, some of the line segments are vertical, reflecting the fact that the formula for $F(x, y)$ may be a fraction whose denominator is sometimes zero.)

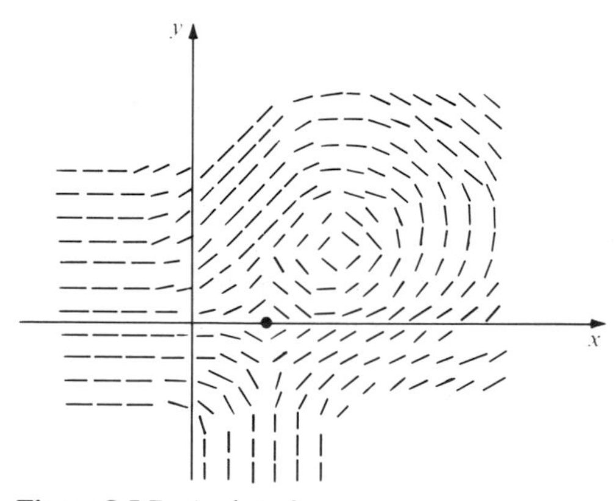

Figure 8.5.7. A plot of a direction field.

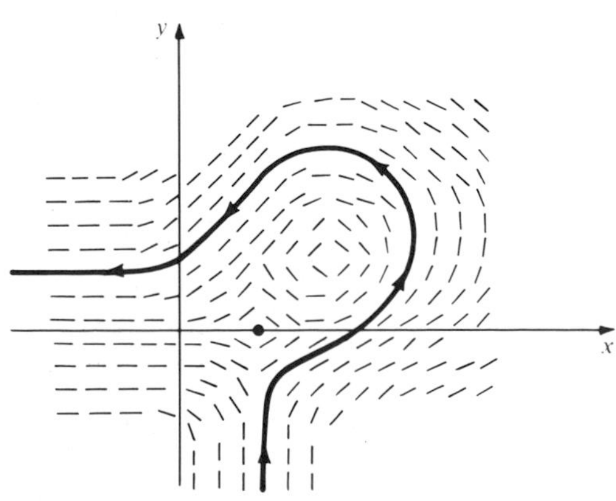

Figure 8.5.8. A solution threads its way through the direction field.

[8] We study such functions in detail beginning with Chapter 14. The material of those later chapters is not needed here.

We saw in Example 5 that differential equations may be given in parametric form

$$\frac{dx}{dt} = g(x, y), \quad \text{and} \quad \frac{dy}{dt} = h(x, y).$$

(so $dy/dx = h(x, y)/g(x, y)$.) Here we seek a parametric curve $(x(t), y(t))$ solving these two equations. From our discussion of parametric curves in Section 2.4, we see that the pair $(g(x, y), h(x, y))$ gives the velocity of the solution curve passing through (x, y). In this formulation, we can interpret Fig. 8.5.7 as a *velocity field*. If one thinks of the motion of a fluid, one can phrase the problem of finding solutions to the above pair of differential equations as follows: given the velocity field of a fluid, find the paths that fluid particles follow. For this reason, a solution curve is often called a *flow line*.

Example 8 Sketch the direction field for the equation $dy/dx = -x/y$ and solve the equation.

Solution Here the slope at (x, y) is $-x/y$. We draw small line segments with these slopes at a number of selected locations to produce Fig. 8.5.9.

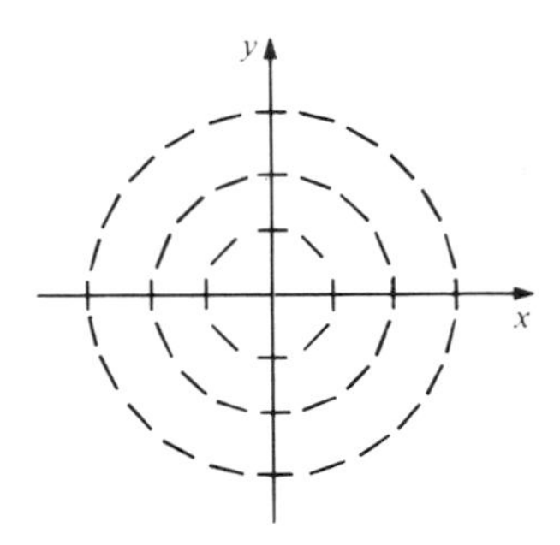

Figure 8.5.9. The direction field for $y' = -x/y$.

The equation is separable:

$$y\,dy = -x\,dx,$$

$$\frac{y^2}{2} = -\frac{x^2}{2} + C,$$

$$y^2 + x^2 = 2C.$$

Thus any solution is a circle and the solutions taken together form a family of circles. This is consistent with the direction field. ▲

When a numerical technique is called for, the direction field idea suggests a simple method. This procedure, called the *Euler method*, replaces the actual solution curve by a polygonal line and follows the direction field by moving a short distance along a straight line. For $dy/dx = F(x, y)$ we start at (x_0, y_0) and break up the interval $[x_0, x_0 + a]$ into n steps $x_0, x_1 = x_0 + a/n$, $x_2 = x_0 + 2a/n, \ldots, x_n = x_0 + a$. Now we recursively define

$$y_1 = F(x_0, y_0)\frac{a}{n} + y_0$$

$$y_2 = F(x_1, y_1)\frac{a}{n} + y_1$$

$$\vdots \qquad \vdots$$

$$y_n = F(x_{n-1}, y_{n-1})\frac{a}{n} + y_{n-1};$$

that is,

$$y_i - y_{i-1} = \left[\frac{dy}{dx}(x_{i-1}, y_{i-1}) \right](x_i - x_{i-1}), \qquad i = 1, 2, \ldots, n$$

to produce the desired approximate solution (the polygonal curve shown in Fig. 8.5.10).

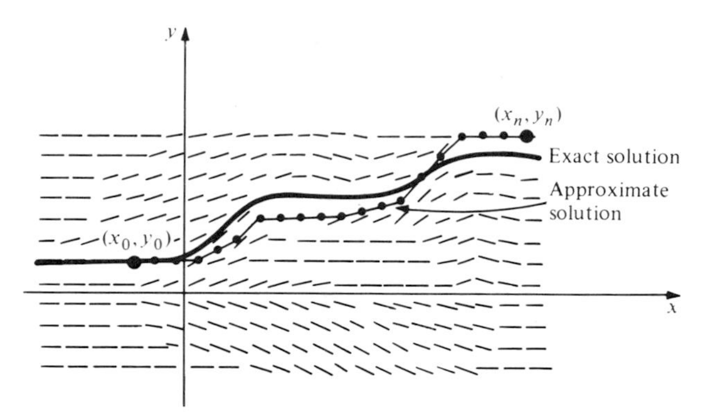

Figure 8.5.10. The Euler method for numerically solving differential equations.

Example 9 Solve the equation $dy/dx = x + \cos y$, $y(0) = 0$ from $x = 0$ to $x = \pi/4$ using a ten-step Euler method; that is, find $y(\pi/4)$ approximately.

Solution The recursive procedure is summarized below. It is helpful to record the data in a table as you proceed.[9] Here $x_0 = 0$, $y_0 = 0$, $a = \pi/4$, $n = 10$; thus $h = a/n = \pi/40 = 0.0785398$.

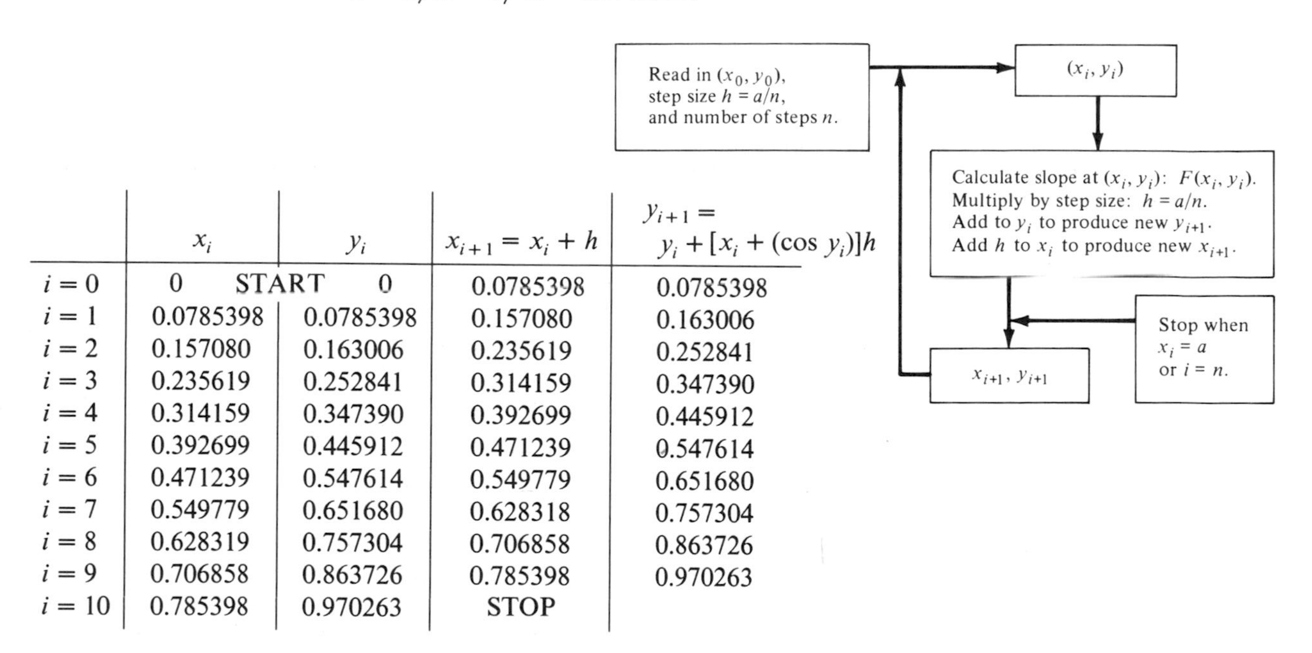

	x_i	y_i	$x_{i+1} = x_i + h$	$y_{i+1} = y_i + [x_i + (\cos y_i)]h$
$i = 0$	0 START	0	0.0785398	0.0785398
$i = 1$	0.0785398	0.0785398	0.157080	0.163006
$i = 2$	0.157080	0.163006	0.235619	0.252841
$i = 3$	0.235619	0.252841	0.314159	0.347390
$i = 4$	0.314159	0.347390	0.392699	0.445912
$i = 5$	0.392699	0.445912	0.471239	0.547614
$i = 6$	0.471239	0.547614	0.549779	0.651680
$i = 7$	0.549779	0.651680	0.628318	0.757304
$i = 8$	0.628319	0.757304	0.706858	0.863726
$i = 9$	0.706858	0.863726	0.785398	0.970263
$i = 10$	0.785398	0.970263	STOP	

Thus $y(\pi/4) \approx 0.970263$. ▲

[9] The Euler or related methods are particularly easy to use with a programmable calculator. In practice, the Euler method is not the most accurate or efficient. Usually the Runge–Kutta or predictor–corrector method is more accurate. (For details and comparative error analyses, a book such as C. W. Gear, *Numerical Initial Value Problems in Ordinary Differential Equations*, Prentice-Hall Englewood Cliffs, N.J. (1971) should be consulted.)

Direction fields can also be used to derive some qualitative information without solving the differential equation.

Example 10 Sketch the direction field for the equation $dy/dx = y^2 - 3y + 2$ and use it to find $\lim_{x\to\infty} y(x)$ geometrically for a solution satisfying $1 < y(0) < 2$.

Solution First we factor $y^2 - 3y + 2 = (y-1)(y-2)$. Thus $y^2 - 3y + 2$ is negative on the interval $(1,2)$ and positive on the intervals $(2,\infty)$ and $(-\infty,1)$. The direction field in the xy plane, which is independent of x, can now be plotted, as in Fig. 8.5.11.

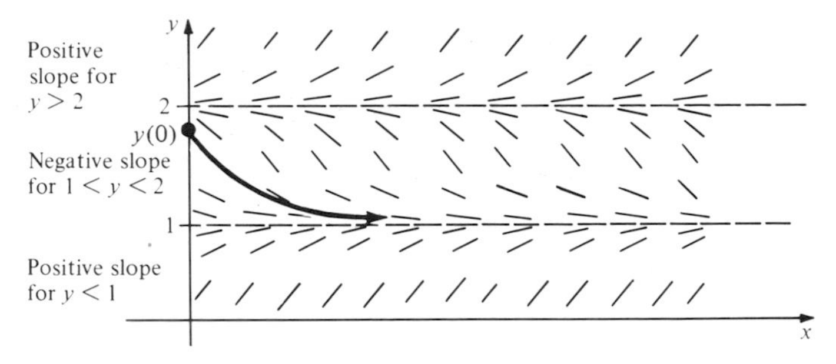

Figure 8.5.11. Direction field for the equation $dy/dx = y^2 - 3y + 2$.

If $1 < y(0) < 2$, then $y(x)$ must thread its way through the direction field, always remaining tangent to it. From Fig. 8.5.11, it is clear that a solution with initial condition $y(0)$ lying in the interval $(1,2)$ is pushed downward, flattens out and becomes asymptotic to the line $y = 1$. Thus $\lim_{x\to\infty} y(x) = 1$. ▲

Supplementary Remark. Figure 8.5.11 enables us to see that the *equilibrium* solution $y = 1$ is stable; in other words, any initial condition $y(0)$ close to 1 gives a solution which remains close to 1 for all x. Compare this with the discussion of stable oscillations in the Supplement to Section 8.1. Likewise, the solution $y = 2$ is unstable. Again, consult the discussion of unstable equilibria in the Supplement to Section 8.3.

Exercises for Section 8.5

In Exercises 1–12 find the solution of the given differential equation satisfying the stated conditions (express your answer implicitly if necessary).

1. $\dfrac{dy}{dx} = \cos x$, $y(0) = 1$.
2. $\dfrac{dy}{dx} = y\cos x$, $y(0) = 1$.
3. $\dfrac{dy}{dx} = 2xy - 2y + 2x - 2$, $y(1) = 0$.
4. $y\dfrac{dy}{dx} = x$, $y(0) = 1$.
5. $\dfrac{1}{y}\dfrac{dy}{dx} = \dfrac{1}{x}$, $y(1) = -2$
6. $x\dfrac{dy}{dx} = \sqrt{1-y^2}$, $y(1) = 0$.
7. $\dfrac{dy}{dx} = \dfrac{xe^{-y}}{(x^2+1)y}$, $y(0) = 1$.
8. $\dfrac{dy}{dx} = y^3\dfrac{\sin x}{1+8y^4}$, $y(0) = 1$.
9. $\dfrac{dy}{dx} = \dfrac{1+y}{1+x}$, $y(0) = 1$.
10. $\dfrac{dy}{dx} = 3xy - x$.
11. $\dfrac{dy}{dx} = \cos x - y\cos x$, $y(0) = 2$.
12. $e^y\left(\dfrac{dy}{dx}\right) = 1 + e^{2y} - xe^{2y} - x$, $y(0) = 1$.
13. The current I in an electric circuit is described by the equation $3(dI/dt) + 8I = 10$, and the initial current is $I_0 = 2.1$ at $t = 0$. Sketch the graph of I as a function of time.
14. Repeat Exercise 13 for $I_0 = 0.3$.
15. *Capacitor equation*. The equation $R(dQ/dt) + Q/C = E$ describes the charge Q on a capacitor, where R, C, and E are constants. (See Fig. 8.5.12.) (a) Find Q as a function of time if $Q = 0$ at $t = 0$. (b) How long does it take for Q to attain 99% of its limiting charge?

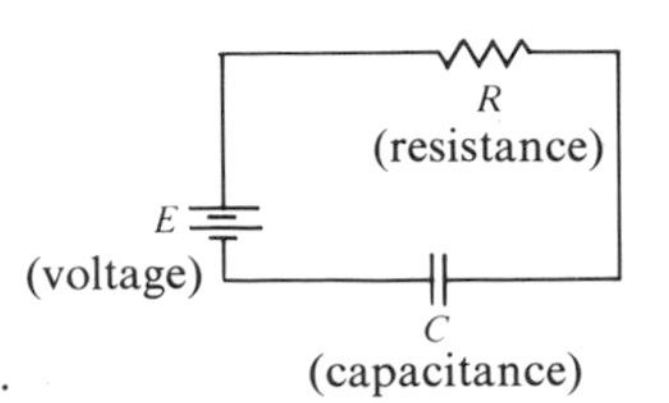

Figure 8.5.12. A circuit with a charging capacitor.

16. Repeat Exercise 15 for $Q = 3.1$ at $t = 0$ and $R = 2$, $E = 10$, and $C = 2$.
17. Verify directly that $x = b/r$, $y = s/c$ is a solution of the predator–prey equations (see Example 5).
18. (a) Verify graphically that the equation $y - \ln y = c$ has exactly two positive solutions if $c < -1$.
(b) What does this have to do with Fig. 8.5.4?
19. *Logistic law of population growth*. If a population can support only P_0 members, the rate of growth of the population may be given by $dP/dt = kP(P_0 - P)$. This modification of the law of growth $dP/dt = \alpha P$ discussed in Section 8.2 is called the logistic law, or the Verhulst law. Solve this equation and show that P tends to P_0 as $t \to \infty$. *Hint*: In solving the equation you may wish to use the identity

$$\frac{1}{P(P_0 - P)} = \frac{1}{P_0}\left(\frac{1}{P} + \frac{1}{P_0 - P}\right).$$

20. *Chemical reaction rates*. Chemical reactions often proceed at a rate proportional to the concentrations of each reagent. For example, consider a reaction of the type $2A + B \to C$ in which two molecules of A and one of B combine to produce one molecule of C. Concentrations are measured in moles per liter, where a mole is a definite number (6×10^{23}) of molecules. Let the concentrations of A, B, and C at time t be a, b, c, and suppose that $c = 0$ at $t = 0$. Since no molecules are destroyed, $b_0 - b = (a_0 - a)/2$, where a_0 and b_0 are the values of a and b at $t = 0$. The rate of change of a is given by $da/dt = ka^2b$ for a constant k. Solve this equation. [*Hint*: You will need to make up an identity like the one used in Exercise 19.]
21. In Example 6, verify that the solution becomes straight as T_0 increases to ∞.
22. *Suspension bridge*. The function $y(x)$ describing a suspended cable which weighs m kilograms per meter and which is subject to a tension T_0 satisfies $dy/dx = (m/T_0)x$. Verify that the cable hangs in a parabolic shape.
23. (a) Find a differential equation satisfied by the family of hyperbolas $xy = k$ for various constants k.
(b) Find a differential equation satisfied by the orthogonal trajectories to the hyperbolas $xy = k$, solve it, and sketch the resulting family of curves.
24. (a) Find a differential equation satisfied by the family of ellipses, $x^2 + 4y^2 = k$, k a constant.
(b) Find a differential equation satisfied by the orthogonal trajectories to this family of ellipses, solve it and sketch.
25. (a) Sketch the family of cubics $y = cx^3$, when c is constant.
(b) Find a differential equation satisfied by this family.
(c) Find a differential equation for the orthogonal family and solve it.
26. Repeat Exercise 25 for the family of cubics $y = x^3 - dx$, where d is constant.
27. (a) Sketch the direction field for the equation $dy/dx = 2y/x$. (b) Solve this equation.
28. Sketch the direction field for the equation $y' = -y/x$. Solve the equation and show that the solutions are consistent with your direction field.
29. Use a ten-step Euler method to find y approximately at $x = 1$ if $dy/dx = y - x^2$ and $y(0) = 1$.
30. If $y' = x + \tan y$, $y(0) = 0$, find $y(1)$ approximately using a ten-step Euler procedure.
31. Find an approximate solution for $y(1)$ if $y' = x\sqrt{1 + y^4}$ and $y(0) = 0$ using a fifteen-step Euler method.
32. Redo Example 11 using a twenty-step Euler method, compare the answers and discuss.

In Exercises 33–36, sketch the direction field of the given equation and use it to find $\lim_{x\to\infty} y(x)$ geometrically for the given $y(0)$.

33. $\dfrac{dy}{dx} = y^2 - 7y + 12$, $y(0) = 3.5$.
34. $\dfrac{dy}{dx} = y^2 + y - 2$, $y(0) = -2$.
35. $\dfrac{dy}{dx} = y^3 - 6y^2 + 11y - 6$, $y(0) = 1.5$.
36. $\dfrac{dy}{dx} = y^3 - 4y^2 + 3y$, $y(0) = 2$.
37. Suppose that $y = f(x)$ solves the equation $dy/dx = e^x y^2 + 4xy^5$, $f(0) = 1$. Calculate $f'''(0)$.
38. Suppose $y'' + 3(y')^3 + 8e^x y^2 = 5\cos x$, $y'(0) = 1$, and $y''(0) = 2$. Calculate $y'''(0)$.

★39. Consider a family of curves defined by a separable equation $dy/dx = g(x)h(y)$. Express the family of orthogonal trajectories implicitly in terms of integrals.

★40. Show by a graphical argument that any straight line through $(b/r, s/c)$ meets the curve $cy - s\ln y = b\ln x - rx + C$ (for $C > 0$) at exactly two points. (See Example 5).

8.6 Linear First-Order Equations

First-order equations which are linear in the unknown function can be solved explicitly.

We have seen that separable equations can be solved directly by integration. There are a few other classes of differential equations that can be solved by reducing them to integration after a suitable transformation. We shall treat one such class now. (Other classes are discussed in Sections 12.7 and 18.3.)

We consider equations that are linear in the unknown function y:

$$\frac{dy}{dx} = P(x)y + Q(x) \tag{1}$$

for given functions P and Q of x. If Q is absent, equation (1) becomes

$$\frac{dy}{dx} = P(x)y \tag{2}$$

which is separable:

$$\frac{1}{y}\,dy = P(x)\,dx,$$

$$\ln|y| = \int P(x)\,dx + C,$$

$$|y| = \exp(C)\exp\left(\int P(x)\,dx\right).$$

Choosing $C = 0$ and $y > 0$ gives the particular solution

$$y = \exp\left(\int P(x)\,dx\right). \tag{3}$$

Now we use the solution (3) of equation (2) to help us simplify equation (1). If y solves equation (1), we divide it by the function (3), obtaining a new function

$$w = y\exp\left(-\int P(x)\,dx\right) \tag{4}$$

which turns out to satisfy a *simpler* equation. By the product and chain rules, we get

$$\frac{dw}{dx} = \frac{dy}{dx}\exp\left(-\int P(x)\,dx\right) - yP(x)\exp\left(\int -P(x)\,dx\right)$$

$$= \left[P(x)y + Q(x)\right]\exp\left(-\int P(x)\,dx\right) - P(x)y\exp\left(\int -P(x)\,dx\right).$$

The terms involving y cancel, leaving

$$\frac{dw}{dx} = Q(x)\exp\left(-\int P(x)\,dx\right).$$

The right-hand side is a function of x alone, so we may integrate:

$$w = \int Q(x)\left[\exp\left(-\int P(x)\,dx\right)\right]dx + C. \tag{5}$$

Combining formulas (4) and (5) gives the general solution y of equation (1), written out explicitly in the following box.

Linear First-Order Equations

The general solution of

$$\frac{dy}{dx} = P(x)y + Q(x)$$

is

$$y = \exp\left(\int P(x)\,dx\right)\left\{\int Q(x)\left[\exp\left(-\int P(x)\,dx\right)\right]dx + C\right\}, \tag{6}$$

where C is an arbitrary constant.

One may verify by direct substitution that the expression (6) for y in this display solves equation (1). Instead of memorizing *formula* (6) for the solution, it may be easier to remember the *method*, as summarized in the following box.

Method for Solving $dy/dx = P(x)y + Q(x)$

1. Calculate $\int P(x)\,dx$, dropping the integration constant.
2. Transpose $P(x)y$ to the left side: $\dfrac{dy}{dx} - P(x)y = Q(x)$.
3. Multiply the equation by $\exp(-\int P(x)\,dx)$.
4. The left-side of the equation should now be a derivative:
   $$\frac{d}{dx}\left[y\exp\left(-\int P(x)\,dx\right)\right].$$
   Check to make sure.
5. Integrate both sides, keeping the constant of integration.
6. Solve the resulting equation for y.
7. Use the initial condition, if given, to solve for the integration constant.

The expression $\exp(-\int P(x)\,dx)$ is called an *integrating factor*, since multiplication by this term enables us to solve the equation by integration.

Example 1 Solve $dy/dx = xy + x$.

Solution We follow the procedure in the preceding box.

1. $P(x) = x, \quad \text{so} \quad \int P(x)\,dx = \dfrac{x^2}{2}.$

2. $\dfrac{dy}{dx} - xy = x$

(transposing xy to the left-hand side).

3. $\exp\left(-\dfrac{x^2}{2}\right)\left\{\dfrac{dy}{dx} - xy\right\} = \exp\left(-\dfrac{x^2}{2}\right)x$

(multiplying by $\exp(-x^2/2)$).

4. $\dfrac{d}{dx}\left\{y\exp\left(-\dfrac{x^2}{2}\right)\right\} = \dfrac{dy}{dx}\exp\left(-\dfrac{x^2}{2}\right) - yx\exp\left(-\dfrac{x^2}{2}\right)$

which equals the left-hand side in step 3. Thus the equation is now

$$\frac{d}{x}\left\{y\exp\left(-\frac{x^2}{2}\right)\right\} = x\exp\left(-\frac{x^2}{2}\right).$$

5. Integration, using the substitution $u = -x^2/2$, yields

$$y\exp\left(-\frac{x^2}{2}\right) = \int x\exp\left(-\frac{x^2}{2}\right)dx = -\exp\left(-\frac{x^2}{2}\right) + C.$$

6. Solving for y,

$$y = C\exp\left(\frac{x^2}{2}\right) - 1. \blacktriangle$$

Example 2 Solve the following equations with the stated initial conditions.

(a) $y' = e^{-x} - y$, $y(0) = 1$;
(b) $y' = \cos^2 x - (\tan x)y$, $y(0) = 1$.

Solution (a) The integrating factor is $\exp(-\int P(x)\,dx) = \exp(-\int -\,dx) = \exp(x)$. Thus

$$e^x(y' + y) = 1,$$

$$\frac{d}{dx}(e^x y) = 1,$$

$$e^x y = x + C.$$

Since $y(0) = 1$, $C = 1$, so $y = (x+1)e^{-x}$.
(b) The integrating factor is $\exp(\int \tan x\,dx) = \exp(-\ln\cos x) = 1/\cos x$. (This is valid only if $\cos x > 0$, but since our initial condition is $x = 0$ where $\cos x = 1$, this is justified.) Thus

$$\frac{1}{\cos x}\left[y' + (\tan x)y\right] = \cos x,$$

$$\frac{d}{dx}\left[\frac{y}{\cos x}\right] = \cos x,$$

$$\frac{y}{\cos x} = \sin x + C.$$

Since $y = 1$ when $x = 0$, $C = 1$. Thus $y = \cos x \sin x + \cos x$. It may be verified that this solution is valid for all x. ▲

Example 3 (**Electric circuits**) In Example 4, Section 8.5, replace E by the sinusoidal voltage

$$E = E_0 \sin\omega t$$

with L, R, and E_0 constants, and solve the resulting equation.

Solution The equation is

$$\frac{dI}{dt} = -\frac{RI}{L} + \frac{E_0}{L}\sin\omega t.$$

We follow the procedure in the preceding box with x replaced by t and y by I:

1. $P(t) = -\dfrac{R}{L}$, a constant, so $\int P(t)\,dt = -tR/L$.

2. $\dfrac{dI}{dt} + \dfrac{RI}{L} = \dfrac{E_0}{L}\sin\omega t.$

3. $\left[\exp\left(\dfrac{tR}{L}\right)\right]\left(\dfrac{dI}{dt} + \dfrac{RI}{L}\right) = \dfrac{E_0}{L}\exp\left(\dfrac{tR}{L}\right)\sin\omega t.$

4. $\frac{d}{dt}\left\{\exp\left(\frac{tR}{L}\right)I\right\} = \frac{E_0}{L}\exp\left(\frac{tR}{L}\right)\sin\omega t.$

5. $\exp\left(\frac{tR}{L}\right)I = \frac{E_0}{L}\int \exp\left(\frac{tR}{L}\right)\sin\omega t\,dt.$

This integral may be evaluated by the method of Example 4, Section 7.4, namely integration by parts twice. One gets

$$\exp\left(\frac{tR}{L}\right)I = \frac{E_0}{L}\left\{\frac{e^{tR/L}}{(R/L)^2+\omega^2}\left(\frac{R}{L}\sin\omega t - \omega\cos\omega t\right)\right\} + C.$$

6. Solving for I,

$$I = \frac{E_0}{L}\,\frac{1}{(R/L)^2+\omega^2}\left(\frac{R}{L}\sin\omega t - \omega\cos\omega t\right) + Ce^{-tR/L}.$$

The constant C is determined by the value of I at $t = 0$. This expression for I contains an oscillatory part, oscillating with the same frequency ω as the driving voltage (but with a phase shift; see Exercise 10) and a transient part $Ce^{-tR/L}$ which decays to zero as $t \to \infty$. (See Fig. 8.6.1.) ▲

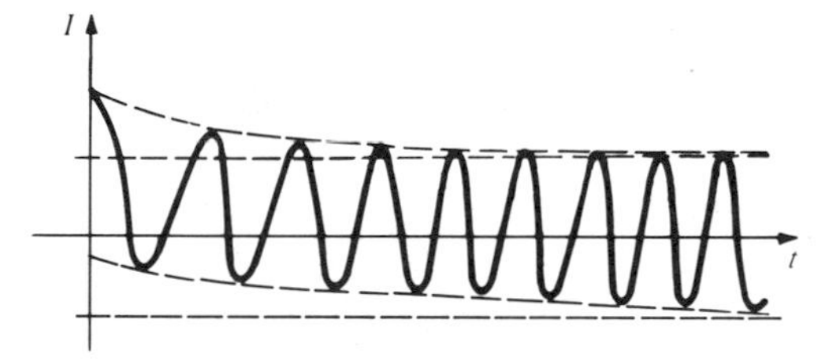

Figure 8.6.1. The graph of the solution of a sinusoidally forced electric circuit containing a resistor and an inductor.

Example 4 **(Pollution)** A small lake contains 4×10^7 liters of pure water at $t = 0$. A polluted stream carries 0.67 liter of pollutant and 10 liters of water into the lake per second. (Assume that this mixes instantly with the lake water.) Meanwhile, 10.67 liters per second of the lake flow out in a drainage stream. Find the amount of pollutant in the lake as a function of time. What is the limiting value?

Solution Let $y(t)$ denote the amount of pollutant in liters in the lake at time t. The amount of pollutant in one liter of lake water is thus

$$\frac{y(t)}{4 \times 10^7}.$$

The rate of change of $y(t)$ is the rate at which pollutant flows out, which is $-10.67y(t)/4 \times 10^7 = -2.67 \times 10^{-7}y(t)$ liters per second, plus the rate at which it flows in, which is 0.67 liter per second. Thus

$$y' = -2.67 \times 10^{-7}y + 0.67.$$

The solution of $dy/dt = ay + b$, $y(0) = 0$, is found using the integrating factor e^{-at}:

$$e^{-at}(y' - ay) = be^{-at},$$

$$\frac{d}{dt}(e^{-at}y) = be^{-at},$$

$$e^{-at}y = -\frac{b}{a}e^{-at} + C.$$

Since $y = 0$ at $t = 0$, $C = b/a$. Thus

$$e^{-at}y = \frac{b}{a}(1 - e^{-at}),$$

$$y = \frac{b}{a}(e^{at} - 1).$$

Here $a = -2.67 \times 10^{-7}$ and $b = 0.67$, so $y = (1 - e^{-2.67 \times 10^{-7}t})(2.51 \times 10^6)$ liters. For t small, y is relatively small; but for larger t, y approaches the (steady-state) catastrophic value of 2.51×10^6 liters, that is, the lake is well over half pollutant. (See Exercise 13 to find out how long this takes.) ▲

Example 5 **(Falling object in a resisting medium)** The downward force acting on a body of mass m falling in air is mg, where g is the gravitational constant. The force of air resistance is γv, where γ is constant of proportionality and v is the downward speed. If a body is released from rest, find its speed as a function of time t. (Assume it is released from a great enough height so that it has not hit the ground by time t.)

Solution Since mass times acceleration is force, and acceleration is the time derivative of velocity, we have the equation $m(dv/dt) = mg - \gamma v$ or, equivalently, $dv/dt = -(\gamma/m)v + g$ which is a linear first-order equation. Its general solution is

$$v = e^{-\gamma t/m}\int ge^{\gamma t/m}\,dt = e^{-\gamma t/m}\left(\frac{mg}{\gamma}e^{\gamma t/m} + C\right).$$

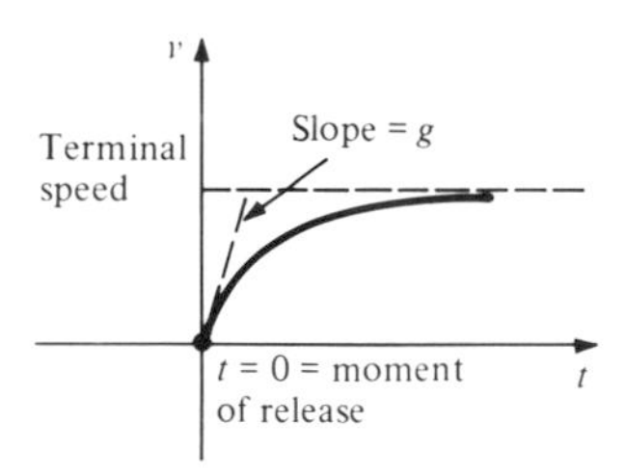

Figure 8.6.2. The speed of an object moving in a resisting medium.

If $v = 0$ when $t = 0$, C must be $-mg/\gamma$, so $v = (mg/\gamma)(1 - e^{-\gamma t/m})$. Note that as $t \to \infty$, $e^{-\gamma t/m} \to 0$ and so $v \to mg/\gamma$ the *terminal* speed. (See Fig. 8.6.2.) For small t the velocity is approximately gt, which is what it would be if there were no air resistance. As t increases, the air resistance slows the velocity and a terminal velocity is approached. ▲

Example 6 **(Rocket propulsion)** A rocket with an initial mass M_0 (kilograms) blasts off at time $t = 0$ (Fig. 8.6.3). The mass decreases with time because the fuel is being

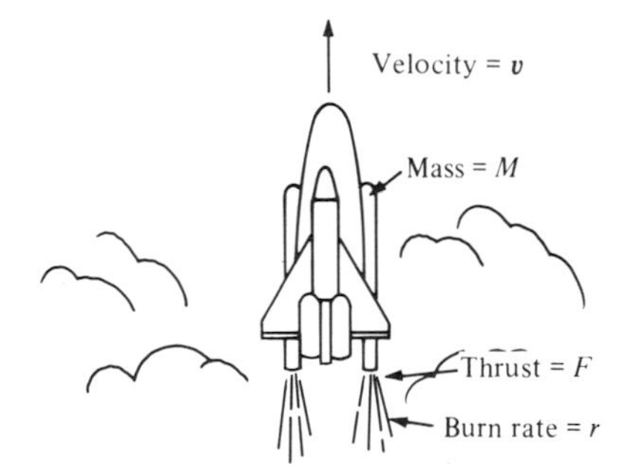

Figure 8.6.3. A rocket blasting off at $t = 0$.

spent at a constant burn rate r (kilograms per second). Thus, the mass at time t is $M = M_0 - rt$. If the thrust is a constant force F, and the velocity is v, Newton's second law gives

$$\frac{d}{dt}(Mv) = F - Mg, \tag{7}$$

where $g = 9.8$ is the gravitational constant. (We neglect air resistance and assume the motion to be vertical.)

(a) Solve equation (7).
(b) If the mass of the rocket at burnout is M_1, compute the velocity at burnout.

Solution (a) Substituting $M = M_0 - rt$ into equation (7) gives

$$\frac{d}{dt}\left[(M_0 - rt)v\right] = F - (M_0 - rt)g.$$

Although this is a linear equation in v, it is already in a form which we can directly integrate:

$$(M_0 - rt)v = \int (F - g(M_0 - rt))\,dt + C$$

$$= Ft - g\left(M_0 t - \frac{rt^2}{2}\right) + C.$$

Solving for v,

$$v = \frac{Ft}{M_0 - rt} - \frac{g}{M_0 - rt}\left(M_0 t - \frac{rt^2}{2}\right) + \frac{C}{(M_0 - rt)}.$$

Since $v = 0$ at $t = 0$, $C = 0$, so the solution is

$$v = \frac{Ft}{M_0 - rt} - \frac{g}{M_0 - rt}\left(M_0 t - \frac{rt^2}{2}\right).$$

(b) At burnout, $M_0 - rt = M_1$, so

$$v = \frac{F(M_0 - M_1)}{rM_1} - \frac{g}{M_1}\left[M_0\left(\frac{M_0 - M_1}{r}\right) - \frac{(M_0 - M_1)^2}{2r}\right]$$

$$= \frac{F(M_0 - M_1)}{rM_1} - \frac{g(M_0^2 - M_1^2)}{2rM_1} = \frac{M_0 - M_1}{rM_1}\left[F - g\left(\frac{M_0 + M_1}{2}\right)\right]$$

is the velocity at burnout. ▲

Exercises for Section 8.6

In Exercises 1–4, solve the given differential equation by the method of this section.

1. $\dfrac{dy}{dx} = \dfrac{y}{1-x} + \dfrac{2}{1-x} + 3.$
2. $\dfrac{dy}{dx} = y\sin x - 2\sin x.$
3. $\dfrac{dy}{dx} = x^3 y - x^3.$
4. $x\left(\dfrac{dy}{dx}\right) = y + x\ln x.$

In Exercises 5–8, solve the given equation with the stated conditions.

5. $y' = y\cos x + 2\cos x$, $y(0) = 0$.
6. $y' = \dfrac{y}{x} + x$, $y(1) = 1$.
7. $xy' = e^x - y$, $y(1) = 0$.
8. $y' = y + \cos 5x$, $y(0) = 0$.

9. Rework Example 3 assuming that the voltage is $E = E_0 \sin\omega t + E_1$, i.e., a sinusoidal plus a constant voltage.
10. In Example 3, use the method on p. 373 of Section 8.1 to determine the amplitude and phase of the oscillatory part:

$$\frac{E_0}{L}\cdot\frac{1}{(R/L)^2 + \omega^2}\left\{\frac{R}{L}\sin\omega t - \omega\cos\omega t\right\}.$$

11. The equation $R(dI/dt) + (I/C) = E$ describes the current I in a circuit containing a resistor with resistance R (a constant) and a capacitor with capacitance C (a constant) as shown in Fig. 8.6.4. If $E = E_0$, a constant, and $I = 0$ at $t = 0$, find I as a function of time. Discuss your solution.

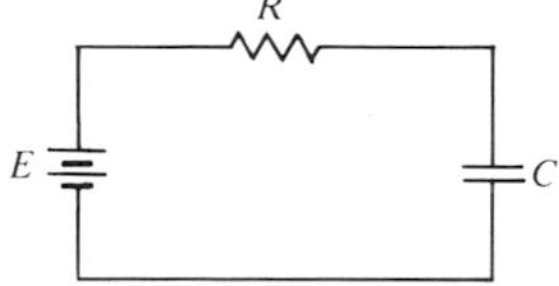

Figure 8.6.4. A resistor and a capacitor in an electric circuit.

12. Repeat Exercise 11 for the case $I = I_0$ at $t = 0$.
13. In Example 4, show that the lake will reach 90% of its limiting pollution value within 3.33 months.
14. *Mixing problem.* A lake contains 5×10^8 liters of water, into which is dissolved 10^4 kilograms of salt at $t = 0$. Water flows into the lake at a rate of 100 liters per second and contains 1% salt; water flows out of the lake at the same rate. Find the amount of salt in the lake as a function of time. When is 90% of the limiting amount of salt reached?

15. *Two-stage mixing*. Suppose the pollutants from the lake in Example 4 empty into a second smaller lake rather than the stream. At $t = 0$, the smaller lake contains 10^7 liters of pure water. The second lake has an exit stream carrying the same volume of fluid as entered. Find the amount of pollution in the second lake as a function of time.
16. Repeat Exercise 15 with the size and flow rates of the lake in Example 4 replaced by those in Exercise 14.
17. The terminal speed of a person in free fall in air is about 64 meters per second (see Example 5). How long does it take to reach 90% of terminal speed? How far has the person fallen in this time? ($g = 9.8$ meters per second^2).
18. The terminal speed of a person falling with a parachute is about 6.3 meters per second (see Example 5). How long does it take to reach 90% of terminal speed? How far has the person fallen in this time? ($g = 9.8$ meters per second^2).
19. *Falling object with drag resistance*. Redo Example 5 assuming that the resistance is proportional to the square of the velocity.
20. An object of mass 10 kilograms is dropped from a balloon. The force of air resistance is $0.07v$, where v is the velocity. What is the object's velocity as a function of time? How far has the object traveled before it is within 10% of its terminal velocity?
21. What is the acceleration of the rocket in Example 6 just before burnout?
22. How high is the rocket in Example 6 at burnout?
23. Sketch the direction field for the equation $dy/dx = y + 2x$ and solve the equation.
24. Sketch the direction field for the equation $dy/dx = -3y + x$ and solve the equation.

★25. Assuming that $P(x)$ and $Q(x)$ are continuous functions of x, prove that the problem $y' = P(x)y + Q(x)$, $y(0) = y_0$ has *exactly* one solution.

26. Express the solution of the equation $y' = xy + 1$, $y(0) = 1$ in terms of an integral.

★27. *Bernoulli's equation*. This equation has the form $dy/dx = P(x)y + Q(x)y^n$, $n = 2, 3, 4, \ldots$.
   (a) Show that the equation satisfied by $w = y^{1-n}$ is linear in w.
   (b) Use (a) to solve the equation $x(dy/dx) = x^4y^3 - y$.

★28. *Riccati equation*. This equation is $dy/dx = P(x) + Q(x)y + R(x)y^2$.
   (a) Let $y_1(x)$ be a known solution (found by inspection). Show that the general solution is $y(x) = y_1(x) + w(x)$, where w satisfies the Bernoulli equation (see Exercise 27) $dw/dx = [Q(x) + 2R(x)y_1(x)]w + R(x)w^2$.
   (b) Use (a) to solve the Riccati equation $y' = y/x + x^3y^2 - x^5$, taking $y_1(x) = x$.

★29. Redo the rocket propulsion Example 6 adding air resistance proportional to velocity.

★30. Solve the equation $dy/dt = -\lambda y + r$, where λ and r are constants. Write a two page report on how this equation was used to study the Van Meegeren art forgeries which were done during World War II. (See M. Braun, *Differential Equations and their Applications*, Third Edition, Springer-Verlag, New York (1983), Section 1.3).

Review Exercises for Chapter 8

Solve the differential equations with the given conditions in Exercises 1–22.

1. $\dfrac{dy}{dt} = 3y$, $y(0) = 1$.
2. $\dfrac{dy}{dt} = y$, $y(0) = 1$.
3. $\dfrac{d^2y}{dt^2} + 3y = 0$, $y(0) = 0$, $y'(0) = 1$.
4. $\dfrac{d^2y}{dt^2} + 9y = 0$, $y(0) = 2$; $y'(0) = 0$.
5. $\dfrac{dy}{dt} = 3y + 1$, $y(0) = 1$.
6. $\dfrac{dy}{dt} = (\cos t)y + \cos t$, $y(0) = 0$.
7. $\dfrac{dy}{dt} = t^3y^2$, $y(0) = 1$.
8. $\dfrac{dy}{dt} + 10y = 0$, $y(0) = 1$.
9. $f' = 4f$; $f(0) = 1$.
10. $f' = -4f$, $f(0) = 1$.
11. $f'' = 4f$; $f(0) = 1$, $f'(0) = 1$.
12. $f'' = -4f$; $f(0) = 1$, $f'(0) = 1$.
13. $\dfrac{d^2x}{dt^2} + x = 0$; $x = 1$ when $t = 0$, $x = 0$ when $t = \pi/4$.
14. $\dfrac{d^2x}{dt^2} + 6x = 0$; $x = 1$ when $t = 0$, $x = 6$ when $t = 1$.
15. $\dfrac{d^2x}{dt^2} - 9x = 0$, $x(0) = 0$, $x'(0) = 1$.
16. $\dfrac{d^2x}{dt^2} + 16x = 0$, $x(0) = 1$, $x'(0) = -1$.
17. $\dfrac{dy}{dx} = e^{x+y}$, $y(0) = 1$.
18. $\dfrac{dx}{dt} = 8e^{3y}$, $y(0) = 1$.
19. $\dfrac{dx}{dt} = -4x$; $x = 1$ when $t = 0$.

20. $\frac{dy}{dx} = y/\ln y,\ y(1) = e$

21. $\frac{dy}{dt} = \frac{y}{t-1} + \frac{1}{t-1},\ y(0) = 0.$

22. $\frac{dx}{dt} = \frac{x}{8-t} + \frac{3}{8-t},\ x(0) = 1.$

23. Solve for $g(t)$: $3(d^2/dt^2)g(t) = -7g(t)$, $g(0) = 1$, $g'(0) = -2$. Find the amplitude and phase of $g(t)$. Sketch.

24. Solve for $z = f(t)$: $d^2z/dt^2 + 5z = 0$, $f(0) = -3$, $f'(0) = 4$. Find the amplitude and phase of $f(t)$. Sketch.

25. Sketch the graph of y as a function of x if $-d^2y/dx^2 = 2y$; $y = \frac{1}{2}$, and $dy/dx = \frac{1}{2}$ when $x = 0$.

26. Sketch the solution of $d^2x/dt^2 + 9x = 0$, where $x(0) = 1$, $x'(0) = 0$.

27. Sketch the graph of the solution of $y' = -8y$, $y = 1$ when $x = 0$.

28. Sketch the graph of $y = f(x)$ if $f' = 2f + 3$, and $f(0) = 0$.

29. Sketch the graph of the solution to $dx/dt = -x + 3$, $x(0) = 0$ and compute $\lim_{t\to\infty} x(t)$.

30. Sketch the graph of the solution to $dx/dt = -2x + 2$, $x(0) = 0$ and compute $\lim_{t\to\infty} x(t)$.

31. Solve $d^2x/dt^2 = dx/dt$, $x(0) = 1$, $x'(0) = 1$ by letting $y = dx/dt$.

32. Solve $d^2x/dt^2 = [1/(1+t)]dx/dt$, $x(0) = 1$, $x'(0) = 1$ by letting $y = dx/dt$.

33. Solve $d^2y/dx^2 + dy/dx = x$, $y(0) = 0$, $y'(0) = 1$. [*Hint*: Let $w = dy/dx$].

34. Solve $y'' + 3yy' = 0$ for $y(x)$ if $y(0) = 1$, $y'(0) = 2$. [*Hint*: $yy' = (y^2/2)'$.]

35. Solve $d^2y/dx^2 = 25y$, $y(0) = 0$, $y(1) = 1$.

36. Solve $d^2y/dx^2 = 36y$, $y(0) = 1$, $y(1) = 0$.

Differentiate the functions in Exercises 37–44.

37. $\sinh(3x^2)$

38. $\tanh(x^3 + x)$

39. $\cosh^{-1}(x^2 + 1)$

40. $\tanh^{-1}(x^4 - 1)$

41. $(\sinh^{-1}x)(\cosh 3x)$

42. $(\cosh^{-1}3x)(\tanh x^2)$

43. $\exp(1 - \cosh^{-1}(3x))$

44. $3(\cosh^{-1}(5x^2) + 1)$

Calculate the integrals in Exercise 45–50.

45. $\int \frac{\cosh x}{1 + \sinh^2 x}\,dx$

46. $\int \operatorname{sech}^2 x \tanh x\sqrt{2 + \tanh^2 x}\,dx$

47. $\int \frac{dx}{9 - x^2}$

48. $\int \frac{dx}{\sqrt{x^2 + 4}}$

49. $\int x \sinh x\,dx$

50. $\int x \cosh x\,dx$

51. A weight of 5 grams hangs on a spring with spring constant $k = 2.1$. Find the displacement $x(t)$ of the mass if $x(0) = 1$ and $x'(0) = 0$.

52. A weight hanging on a spring oscillates with a frequency of two cycles per second. Find the displacement $x(t)$ of the mass if $x(0) = 1$ and $x'(0) = 0$.

53. An observer sees a weight of 10 grams on a spring undergoing the motion $x(t) = 10\sin(8t)$.
(a) What is the spring constant?
(b) What force acts on the weight at $t = \pi/16$?

54. A 3-foot metal rod is suspended horizontally from a spring, as shown in Fig. 8.R.1. The rod bobs up and down around the equilibrium point, 5 feet from the ground, and an amplitude of 1 foot and a frequency of two bobs per second. What is the maximum length of its shadow? How fast is its shadow changing length when the rod passes the middle of its bob?

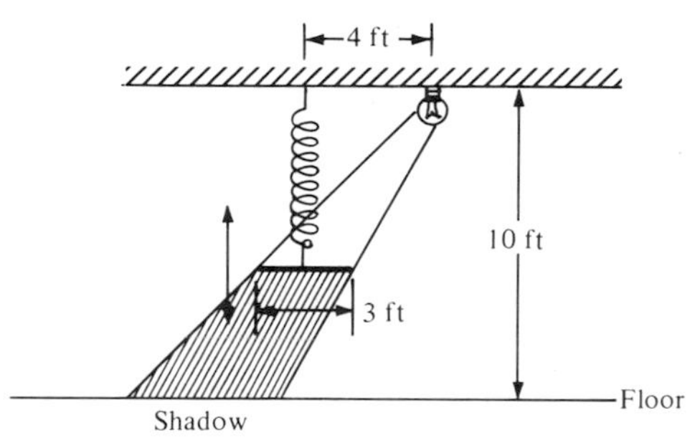

Figure 8.R.1. Study the movement of the shadow of the bobbing rod.

55. *Simple Harmonic Motion with Damping*. Consider the equation $x'' + 2\beta x' + \omega^2 x = 0$, where $0 < \beta < \omega$. (a) Show that $y = e^{\beta t}x$ satisfies a harmonic oscillator equation. (b) Show that the solution is of the form $x = e^{-\beta t}(A\cos\omega_1 t + B\sin\omega_1 t)$ where $\omega_1 = \sqrt{\omega^2 - \beta^2}$, and A and B are constants. (c) Solve $x'' + 2x' + 4x = 0$; $x(0) = 1$, $x'(0) = 0$, and sketch.

56. *Forced oscillations*.
(a) Show that a solution of the differential equation $x'' + 2\beta x' + \omega^2 x = f_0\cos\omega_0 t$ is

$$x_1(t) = \frac{f_0}{\left(\omega^2 - \omega_0^2\right)^2 + 4\omega_0^2\beta^2}\left[2\omega_0\beta\sin\omega_0 t + \left(\omega^2 - \omega_0^2\right)\cos\omega_0 t\right].$$

(b) Show that the general solution is $x(t) = x_1(t) + x_0(t)$ where x_0 is the solution found in Exercise 55.
(c) *Resonance*. Show that the "amplitude" $f_0/[(\omega^2 - \omega_0^2)^2 + 4\omega_0^2\beta^2]$ of the solution is largest when ω_0 is near ω (the natural frequency) for β (the friction constant) small, by maximizing the amplitude for fixed f_0, ω,

β and variable ω_0. (This is the phenomenon responsible for the Tacoma bridge disaster . . . somewhat simplified of course; see Section 12.7 for further information.)

57. If a population doubles every 10 years and is now 100,000, how long will it take to reach 10 million?
58. The half-life of a certain radioactive substance is 15,500 years. What percentage will have decayed after 50,000 years?
59. A certain radioactive substance decreases at a rate of 0.00128% per year. What is its half-life?
60. (a) The population of the United States in 1900 was about 76 million; in 1910, about 92 million. Assume that the population growth is uniform, so $f(t) = e^{\gamma t}f(1900)$, t in years after 1900. (i) Show that $\gamma \approx 0.0191$. (ii) What would have been a reasonable prediction for the population in 1960? In 1970? (iii) At this rate, how long does it take for the population to double?
(b) The actual U.S. population in 1960 was about 179 million and in 1970 about 203 million. (i) By what fraction did the "growth rate factor" ("γ") change between 1900–1910 and 1960–1970? (ii) Compare the percentage increase in population from 1900 to 1910 with the percentage increase from 1960 to 1970.
61. If an object cools from 100°C to 80°C in an environment of 18°C in 8 minutes, how long will it take to cool from 100° to 50°C?
62. "Suppose the pharaohs had built nuclear energy plants. They might have elected to store the resulting radioactive wastes inside the pyramids they built. Not a bad solution, considering how well the pyramids have lasted. But plutonium-239 stored in the oldest of them—some 4600 years ago—would today still exhibit 88 percent of its initial radioactivity." (see G. Hardin, "The Fallibility Factor," *Skeptic* **14** (1976): 12.)
(a) What is the half-life of plutonium?
(b) How long will it take for plutonium stored today to have only 1% of its present radioactivity? How long for $\frac{1}{1000}$?
63. (a) The oil consumption rate satisfies the equation $C(t) = C_0e^{rt}$, where C_0 is the consumption rate at $t = 0$ (number of barrels per year) and r is a constant. If the consumption rate is $C_0 = 2.5 \times 10^{10}$ barrels per year in 1976 and $r = 0.06$, how long will it take before 2×10^{12} barrels (the total world's supply) are used up?
(b) As the fuel is almost used up, the prices will probably skyrocket and other sources of energy will be turned to. Let $S(t)$ be the supply left at time t. Assume that $dS/dt = -\alpha S$, where α is a constant (the panic factor). Find $S(t)$.
64. (a) A bank advertises "5% interest on savings—but you earn more because it is compounded continuously." The formula for computing the amount $M(t)$ of money in an account t days after $M(0)$ dollars is deposited (and left untouched) is $M(t) = M(0)e^{0.05t/365}$. What is the percentage increase on an amount $M(0)$ left untouched for 1 year?
(b) A bank wants to compute its interest by the method in part (a), but it wants to give only \$5 interest on each \$100 that is left untouched for 1 year. How must it change the formula for that to occur?
65. A certain electric circuit is governed by the equation $L dI/dt + RI = E$, where E, R, and L are constants. Graph the solution if $I_0 < E/R$.
66. If a savings account containing P dollars grows at a rate $dP/dt = rP + W$ (interest with continuous deposits), find P in terms of its value P_0 at $t = 0$.
67. (a) Sketch the direction field for the equation $y' = -9x/y$. Solve the equation exactly.
(b) Find the orthogonal trajectories for the solutions in (a).
68. Consider the predator–prey model in Example 5, Section 8.5. Solve this explicitly if $r = 0$ (i.e., ignore deaths of the prey).
69. Suppose your car radiator holds 4 gallons of fluid two thirds of which is water and one third is old antifreeze. The mixture begins flowing out at a rate of $\frac{1}{2}$ gallon per minute while fresh water is added at the same rate. How long does it take for the mixture to be 95% fresh water? Is it faster to wait until the radiator has drained before adding fresh water?
70. An object falling freely with air resistance has a terminal speed of 20 meters per sec. Find a formula for its velocity as a function of time.
71. The current I in a certain electric circuit is governed by $dI/dt = -3I + 2\sin(\pi t)$, and $I = 1$ at $t = 0$. Find the solution.
72. In Example 6, Section 8.6, suppose that the burnout mass is 10% of the initial mass M_0, the burnout time is 3 minutes, and the rocket thrust is $3M_0g$. Calculate the acceleration of the rocket just before burnout in terms of M_0 and g.
73. Sketch the direction field for the equation $y' = 3y + 4$ and solve it.
74. Sketch the direction field for the equation $y' = -4y + 1$ and solve it.
75. Let $x(t)$ be the solution of $dx/dt = x^2 - 5x + 4$, $x(0) = 3$. Find $\lim_{t\to\infty} x(t)$.
76. Let x satisfy $dx/dt = x^3 - 4x^2 + 3x$, $x(0) = 2$. Find $\lim_{t\to\infty} x(t)$.
77. Test the accuracy of the Euler method by using a ten-step Euler method on the problem of finding $y(1)$ if $y' = y$ and $y(0) = 1$. Compare your answer with the exact solution and with a twenty-step Euler method.

78. Solve $y' = \csc y$ approximately for $0 \leqslant x \leqslant 1$ with $y(0) = 1$ using a ten-step Euler method.

79. Numerically solve for $y(2)$ if $y' = y^2$ and $y(0) = 1$, using a twenty-step Euler method. Do you detect some numerical trouble? What do you think is going wrong?

80. Solve for $y(1)$ if $y' = \cos(x + y)$ and $y(0) = 0$, using a ten-step Euler method.

81. Solve $y' = ay + b$, given constants a and b, by
(a) introducing $w = y + b/a$ and a differential equation for w;
(b) treating it as a separable equation; and
(c) treating it as a linear equation.
Are your answers the same?

★82. Let $x(t)$ be the solution of $dx/dt = -x + 3\sin \pi t$, $x(0) = 1$. Find α, ω and θ such that $\lim_{t\to\infty}[x(t) - \alpha\cos(\omega t + \theta)] = 0$.

★83. *Simple pendulum.* The equation of motion for a simple pendulum (see Fig. 8.R.2) is

$$\frac{d^2\theta}{dt^2} = -\frac{g}{L}\sin\theta.$$

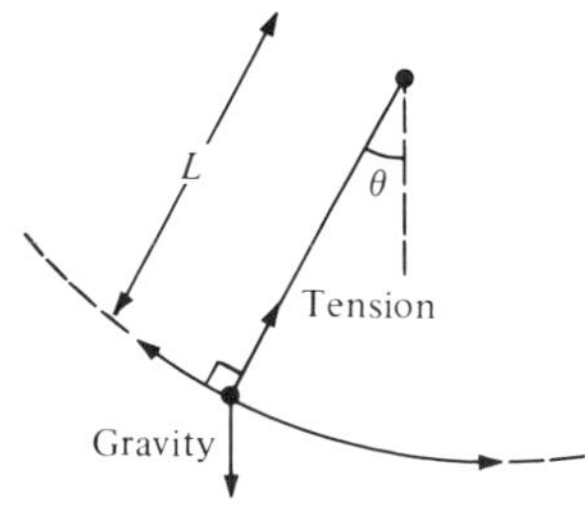

Figure 8.R.2. A simple pendulum.

(a) Let $w(t) = d\theta/dt$. Show that

$$\frac{d}{dt}\left(\frac{w^2}{2}\right) = \frac{g}{L}\frac{d}{dt}\cos\theta,$$

and so

$$\frac{w^2}{2} = \frac{g}{L}(\cos\theta - \cos\theta_0),$$

where $w = 0$ when $\theta = \theta_0$ (the maximum value of θ).

(b) Conclude that θ is implicitly determined by

$$\int_{\theta_0}^{\theta} \frac{d\theta}{\sqrt{\cos\theta - \cos\theta_0}} = \sqrt{\frac{2g}{L}}\, t.$$

(c) Show that the period of oscillation is

$$T = 2\sqrt{\frac{2L}{g}} \int_0^{\theta_0} \frac{d\theta}{\sqrt{\cos\theta - \cos\theta_0}}$$

$$= 4\sqrt{\frac{L}{g}} \int_0^{\pi/2} \frac{d\phi}{\sqrt{1 - k^2\sin^2\phi}},$$

where $k = \sin(\theta_0/2)$. [*Hint*: Write $\cos\theta = 1 - 2\sin^2(\theta/2)$.] The last integral is called an *elliptic integral of the first kind*, and cannot be evaluated explicitly.

(d) How does the answer in (c) compare with the prediction from linearized oscillations for θ_0 small?

★84. A photographer dips a thermometer into a developing solution to determine its temperature in degrees Centigrade. The temperature $\theta(t)$ registered by the thermometer satisfies a differential equation $d\theta/dt = -k(\theta - \bar{\theta})$, where $\bar{\theta}$ is the true temperature of the solution and k is a constant. How can the photographer determine when θ is correct to within 0.1°C?

★85. (a) Find $y = f(x)$ so that

$$\int_a^b f(x)\,dx = \int_a^b \sqrt{1 + [f'(x)]^2}\, dx$$

for all a and b.

(b) What geometric problem leads to the problem (a).

Chapter 9

Applications of Integration

Many physical and geometric quantities can be expressed as integrals.

Our applications of integration in Chapter 4 were limited to area, distance–velocity, and rate problems. In this chapter, we will see how to use integrals to set up problems involving volumes, averages, centers of mass, work, energy, and power. The techniques developed in Chapter 7 make it possible to solve many of these problems completely.

9.1 Volumes by the Slice Method

The volume of a solid region is an integral of its cross-sectional areas.

By thinking of a region in space as being composed of "infinitesimally thin slices," we shall obtain a formula for volumes in terms of the areas of slices. In this section, we apply the formula in a variety of special cases. Further methods for calculating volumes will appear when we study multiple integration in Chapter 17.

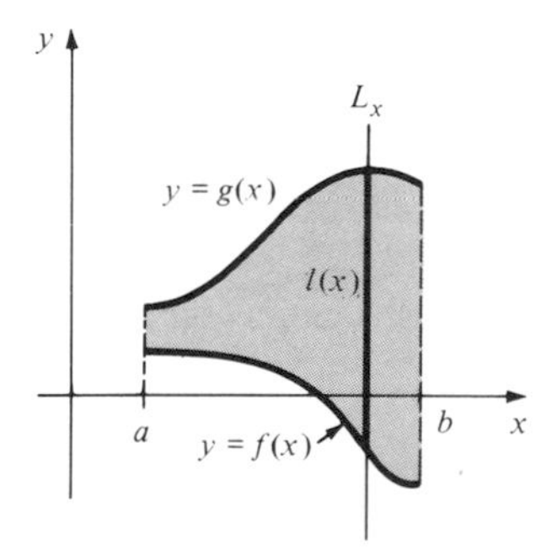

Figure 9.1.1. The area of the shaded region is $\int_a^b l(x)\,dx$.

We will develop the slice method for volumes by analogy with the computation of areas by integration. If f and g are functions with $f(x) \leqslant g(x)$ on $[a, b]$, then the area between the graphs of f and g is $\int_a^b [g(x) - f(x)]\,dx$ (see Section 4.6). We recall the infinitesimal argument for this formula. Think of the region as being composed of infinitesimally thin strips obtained by cutting with lines perpendicular to the x axis. Denote the vertical line through x by L_x; the intersection of L_x with the region between the graphs has length $l(x) = g(x) - f(x)$, and the corresponding "infinitesimal rectangle" with thickness dx has area $l(x)\,dx$ (= height × width) (see Fig. 9.1.1). The area of the entire region, obtained by "summing" the infinitesimal areas, is

$$\int_a^b l(x)\,dx = \int_a^b [g(x) - f(x)]\,dx.$$

Given a region surrounded by a closed curve, we can often use the same formula $\int_a^b l(x)\,dx$ to find its area. To implement this, we position it conveniently with respect to the axes and determine a and b by noting where the ends of the region are. We determine $l(x)$ by using the geometry of the situation at hand. This is done for a disk of radius r in

Fig. 9.1.2. We may evaluate the integral $\int_{-r}^{r} 2\sqrt{r^2 - x^2}\, dx$ by using integral tables to obtain the answer πr^2, in agreement with elementary geometry. (One can also readily evaluate integrals of this type by using the substitution $x = r\cos\theta$.)

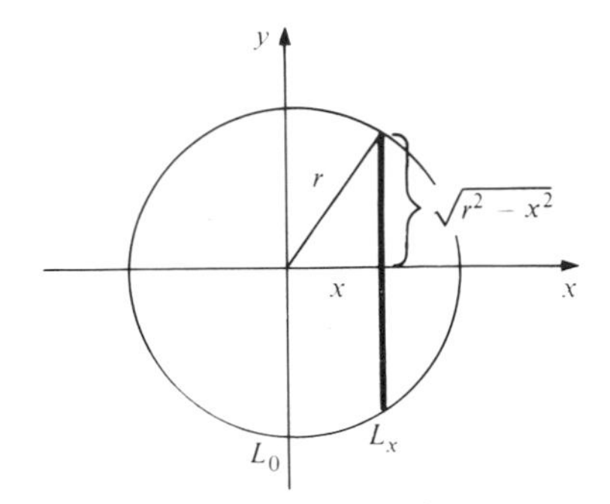

Figure 9.1.2. Area of the disk $= \int_{-r}^{r} 2\sqrt{r^2 - x^2}\, dx$.

To find the volume of a *solid* region, we imagine it sliced by a family of parallel *planes*: The plane P_x is perpendicular to a fixed x axis in space at a distance x from a reference point (Fig. 9.1.3).

The plane P_x cuts the solid in a plane region R_x; the corresponding "infinitesimal piece" of the solid is a slab whose base is a region R_x and whose thickness is dx (Fig. 9.1.4). The volume of such a cylinder is equal to the area

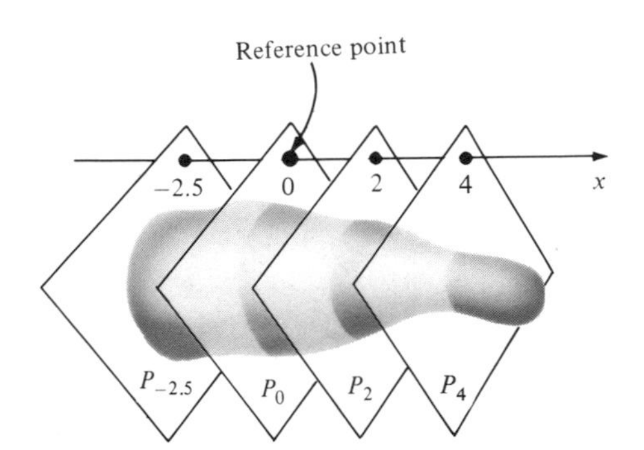

Figure 9.1.3. The plane P_x is at distance x from P_0.

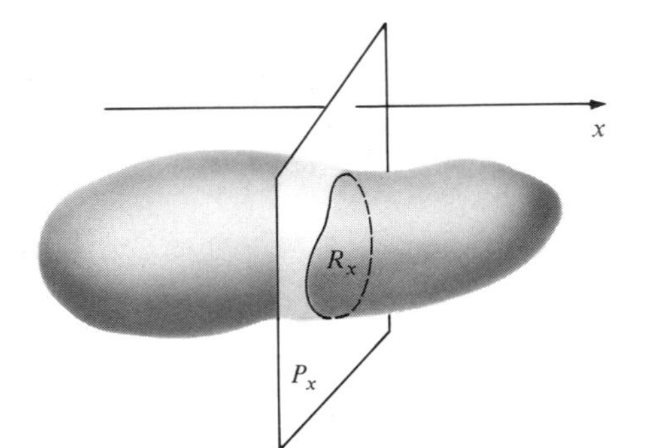

Figure 9.1.4. An infinitesimally thin slice of a solid.

of the base R_x times the thickness dx. If we denote the area of R_x by $A(x)$, then this volume is $A(x)\,dx$. Thus the volume of the entire solid, obtained by summing, is the integral $\int_a^b A(x)\,dx$, where the limits a and b are determined by the ends of the solid.

The Slice Method

Let S be a solid and P_x be a family of parallel planes such that:

1. S lies between P_a and P_b;
2. the area of the slice of S cut by P_x is $A(x)$.

Then the volume of S is equal to

$$\int_a^b A(x)\,dx.$$

The slice method can also be justified using step functions. We shall see how to do this below.

In simple cases, the areas $A(x)$ can be computed by elementary geometry. For more complicated problems, it may be necessary to do a preliminary integration to find the $A(x)$'s themselves.

Example 1 Find the volume of a ball[1] of radius r.

Solution Draw the ball above the x axis as in Fig. 9.1.5.

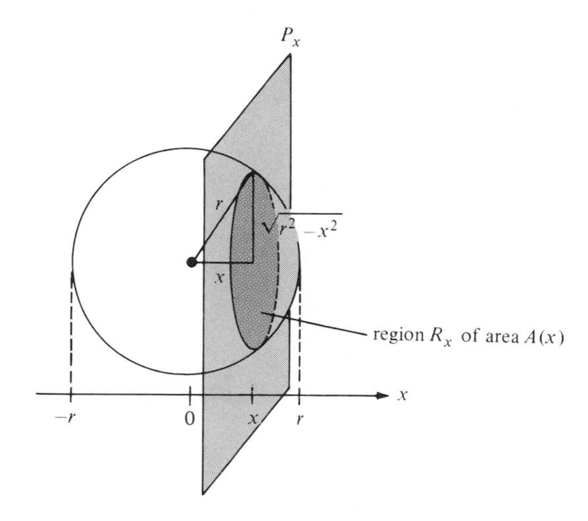

Figure 9.1.5. The area of the slice at x of a ball of radius r is $A(x) = \pi(r^2 - x^2)$.

Let the plane P_0 pass through the center of the ball. The ball lies between P_{-r} and P_r, and the slice R_x is a disk of radius $\sqrt{r^2 - x^2}$. The area of the slice is $\pi \times (\text{radius})^2$; i.e., $A(x) = \pi(\sqrt{r^2 - x^2}\,)^2 = \pi(r^2 - x^2)$. Thus the volume is

$$\int_{-r}^{r} A(x)\,dx = \int_{-r}^{r} \pi(r^2 - x^2)\,dx = \pi\left(r^2x - \frac{x^3}{3}\right)\Bigg|_{-r}^{r} = \frac{4}{3}\pi r^3. \blacktriangle$$

Example 2 Find the volume of the conical solid in Fig. 9.1.6. (The base is a circle.)

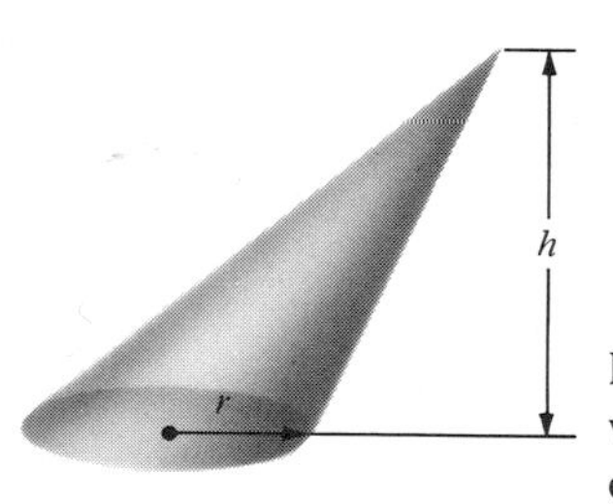

Figure 9.1.6. Find the volume of this oblique circular cone.

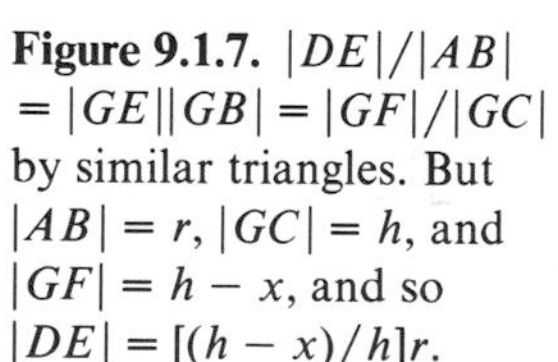

Figure 9.1.7. $|DE|/|AB| = |GE||GB| = |GF|/|GC|$ by similar triangles. But $|AB| = r$, $|GC| = h$, and $|GF| = h - x$, and so $|DE| = [(h - x)/h]r$.

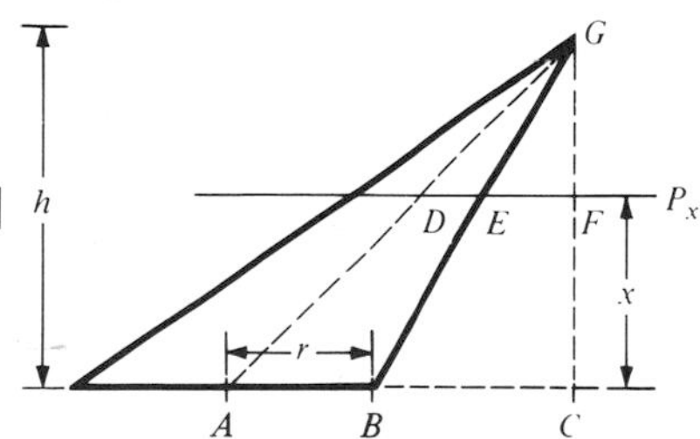

Solution We let the x axis be vertical and choose the family P_x of planes such that P_0 contains the base of the cone and P_x is at distance x above P_0. Then the cone lies between P_0 and P_h, and the plane section by P_x is a disk with radius $[(h - x)/h]r$ and area $\pi[(h - x)/h]^2r^2$ (see Fig. 9.1.7). By the slice method,

[1] A *sphere* is the set of points in space at a fixed distance from a point. A *ball* is the solid region enclosed by a sphere, just as a disk is the plane region enclosed by a circle.

the volume is

$$\int_0^h A(x)\,dx = \int_0^h \pi \frac{(h-x)^2}{h^2} r^2\,dx = \frac{\pi r^2}{h^2}\int_0^h (h^2 - 2xh + x^2)\,dx$$

$$= \frac{\pi r^2}{h^2}\left[(h^2x - hx^2) + \frac{x^3}{3}\right]\Big|_0^h = \frac{1}{3}\pi r^3 h. \ \blacktriangle$$

Example 3 Find the volume of the solid W shown in Fig. 9.1.8. It can be thought of as a wedge-shaped piece of a cylindrical tree of radius r obtained by making two saw cuts to the tree's center, one horizontally and one at an angle θ.

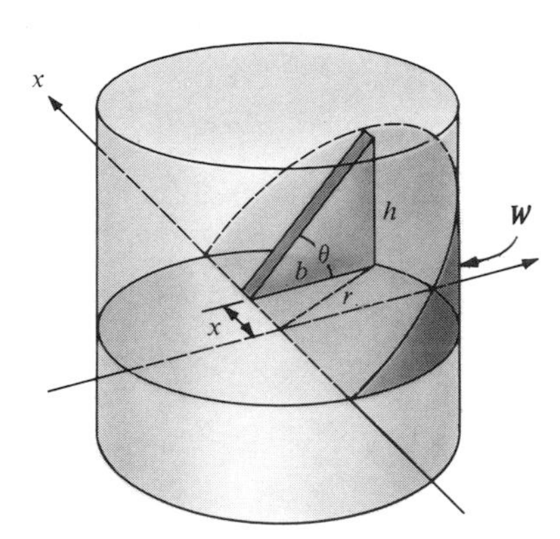

Figure 9.1.8. Find the volume of the wedge W.

Solution With the setup in Fig. 9.1.8, we slice W by planes to produce triangles R_x of area $A(x)$ as shown. The base b of the triangle is $b = \sqrt{r^2 - x^2}$, and its height is $h = b\tan\theta = \sqrt{r^2 - x^2}\tan\theta$. Thus, $A(x) = \frac{1}{2}bh = \frac{1}{2}(r^2 - x^2)\tan\theta$. Hence, the volume is

$$\int_{-r}^{r} A(x)\,dx = \int_{-r}^{r} \frac{1}{2}(r^2 - x^2)\tan\theta\,dx = \frac{1}{2}(\tan\theta)\left(r^2x - \frac{x^2}{3}\right)\Big|_{-r}^{r}$$

$$= \frac{1}{2}(\tan\theta)\left(2r^3 - \frac{2r^3}{3}\right) = \frac{2r^3}{3}\tan\theta.$$

Notice that even though we started with a region with a circular boundary, π does not occur in the answer! ▲

Example 4 A ball of radius r is cut into three pieces by parallel planes at a distance of $r/3$ on each side of the center. Find the volume of each piece.

Solution The middle piece lies between the planes $P_{-r/3}$ and $P_{r/3}$ of Example 1, and the area function is $A(x) = \pi(r^2 - x^2)$ as before, so the volume of the middle piece is

$$\int_{-r/3}^{r/3} \pi(r^2 - x^2)\,dx = \pi\left(r^2x - \frac{x^3}{3}\right)\Big|_{-r/3}^{r/3}$$

$$= \pi\left(\frac{r^3}{3} - \frac{r^3}{81} + \frac{r^3}{3} - \frac{r^3}{81}\right) = \frac{52}{81}\pi r^3.$$

This leaves a volume of $(\frac{4}{3} - \frac{52}{81})\pi r^3 = \frac{56}{81}\pi r^3$ to be divided between the two outside pieces. Since they are congruent, each of them has volume $\frac{28}{81}\pi r^3$. (You may check this by computing $\int_{r/3}^{r}\pi(r^2 - x^2)\,dx$.) ▲

One way to construct a solid is to take a plane region R, as shown in Fig. 9.1.9 and revolve it around the x axis so that it sweeps out a solid region S. Such solids are common in woodworking shops (lathe-tooled table legs), in pottery

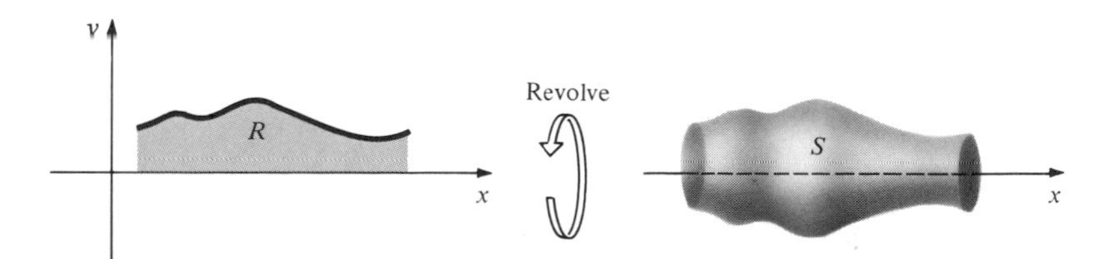

Figure 9.1.9. S is the solid of revolution obtained by revolving the plane region R about the x axis.

studios (wheel-thrown pots), and in nature (unicellular organisms).[2] They are called *solids of revolution* and are said to have *axial symmetry*.

Suppose that region R is bounded by the lines $x = a$, $x = b$, and $y = 0$, and by the graph of the function $y = f(x)$. To compute the volume of S by the slice method, we use the family of planes perpendicular to the x axis, with P_0 passing through the origin. The plane section of S by P_x is a circular disk of radius $f(x)$ (see Fig. 9.1.10), so its area $A(x)$ is $\pi[f(x)]^2$. By the basic formula of the slice method, the volume of S is

$$\int_a^b A(x)\,dx = \int_a^b \pi[f(x)]^2\,dx = \pi\int_a^b [f(x)]^2\,dx.$$

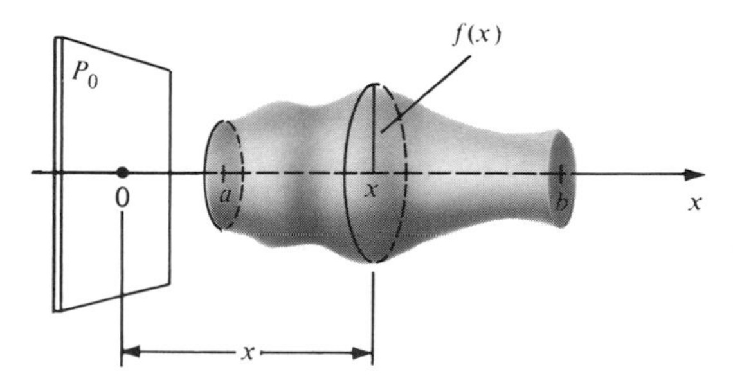

Figure 9.1.10. The volume of a solid of revolution obtained by the disk method.

We use the term "disk method" for this special case of the slice method since the slices are disks.

> **Volume of a Solid of Revolution: Disk Method**
>
> The volume of the solid of revolution obtained by revolving the region under the graph of a (non-negative) function $f(x)$ on $[a, b]$ about the x axis is
>
> $$\pi\int_a^b [f(x)]^2\,dx.$$

Example 5 The region under the graph of x^2 on $[0, 1]$ is revolved about the x axis. Sketch the resulting solid and find its volume.

Solution The solid, which is shaped something like a trumpet, is sketched in Fig. 9.1.11.

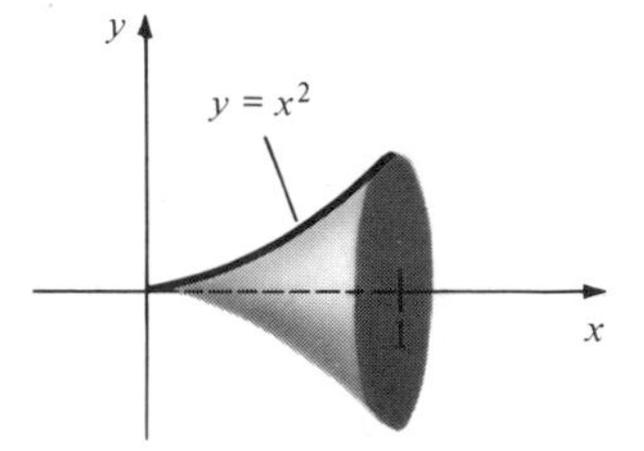

Figure 9.1.11. The volume of this solid of revolution is $\pi\int_0^1 (x^2)^2\,dx$.

[2] See D'Arcy Thompson, *On Growth and Form*, abridged edition, Cambridge University Press (1969).

According to the disk method, its volume is

$$\pi\int_0^1 (x^2)^2\,dx = \pi\int_0^1 x^4\,dx = \frac{\pi x^5}{5}\bigg|_0^1 = \frac{\pi}{5}\,. \ \blacktriangle$$

Example 6 The region between the graphs of $\sin x$ and x on $[0,\pi/2]$ is revolved about the x axis. Sketch the resulting solid and find its volume.

Solution The solid is sketched in Fig. 9.1.12. It has the form of a hollowed-out cone.

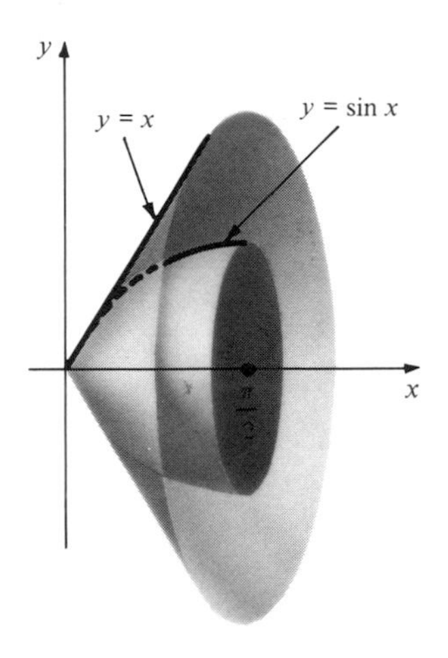

Figure 9.1.12. The region between the graphs of $\sin x$ and x is revolved about the x axis.

The volume is that of the cone minus that of the hole. The cone is obtained by revolving the region under the graph of x on $[0,1]$ about the axis, so its volume is

$$\pi\int_0^{\pi/2} x^2\,dx = \frac{\pi^4}{24}\,.$$

The hole is obtained by revolving the region under the graph of $\sin x$ on $[0,\pi/2]$ about the x axis, so its volume is

$$\pi\int_0^{\pi/2}\sin^2 x\,dx = \int_0^{\pi/2}\frac{1-\cos 2x}{2}\,dx \qquad (\text{since } \cos 2x = 1-2\sin^2 x)$$

$$= \pi\left(\frac{x}{2}-\frac{1}{4}\sin 2x\right)\bigg|_0^{\pi/2}$$

$$= \pi\left(\frac{\pi}{4}-0-0+0\right) = \frac{\pi^2}{4}\,.$$

Thus the volume of our solid is $\pi^4/24 - \pi^2/4 \approx 1.59$. ▲

The volume of the solid obtained by rotating the region between the graphs of two functions f and g (with $f(x) \leqslant g(x)$ on $[a,b]$) can be done as in Example 6 or by the *washer method* which proceeds as follows. In Fig. 9.1.13, the volume of the shaded region (the "washer") is the area × thickness. The area of the washer is the area of the complete disk minus that of the hole. Thus, the washer's volume is

$$\left(\pi[g(x)]^2 - \pi[f(x)]^2\right)dx.$$

Thus, the total volume is

$$\pi\int_a^b\left([g(x)]^2 - [f(x)]^2\right)dx.$$

The reader should notice that this method gives the same answer as one finds by using the method of Example 6.

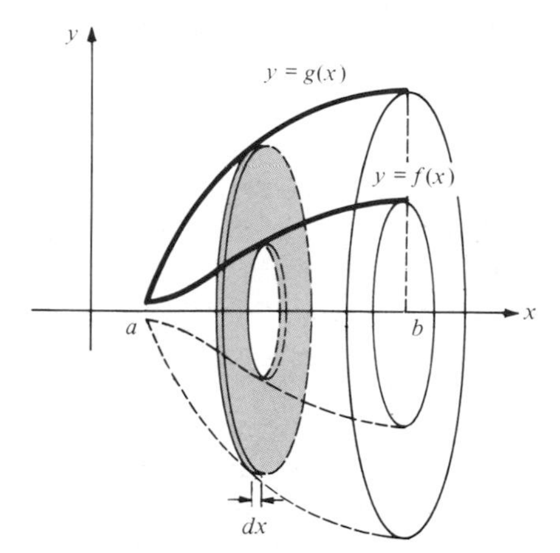

Figure 9.1.13. The washer method.

Our formula for volumes by the slice method was introduced via infinitesimals. A more rigorous argument for the formula is based on the use of upper and lower sums.[3] To present this argument, we first look at the case where S is composed of n cylinders, as in Fig. 9.1.14.

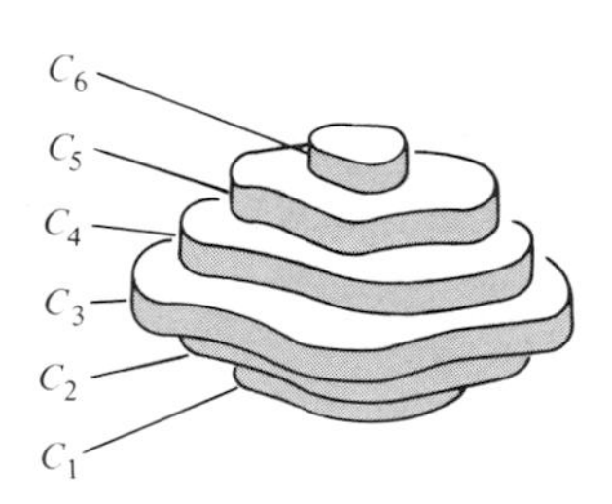

Figure 9.1.14. A "stepwise cylindrical" solid.

If the ith cylinder C_i lies between the planes $P_{x_{i-1}}$ and P_{x_i} and has cross-sectional area k_i, then the function $A(x)$ is a step function on the interval $[x_0, x_n]$; in fact, $A(x) = k_i$ for x in (x_{i-1}, x_i). The volume of C_i is the product of its base area k_i by its height $\Delta x_i = x_i - x_{i-1}$, so the volume of the total figure is $\sum_{i=1}^{n} k_i \Delta x_i$; but this is just the integral $\int_{x_0}^{x_n} A(x)\,dx$ of the step function $A(x)$. We conclude that if S is a "stepwise cylindrical" solid between the planes P_a and P_b, then

$$\text{volume } S = \int_a^b A(x)\,dx.$$

If S is a reasonably "smooth" solid region, we expect that it can be squeezed arbitrarily closely between stepwise cylindrical regions on the inside and outside. Specifically, for every positive number ε, there should be a stepwise cylindrical region S_i inside S and another such region S_o outside S such that (volume S_o) − (volume S_i) $< \varepsilon$. If $A_i(x)$ and $A_o(x)$ are the corresponding functions, then A_i and A_o are step functions, and we have the inequality $A_i(x) \leqslant A(x) \leqslant A_o(x)$, so

$$\text{volume } S_i = \int_a^b A_i(x)\,dx \leqslant \int_a^b A(x)\,dx \leqslant \int_a^b A_o(x)\,dx = \text{volume } S_o .$$

Since S encloses S_i and S_o encloses S, volume $S_i \leqslant$ volume $S \leqslant$ volume S_o. Thus the numbers (volume S) and $\int_a^b A(x)\,dx$ both belong to the same interval [(volume S_i), (volume S_o)], which has length less than ε. It follows that the difference between (volume S) and $\int_a^b A(x)\,dx$ is less than any positive number ε; the only way this can be so is if the two numbers are equal.

Supplement to Section 9.1: Cavalieri's Delicatessen

The idea behind the slice method goes back, beyond the invention of calculus, to Francesco Bonaventura Cavalieri (1598–1647), a student of Galileo and then professor at the University of Bologna. An accurate report of the events leading to Cavalieri's discovery is not available, so we have taken the liberty of inventing one.

[3] Even this justification, as we present it, is not yet completely satisfactory. For example, do we get the same answer if we slice the solid a different way? The answer is yes, but the proof uses multiple integrals (see Chapter 17).

Cavalieri's delicatessen usually produced bologna in cylindrical form, so that the volume would be computed as $\pi \cdot \text{radius}^2 \cdot \text{length}$. One day, the casings were a bit weak, and the bologna came out with odd bulges. The scale was not working that day, either, so the only way to compute the price of the bologna was in terms of its volume.

Cavalieri took his best knife and sliced the bologna into n very thin slices, each of thickness Δx, and measured the radii $r_1, r_2, \ldots, r_n$ of the slices (fortunately, they were round). He then estimated the volume to be $\sum_{i=1}^{n} \pi r_i^2 \Delta x_i$, the sum of the volumes of the slices.

Cavalieri was moonlighting from his regular job as a professor at the University of Bologna. That afternoon, he went back to his desk and began the book "Geometria indivisibilium continuorum nova quandum ratione promota" ("Geometry shows the continuous indivisibility between new rations and getting promoted"), in which he stated what is now known[4] as *Cavalieri's principle*:

If two solids are sliced by a family of parallel planes in such a way that corresponding sections have equal areas, then the two solids have the same volume.

The book was such a success that Cavalieri sold his delicatessen and retired to a life of occasional teaching and eternal glory.

[4] Honest!

Exercises for Section 9.1

In Exercises 1–4, use the slice method to find the volume of the indicated solid.

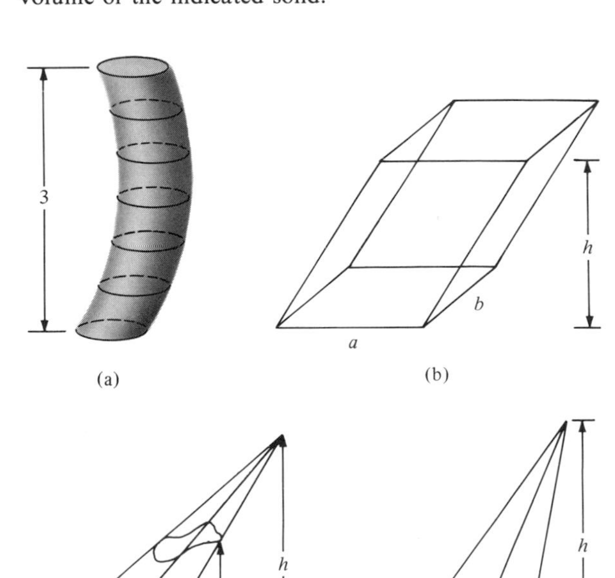

Figure 9.1.15. The solids for Exercises 1–4.

1. The solid in Fig. 9.1.15(a); each plane section is a circle of radius 1.
2. The parallelepiped in Fig. 9.1.15(b); the base is a rectangle with sides a and b.
3. The solid in Fig. 9.1.15(c); the base is a figure of area A and the figure at height x has area $A_x = [(h - x)/h]^2 A$.
4. The solid in Fig. 9.1.15(d); the base is a right triangle with sides b and l.
5. Find the volume of the tent in Fig. 9.1.16. The plane section at height x above the base is a square of side $\frac{1}{6}(6 - x)^2 - \frac{1}{6}$. The height of the tent is 5 feet.

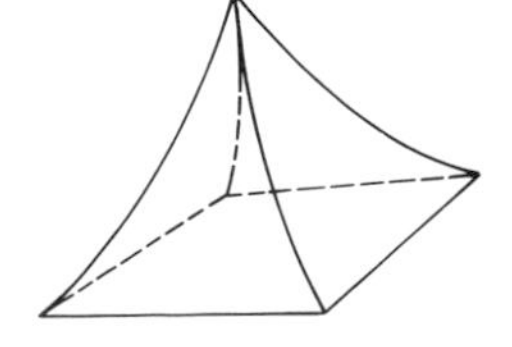

Figure 9.1.16. Find the volume of this tent.

6. What would the volume of the tent in the previous exercise be if the base and cross sections were equilateral triangles instead of squares (with the same side lengths)?
7. The base of a solid S is the disk in the xy plane with radius 1 and center $(0, 0)$. Each section of S cut by a plane perpendicular to the x axis is an equilateral triangle. Find the volume of S.

8. A plastic container is to have the shape of a truncated pyramid with upper and lower bases being squares of side length 10 and 6 centimeters, respectively. How high should the container be to hold exactly one liter (= 1000 cubic centimeters)?
9. The conical solid in Fig. 9.1.6 is to be cut by horizontal planes into four pieces of equal volume. Where should the cuts be made? [*Hint:* What is the volume of the portion of the cone above the plane P_x?]
10. The tent in Exercise 5 is to be cut into two pieces of equal volume by a plane parallel to the base. Where should the cut be made?
    (a) Express your answer as the root of a fifth-degree polynomial.
    (b) Find an approximate solution using the method of bisection.
11. A wedge is cut in a tree of radius 0.5 meter by making two cuts to the tree's center, one horizontal and another at an angle of 15° to the first. Find the volume of the wedge.
12. A wedge is cut in a tree of radius 2 feet by making two cuts to the tree's center, one horizontal and another at an angle of 20° to the first. Find the volume of the wedge.
13. Find the volume of the solid in Fig. 9.1.17(a).
14. Find the volume of the solid in Fig. 9.1.17(b).

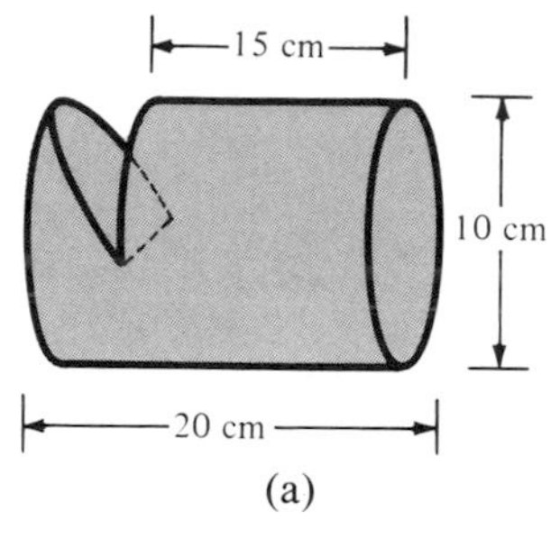

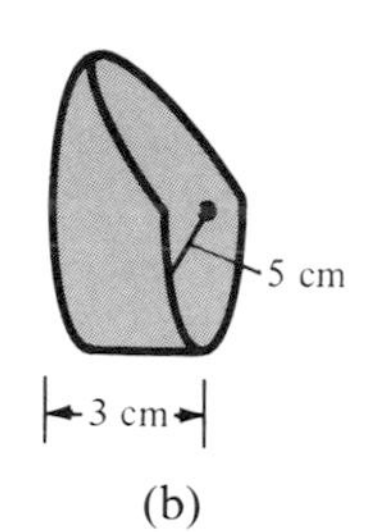

Figure 9.1.17. Find the volumes of these solids.

In Exercises 15–26, find the volume of the solid obtained by revolving each of the given regions about the x axis and sketch the region.

15. The region under the graph of $3x + 1$ on $[0, 2]$.
16. The region under the graph of $2 - (x - 1)^2$ on $[0, 2]$.
17. The region under the graph of $\cos x + 1$ on $[0, 2\pi]$.
18. The region under the graph of $\cos 2x$ on $[0, \pi/4]$.
19. The region under the graph of $x(x - 1)^2$ on $[1, 2]$.
20. The region under the graph of $\sqrt{4 - 4x^2}$ on $[0, 1]$.
21. The semicircular region with center $(a, 0)$ and radius r (assume that $0 < r < a, y \geqslant 0$).
22. The region between the graphs of $\sqrt{3 - x^2}$ and $5 + x$ on $[0, 1]$.
23. The square region with vertices $(4, 6)$, $(5, 6)$, $(5, 7)$, and $(4, 7)$.
24. The region in Exercise 23 moved 2 units upward.
25. The region in Exercise 23 rotated by 45° around its center.
26. The triangular region with vertices $(1, 1)$, $(2, 2)$, and $(3, 1)$.

★27. A vase with axial symmetry has the cross section shown in Fig. 9.1.18 when it is cut by a plane through its axis of symmetry. Find the volume of the vase to the nearest cubic centimeter.

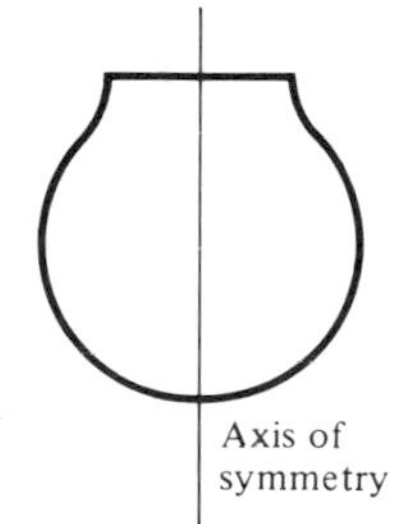

Figure 9.1.18. Cross section of a vase.

★28. A right circular cone of base radius r and height 14 is to be cut into three equal pieces by parallel planes which are parallel to the base. Where should the cuts be made?

★29. Find the formula for the volume of a doughnut with outside radius R and a hole of radius r.

★30. Use the fact that the area of a disk of radius r is $\pi r^2 = \int_{-r}^{r} 2\sqrt{r^2 - x^2}\, dx$ to compute the area inside the ellipse $y^2/4 + x^2 = r^2$.

★31. Prove Cavalieri's principle.

★32. Using Cavalieri's principle, without integration, find a relation between the volumes of:
    (a) a hemisphere of radius 1;
    (b) a right circular cone of base radius 1 and height 1;
    (c) a right circular cylinder of base radius 1 and height 1.

[*Hint:* Consider two of the solids side by side as a single solid. The sum of two volumes will equal the third.]

9.2 Volumes by the Shell Method

A solid of revolution about the y axis can be regarded as composed of cylindrical shells.

In the last section, we computed the volume of the solid obtained by revolving the region under the graph of a function about the x axis. Another way to obtain a solid S is to revolve the region R under the graph of a non-negative function $f(x)$ on $[a,b]$ about the y axis as shown in Fig. 9.2.1. We assume that $0 \leqslant a < b$.

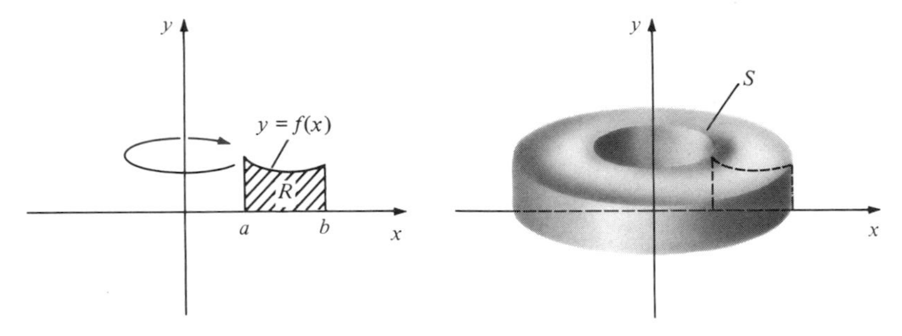

Figure 9.2.1. The solid S is obtained by revolving the plane region R about the y axis.

To find the volume of S, we use the method of infinitesimals. (Another argument using step functions is given at the end of the section.) If we rotate a strip of width dx and height $f(x)$ located at a distance x from the axis of rotation, the result is a cylindrical shell of radius x, height $f(x)$, and thickness dx. We may "unroll" this shell to get a flat rectangular sheet whose length is $2\pi x$, the circumference of the cylindrical shell (see Fig. 9.2.2). The volume of the sheet is thus the product of its area $2\pi x f(x)$ and its thickness dx. The total volume of the solid, obtained by summing the volumes of the infinitesimal shells, is the integral $\int_a^b 2\pi x f(x)\,dx$. If we revolve the region between the graphs of $f(x)$ and $g(x)$, with $f(x) \leqslant g(x)$ on $[a,b]$, the height is $g(x) - f(x)$, and so the volume is $2\pi\int_a^b x[g(x) - f(x)]\,dx$.

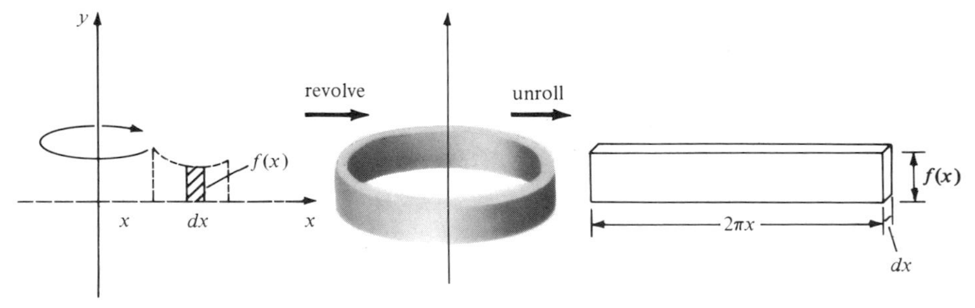

Figure 9.2.2. The volume of the cylindrical shell is $2\pi x f(x)\,dx$.

Example 1 The region under the graph of x^2 on [0, 1] is revolved about the y axis. Sketch the resulting solid and find its volume.

Solution The solid, in the shape of a bowl, is sketched in Fig. 9.2.3. Its volume is

$$2\pi\int_0^1 x\cdot x^2\,dx = 2\pi\int_0^1 x^3\,dx = 2\pi\left.\frac{x^4}{4}\right|_0^1 = \frac{\pi}{2}\,. \blacktriangle$$

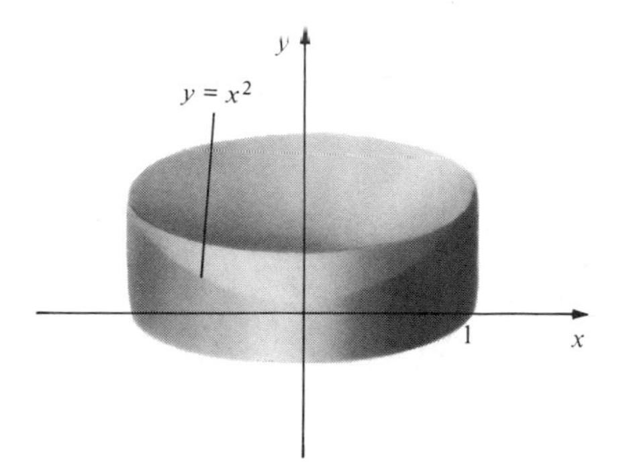

Figure 9.2.3. Find the volume of the "bowl-like" solid.

Volume of a Solid of Revolution: Shell Method

The volume of the solid of revolution obtained by revolving about the y axis the region under the graph of a (non-negative) function $f(x)$ on $[a,b]$ $(0 \leqslant a < b)$ is

$$2\pi \int_a^b xf(x)\,dx.$$

If the region between the graphs of $f(x)$ and $g(x)$ is revolved, the volume is

$$2\pi \int_a^b x[\,g(x) - f(x)\,]\,dx.$$

Example 2 Find the capacity of the bowl in Example 1.

Solution The capacity of the bowl is the volume of the region obtained by rotating the region between the curves $y = x^2$ and $y = 1$ on $[0,1]$ around the y axis.

By the second formula in the box above, with $f(x) = x^2$ and $g(x) = 1$, the volume is

$$2\pi \int_0^1 x(1 - x^2)\,dx = 2\pi \int_0^1 (x - x^3)\,dx = 2\pi\left(\frac{x^2}{2} - \frac{x^4}{4}\right)\Big|_0^1 = \frac{\pi}{2}. \;\blacktriangle$$

We could have found the capacity in Example 2 by subtracting the result of Example 1 from the volume of the right circular cylinder with radius 1 and height 1, namely $\pi r^2 h = \pi$. Another way to find the capacity is by the slice method, using y as the independent variable. The slice at height y is a disk of radius $x = \sqrt{y}$, so the volume is

$$\int_0^1 \pi\left(\sqrt{y}\right)^2 dy = \int_0^1 \pi y\,dy = \frac{1}{2}\pi y^2\Big|_0^1 = \frac{\pi}{2}.$$

Example 3 Sketch and find the volume of the solid obtained by revolving each of the following regions about the y axis: (a) the region under the graph of e^x on $[1,3]$; (b) the region under the graph of $2x^3 + 5x + 1$ on $[0,1]$.

Solution (a) Volume $= 2\pi\int_1^3 xe^x\,dx$. This integral may be evaluated by integration by parts to give $2\pi(xe^x|_1^3 - \int_1^3 e^x\,dx) = 2\pi[e^x(x-1)]|_1^3 = 4\pi e^3$. (see Fig. 9.2.4(a)).

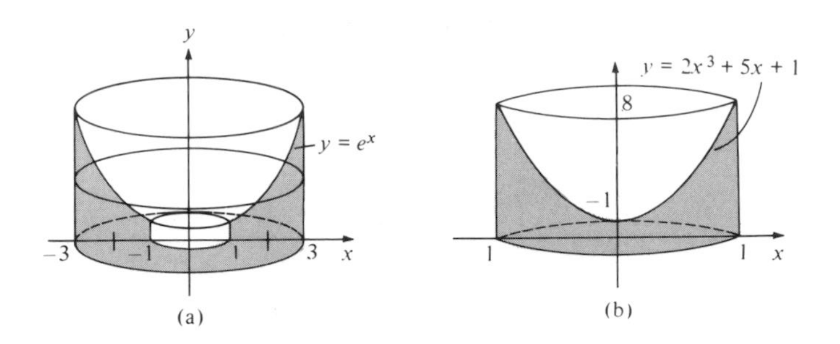

Figure 9.2.4. Find the volume of the shaded solids.

(b) Volume $= 2\pi\int_0^1 x(2x^3+5x+1)\,dx = 2\pi\left[\dfrac{2x^5}{5}+\dfrac{5x^3}{3}+\dfrac{x^2}{2}\right]\Bigg|_0^1 = \dfrac{77}{15}\pi.$

(See Fig. 9.2.4(b)). ▲

Example 4 Find the volume of the "flying saucer" obtained by rotating the region between the curves $y = -\frac{1}{4}(1-x^4)$ and $y = \frac{1}{6}(1-x^6)$ on $[0,1]$ about the y axis.

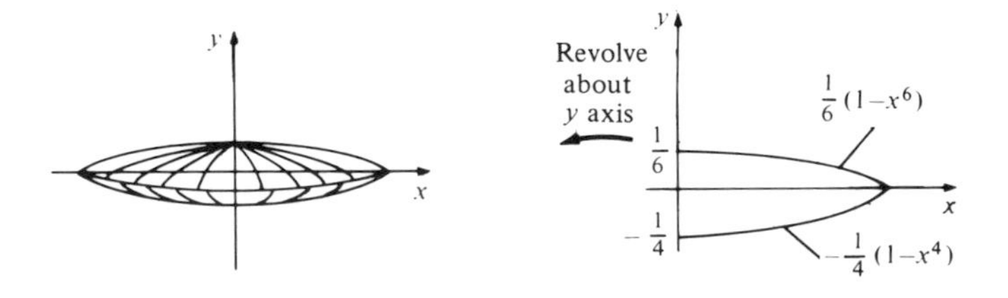

Figure 9.2.5. The flying saucer.

Solution See Fig. 9.2.5. The height of the shell at radius x is $\frac{1}{6}(1-x^6)+\frac{1}{4}(1-x^4) = (5/12)-(x^6/6)-(x^4/4)$, so the volume is

$$2\pi\int_0^1 x\left(\frac{5}{12}-\frac{x^6}{6}-\frac{x^4}{4}\right)dx = 2\pi\left(\frac{5}{24}x^2-\frac{x^8}{48}-\frac{x^6}{24}\right)\Bigg|_0^1$$

$$= 2\pi\left(\frac{5}{24}-\frac{1}{48}-\frac{1}{24}\right) = \frac{7\pi}{24}. \ ▲$$

Example 5 A hole of radius r is drilled through the center of a ball of radius R. How much material is removed?

Solution See Figure 9.2.6. The shell at distance x from the axis of the hole has height

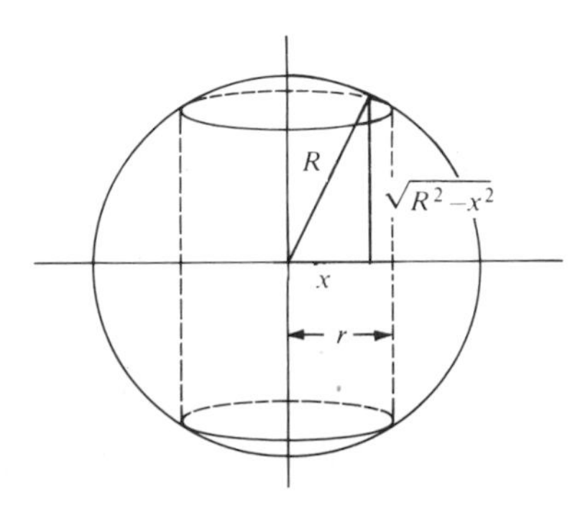

Figure 9.2.6. A ball with a hole drilled through it.

$2\sqrt{R^2 - x^2}$. The shells removed have x running from 0 to r, so their total volume is

$$2\pi\int_0^r 2x\sqrt{R^2 - x^2}\,dx = 2\pi\int_0^{r^2}\sqrt{R^2 - u}\,du = 2\pi\left[-\frac{2}{3}(R^2 - u)^{3/2}\Big|_0^{r^2}\right]$$

$$= \frac{4}{3}\pi\left[R^3 - (R^2 - r^2)^{3/2}\right]$$

$$= \frac{4}{3}\pi R^3\left[1 - \left(1 - \frac{r^2}{R^2}\right)^{3/2}\right].$$

Notice that if we set $r = R$, we get $\frac{4}{3}\pi R^3$; we then recover the formula for the volume of the ball, computed by the shell method. ▲

Example 6 The disk with radius 1 and center $(4,0)$ is revolved around the y axis. Sketch the resulting solid and find its volume.

Solution The doughnut-shaped solid is shown in Fig. 9.2.7.[5] We observe that if the solid is sliced in half by a plane through the origin perpendicular to the y axis, the top half is the solid obtained by revolving about the y axis the region under the semicircle $y = \sqrt{1 - (x-4)^2}$ on the interval $[3,5]$.

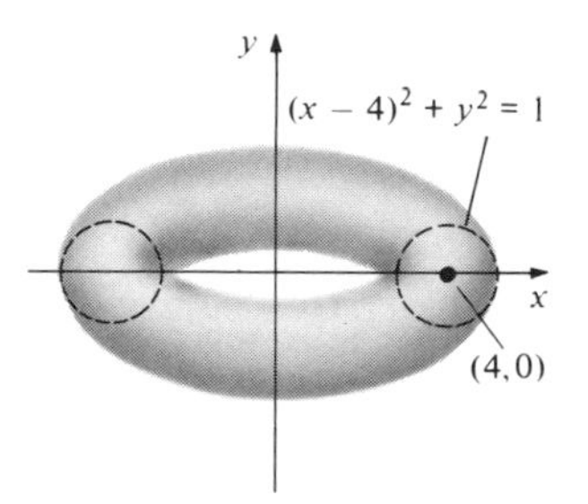

Figure 9.2.7. The disk $(x-4)^2 + y^2 \leqslant 1$ is revolved about the y axis.

The volume of that solid is

$$2\pi\int_3^5 x\sqrt{1 - (x-4)^5}\,dx$$

$$= 2\pi\int_{-1}^1 (u+4)\sqrt{1 - u^2}\,du \qquad (u = x - 4)$$

$$= 2\pi\int_{-1}^1 \sqrt{1 - u^2}\,u\,du + 8\pi\int_{-1}^1 \sqrt{1 - u^2}\,du.$$

Now $\int_{-1}^1\sqrt{1 - u^2}\,u\,du = 0$ because the function $f(u) = \sqrt{1 - u^2}\,u$ is odd: $f(-u) = -f(u)$ so that $\int_{-1}^0 f(u)\,du$ is exactly the negative of $\int_0^1 f(u)\,du$.

On the other hand, $\int_{-1}^1\sqrt{1 - u^2}\,du$ is just the area of a semicircular region of radius 1—that is, $\pi/2$—so the volume of the upper half of the doughnut is $8\pi \cdot (\pi/2) = 4\pi^2$, and the volume of the entire doughnut is twice that, or $8\pi^2$. (Notice that this is equal to the area π of the rotated disk times the circumference 8π of the circle traced out by its center $(4,0)$.) ▲

[5] Mathematicians call this a *solid torus*. The surface of this solid (an "inner tube") is a *torus*.

We conclude this section with a justification of the shell method using step functions. Consider again the solid S in Fig. 9.2.1. We break the region R into thin vertical strips and rotate them into shells, as in Fig. 9.2.8.

What is the volume of such a shell? Suppose for a moment that f has the

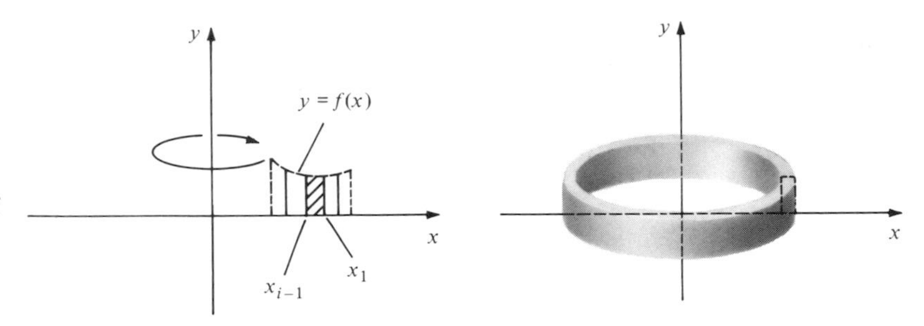

Figure 9.2.8. The volume of a solid of revolution obtained by the shell method.

constant value k_i on the interval (x_{i-1}, x_i). Then the shell is the "difference" of two cylinders of height k_i, one with radius x_i and one with radius x_{i-1}. The volume of the shell is, therefore, $\pi x_i^2 k_i - \pi x_{i-1}^2 k_i = \pi k_i(x_i^2 - x_{i-1}^2)$; we may observe that this last expression is $\int_{x_{i-1}}^{x_i} 2\pi k_i x\,dx$.

If f is a step function on $[a,b]$, with partition $(x_0, \ldots, x_n)$ and $f(x) = k_i$ on (x_{i-1}, x_i), then the volume of the collection of n shells is

$$\sum_{i=1}^{n} \int_{x_{i-1}}^{x_i} 2\pi k_i x\,dx;$$

but $k_i = f(x)$ on (x_{i-1}, x_i), so this is

$$\sum_{i=1}^{n} \int_{x_{i-1}}^{x_i} 2\pi x f(x)\,dx,$$

which is simply $\int_a^b 2\pi x f(x)\,dx$. We now have the formula

$$\text{volume} = 2\pi \int_a^b x f(x)\,dx,$$

which is valid whenever $f(x)$ is a step function on $[a,b]$. To show that the same formula is valid for general f, we squeeze f between step functions above and below using the same argument we used for the slice method.

Exercises for Section 9.2

In Exercises 1–12, find the volume of the solid obtained by revolving each of the following regions about the y axis and sketch the region.

1. The region under the graph of $\sin x$ on $[0,\pi]$.
2. The region under the graph of $\cos 2x$ on $[0,\pi/4]$.
3. The region under the graph of $2-(x-1)^2$ on $[0,2]$.
4. The region under the graph of $\sqrt{4-4x^2}$ on $[0,2]$.
5. The region between the graphs of $\sqrt{3-x^2}$ and $5+x$ on $[0,1]$.
6. The region between the graphs of $\sin x$ and x on $[0,\pi/2]$.
7. The circular region with center $(a,0)$ and radius r $(0<r<a)$.
8. The circular region with radius 2 and center $(6,0)$.
9. The square region with vertices $(4,6)$, $(5,6)$, $(5,7)$, and $(4,7)$.
10. The region in Exercise 9 moved 2 units upward.
11. The region in Exercise 9 rotated by 45° around its center.
12. The triangular region with vertices $(1,1)$, $(2,2)$, and $(3,1)$.

13. The region under the graph of $\sqrt{x}$ on $[0,1]$ is revolved around the y axis. Sketch the resulting solid and find its volume. Relate the result to Example 5 of the previous section.
14. Find the volume in Example 4 by the slice method.
15. A cylindrical hole of radius $\frac{1}{2}$ is drilled through the center of a ball of radius 1. Use the shell method to find the volume of the resulting solid.

16. Find the volume in Exercise 15 by the slice method.
17. Find the volume of the solid torus obtained by rotating the disk $(x-3)^2 + y^2 \leqslant 4$ about the y axis.
18. Find the volume of the solid torus obtained by rotating the disk $x^2 + (y-5)^2 \leqslant 9$ about the x axis.
19. A spherical shell of radius r and thickness h is, by definition, the region between two concentric spheres of radius $r - h/2$ and $r + h/2$.
 (a) Find a formula for the volume $V(r,h)$ of a spherical shell of radius r and thickness h.
 (b) For fixed r, what is $(d/dh)V(r,h)$ when $h = 0$? Interpret your result in terms of the surface area of the sphere.
20. In Exercise 19, find $(d/dr)V(r,h)$ when h is held fixed. Give a geometric interpretation of your answer.

★21. (a) Find the volume of the solid torus $T_{a,b}$ obtained by rotating the disk with radius a and center $(b,0)$ about the y axis, $0 < a < b$.
 (b) What is the volume of the region between the solid tori $T_{a,b}$ and $T_{a+h,b}$, assuming $0 < a+h < b$?
 (c) Using the result in (b), guess a formula for the area of the torus which is the surface of $T_{a,b}$. (Compare Exercise 19).

★22. Let $f(x)$ and $g(y)$ be inverse functions with $f(a) = \alpha$, $f(b) = \beta$, $0 \leqslant a < b$, $0 \leqslant \alpha < \beta$. Show that

$$2\pi \int_{\alpha}^{\beta} y g(y)\, dy = b\pi\beta^2 - a\pi\alpha^2 - \pi \int_a^b [f(x)]^2\, dx.$$

Interpret this statement geometrically.

★23. Use Exercise 22 to compute the volume of the solid obtained by revolving the graph $y = \cos^{-1}x$, $0 \leqslant x \leqslant 1$, about the x axis.

9.3 Average Values and the Mean Value Theorem for Integrals

The average height of a region under a graph is its area divided by the length of the base.

The average value of a function on an interval will be defined in terms of an integral, just as the average or mean of a list $a_1, \dots, a_n$ of n numbers is defined in terms of a sum as $(1/n)\sum_{i=1}^{n} a_i$.

If a grain dealer buys wheat from n farmers, buying b_i bushels from the ith farmer at the price of p_i dollars per bushel, the average price is determined not by taking the simple average of the p_i's, but rather by the "weighted average":

$$p_{\text{average}} = \frac{\sum_{i=1}^{n} p_i b_i}{\sum_{i=1}^{n} b_i} = \frac{\text{total dollars}}{\text{total bushels}}.$$

If a cyclist changes speed intermittently, travelling at v_1 miles per hour from t_0 to t_1, v_2 miles per hour from t_1 to t_2, and so on up to time t_n, then the average speed for the trip is

$$v_{\text{average}} = \frac{\sum_{i=1}^{n} v_i(t_i - t_{i-1})}{\sum_{i=1}^{n}(t_i - t_{i-1})} = \frac{\text{total miles}}{\text{total hours}}.$$

If, in either of the last two examples, the b_i's or $(t_i - t_{i-1})$'s are all equal, then the average value is simply the usual average of the p_i's or the v_i's.

If f is a step function on $[a,b]$ and we have a partition $(x_0, x_1, \dots, x_n)$ with $f(x) = k_i$ on (x_{i-1}, x_i), then the *average value* of f on the interval $[a,b]$ is defined to be

$$\overline{f(t)}_{[a,b]} = \frac{\sum_{i=1}^{n} k_i \Delta x_i}{\sum_{i=1}^{n} \Delta x_i}. \tag{1}$$

In other words, each interval is weighted by its length.

How can we define the average value of a function which is not a step function? For instance, it is common to talk of the average temperature at a place on earth, although the temperature is not a step function. We may rewrite (1) as

$$\overline{f(x)}_{[a,b]} = \frac{\int_a^b f(x)\,dx}{b-a}, \tag{2}$$

and this leads us to adopt formula (2) as the definition of the average value for any integrable function f, not just a step function.

Average Value

If the function f has an integral on $[a,b]$, then the average value $\overline{f(x)}_{[a,b]}$ of f on $[a,b]$ is defined by the formula

$$\overline{f(x)}_{[a,b]} = \frac{1}{b-a}\int_a^b f(x)\,dx.$$

Example 1 Find the average value of $f(x) = x^2$ on $[0,2]$.

Solution By definition, we have

$$\overline{x^2}_{[0,2]} = \frac{1}{2-0}\int_0^2 x^2\,dx = \frac{1}{2}\cdot\frac{1}{3}x^3\Big|_0^2 = \frac{4}{3}. \ \blacktriangle$$

Example 2 Show that if $v = f(t)$ is the velocity of a moving object, then the definition of $\bar{v}_{[a,b]}$ agrees with the usual notion of average velocity.

Solution By the definition,

$$\bar{v}_{[a,b]} = \frac{1}{b-a}\int_a^b v\,dt;$$

but $\int_a^b v\,dt$ is the distance travelled between $t = a$ and $t = b$, so $\bar{v}_{[a,b]} =$ (distance travelled)/(time of travel), which is the usual definition of average velocity. ▲

Example 3 Find the average value of $\sqrt{1-x^2}$ on $[-1,1]$.

Solution By the formula for average values, $\overline{\sqrt{1-x^2}}_{[-1,1]} = (\int_{-1}^{1}\sqrt{1-x^2}\,dx)/2$; but $\int_{-1}^{1}\sqrt{1-x^2}\,dx$ is the area of the upper semicircle of $x^2+y^2=1$, which is $\frac{1}{2}\pi$, so $\overline{\sqrt{1-x^2}}_{[-1,1]} = \pi/4 \approx 0.785$. ▲

Example 4 Find $\overline{x^2\sin x^3}_{[0,\pi]}$.

Solution

$$\begin{aligned}\overline{x^2\sin(x^3)}_{[0,\pi]} &= \frac{1}{\pi}\int_0^\pi x^2\sin x^3\,dx \\ &= \frac{1}{\pi}\int_0^{\pi^3}\sin u\,\frac{du}{3} \qquad \text{(substituting } u = x^3\text{)} \\ &= \frac{1}{3\pi}(-\cos u)\Big|_0^{\pi^3} \\ &= \frac{1}{3\pi}(1-\cos\pi^3) \approx 0.0088. \ \blacktriangle\end{aligned}$$

We may rewrite the definition of the average value in the form

$$\int_a^b f(x)\,dx = \overline{f(x)}_{[a,b]}(b-a),$$

and the right-hand side can be interpreted as the integral of a constant function:

$$\int_a^b f(x)\,dx = \int_a^b \overline{f(x)}_{[a,b]}\,dx.$$

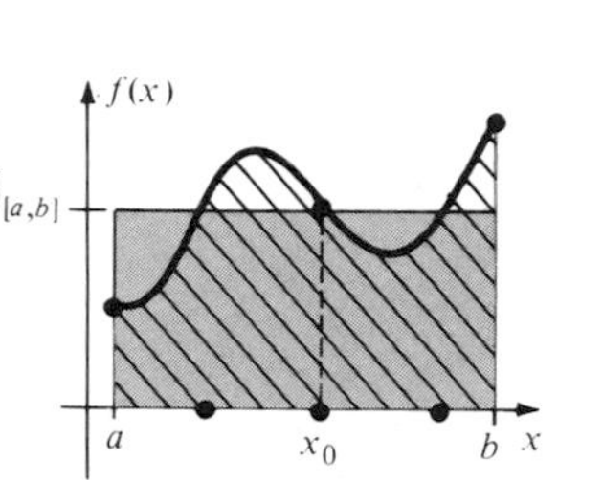

Figure 9.3.1. The average value is defined so that the area of the rectangle equals the area under the graph. The dots on the x axis indicate places where the average value is attained.

Geometrically, the average value is the height of the rectangle with base $[a,b]$ which has the same area as the region under the graph of f (see Fig. 9.3.1). Physically, if the graph of f is a picture of the surface of wavy water in a narrow channel, then the average value of f is the height of the water when it settles.

An important property of average values is given in the following statement:

If $m \leqslant f(x) \leqslant M$ for all x in $[a,b]$, then $m \leqslant \overline{f(x)}_{[a,b]} \leqslant M$.

Indeed, the integrals $\int_a^b m\,dx$ and $\int_a^b M\,dx$ are lower and upper sums for f on $[a,b]$, so

$$m(b-a) \leqslant \int_a^b f(x)\,dx \leqslant M(b-a).$$

Dividing by $(b-a)$ gives the desired result.

By the extreme value theorem (Section 3.5), $f(x)$ attains a minimum value m and a maximum value M on $[a,b]$. Then $m \leqslant f(x) \leqslant M$ for x in $[a,b]$, so $\overline{f(x)}_{[a,b]}$ lies between m and M, by the preceding proposition. By the first version of the intermediate value theorem (Section 3.1), applied to the interval between the points where $f(x) = m$ and $f(x) = M$, we conclude that there is a x_0 in this interval (and thus in $[a,b]$), such that $f(x_0) = \overline{f(x)}_{[a,b]}$.

In other words, we have proved that the average value of a continuous function on an interval is always attained somewhere on the interval. This result is known as the *mean value theorem for integrals.*

Mean Value Theorem for Integrals

Let f be continuous on $[a,b]$. Then there is a point x_0 in (a,b) such that

$$f(x_0) = \frac{1}{b-a}\int_a^b f(x)\,dx.$$

Notice that in Fig. 9.3.1, the mean value is attained at three different points.

Example 5 Give another proof of the mean value theorem for integrals by using the fundamental theorem of calculus and the mean value theorem for derivatives.

Solution Let f be continuous on $[a,b]$, and define $F(x) = \int_a^x f(s)\,ds$. By the fundamental theorem of calculus (alternative version), $F'(x) = f(x)$ for x in (a,b). (Exercise 29 asks you to verify that F is continuous at a and b—we accept it here.) By the mean value theorem for derivatives, there is some x_0 in (a,b) such that

$$F'(x_0) = \frac{F(b)-F(a)}{b-a}.$$

Substituting for F and F' in terms of f, we have

$$f(x_0) = \frac{\int_a^b f(x)\,dx - \int_a^a f(x)\,dx}{b-a} = \frac{\int_a^b f(x)\,dx}{b-a} = \overline{f(x)}_{[a,b]},$$

which establishes the mean value theorem for integrals. ▲

Exercises for Section 9.3

In Exercises 1–4, find the average value of the given function on the given interval.

1. x^3 on $[0,1]$
2. x^2+1 on $[1,2]$
3. $x/(x^2+1)$ on $[1,2]$
4. $\cos^2 x \sin x$ on $[0,\pi/2]$

Calculate each of the average values in Exercises 5–16.

5. $\overline{x^3}_{[0,2]}$
6. $\overline{z^3+z^2+1}_{[1,2]}$
7. $\overline{1/(1+t^2)}_{[-1,1]}$
8. $\overline{[(x^3+x-2)/(x^2+1)]}_{[-1,1]}$
9. $\overline{\sin^{-1}x}_{[0,1]}$
10. $\overline{\sin^{-1}x}_{[-1/2,0]}$
11. $\overline{\sin x \cos 2x}_{[0,\pi/2]}$
12. $\overline{(x^2+x-1)\sin x}_{[0,\pi/4]}$
13. $\overline{x^3+\sqrt{1/x}}_{[1,3]}$
14. $\overline{\sqrt{1-t^2}}_{[0,1]}$
15. $\overline{\sin^2 x}_{[0,\pi]}$
16. $\overline{\ln x}_{[1,e]}$

17. What was the average temperature in Goose Brow on June 13, 1857? (See Fig. 9.3.2).

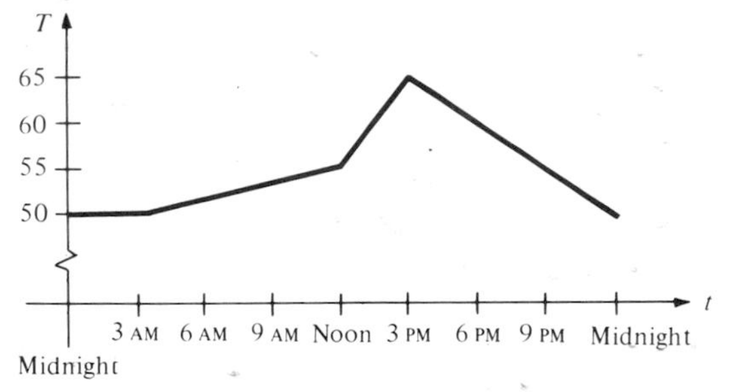

Figure 9.3.2. Temperature in Goose Brow on June 13, 1857.

18. Find the average temperature in Goose Brow (Fig. 9.3.2) during the periods midnight to 3P.M. and 3P.M. to midnight. How is the average over the whole day related to these numbers?

19. (a) Find $\overline{t^2+3t+2}_{[0,x]}$ as a function of x. (b) Evaluate this function of x for $x=0.1$, 0.01, 0.0001. Try to explain what is happening.

20. Find $\overline{\cos\theta}_{[\pi,\pi+\theta]}$ as a function of θ and evaluate the limit as $\theta \to 0$.

21. Show that if $\overline{f'(x)}_{[a,b]} = 0$ then $f(b) = f(a)$.

22. Show that if $a < b < c$, then

$$\overline{f(t)}_{[a,c]} = \left(\frac{b-a}{c-a}\right)\overline{f(t)}_{[a,b]} + \left(\frac{c-b}{c-a}\right)\overline{f(t)}_{[b,c]}.$$

★23. How is the average of $f(x)$ on $[a,b]$ related to that of $f(x)+k$ for a constant k? Explain the answer in terms of a graph.

★24. If $f(x) = g(x) + h(x)$ on $[a,b]$, show that the average of f on $[a,b]$ is the sum of the averages of g and h on $[a,b]$.

★25. Suppose that f' exists and is continuous on $[a,b]$. Prove the mean value theorem for derivatives from the mean value theorem for integrals.

★26. Let f be defined on the real line and let $a(x) = \overline{f(x)}_{[0,x]}$.
   (a) Derive the formula $a'(x) = (1/x)[f(x) - a(x)]$.
   (b) Interpret the formula in the cases $f(x) = a(x)$, $f(x) < a(x)$, and $f(x) > a(x)$.
   (c) When baseball players strike out, it lowers their batting average more at the beginning of the season than at the end. Explain why.

★27. The *geometric mean* of the positive numbers $a_1, \cdots, a_n$ is the nth root of the product $a_1 \cdots a_n$. Define the geometric mean of a positive function $f(x)$ on $[a,b]$. [*Hint:* Use logarithms.]

★28. (a) Use the idea of Exercise 27 to prove the arithmetic–geometric mean inequality (see Example 12, Section 3.5). [*Hint:* Use the fact that e^x is concave upwards.] (b) Generalize from numbers to functions.

★29. If f is continuous on $[a,b]$ and $F(x) = \int_a^x f(s)\,ds$, verify directly using the definition of continuity in Section 11.1 that F is continuous on $[a,b]$.

★30. (a) At what point of the interval $[0,a]$ is the average value of e^x achieved?
   (b) Denote the expression found in part (a) by $p(a)$. Evaluate $p(a)$ for $a = 1$, 10, 100, 1000 and $a = 0.1$, 0.01, 0.0001, and 0.000001. Be sure that your answers are reasonable.
   (c) Guess the limits $\lim_{a\to 0} p(a)/a$ and $\lim_{a\to\infty} p(a)/a$.

9.4 Center of Mass

The center of mass of a region is the point where it balances.

An important problem in mechanics, which was considered by Archimedes, is to locate the point on which a plate of some given irregular shape will balance (Fig. 9.4.1). This point is called the center of mass, or center of gravity, of the plate. The center of mass can also be defined for solid objects, and its applications range from theoretical physics to the problem of arranging wet towels to spin in a washing machine.

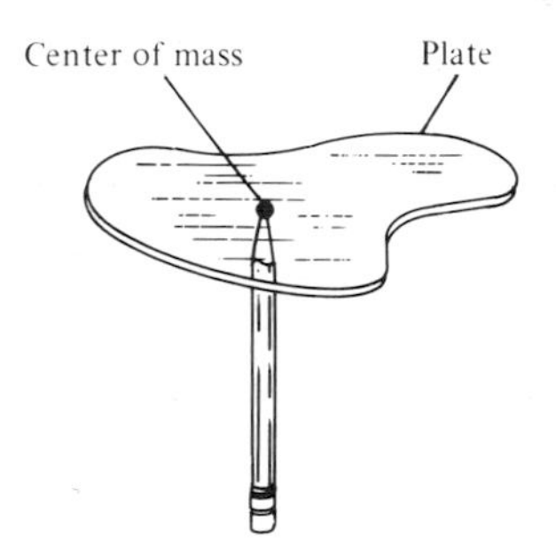

Figure 9.4.1. The plate balances when supported at its center of mass.

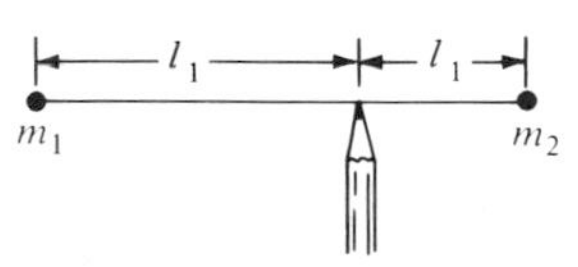

Figure 9.4.2. The support is at the center of mass when $m_1 l_1 = m_2 l_2$.

To give a mathematical definition of the center of mass, we begin with the ideal case of two point masses, m_1 and m_2, attached to a light rod whose mass we neglect. (Think of a see-saw.) If we support the rod (see Fig. 9.4.2) at a point which is at distance l_1 from m_1 and distance l_2 from m_2, we find that the rod tilts down at m_1 if $m_1 l_1 > m_2 l_2$ and down at m_2 if $m_1 l_1 < m_2 l_2$. It balances when

$$m_1 l_1 = m_2 l_2 . \tag{1}$$

One can derive this balance condition from basic physical principles, or one may accept it as an experimental fact; we will not try to prove it here, but rather study its consequences.

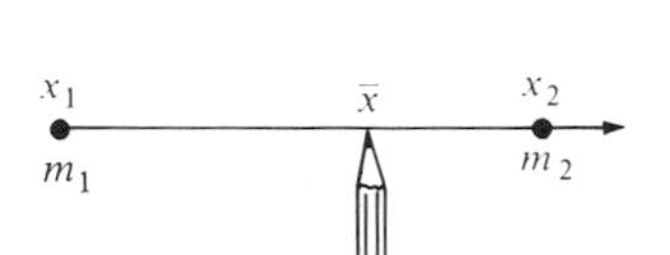

Figure 9.4.3. The center of mass is at $\bar{x}$ if $m_1(\bar{x} - x_1) = m_2(x_2 - \bar{x})$.

Suppose that the rod lies along the x axis, with m_1 at x_1 and m_2 at x_2. Let $\bar{x}$ be the position of the center of mass. Comparing Figs. 9.4.2 and 9.4.3, we see that $l_1 = \bar{x} - x_1$ and $l_2 = x_2 - \bar{x}$, so formula (1) may be rewritten as $m_1(\bar{x} - x_1) = m_2(x_1 - \bar{x})$. Solving for $\bar{x}$ gives the explicit formula

$$\bar{x} = \frac{m_1 x_1 + m_2 x_2}{m_1 + m_2} . \tag{2}$$

We may observe that the position of the center of mass is just the *weighted average* of the positions of the individual masses. This suggests the following generalization.

Center of Mass on the Line

If n masses, $m_1, m_2, \ldots, m_n$, are placed at the points $x_1, x_2, \ldots, x_n$, respectively, their center of mass is located at

$$\bar{x} = \frac{\sum_{i=1}^{n} m_i x_i}{\sum_{i=1}^{n} m_i} . \tag{3}$$

We may accept formula (3), as we did formula (1), as a physical fact, or we may derive it (see Example 1) from formula (2) and the following principle, which is also accepted as a general physical fact.

Consolidation Principle

If a body B is divided into two parts, B_1 and B_2, with masses M_1 and M_2, then the center of mass of the body B is located as if B consisted of two point masses: M_1 located at the center of mass of B_1, and M_2 located at the center of mass of B_2.

Example 1 Using formula (2) and the consolidation principle, derive formula (3) for the case of three masses.

Solution We consider the body B consisting of m_1, m_2, and m_3 as divided into B_1, consisting of m_1 and m_2, and B_2, consisting of m_3 alone. (See Fig. 9.4.4.)

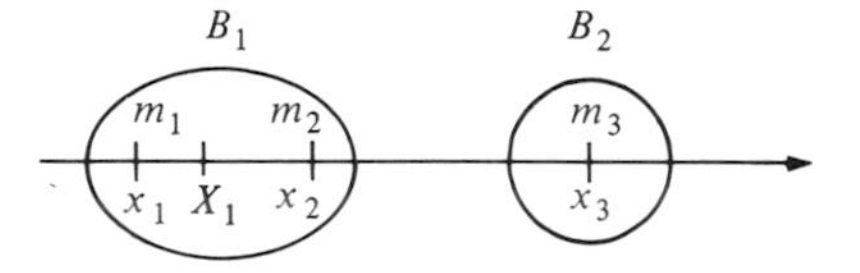

Figure 9.4.4. Center of mass of three points by the consolidation principle.

By formula (2) we know that X_1, the center of mass of B_1, is given by

$$X_1 = \frac{m_1x_1 + m_2x_2}{m_1 + m_2}.$$

The mass M_1 of B_1 is $m_1 + m_2$. The body B_2 has center of mass at $X_2 = x_3$ and mass $M_2 = m_3$. Applying formula (2) once again to the point masses M_1 at X_1 and M_2 at X_2 gives the center of mass $\bar{x}$ of B by the consolidation principle:

$$\bar{x} = \frac{M_1X_1 + M_2X_2}{M_1 + M_2} = \frac{(m_1 + m_2)\left(\dfrac{m_1x_1 + m_2x_2}{m_1 + m_2}\right) + m_3x_3}{(m_1 + m_2) + m_3}$$

$$= \frac{m_1x_1 + m_2x_2 + m_3x_3}{m_1 + m_2 + m_3},$$

which is exactly formula (3) for $n = 3$. ▲

Example 2 Masses of 10, 20, and 25 grams are located at $x_1 = 0$, $x_2 = 5$, and $x_3 = 12$ centimeters, respectively. Locate the center of mass.

Solution Using formula (3), we have

$$\bar{x} = \frac{10(0) + 20(5) + 25(12)}{10 + 20 + 25} = \frac{400}{55} = \frac{80}{11} \approx 7.27 \text{ centimeters. } \blacktriangle$$

Now let us study masses in the plane. Suppose that the masses $m_1, m_2, \ldots, m_n$ are located at the points $(x_1, y_1), \ldots, (x_n, y_n)$. We imagine the masses as being attached to a weightless card, and we seek a point $(\bar{x}, \bar{y})$ on the card where it will balance. (See Fig. 9.4.5.)

To locate the center of mass $(\bar{x}, \bar{y})$, we note that a card which balances on the *point* $(\bar{x}, \bar{y})$ will certainly balance along any *line* through $(\bar{x}, \bar{y})$. Take, for instance, a line parallel to the y axis (Fig. 9.4.6). The balance along this line will not be affected if we move each mass parallel to the line so that m_1, m_2, m_3, and m_4 are lined up parallel to the x axis (Fig. 9.4.7).

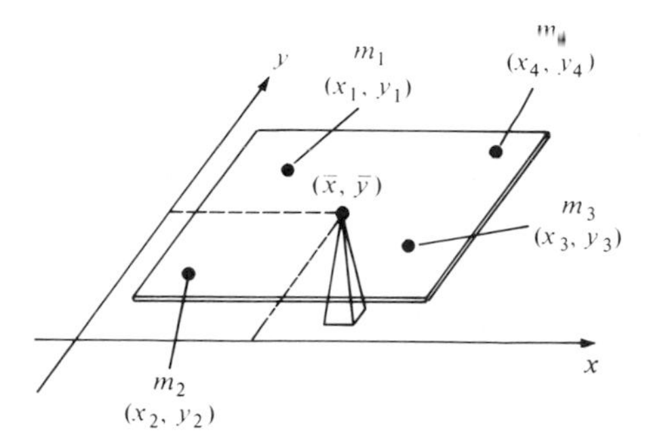

Figure 9.4.5. The card balances at the center of mass.

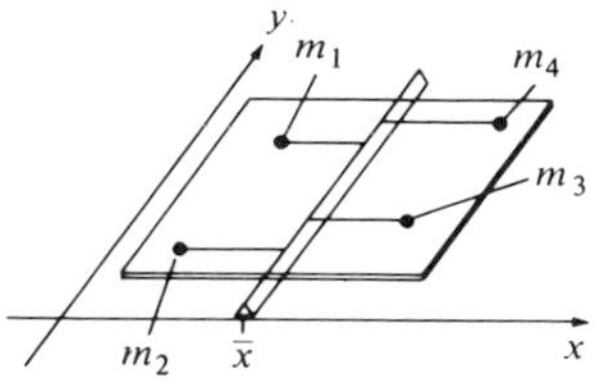

Figure 9.4.6. If the card balances at a point, it balances along any line through that point.

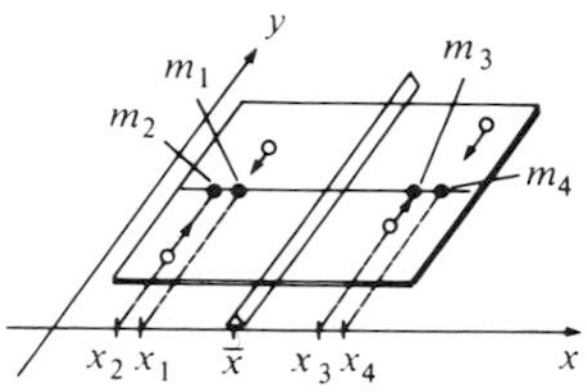

Figure 9.4.7. Moving the masses parallel to a line does not affect the balance along this line.

Now we can apply the balance equation (3) for masses in a line to conclude that the x component $\bar{x}$ of the center of mass is equal to the weighted average

$$\bar{x} = \frac{\sum_{i=1}^{n} m_i x_i}{\sum_{i=1}^{n} m_i}$$

of the x components of the point masses.

Repeating the construction for a balance line parallel to the x axis (we urge you to draw versions of Figs. 9.4.6 and 9.4.7 for this case), and applying formula (3) to the masses as lined up parallel to the y axis, we conclude that

$$\bar{y} = \frac{\sum_{i=1}^{n} m_i y_i}{\sum_{i=1}^{n} m_i}.$$

These two equations completely determine the position of the center of mass.

Center of Mass in the Plane

If n masses, $m_1, m_2, \ldots, m_n$, are placed at the n points (x_1, y_1), $(x_2, y_2), \ldots, (x_n, y_n)$, respectively, then their center of mass is located at $(\bar{x}, \bar{y})$, where

$$\bar{x} = \frac{\sum_{i=1}^{n} m_i x_i}{\sum_{i=1}^{n} m_i} \quad \text{and} \quad \bar{y} = \frac{\sum_{i=1}^{n} m_i y_i}{\sum_{i=1}^{n} m_i}. \tag{4}$$

Example 3 Masses of 10, 15, and 30 grams are located at $(0,1)$, $(1,1)$, and $(1,0)$. Find their center of mass.

Solution Applying formula (4), with $m_1 = 10$, $m_2 = 15$, $m_3 = 30$, $x_1 = 0$, $x_2 = 1$, $x_3 = 1$, $y_1 = 1$, $y_2 = 1$, and $y_3 = 0$, we have

$$\bar{x} = \frac{10 \cdot 0 + 15 \cdot 1 + 30 \cdot 1}{10 + 15 + 30} = \frac{9}{11}$$

and

$$\bar{y} = \frac{10 \cdot 1 + 15 \cdot 1 + 30 \cdot 0}{10 + 15 + 30} = \frac{5}{11},$$

so the center of mass is located at $(\frac{9}{11}, \frac{5}{11})$. ▲

Example 4 Particles of mass 1, 2, 3, and 4 are located at successive vertices of a unit square. How far from the center of the square is the center of mass?

Solution We take the vertices of the square to be (0, 0), (1, 0), (1, 1), and (0, 1). (See Fig. 9.4.8.) The center is at $(\frac{1}{2}, \frac{1}{2})$ and the center of mass is located by formula (4):

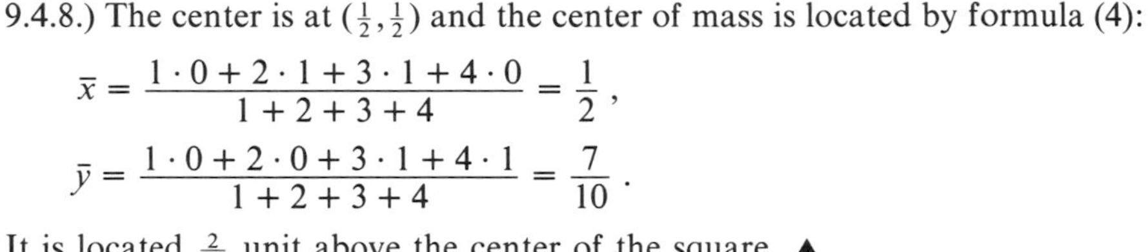

$$\bar{x} = \frac{1 \cdot 0 + 2 \cdot 1 + 3 \cdot 1 + 4 \cdot 0}{1 + 2 + 3 + 4} = \frac{1}{2},$$

$$\bar{y} = \frac{1 \cdot 0 + 2 \cdot 0 + 3 \cdot 1 + 4 \cdot 1}{1 + 2 + 3 + 4} = \frac{7}{10}.$$

It is located $\frac{2}{10}$ unit above the center of the square. ▲

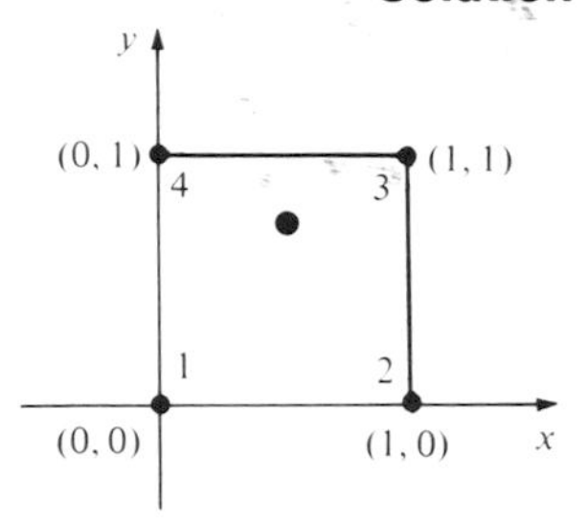

Figure 9.4.8. The center of mass of these four weighted points is located at $(\frac{1}{2}, \frac{7}{10})$.

We turn now from the study of center of mass for point masses to that for flat plates of various shapes.

A flat plate is said to be of *uniform density* if there is a constant ρ such that the mass of any piece of the plate is equal to ρ times the area of the piece. The number ρ is called the *density* of the plate. We represent a plate of uniform density by a region R in the plane; we will see that the value of ρ is unimportant as far as the center of mass is concerned.

A line l is called an *axis of symmetry* for the region R if the region R is taken into itself when the plane is flipped 180° around l (or, equivalently, reflected across l). For example, a square has four different axes of symmetry, a nonsquare rectangle two, and a circle infinitely many (see Fig. 9.4.9). Since a region will obviously balance along an axis of symmetry l, the center of mass must lie somewhere on l.

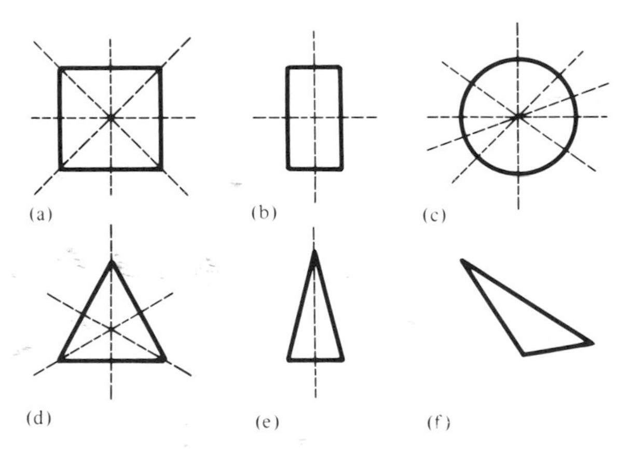

Figure 9.4.9. The axes of symmetry of various geometric figures.

> **Symmetry Principle**
>
> If l is an axis of symmetry for the plate R of uniform density, then the center of mass of R lies on l.

If a plate admits more than one axis of symmetry, then the center of mass must lie on all the axes. In this case, we can conclude that the center of mass lies at the point of intersection of the axes of symmetry. Looking at parts (a) through (d) of Fig. 9.4.9, we see that in each case the center of mass is located at the "geometric center" of the figure. In case (e), we know only that the center of mass is on the altitude; in case (f), symmetry cannot be applied to determine the center of mass.

Using infinitesimals, we shall now derive formulas for the center of mass of the region under the graph of a function f, with uniform density ρ. As we did when computing areas, we think of the region under the graph of f on $[a,b]$ as being composed of "infinitely many rectangles of infinitesimal width." The rectangle at x with width dx has area $f(x)\,dx$ and mass $\rho f(x)\,dx$; its center of mass is located at $(x,\frac{1}{2}f(x))$ (by the symmetry principle) (see Fig. 9.4.10). [The center of mass is "really" at $(x+\frac{1}{2}dx,\frac{1}{2}f(x))$ but since the region is infinitesimally thin, we use $(x,\frac{1}{2}f(x))$—a more careful argument is given in the supplement to this section.]

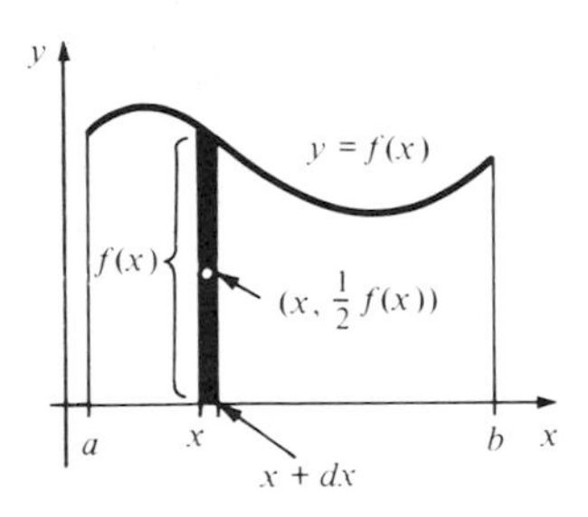

Figure 9.4.10. The "infinitesimal rectangle" has mass $\rho f(x)\,dx$ and center of mass at $(x,\frac{1}{2}f(x))$.

Now we apply the consolidation principle, but instead of summing, we replace the sums in formula (4) by integrals and arrive at the following result:

$$\bar{x}=\frac{\int_a^b x\rho f(x)\,dx}{\int_a^b \rho f(x)\,dx}=\frac{\int_a^b xf(x)\,dx}{\int_a^b f(x)\,dx},\quad\text{and}$$

$$\bar{y}=\frac{\int_a^b \frac{1}{2}f(x)\rho f(x)\,dx}{\int_a^b \rho f(x)\,dx}=\frac{\frac{1}{2}\int_a^b[f(x)]^2\,dx}{\int_a^b f(x)\,dx};\ (\rho\text{ cancels since it is constant}).$$

Since the center of mass depends only upon the region in the plane, and not upon the density ρ, we usually refer to $(\bar{x},\bar{y})$ simply as the *center of mass of the region.*

Center of Mass of the Region under a Graph

The center of mass of a plate of uniform density represented by the region under the graph of a (non-negative) function $f(x)$ on $[a,b]$ is located at $(\bar{x},\bar{y})$, where

$$\bar{x}=\frac{\int_a^b xf(x)\,dx}{\int_a^b f(x)\,dx}\quad\text{and}\quad\bar{y}=\frac{\frac{1}{2}\int_a^b[f(x)]^2\,dx}{\int_a^b f(x)\,dx}.\tag{5}$$

Example 5 Find the center of mass of the region under the graph of x^2 from 0 to 1.

Solution By formulas (5), with $f(x)=x^2$, $a=0$, and $b=1$,

$$\bar{x}=\frac{\int_0^1 x^3\,dx}{\int_0^1 x^2\,dx}=\frac{1/4}{1/3}=\frac{3}{4}\quad\text{and}\quad\bar{y}=\frac{\frac{1}{2}\int_0^1 x^4\,dx}{\int_0^1 x^2\,dx}=\frac{1/10}{1/3}=\frac{3}{10},$$

so the center of mass is located at $(\frac{3}{4},\frac{3}{10})$. (See Fig. 9.4.11.) (You can verify this result experimentally by cutting a figure out of stiff cardboard and seeing where it balances.) ▲

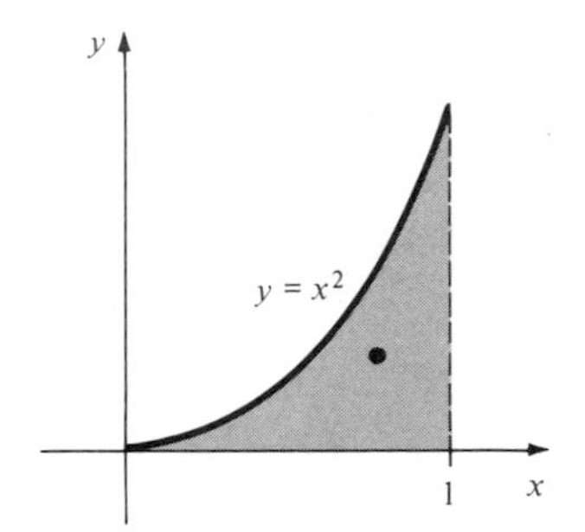

Figure 9.4.11. The center of mass of the shaded region is located at $(\frac{3}{4}, \frac{3}{10})$.

Example 6 Find the center of mass of a semicircular region of radius 1.

Solution We take the region under the graph of $\sqrt{1-x^2}$ on $[-1, 1]$. Since the y axis is an axis of symmetry, the center of mass must lie on this axis; that is, $\bar{x} = 0$. (You can also calculate $\int_{-1}^{1} x\sqrt{1-x^2}\,dx$ and find it to be zero.) By equation (5),

$$\bar{y} = \frac{\int_{-1}^{1}(1-x^2)\,dx}{\int_{-1}^{1}\sqrt{1-x^2}\,dx}.$$

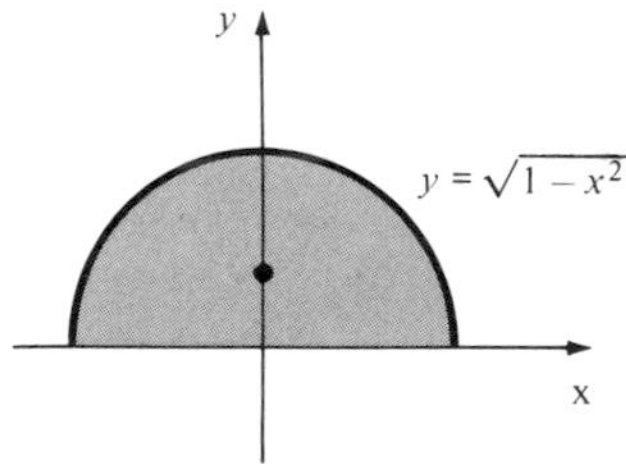

Figure 9.4.12. The center of mass of the semicircular region is located at $(0, 4/3\pi)$.

The denominator is the area $\pi/2$ of the semicircle. The numerator is $\frac{1}{2}\int_{-1}^{1}(1-x^2)\,dx = \frac{1}{2}[x - x^3/3]\big|_{-1}^{1} = \frac{2}{3}$, so

$$\bar{y} = \frac{2/3}{\pi/2} = \frac{4}{3\pi} \approx 0.42,$$

and so the center of mass is located at $(0, 4/3\pi)$ (see Fig. 9.4.12). ▲

Using the consolidation principle, we can calculate the center of mass of a region which is *not* under a graph by breaking it into simpler regions, as we did for areas in Section 4.

Example 7 Find the center of mass of the region consisting of a disk of radius 1 centered at the origin and the region under the graph of $\sin x$ on $(2\pi, 3\pi)$.

Solution The center of mass of the disk is at $(0, 0)$, since the x and y axes are both axes of symmetry. For the region under the graph of $\sin x$ on $[2\pi, 3\pi]$, the line $x = \frac{5}{2}\pi$ is an axis of symmetry. To find the y coordinate of the center of mass, we use formula (5) and the identity $\sin^2 x = (1 - \cos 2x)/2$ to obtain

$$\frac{\frac{1}{2}\int_{2\pi}^{3\pi}\sin^2 x\,dx}{\int_{2\pi}^{3\pi}\sin x\,dx} = \frac{\frac{1}{2}(x/2 - \sin 2x/4)\big|_{2\pi}^{3\pi}}{-\cos x\big|_{2\pi}^{3\pi}} = \frac{(1/2)\cdot\pi/2}{2} = \frac{\pi}{8} \approx 0.393.$$

(Notice that this region is more "bottom heavy" than the semicircular region.)

By the consolidation principle, the center of mass of the total figure is the same as one consisting of two points: one at $(0, 0)$ with mass $\rho\pi$ and one at $(\frac{5}{2}\pi, \pi/8)$ with mass 2ρ. The center of mass is, therefore, at $(\bar{x}, \bar{y})$, where

$$\bar{x} = \frac{\rho\pi\cdot 0 + 2\rho\cdot\frac{5}{2}\pi}{\rho\pi + \rho 2} = \frac{5\pi}{2+\pi} \quad \text{and} \quad \bar{y} = \frac{\rho\pi\cdot 0 + 2\rho\cdot\pi/8}{\rho\pi + \rho 2} = \frac{\pi/4}{2+\pi}.$$

Hence $(\bar{x}, \bar{y})$ is approximately $(3.06, 0.15)$ (see Fig. 9.4.13). ▲

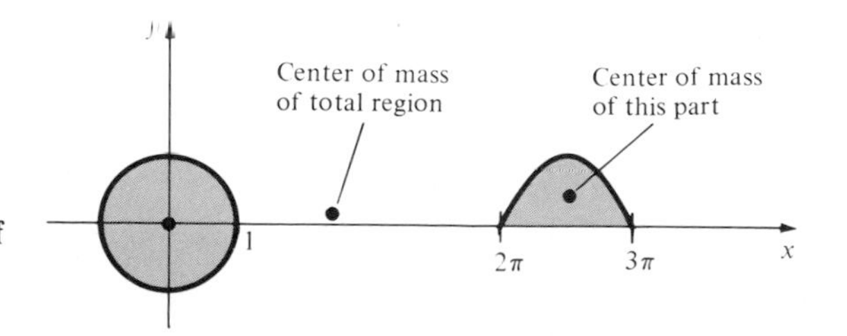

Figure 9.4.13. The center of mass is found by the consolidation principle.

Supplement to Section 9.4: A Derivation of the Center of Mass Formula (5) Using Step Functions

We begin by considering the case in which f is a step function on $[a,b]$ with $f(x) \geqslant 0$ for x in $[a,b]$. Let R be the region under the graph of f and let $(x_0, \ldots, x_n)$ be a partition of $[a,b]$ such that f is a constant k_i on (x_{i-1}, x_i). Then R is composed of n rectangles $R_1, \ldots, R_n$ of areas $k_i(x_i - x_{i-1}) = k_i\Delta x_i$ and masses $\rho k_i \Delta x_i = m_i$. By the symmetry principle, the center of mass of R_i is located at $(\bar{x}_i, \bar{y}_i)$, where $\bar{x}_i = \frac{1}{2}(x_{i-1} + x_i)$ and $\bar{y}_i = \frac{1}{2}k_i$. (See Fig. 9.4.14.)

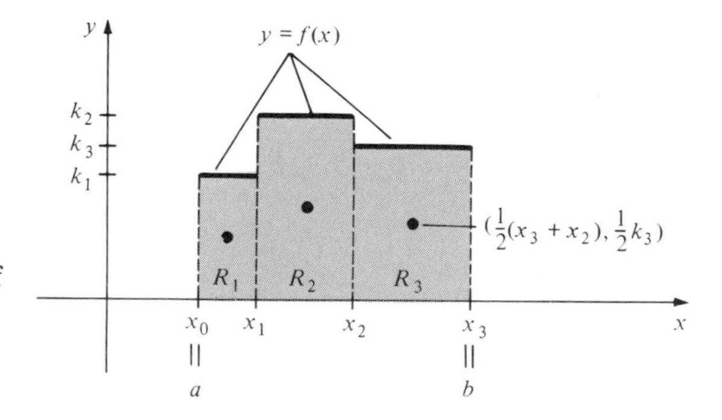

Figure 9.4.14. The center of mass of the shaded region is obtained by the consolidation principle.

Now we use the consolidation principle, extended to a decomposition into n pieces, to conclude that the center of mass of R is the same as the center of mass of masses $m_1, \ldots, m_n$ placed at the points $(\bar{x}_1, \bar{y}_1), \ldots, (\bar{x}_n, \bar{y}_n)$. By formula (4), we have, first of all,

$$\bar{x} = \frac{\sum_{i=1}^{n} m_i \bar{x}_i}{\sum_{i=1}^{n} m_i} = \frac{\sum_{i=1}^{n} \rho k_i \Delta x_i \left[\frac{1}{2}(x_{i-1} + x_i)\right]}{\sum_{i=1}^{n} \rho k_i \Delta x_i}.$$

We wish to rewrite the numerator and denominator as integrals, so that we can eventually treat the case where f is not a step function. The denominator is easy to handle. Factoring out ρ gives $\rho\sum_{i=1}^{n} k_i \Delta x_i$, which we recognize as $\rho\int_a^b f(x)\,dx$, the total mass of the plate. The numerator of $\bar{x}$ equals

$$\frac{1}{2}\rho\sum_{i=1}^{n} k_i(x_i - x_{i-1})(x_i + x_{i-1}) = \frac{1}{2}\rho\sum_{i=1}^{n} k_i(x_i^2 - x_{i-1}^2).$$

We notice that $k_i(x_i^2 - x_{i-1}^2) = \int_{x_{i-1}}^{x_i} 2k_i x\,dx$, which we can also write as $\int_{x_{i-1}}^{x_i} 2xf(x)\,dx$, since $f(x) = k_i$ on (x_{i-1}, x_i). Now the numerator of $\bar{x}$ becomes

$$\frac{1}{2}\rho\sum_{i=1}^{n}\int_{x_{i-1}}^{x_i} 2xf(x)\,dx = \frac{1}{2}\rho\int_a^b 2xf(x)\,dx = \rho\int_a^b xf(x)\,dx,$$

and we have

$$\bar{x} = \frac{\rho\int_a^b xf(x)\,dx}{\rho\int_a^b f(x)\,dx} = \frac{\int_a^b xf(x)\,dx}{\int_a^b f(x)\,dx} \qquad \text{(cancelling } \rho\text{).}$$

To find the y coordinate of the center of mass, we use the second half of formula (4):

$$\bar{y} = \frac{\sum_{i=1}^{n} m_i \bar{y}_i}{\sum_{i=1}^{n} m_i} = \frac{\sum_{i=1}^{n} \rho k_i \Delta x_i (\frac{1}{2} k_i)}{\sum_{i=1}^{n} \rho k_i \Delta x_i}$$

The denominator is the total mass $\rho \int_a^b f(x)\,dx$, as before. The numerator is $\frac{1}{2}\rho \sum_{i=1}^{n} k_i^2 \Delta x_i$, and we recognize $\sum_{i=1}^{n} k_i^2 \Delta x_i$ as the integral $\int_a^b [f(x)]^2\,dx$ of the step function $[f(x)]^2$. Thus,

$$\bar{y} = \frac{\frac{1}{2}\rho \int_a^b [f(x)]^2\,dx}{\rho \int_a^b f(x)\,dx} = \frac{\frac{1}{2}\int_a^b [f(x)]^2\,dx}{\int_a^b f(x)\,dx}.$$

We have derived the formulas for $\bar{x}$ and $\bar{y}$ for the case in which $f(x)$ is a step function; however, they make sense as long as $f(x)$, $xf(x)$, and $[f(x)]^2$ are integrable on $[a,b]$. As usual, we carry over the same formula to general f, so formulas (5) are derived.

Exercises for Section 9.4

1. Redo Example 1 by choosing B_1 to consist of m_1 alone and B_2 to consist of m_2 and m_3.
2. Assuming formula (2) and the consolidation principle, derive formula (3) for the case of four masses by dividing the masses into two groups of two masses each.
3. Using formulas (2) and (3) for two and three masses, and the consolidation principle, derive formula (3) for four masses.
4. Assume that you have derived formula (3) from formula (2) and the consolidation principle for n masses. Now derive formula (3) for $n+1$ masses.
5. Masses of 1, 3, 5, and 7 units are located at the points 7, 3, 5, and 1, respectively, on the x axis. Where is the center of mass?
6. Masses of 2, 4, 6, 8, and 10 units are located at the points $x_1 = 0$, $x_2 = 1$, $x_3 = 3$, $x_4 = -1$, and $x_5 = -2$ on the x axis. Locate the center of mass.
7. For each integer i from 1 to 100, a point of mass i is located at the point $x = i$. Where is the center of mass? (See Exercise 41(a), Section 4.1.)
8. Suppose that n equal masses are located at the points $1, 2, 3, \ldots, n$ on a line. Where is their center of mass?

In Exercises 9–12, find the center of mass for the given arrangement of masses.

9. 10 grams at $(1,0)$ and 20 grams at $(1,2)$.
10. 15 grams at $(-3,2)$ and 30 grams at $(4,2)$.
11. 5 grams at $(1,1)$, 8 grams at $(3,2)$, and 10 grams at $(0,0)$.
12. 2 grams at $(4,2)$, 3 grams at $(3,2)$, and 4 grams at $(5,3)$.

13. (a) Equal masses are placed at the vertices of an equilateral triangle whose base is the segment from $(0,0)$ to $(1,0)$. Where is the center of mass?
(b) The mass at $(0,0)$ is doubled. Where is the center of mass now?
14. Masses of 2, 3, 4, and 5 kilograms are placed at the points $(1,2)$, $(1,4)$, $(3,5)$, and $(2,6)$, respectively. Where should a mass of 1 kilogram be placed so that the configuration of five masses has its center of mass at the origin?
15. Verify the consolidation principle for the situation in which four masses in the plane are divided into two groups containing one mass and three masses each. (Assume that formula (5) holds for $n = 3$.)
16. Equal masses are placed at the points (x_1, y_1), (x_2, y_2), and (x_3, y_3). Show that their center of mass is at the intersection point of the medians of the triangle at whose vertices the masses are located.

Find the center of mass of the regions in Exercises 17–22.

17. The region under the graph of $4/x^2$ on $[1,3]$.
18. The region under the graph of $1 + x^2 + x^4$ on $[-1,1]$.
19. The region under the graph of $\sqrt{1-x^2}$ on $[0,1]$.
20. The region under the graph of $\sqrt{1 - x^2/a^2}$ on $[-a,a]$.
21. The triangle with vertices at $(0,0)$, $(0,2)$, and $(4,0)$.
22. The triangle with vertices at $(1,0)$, $(4,0)$, and $(2,3)$.

23. If, in formula (3), we have $a \leqslant x_i \leqslant b$ for all x_i, show that $a \leqslant \bar{x} \leqslant b$ as well. Interpret this statement geometrically.
24. Let a mass m_i be placed at position x_i on a line ($i = 1, \ldots, n$). Show that the function $f(x) = \sum_{i=1}^{n} m_i (x - x_i)^2$ is minimized when x is the center of mass of the n particles.

25. Suppose that masses m_i are located at points x_i on the line and are moving with velocity $v_i = dx_i/dt$ $(i = 1, \ldots, n)$. The *total momentum* of the particles is defined to be $P = m_1v_1 + m_2v_2 + \cdots + m_nv_n$. Show that $P = Mv$, where M is the total mass and v is the velocity of the center of mass (i.e., the rate of change of the position of the center of mass with respect to time).
26. A mass m_i is at position $x_i = f_i(t)$ at time t. Show that if the force on m_i is $F_i(t)$, and $F_1(t) + F_2(t) = 0$, then the center of mass of m_1 and m_2 moves with constant velocity.
27. From a disk of radius 5, a circular hole with radius 2 and center 1 unit from the center of the disk is cut out. Sketch and find the center of mass of the resulting figure.
28. Suppose that $f(x) \leqslant g(x)$ for all x in $[a,b]$. Show that the center of mass of the region between the graphs of f and g on $[a,b]$ is located at $(\bar{x}, \bar{y})$, where

$$\bar{x} = \frac{\int_a^b x[g(x) - f(x)]\,dx}{\int_a^b [g(x) - f(x)]\,dx}, \quad \text{and}$$

$$\bar{y} = \frac{\frac{1}{2}\int_a^b [g(x) + f(x)][g(x) - f(x)]\,dx}{\int_a^b [g(x) - f(x)]\,dx}.$$

29. Find the center of mass of the region between the graphs of $\sin x$ and $\cos x$ on $[0, \pi/4]$. [*Hint:* Find the center of mass of each infinitesimal strip making up the region, or use Exercise 28.]
30. Find the center of mass of the region between the graphs of $-x^4$ and x^2 on $[-1, 1]$. (See the hint in Exercise 29.)
★31. Find the center of mass of the triangular region with vertices (x_1, y_1), (x_2, y_2), and (x_3, y_3). (For convenience, you may assume that $x_1 \leqslant x_2 \leqslant x_3$, $y_1 \leqslant y_3$, and $y_2 \leqslant y_3$.) Compare with Exercise 16.

9.5 Energy, Power, and Work

Energy is the integral of power over time, and work is the integral of force over distance.

Energy appears in various forms and can often be converted from one form into another. For instance, a solar cell converts the energy in light into electrical energy; a fusion reactor, in changing atomic structures, transforms nuclear energy into heat energy. Despite the variety of forms in which energy may appear, there is a common unit of measure for all these forms. In the MKS (meter-kilogram-second) system, it is the *joule*, which equals 1 kilogram meter^2 per second^2.

Energy is an "extensive" quantity. This means the following: the longer a generator runs, the more electrical energy it produces; the longer a light bulb burns, the more energy it consumes. The rate (with respect to time) at which some form of energy is produced or consumed is called the *power* output or input of the energy conversion device. Power is an *instantaneous* or "intensive" quantity. By the fundamental theorem of calculus, we can compute the total energy transformed between times a and b by integrating the power from a to b.

Power and Energy

Power is the rate of change of energy with respect to time:

$$P = \frac{dE}{dt}.$$

The total energy over a time period is the integral of power with respect to time:

$$E = \int_a^b P\,dt.$$

A common unit of measurement for power is the *watt*, which equals 1 joule per second. One *horsepower* is equal to 746 watts. The *kilowatt-hour* is a unit of energy equal to the energy obtained by using 1000 watts for 1 hour (3600 seconds)—that is, 3,600,000 joules.

Example 1 The power output (in watts) of a 60-cycle generator varies with time (measured in seconds) according to the formula $P = P_0\sin^2(120\pi t)$, where P_0 is the maximum power output. (a) What is the total energy output during an hour? (b) What is the average power output during an hour?

Solution (a) The energy output, in joules, is

$$E = \int_0^{3600} P_0 \sin^2(120\pi t)\, dt.$$

Using the formula $\sin^2\theta = (1 - \cos 2\theta)/2$, we find

$$E = \frac{1}{2} P_0 \int_0^{3600} (1 - \cos 240\pi t)\, dt = \frac{1}{2} P_0 \left[t - \frac{1}{240} \sin 240\pi t \right]\Bigg|_0^{3600}$$

$$= \frac{1}{2} P_0 \big[3600 - 0 - (0 - 0) \big] = 1800 P_0 .$$

(b) The average power output is the energy output divided by the time (see Section 9.3), or $1800\ P_0/3600 = \frac{1}{2} P_0$; in this case, half the maximum power output. ▲

A common form of energy is mechanical energy—the energy stored in the movement of a massive object (*kinetic energy*) or the energy stored in an object by virtue of its position (*potential energy*). The latter is illustrated by the energy we can extract from water stored above a hydroelectric power plant.

We accept the following principles from physics:

1. The kinetic energy of a mass m moving with velocity v is $\frac{1}{2} mv^2$.
2. The (gravitational) potential energy of a mass m at a height h is mgh (here g is the gravitational acceleration; $g = 9.8$ meters/(second)^2 = 32 feet/(second)^2.

The total force on a moving object is equal to the product of the mass m and the acceleration $dv/dt = d^2x/dt^2$. The unit of force is the *newton* which is 1 kilogram meter per second^2. If the force depends upon the position of the object, we may calculate the variation of the kinetic energy $K = \frac{1}{2} mv^2$ with position. We have

$$\frac{dK}{dx} = \frac{dK/dt}{dx/dt} = \frac{(d/dt)(\frac{1}{2} mv^2)}{v} = \frac{mv\, dv/dt}{v} = m\frac{dv}{dt} = F.$$

Applying the fundamental theorem of calculus, we find that the change ΔK of kinetic energy as the particle moves from a to b is $\int_a^b F\, dx$. Often we can divide the total force on an object into parts arising from identifiable sources (gravity, friction, fluid pressure). We are led to define the *work* W done by a particular force F on a moving object (even if there are other forces present) as $W = \int_a^b F\, dx$. Note that if the force F is constant, then the work done is simply the product of F with the *displacement* $\Delta x = b - a$. Accordingly, 1 joule equals 1 newton-meter.

Force and Work

The work done by a force on a moving object is the integral of the force with respect to position:

$$W = \int_a^b F\,dx.$$

If the force is constant,

$$\text{Work} = \text{Force} \times \text{Displacement}.$$

If the total force F is a sum $F_1 + \cdots + F_n$, then we have

$$\Delta K = \int_a^b (F_1 + \cdots + F_n)\,dx = \int_a^b F_1\,dx + \cdots + \int_a^b F_n\,dx.$$

Thus the total change in kinetic energy is equal to the sum of the works done by the individual forces.

Example 2 The acceleration of gravity near the earth is $g = 9.8$ meters/(second)^2. How much work does a weight-lifter do in raising a 50-kilogram barbell to a height of 2 meters? (See Figure 9.5.1.)

Figure 9.5.1. How much work did the weight-lifter do?

Solution We let x denote the height of the barbell above the ground. Before and after the lifts, the barbell is stationary, so the net change in kinetic energy is zero. The work done by the weight-lifter must be the *negative* of the work done by gravity. Since the pull of gravity is downward, its force is -9.8 meters per second^2 $\times$ 50 kilograms $= -490$ kilograms $\cdot$ meters per second^2 $= -490$ newtons; $\Delta x = 2$ meters, so the work done by gravity is -980 kilograms $\cdot$ meters^2 per second^2 $= -980$ joules. Thus the work done by the weight-lifter is 980 joules. (If the lift takes s seconds, the average power output is $(980/s)$ watts.) ▲

Example 3 Show that the *power* exerted by a force F on a moving object is Fv, where v is the velocity of the object.

Solution Let E be the energy content. By our formula for work, we have $\Delta E = \int_a^b F\,dx$, so $dE/dx = F$. To compute the power, which is the *time* derivative of E, we use the chain rule:

$$P = \frac{dE}{dt} = \frac{dE}{dx} \cdot \frac{dx}{dt} = Fv.$$

(In pushing a child on a swing, it is most effective to exert your force at the bottom of the swing, when the velocity is greatest.) ▲

Example 4 A pump is to empty the conical tank of water shown in Fig. 9.5.2. How much energy (in joules) is required for the job? (A cubic meter of water has mass 10^3 kilograms.)

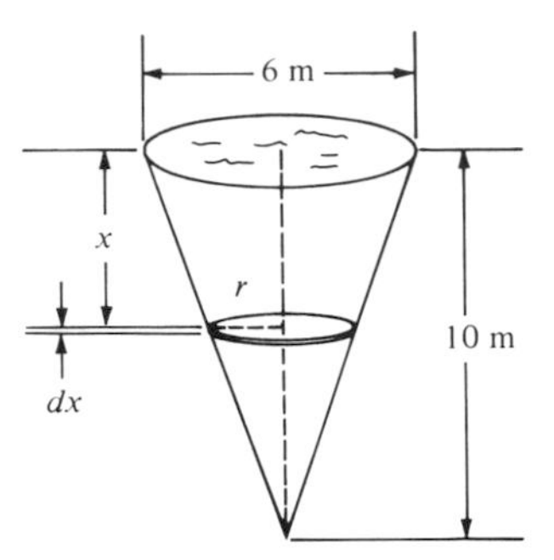

Figure 9.5.2. To calculate the energy needed to empty the tank, we add up the energy needed to remove slabs of thickness dx.

Solution Consider a layer of thickness dx at depth x, as shown in Fig. 9.5.2. By similar triangles, the radius is $r = \frac{3}{10}(10 - x)$, so the volume of the layer is given by $\pi \cdot \frac{9}{100}(10 - x)^2\,dx$ and its mass is $10^3 \cdot \pi \cdot \frac{9}{100}(10 - x)^2\,dx = 90\pi(10 - x)^2\,dx$. To lift this layer x meters to the top of the tank takes $90\pi(10 - x)^2\,dx \cdot g \cdot x$ joules of work, where $g = 9.8$ meters per second^2 is the acceleration due to gravity (see Example 2). Thus, the total work done in emptying the tank is

$$
\begin{aligned}
90g\pi\int_0^{10}(10 - x)^2 x\,dx &= 90g\pi\left[100\frac{x^2}{2} - 20\frac{x^3}{3} + \frac{x^4}{4}\right]\Bigg|_0^{10} \\
&= 90g\pi(10^4)\left[\frac{1}{2} - \frac{2}{3} + \frac{1}{4}\right] \\
&= (90)(9.8)(\pi)(10^4)\left(\frac{1}{12}\right) \\
&\approx 2.3 \times 10^6 \text{ joules. } \blacktriangle
\end{aligned}
$$

Example 5 The pump which is emptying the conical tank in Example 4 has a power output of 10^5 joules per hour (i.e., 27.77 watts). What is the water level at the end of 6 minutes of pumping? How fast is the water level dropping at this time?

Solution The total energy required to pump out the top h meters of water is

$$
\begin{aligned}
90g\pi\int_0^{h}(10 - x)^2 x\,dx &= 90g\pi\left(100\frac{h^2}{2} - 20\frac{h^3}{3} + \frac{h^4}{4}\right) \\
&\approx 2770h^2\left(50 - \frac{20}{3}h + \frac{h^2}{4}\right).
\end{aligned}
$$

At the end of 6 minutes ($\frac{1}{10}$ hour), the pump has produced 10^4 joules of energy, so the water level is h meters from the top, where h is the solution of

$$2770h^2\left(50 - \frac{20}{3}h + \frac{h^2}{4}\right) = 10^4.$$

Solving this numerically by the method of bisection (see Section 3.1) gives $h \approx 0.27$ meter.

At the end of t hours, the total energy output is $10^5 t$ joules, so

$$90g\pi\int_0^{h}(10 - x)^2 x\,dx = 10^5 t,$$

where h is the amount pumped out at time t. Differentiating both sides with respect to t gives

$$90g\pi(10-h)^2h\frac{dh}{dt}=10^5, \quad \text{so} \quad \frac{dh}{dt}=\frac{10^4}{9g\pi(10-h)^2h}.$$

when $h=0.27$, this is 1.41 meters per hour. ▲

Supplement to Section 9.5: Integrating Sunshine

We will now apply the theory and practice of integration to compute the total amount of sunshine received during a day, as a function of latitude and time of year. If we have a horizontal square meter of surface, then the rate at which solar energy is received by this surface—that is, the *intensity* of the solar radiation—is proportional to the sine of the angle A of elevation of the sun above the horizon.[6] Thus the intensity is highest when the sun is directly overhead ($A=\pi/2$) and reduces to zero at sunrise and sunset.

The total energy received on day T must therefore be the product of a constant (which can be determined only by experiment, and which we will ignore) and the integral $E=\int_{t_0(T)}^{t_1(T)}\sin A\,dt$, where t is the time of day (measured in hours from noon) and $t_0(T)$ and $t_1(T)$ are the times of sunrise and sunset on day T. (When the sun is below the horizon, although $\sin A$ is negative, the solar intensity is simply zero.)

We presented a formula for $\sin A$ (formula (1) in the Supplement to Chapter 5, to be derived in the Supplement to Chapter 14), and used it to determine the time of sunset (formula (3) in the Supplement to Chapter 5). The time of sunrise is the negative of the time of sunset, so we have[7]

$$E=\int_{-S}^{S}\sin A\,dt, \tag{1}$$

where

$$\sin A=\cos l\sqrt{1-\sin^2\alpha\cos^2\left(\frac{2\pi T}{365}\right)}\cos\left(\frac{2\pi t}{24}\right)+\sin l\sin\alpha\cos\left(\frac{2\pi T}{365}\right)$$

and

$$S=\frac{24}{2\pi}\cos^{-1}\left[-\tan l\frac{\sin\alpha\cos(2\pi T/365)}{\sqrt{1-\sin^2\alpha\cos^2(2\pi T/365)}}\right]. \tag{2}$$

Here $\alpha\approx 23.5°$ is the inclination of the earth's axis from the perpendicular to the plane of the earth's orbit; l is the latitude of the point where the sunshine is being measured.

The integration will be simpler than you may expect. First of all, we simplify notation by writing k for the expression $\sin\alpha\cos(2\pi T/365)$, which appears so often. Then we have

$$\begin{aligned}E&=\int_{-S}^{S}\left[\cos l\sqrt{1-k^2}\cos\left(\frac{2\pi t}{24}\right)+(\sin l)k\right]dt\\&=\cos l\sqrt{1-k^2}\int_{-S}^{S}\cos\left(\frac{2\pi t}{24}\right)dt+(\sin l)k\int_{-S}^{S}dt.\end{aligned}$$

[6] We will justify this assertion in the Supplement to Chapter 14. We also note that, strictly speaking, it applies only if we neglect absorption by the atmosphere or assume that our surface is at the top of the atmosphere.

[7] All these calculations assume that there is a sunrise and sunset. In the polar regions during the summer, the calculations must be altered (see Exercise 5 below).

Integration gives

$$E = \cos l\sqrt{1-k^2}\left(\frac{24}{2\pi}\sin\frac{2\pi t}{24}\Big|_{-S}^{S}\right) + 2Sk\sin l$$

$$= \cos l\sqrt{1-k^2}\left(\frac{24}{2\pi}\right)\left(\sin\frac{2\pi S}{24} - \frac{\sin 2\pi(-S)}{24}\right) + 2Sk\sin l$$

$$= \frac{24}{\pi}\cos l\sqrt{1-k^2}\sin\frac{2\pi S}{24} + 2Sk\sin l.$$

The expression $\sin(2\pi S/24)$ can be simplified. Using the formula

$$\cos(2\pi S/24) = -(\tan l)\left(k/\sqrt{1-k^2}\right),$$

we get

$$\sin\frac{2\pi S}{24} = \sqrt{1-\cos^2\frac{2\pi S}{24}} = \sqrt{1-\frac{(\tan^2 l)k^2}{1-k^2}}$$

$$= \sqrt{\frac{1-k^2-(\tan^2 l)k^2}{1-k^2}}$$

$$= \sqrt{\frac{1-(1+\tan^2 l)k^2}{1-k^2}} = \sqrt{\frac{1-(\sec^2 l)k^2}{1-k^2}},$$

and so, finally, we get,

$$E = \frac{24}{\pi}\cos l\sqrt{1-(\sec^2 l)k^2} + \frac{24}{\pi}k\sin l\cos^{-1}\left[-\frac{(\tan l)k}{\sqrt{1-k^2}}\right].$$

Since both k and $\sqrt{1-k^2}$ appear, we can do even better by writing $k = \sin D$ (the number D is important in astronomy—it is called the *declination*), and we get

$$E = \frac{24}{\pi}\left[\cos l\sqrt{1-\sec^2 l\sin^2 D} + \sin l\sin D\cos^{-1}(-\tan l\tan D)\right].$$

Since we have already ignored a constant factor in E, we will also ignore the factor $24/\pi$. Incorporating $\cos l$ into the square root, we obtain as our final result

$$E = \sqrt{\cos^2 l - \sin^2 D} + \sin l\sin D\cos^{-1}(-\tan l\tan D), \tag{3}$$

where $\sin D = \sin\alpha\cos(2\pi T/365)$.

Plotting E as a function of l for various values of T leads to graphs like those in Figs. 3.5.4 and 9.5.3.

Example When does the equator receive the most solar energy? The least?

Solution At the equator, $l = 0$, so we have

$$E = \sqrt{1-\sin^2 D} = \sqrt{1-\sin^2\alpha\cos^2\left(\frac{2\pi T}{365}\right)}.$$

We see by inspection that E is largest when $\cos^2(2\pi T/365) = 0$—that is, when $T/365 = \frac{1}{4}$ or $\frac{3}{4}$; that is, on the first days of fall and spring: on these days $E = 1$. We note that E is smallest on the first days of summer and winter, when $\cos^2(2\pi/365) = 1$ and we have $E = \sqrt{1-\sin^2\alpha} = \cos\alpha = \cos 23.5° = 0.917$, or about 92% of the maximum value. ▲

Using this example we can standardize units in which E can be measured. One unit of E is the total energy received on a square meter at the equator on the first day of spring. All other energies may be expressed in terms of this unit.

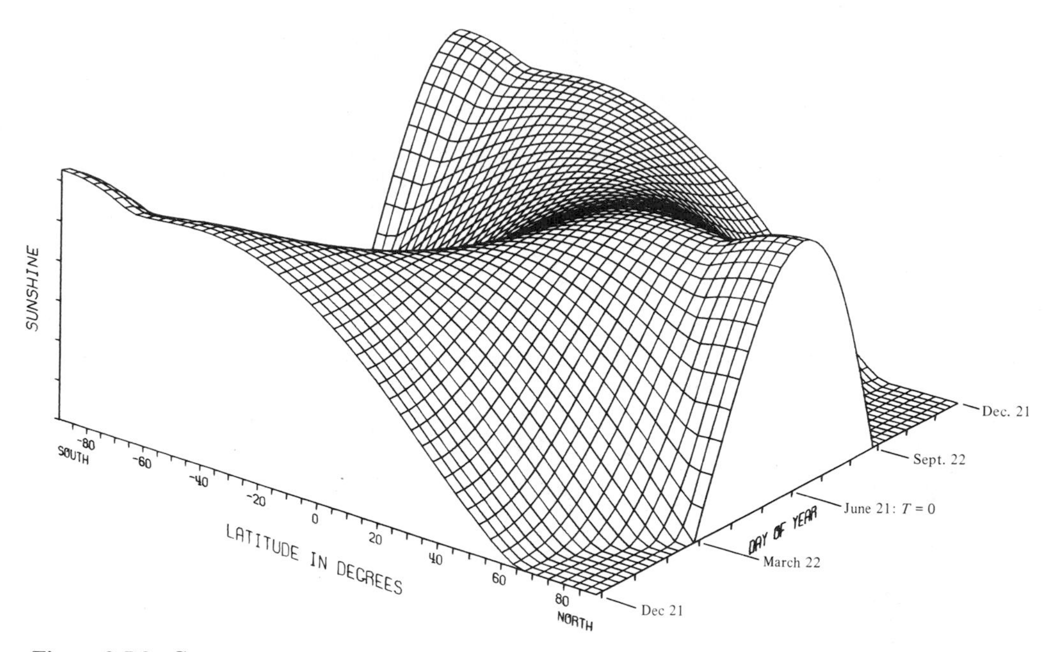

Figure 9.5.3. Computer-generated graph of the daily sunshine intensity on the earth as a function of day of the year and latitude.

Exercises for the Supplement to Section 9.5

1. Compare the solar energy received on June 21 at the Arctic Circle ($l = 90° - \alpha$) with that received at the equator.
2. What would the inclination of the earth need to be in order for E on June 21 to have the same value at the equator as at the latitude $90° - \alpha$?
3. (a) Express the total solar energy received over a whole year at latitude l by using summation notation. (b) Write down an integral which is approximately equal to this sum. Can you evaluate it?
4. Simplify the integral in the solution of Exercise 3(b) for the cases $l = 0$ (equator) and $l = 90° - \alpha$ (Arctic Circle). In each case, one of the two terms in the integrand can be integrated explicitly: find the integral of this term.
5. Find the total solar energy received at a latitude in the polar region on a day on which the sun never sets.
6. How do you think the climate of the earth would be affected if the inclination α were to become: (a) 10°? (b) 40°? (In each case, discuss whether the North Pole receives more or less energy during the year than the equator—see Exercise 5.)
7. Consider equation (3) for E. For $D = \pi/8$, compute dE/dl at $l = \pi/4$. Is your answer consistent with the graph in Fig. 9.5.3 (look in the plane of constant T)?
8. Determine whether a square meter at the equator or at the North Pole receives more solar energy: (a) during the month of February, (b) during the month of April, (c) during the entire year.

Exercises for Section 9.5

1. The power output (in watts) of a 60-cycle generator is $P = 1050\sin^2(120\pi t)$, where t is measured in seconds. What is the total energy output in an hour?
2. A worker, gradually becoming tired, has a power output of $30e^{-2t}$ watts for $0 \leqslant t \leqslant 360$, where t is the time in seconds from the start of a job. How much energy is expended during the job?
3. An electric motor is operating with power $15 + 2\sin(t\pi/24)$ watts, where t is the time in *hours* measured from midnight. How much energy is consumed in one day's operation?
4. The power output of a solar cell is $25\sin(\pi t/12)$ watts, where t is the time in *hours* after 6 A.M. How many joules of energy are produced between 6 A.M. and 6 P.M?

In Exercises 5–8, compute the work done by the given force acting over the given interval.

5. $F = 3x$; $0 \leqslant x \leqslant 1$.
6. $F = k/x^2$; $1 \leqslant x \leqslant 6$ (k a constant).
7. $F = 1/(4 + x^2)$; $0 \leqslant x \leqslant 1$.
8. $F = \sin^3 x \cos^2 x$; $0 \leqslant x \leqslant 2$. [*Hint:* Write $\sin^3 x = \sin x(1 - \cos^2 x)$.]

9. How much power must be applied to raise an object of mass 1000 grams at a rate of 10 meters per second (at the Earth's surface)?
10. The gravitational force on an object at a distance r from the center of the earth is k/r^2, where k is a constant. How much work is required to move the object:
    (a) From $r = 1$ to $r = 10$?
    (b) From $r = 1$ to $r = 1000$?
    (c) From $r = 1$ to $r = 10{,}000$?
    (d) From $r = 1$ to "$r = \infty$"?
11. A particle with mass 1000 grams has position $x = 3t^2 + 4$ meters at time t seconds. (a) What is the kinetic energy at time t? (b) What is the rate at which power is being supplied to the object at time $t = 10$?
12. A particle of mass 20 grams is at rest at $t = 0$, and power is applied at the rate of 10 joules per second. (a) What is the energy at time t? (b) If all the energy is kinetic energy, what is the velocity at time t? (c) How far has the particle moved at the end of t seconds? (d) What is the force on the particle at time t?
13. A force $F(x) = -3x$ newtons acts on a particle between positions $x = 1$ and $x = 0$. What is the change in kinetic energy of the particle between these positions?
14. A force $F(x) = 3x\sin(\pi x/2)$ newtons acts on a particle between positions $x = 0$ and $x = 2$. What is the increase in kinetic energy of the particle between these positions?
15. (a) The power output of an electric generator is $25\cos^2(120\pi t)$ joules per second. How much energy is produced in 1 hour? (b) The output of the generator in part (a) is converted, with 80% efficiency, into the horizontal motion of a 250-gram object. How fast is the object moving at the end of 1 minute?
16. The generator in Exercise 15 is used to lift a 500-kilogram weight and the energy is converted via pulleys with 75% efficiency. How high can it lift the weight in an hour?

Exercises 17–20 refer to Figure 9.5.4.

17. How much energy is required to pump all the water out of the swimming pool?
18. Suppose that a mass equal to that of the water in the pool were moving with kinetic energy equal to the result of Exercise 17. What would its velocity be?
19. Repeat Exercise 17 assuming that the pool is filled with a liquid three times as dense as water.
20. Repeat Exercise 18 assuming that the pool is filled with a liquid three times as dense as water.

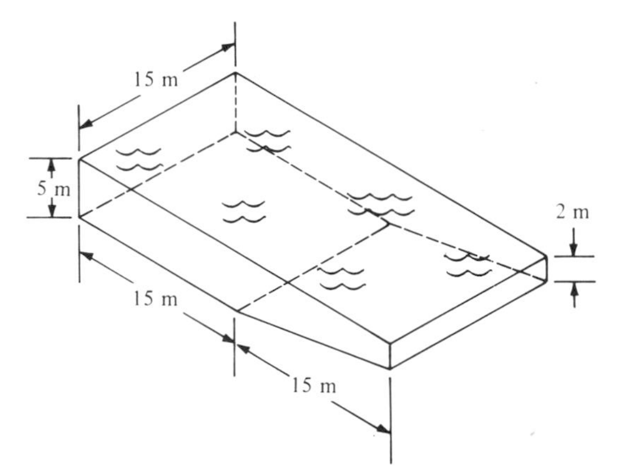

Figure 9.5.4. How much energy is required to empty this pool of water?

21. Suppose that a spring has a natural length of 10 centimeters, and that a force of 3 newtons is required to stretch it to 15 centimeters. How much work is needed to compress the spring to 5 centimeters?
22. If all the energy in the compressed spring in Exercise 21 is used to fire a ball weighing 20 grams, how fast will the ball travel?
23. How much work is required to fill the tank in Figure 9.5.5 with water from ground level?
24. A solid concrete monument is built in the pyramid shape of Fig. 9.5.6. Assume that the concrete weighs 260 pounds per cubic foot. How much work is done in erecting the monument? ($g = 32$ feet per second^2; express your answer in units of foot-pounds^2 per second^2.)

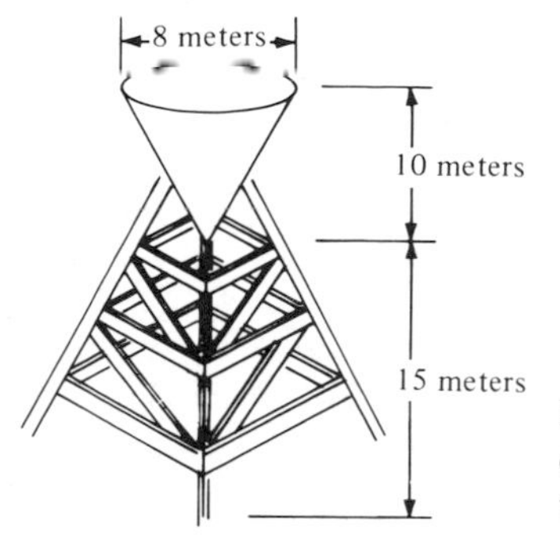

Figure 9.5.5. How much energy is needed to fill this tank?

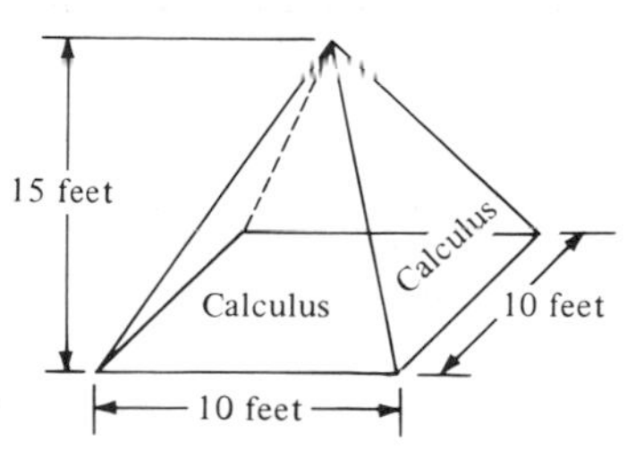

Figure 9.5.6. How much energy is needed to erect this monument?

Review Exercises for Chapter 9

In Exercises 1–4, find the volume of the solid obtained by rotating the region under the given graph about (a) the x axis and (b) the y axis.

1. $y = \sin x,\ 0 \leqslant x \leqslant \pi$
2. $y = 3\sin 2x,\ 0 \leqslant x \leqslant \pi/4$
3. $y = e^x,\ 0 \leqslant x \leqslant \ln 2$
4. $y = 5e^{2x},\ 0 \leqslant x \leqslant \ln 4$

5. A cylindrical hole of radius $\frac{1}{3}$ is drilled through the center of a ball of radius 1. What is the volume of the resulting solid?
6. A wedge is cut in a tree of radius 1 meter by making two cuts to the center, one horizontally, and one at an angle of 20° to the first. Find the volume of the wedge.
7. Find the volume of the "football" whose dimensions are shown in Fig. 9.R.1. The two arcs in the figure are segments of parabolas.

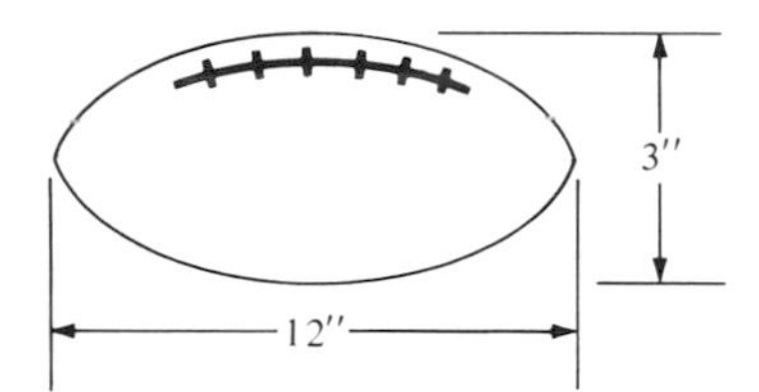

Figure 9.R.1. Find the volume of the football.

8. Imagine the "football" in Fig. 9.R.1, formed by revolving a parabola, to be solid. A hole with radius 1 inch is drilled along the axis of symmetry. How much material is removed?

In Exercises 9–12, find the average value of each function on the stated interval.

9. $1 + t^3,\ 0 \leqslant t \leqslant 1$
10. $t\sin(t^2),\ \pi \leqslant t \leqslant 3\pi/2$
11. $xe^x,\ 0 \leqslant x \leqslant 1$
12. $\dfrac{1}{1+x^2},\ 1 \leqslant x \leqslant 3$

13. If $\int_0^2 f(x)\,dx = 4$, what is the average value of $g(x) = 3f(x)$ on $[0,2]$?
14. If $f(x) = kg(cx)$ on $[a,b]$, how is the average of f on $[a,b]$ related to that of g on $[ac, bc]$?
15. Show that for some x in $[0,\pi]$, $\dfrac{\pi}{2+\cos x}$ is equal to $\displaystyle\int_0^{\pi} \frac{d\theta}{2+\cos\theta}$.
16. (a) Prove that
$$1/\sqrt{2} \leqslant \int_0^1 \left[dt/\sqrt{t^3+1}\,\right] \leqslant 1.$$
(b) Prove that $\int_0^1 [dt/\sqrt{t^3+1}\,] = \sin\theta$ for some θ, $\pi/4 \leqslant \theta \leqslant \pi/2$.

In Exercises 17–22, let μ be the average value of f on $[a,b]$. Then the average value of $[f(x)-\mu]^2$ on $[a,b]$ is called the *variance* of f on $[a,b]$, and the square root of the variance is called the *standard deviation* of f on $[a,b]$ and is denoted σ. Find the average value, variance, and standard deviation of each of the following functions on the interval specified.

17. x^2 on $[0,1]$
18. $3 + x^2$ on $[0,1]$
19. xe^x on $[0,1]$
20. $\sin 2x$ on $[0, 4\pi]$
21. $f(x) = 1$ on $[0,1]$ and 2 on $(1,2]$.
22. $f(x) = \begin{cases} 2 & \text{on } [0,1] \\ 3 & \text{on } (1,2] \\ 1 & \text{on } (2,3] \\ 5 & \text{on } (3,4] \end{cases}$

23. Let the region under the graph of a positive function $f(x)$, $a \leqslant x \leqslant b$, be revolved about the x axis to form a solid S. Suppose this solid has a mass density of $\rho(x)$ grams per cubic centimeter at a distance x along the x axis. (a) Find a formula for the mass of S. (b) If $f(x) = x^2$, $a = 0$ and $b = 1$, and $\rho(x) = (1 + x^4)$, find the mass of S.
24. A rod has linear mass density $\mu(x)$ grams per centimeter at the point x along its length. If the rod extends from $x = a$ to $x = b$, find a formula for the location of the rod's center of mass.

In Exercises 25–28, find the center of mass of the region under the given graph on the given interval.

25. $y = x^4$ on $[0,2]$
26. $y = x^3 + 2$ on $[0,1]$
27. $y = \ln(1+x)$ on $[0,1]$
28. $y = e^x$ on $[1,2]$

29. Find the center of mass of the region between the graphs of $y = x^3$ and $y = -x^2$ between $x = 0$ and $x = 1$ (see Exercise 28, Section 9.4).
30. Find the center of mass of the region composed of the region under the graph $y = \sin x$, $0 \leqslant x \leqslant \pi$, and the circle with center at $(5,0)$ and radius 1.
31. Over a time period $0 \leqslant t \leqslant 6$ (t measured in minutes), an engine is consuming power at a rate of $20 + 5te^{-t}$ watts. What is (a) the total energy consumed? (b) The average power used?
32. Water is being pumped from a deep, irregularly shaped well at a constant rate of $3\frac{1}{2}$ cubic meters per hour. At a certain instant, it is observed that the water level is dropping at a rate of 1.2 meters per hour. What is the cross-sectional area of the well at that depth?
33. A force $F(x) = 30\sin(\pi x/4)$ newtons acts on a particle between positions $x = 2$ and $x = 4$. What is the increase in kinetic energy (in joules) between these positions?
34. The engine in Fig. 9.R.2 is using energy at a rate of 300 joules per second to lift the weight of 600 kilograms. If the engine operates at 60% efficiency, at what speed (meters per second) can it raise the weight?

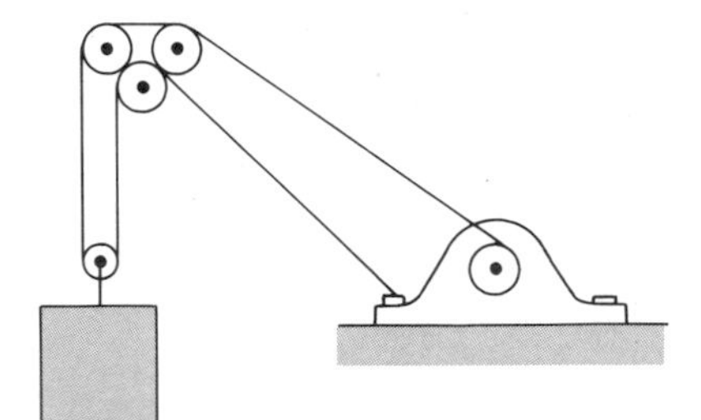

Figure 9.R.2. The engine for Problem 34.

35. Find a formula for the work required to empty a tank of water which is a solid of revolution about a vertical axis of symmetry.
36. How much work is required to empty the tank shown in Fig. 9.R.3? [*Hint:* Use the result of Exercise 35.]

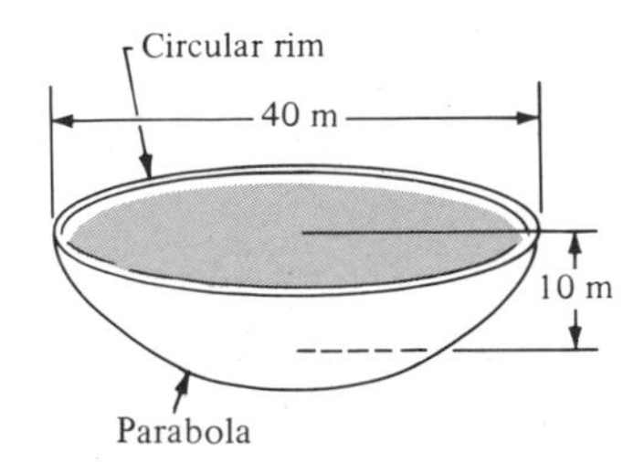

Figure 9.R.3. How much energy is needed to empty the tank?

37. The pressure (force per unit area) at a depth h below the surface of a body of water is given by $p = \rho g h = 9800h$, measured in newtons per square meter. (This formula derives from the fact that the force needed to support a column of water of cross-sectional area A is (volume) $\times$ (density) $\times$ (g) $= Ah\rho g$, so the force per unit area is $\rho g h$, where $\rho = 10^3$ kilograms per cubic meter, and $g = 9.8$ meters per second per second).
    (a) For the dam shown in Fig. 9.R.4(a), show that the total force exerted on it by the water is $F = \frac{1}{2}\int_a^b \rho g[f(x)]^2\,dx$. [*Hint:* First calculate the force exerted on a vertical rectangular slab.]
    (b) Make up a geometric theorem relating F to the volume of a certain solid.
    (c) Find the total force exerted on the dam whose face is shown in Fig. 9.R.4(b).

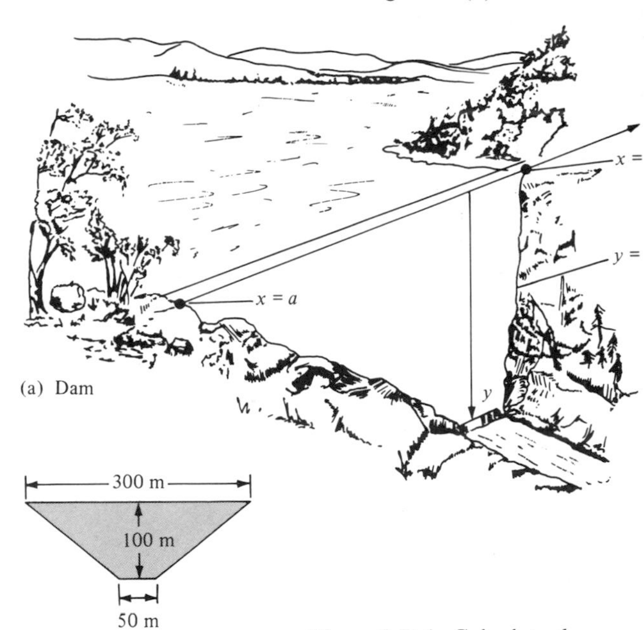

Figure 9.R.4. Calculate the force on the dam.

38. (a) *Pappus' theorem for volumes.* Use the shell method to show that if a region R in the xy plane is revolved around the y axis, the volume of the resulting solid equals the area of R times the circumference of the circle obtained by revolving the center of mass of R around the y axis.
    (b) Use Pappus' theorem to do Exercise 21(a) in Section 9.2.
    (c) Assuming the formula $V = \frac{4}{3}\pi r^3$ for the volume of a ball, use Pappus's theorem to find the center of mass of the semicircular region $x^2 + y^2 \leqslant r^2$, $x \geqslant 0$.

★39. See the instructions for Exercises 17–22.
    (a) Suppose that $f(t)$ is a step function on $[a,b]$, with value k_i on the interval (t_{i-1}, t_i) belong-

ing to a partition $(t_0, t_1, \ldots, t_n)$. Find a formula for the standard deviation of f on $[a, b]$.

(b) Simplify your formula in part (a) for the case when all the Δt_i's are equal.

(c) Show that if the standard deviation of a step function is zero, then the function has the same value on all the intervals of the partition; i.e., the function is constant.

(d) Give a definition for the standard deviation of a list $a_1, \ldots, a_n$ of numbers.

(e) What can you say about a list of numbers if its standard deviation is zero?

★40. (a) Prove, by analogy with the mean value theorem for integrals, the *second mean value theorem*: If f and g are continuous on $[a, b]$ and $g(x) \geqslant 0$, for x in $[a, b]$, then there is a point t_0 in $[a, b]$ such that

$$\int_a^b f(t)g(t)\,dt = f(t_0)\int_a^b g(t)\,dt.$$

(b) Show that the mean value theorem for integrals is a special case of the result in part (a).

(c) Show by example that the conclusion of part (a) is false without the assumption that $g(t) \geqslant 0$.

★41. Show that if f is an increasing continuous function on $[a, b]$, the mean value theorem for integrals implies the conclusion of the intermediate value theorem.

★42. Show that in the context of Exercise 35, the work needed to empty the tank equals Mgh, where M is the total mass of water in the tank and h is the distance of the center of mass of the tank below the top of the tank.

★43. Let $f(x) > 0$ for x in $[a, b]$. Find a relation between the average value of the logarithmic derivative $f'(x)/f(x)$ of f on $[a, b]$ and the values of f at the endpoints.

Chapter 10

Further Techniques and Applications of Integration

Some simple geometric problems require advanced methods of integration.

Besides the basic methods of integration associated with reversing the differentiation rules, there are special methods for integrands of particular forms. Using these methods, we can solve some interesting length and area problems.

10.1 Trigonometric Integrals

The key to evaluating many integrals is a trigonometric identity or substitution.

The integrals treated in this section fall into two groups. First, there are some purely trigonometric integrals that can be evaluated using trigonometric identities. Second, there are integrals involving quadratic functions and their square roots which can be evaluated using trigonometric substitutions.

We begin by considering integrals of the form

$$\int \sin^m x \cos^n x \, dx,$$

where m and n are integers. The case $n = 1$ is easy, for if we let $u = \sin x$, we find

$$\int \sin^m x \cos x \, dx = \int u^m \, du = \frac{u^{m+1}}{m+1} + C = \frac{\sin^{m+1}(x)}{m+1} + C$$

(or $\ln|\sin x| + C$, if $m = -1$). The case $m = 1$ is similar:

$$\int \sin x \cos^n x \, dx = -\frac{\cos^{n+1}(x)}{n+1} + C$$

(or $-\ln|\cos x| + C$, if $n = -1$). If either m or n is odd, we can use the identity $\sin^2 x + \cos^2 x = 1$ to reduce the integral to one of the types just treated.

Example 1 Evaluate $\int \sin^2 x \cos^3 x \, dx$.

Solution $\int \sin^2 x \cos^3 x \, dx = \int \sin^2 x \cos^2 x \cos x \, dx = \int (\sin^2 x)(1 - \sin^2 x)\cos x \, dx$, which can be integrated by the substitution $u = \sin x$. We get

$$\int u^2(1 - u^2) \, du = \frac{u^3}{3} - \frac{u^5}{5} + C = \frac{1}{3}\sin^3 x - \frac{1}{5}\sin^5 x + C. \blacktriangle$$

If $m = 2k$ and $n = 2l$ are both even, we can use the half-angle formulas $\sin^2 x = (1 - \cos 2x)/2$ and $\cos^2 x = (1 + \cos 2x)/2$ to write

$$\int \sin^{2k}x \cos^{2l}x\,dx = \int \left(\frac{1 - \cos 2x}{2}\right)^k \left(\frac{1 + \cos 2x}{2}\right)^l dx$$

$$= \frac{1}{2}\int \left(\frac{1 - \cos y}{2}\right)^k \left(\frac{1 + \cos y}{2}\right)^l dy,$$

where $y = 2x$. Multiplying this out, we are faced with a sum of integrals of the form $\int \cos^m y\,dy$, with m ranging from zero to $k + l$. The integrals for odd m can be handled by the previous method; to those with even m we apply the half-angle formula once again. The whole process is repeated as often as necessary until everything is integrated.

Example 2 Evaluate $\int \sin^2 x \cos^2 x\,dx$.

Solution

$$\int \sin^2 x \cos^2 x\,dx = \int \left(\frac{1 - \cos 2x}{2}\right)\left(\frac{1 + \cos 2x}{2}\right) dx$$

$$= \frac{1}{4}\int (1 - \cos^2 2x)\,dx = \frac{x}{4} - \frac{1}{4}\int \cos^2 2x\,dx$$

$$= \frac{x}{4} - \frac{1}{4}\int \frac{1 + \cos 4x}{2}\,dx = \frac{x}{4} - \frac{x}{8} - \frac{1}{8}\int \cos 4x\,dx$$

$$= \frac{x}{8} - \frac{\sin 4x}{32} + C. \blacktriangle$$

Trigonometric Integrals

To evaluate $\int \sin^m x \cos^n x\,dx$:

1. If m is odd, write $m = 2k + 1$, and

$$\int \sin^m x \cos^n x\,dx = \int \sin^{2k}x \cos^n x \sin x\,dx$$

$$= \int (1 - \cos^2 x)^k \cos^n x \sin x\,dx.$$

Now integrate by substituting $u = \cos x$.

2. If n is odd, write $n = 2l + 1$, and

$$\int \sin^m x \cos^n x\,dx = \int \sin^m x \cos^{2l}x \cos x\,dx$$

$$= \int \sin^m x (1 - \sin^2 x)^l \cos x\,dx.$$

Now integrate by substituting $u = \sin x$.

3. (a) If m and n are even, write $m = 2k$ and $n = 2l$ and

$$\int \sin^{2k}x \cos^{2l}x\,dx = \int \left(\frac{1 - \cos 2x}{2}\right)^k \left(\frac{1 + \cos 2x}{2}\right)^l dx.$$

Substitute $y = 2x$. Expand and apply step 2 to the odd powers of $\cos y$.

(b) Apply step 3(a) to the even powers of $\cos y$ and continue until the integration is completed.

Example 3 Evaluate: (a) $\int_0^{2\pi}\sin^4x\cos^2x\,dx$ (b) $\int(\sin^2x+\sin^3x\cos^2x)\,dx$. (c) $\int\tan^3\theta\sec^3\theta\,d\theta$.

Solution (a) Substitute $\sin^2x=(1-\cos 2x)/2$ and $\cos^2x=(1+\cos 2x)/2$ to get

$$\begin{aligned}\int\sin^4x\cos^2x\,dx &= \int\frac{(1-\cos 2x)^2}{4}\,\frac{(1+\cos 2x)}{2}\,dx\\ &= \frac{1}{8}\int(1-2\cos 2x+\cos^2 2x)(1+\cos 2x)\,dx\\ &= \frac{1}{8}\int(1-\cos 2x-\cos^2 2x+\cos^3 2x)\,dx\\ &= \frac{1}{16}\int(1-\cos y-\cos^2y+\cos^3y)\,dy,\end{aligned}$$

where $y=2x$. Integrating the last two terms gives

$$\int\cos^2y\,dy=\int\frac{(1+\cos 2y)}{2}\,dy=\frac{y}{2}+\frac{\sin 2y}{4}+C$$

and

$$\int\cos^3y\,dy=\int(1-\sin^2y)\cos y\,dy=\sin y-\frac{\sin^3y}{3}+C.$$

Thus

$$\begin{aligned}\int\sin^4x\cos^2x\,dx &= \frac{1}{16}\left(y-\sin y-\frac{y}{2}-\frac{\sin 2y}{4}+\sin y-\frac{\sin^3y}{3}\right)+C\\ &= \frac{1}{16}\left(\frac{y}{2}-\frac{\sin 2y}{4}-\frac{\sin^3y}{3}\right)+C\\ &= \frac{1}{16}\left(x-\frac{\sin 4x}{4}-\frac{\sin^3 2x}{3}\right)+C,\end{aligned}$$

and so $\int_0^{2\pi}\sin^4x\cos^2x\,dx=\frac{\pi}{8}$.

(b)
$$\begin{aligned}\int(\sin^2x+\sin^3x\cos^2x)\,dx &= \int\sin^2x\,dx+\int\sin^3x\cos^2x\,dx\\ &= \int\left(\frac{1-\cos 2x}{2}\right)dx\\ &\quad+\int(1-\cos^2x)\cos^2x\sin x\,dx\\ &= \frac{x}{2}-\frac{\sin 2x}{4}-\int(1-u^2)u^2\,du \qquad (u=\cos x)\\ &= \frac{x}{2}-\frac{\sin 2x}{4}-\frac{\cos^3x}{3}+\frac{\cos^5x}{5}+C.\end{aligned}$$

(c) *Method* 1. Rewrite in terms of $\sec\theta$ and its derivative $\tan\theta\sec\theta$:

$$\begin{aligned}\int\tan^3\theta\sec^3\theta\,d\theta &= \int(\tan\theta\sec\theta)(\tan^2\theta\sec^2\theta)\,d\theta\\ &= \int(\tan\theta\sec\theta)(\sec^2\theta-1)\sec^2\theta\,d\theta \qquad (1+\tan^2\theta=\sec^2)\\ &= \int(u^2-1)u^2\,du \qquad (u=\sec\theta)\\ &= \frac{u^5}{5}-\frac{u^3}{3}+C=\frac{\sec^5\theta}{5}-\frac{\sec^3\theta}{3}+C.\end{aligned}$$

Method 2. Convert to sines and cosines:

$$\int \tan^3\theta \sec^3\theta \, d\theta = \int \frac{\sin^3\theta}{\cos^6\theta} \, d\theta = \int \frac{\sin\theta(1-\cos^2\theta)}{\cos^6\theta} \, d\theta$$

$$= -\int \frac{1-u^2}{u^6} \, du \qquad (u = \cos\theta)$$

$$= \frac{u^{-5}}{5} - \frac{u^{-3}}{3} + C = \frac{\sec^5\theta}{5} - \frac{\sec^3\theta}{3} + C. \ \blacktriangle$$

Certain other integration problems yield to the use of the addition formulas:

$$\sin(x+y) = \sin x \cos y + \cos x \sin y, \tag{1a}$$

$$\cos(x+y) = \cos x \cos y - \sin x \sin y \tag{1b}$$

and the product formulas:

$$\sin x \cos y = \tfrac{1}{2}\left[\sin(x-y) + \sin(x+y)\right], \tag{2a}$$

$$\sin x \sin y = \tfrac{1}{2}\left[\cos(x-y) - \cos(x+y)\right], \tag{2b}$$

$$\cos x \cos y = \tfrac{1}{2}\left[\cos(x-y) + \cos(x+y)\right]. \tag{2c}$$

Example 4 Evaluate (a) $\int \sin x \cos 2x \, dx$ and (b) $\int \cos 3x \cos 5x \, dx$.

Solution (a) $\displaystyle\int \sin x \cos 2x \, dx = \frac{1}{2}\int (\sin 3x - \sin x) \, dx = -\frac{\cos 3x}{6} + \frac{\cos x}{2} + C.$

(see product formula (2a)).

(b) $\displaystyle\int \cos 3x \cos 5x \, dx = \frac{1}{2}\int (\cos 8x + \cos 2x) \, dx = \frac{\sin 8x}{16} + \frac{\sin 2x}{4} + C$

(see product formula (2c)). ▲

Example 5 Evaluate $\int \sin ax \sin bx \, dx$, where a and b are constants.

Solution If we use identity (2b), we get

$$\int \sin ax \sin bx \, dx = \frac{1}{2}\int \left[\cos(a-b)x - \cos(a+b)x\right] dx$$

$$= \begin{cases} \dfrac{1}{2}\dfrac{\sin(a-b)x}{a-b} - \dfrac{1}{2}\dfrac{\sin(a+b)x}{a+b} + C & \text{if } a \neq \pm b, \\[2mm] \dfrac{x}{2} - \dfrac{1}{4a}\sin 2ax + C, & \text{if } a = b, \\[2mm] \dfrac{1}{4a}\sin 2ax - \dfrac{x}{2} + C, & \text{if } a = -b. \end{cases}$$

[The difference between the case $a \neq \pm b$ and the other two should be noted. The first case is "pure oscillation" in that it consists of two sine terms. The others contain the nonoscillating linear term $x/2$, called a *secular term*. This example is related to the phenomena of *resonance*: when an oscillating system is subjected to a sinusoidally varying force, the oscillation will build up indefinitely if the force has the same frequency as the oscillator. See Review Exercise 56, Chapter 8 and the discussion in the last part of Section 12.7, following equation (14).] ▲

Many integrals containing factors of the form $\sqrt{a^2 \pm x^2}$, $\sqrt{x^2 - a^2}$, or $a^2 + x^2$ can be evaluated or simplified by means of trigonometric substitutions. In order to remember what to substitute, it is useful to draw the appropriate right-angle triangle, as in the following box.

Trigonometric Substitutions

1. If $\sqrt{a^2 - x^2}$ occurs, try $x = a\sin\theta$; then $dx = a\cos\theta\, d\theta$ and $\sqrt{a^2 - x^2} = a\cos\theta$; ($a > 0$ and θ is an acute angle).

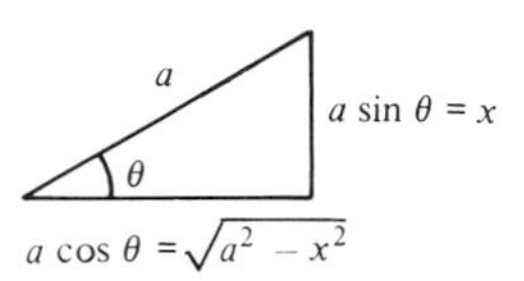

2. If $\sqrt{x^2 - a^2}$ occurs, try $x = a\sec\theta$; then $dx = a\tan\theta\sec\theta\, d\theta$ and $\sqrt{x^2 - a^2} = a\tan\theta$.

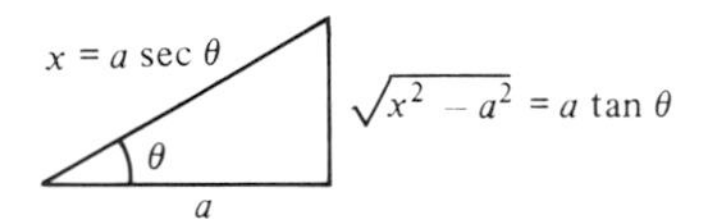

3. If $\sqrt{a^2 + x^2}$ or $a^2 + x^2$ occurs, try $x = a\tan\theta$; then $dx = a\sec^2\theta\, d\theta$ and $\sqrt{a^2 + x^2} = a\sec\theta$ (one can also use $x = a\sinh\theta$; then $\sqrt{a^2 + x^2} = a\cosh\theta$).

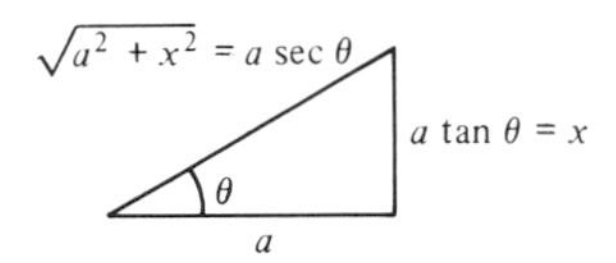

Example 6 Evaluate: (a) $\int \frac{\sqrt{9 - x^2}}{x^2}\, dx$, (b) $\int \frac{dx}{\sqrt{4x^2 - 1}}$.

Solution (a) Let $x = 3\sin\theta$, so $\sqrt{9 - x^2} = 3\cos\theta$. Thus $dx = 3\cos\theta\, d\theta$ and

$$
\begin{aligned}
\int \frac{\sqrt{9 - x^2}}{x^2}\, dx &= \int \frac{3\cos\theta}{9\sin^2\theta} 3\cos\theta\, d\theta \\
&= \int \frac{\cos^2\theta}{\sin^2\theta}\, d\theta = \int \frac{1 - \sin^2\theta}{\sin^2\theta}\, d\theta \\
&= \int (\csc^2\theta - 1)\, d\theta = -\cot\theta - \theta + C \\
&= -\frac{\sqrt{9 - x^2}}{x} - \sin^{-1}\left(\frac{x}{3}\right) + C.
\end{aligned}
$$

In the last line, we used the first figure in the preceding box to get the identity $\cot\theta = \sqrt{a^2 - x^2}/x$ with $a = 3$.

(b) Let $x = \frac{1}{2}\sec\theta$, so $dx = \frac{1}{2}\tan\theta\sec\theta\, d\theta$, and $\sqrt{4x^2 - 1} = \tan\theta$. Thus

$$\int \frac{dx}{\sqrt{4x^2-1}} = \frac{1}{2}\int \frac{\tan\theta\sec\theta}{\tan\theta}\,d\theta = \frac{1}{2}\int \sec\theta\, d\theta.$$

Here is a trick[1] for evaluating $\int \sec\theta$:

$$\int \sec\theta\, d\theta = \int \sec\theta\,\frac{\sec\theta + \tan\theta}{\sec\theta + \tan\theta}\,d\theta = \int \frac{\sec^2\theta + \sec\theta\tan\theta}{\sec\theta + \tan\theta}\,d\theta$$
$$= \ln|\sec\theta + \tan\theta| + C \qquad (\text{substituting } u = \sec\theta + \tan\theta).$$

Thus

$$\int \frac{dx}{\sqrt{4x^2-1}} = \frac{1}{2}\ln|2x + \sqrt{4x^2-1}\,| + C$$

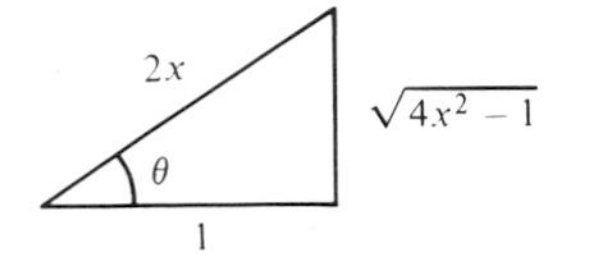

Figure 10.1.1. Geometry of the substitution $x = \frac{1}{2}\sec\theta$.

(see Fig. 10.1.1).

If you have studied the hyperbolic functions you should note that this integral can also be evaluated by means of the formula $\int[du/\sqrt{u^2-1}\,] = \cosh^{-1}u + C$. ▲

These examples show that trigonometric substitutions work quite well in the presence of algebraic integrands involving square roots. You should also keep in mind the possibility of a simple algebraic substitution or using the direct integration formulas involving inverse trigonometric and hyperbolic functions.

Example 7 Evaluate:

$$\text{(a)}\ \int \frac{x}{\sqrt{4-x^2}}\,dx; \qquad \text{(b)}\ \int \frac{1}{\sqrt{x^2-4}}\,dx; \qquad \text{(c)}\ \int \frac{x^2}{\sqrt{4-x^2}}\,dx.$$

Solution (a) Let $u = 4 - x^2$, so $du = -2x\,dx$. Thus

$$\int \frac{x}{\sqrt{4-x^2}}\,dx = -\frac{1}{2}\int \frac{du}{\sqrt{u}} = -\sqrt{u} + C = -\sqrt{4-x^2} + C$$

(no trigonometric function appears).

$$\text{(b)}\ \int \frac{1}{\sqrt{x^2-4}}\,dx = \int \frac{du}{\sqrt{u^2-1}} \qquad \left(u = \frac{x}{2}\right)$$
$$= \cosh^{-1}u + C = \cosh^{-1}\left(\frac{x}{2}\right) + C$$
$$= \ln\left(\frac{x}{2} + \sqrt{\frac{x^2}{4} - 1}\right) + C \qquad (\text{see p. 396}).$$

You may use the method of Example 6(b) if you are not familiar with hyperbolic functions.

(c) To evaluate $\int(x^2/\sqrt{4-x^2}\,)dx$, let $x = 2\sin\theta$; then $dx = 2\cos\theta\, d\theta$, and $\sqrt{4-x^2} = 2\cos\theta$. Thus

$$\int \frac{x^2}{\sqrt{4-x^2}}\,dx = \int \frac{4\sin^2\theta}{2\cos\theta}\cdot 2\cos\theta\, d\theta = 4\int \sin^2\theta\, d\theta$$
$$= 4\int \frac{1-\cos 2\theta}{2}\,d\theta = 2\theta - \sin 2\theta + C$$
$$= 2\theta - 2\sin\theta\cos\theta + C.$$

[1] The same trick shows that $\int \operatorname{csch}\theta\, d\theta = -\ln|\operatorname{csch}\theta + \coth\theta| + C$.

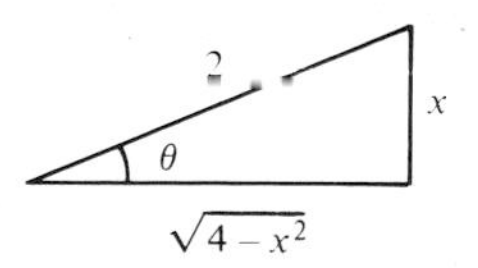

Figure 10.1.2. Geometry of the substitution $x = 2\sin\theta$.

From Fig. 10.1.2 we get

$$\int \frac{x^2}{\sqrt{4-x^2}}\,dx = 2\sin^{-1}\left(\frac{x}{2}\right) - 2\left(\frac{x}{2}\right)\left(\frac{\sqrt{4-x^2}}{2}\right) + C$$

$$= 2\sin^{-1}\frac{x}{2} - \frac{1}{2}x\sqrt{4-x^2} + C. \blacktriangle$$

Completing the square can be useful in simplifying integrals involving the expression $ax^2 + bx + c$. The following two examples illustrate the method.

Example 8 Evaluate $\displaystyle\int \frac{dx}{\sqrt{10+4x-x^2}}$.

Solution To complete the square, write $10 + 4x - x^2 = -(x+a)^2 + b$; solving for a and b, we find $a = -2$ and $b = 14$, so $10 + 4x - x^2 = -(x-2)^2 + 14$. Hence

$$\int \frac{dx}{\sqrt{10+4x-x^2}} = \int \frac{dx}{\sqrt{14-(x-2)^2}} = \int \frac{du}{\sqrt{14-u^2}},$$

where $u = x - 2$. This integral is $\sin^{-1}(u/\sqrt{14}) + C$, so our final answer is

$$\sin^{-1}\left(\frac{x-2}{\sqrt{14}}\right) + C. \blacktriangle$$

Completing the Square

If an integral involves $ax^2 + bx + c$, complete the square and then use a trigonometric substitution or some other method to evaluate the integral.

Example 9 Evaluate (a) $\displaystyle\int \frac{dx}{x^2+x+1}$; (b) $\displaystyle\int \frac{dx}{\sqrt{x^2+x+1}}$.

Solution (a)

$$\int \frac{dx}{x^2+x+1} = \int \frac{dx}{(x+1/2)^2+3/4}$$

$$= \int \frac{du}{u^2+3/4} \qquad \left(u = x + \frac{1}{2}\right)$$

$$= \frac{1}{\sqrt{3/4}}\tan^{-1}\left(\frac{u}{\sqrt{3/4}}\right) + C$$

$$= \frac{2}{\sqrt{3}}\tan^{-1}\left(\frac{2x+1}{\sqrt{3}}\right) + C.$$

(b)

$$\int \frac{dx}{\sqrt{x^2+x+1}} = \int \frac{du}{\sqrt{u^2+3/4}} \qquad \left(u = x + \frac{1}{2}\right)$$

$$= \ln\left|u + \sqrt{u^2+3/4}\,\right| + C$$

$$= \ln\left|x + \frac{1}{2} + \sqrt{(x+1/2)^2+3/4}\,\right| + C$$

$$= \ln\left|x + \frac{1}{2} + \sqrt{x^2+x+1}\,\right| + C. \blacktriangle$$

Applications of the kind encountered in earlier chapters may involve integrals of the type in this section. Here is an example.

Example 10 Find the average value of $\sin^2x\cos^2x$ on the interval $[0, 2\pi]$.

Solution By definition, the average value is the integral divided by the length of the interval:

$$\frac{1}{2\pi}\int_0^{2\pi}\sin^2x\cos^2x\,dx.$$

By Example 2, $\int\sin^2x\cos^2x\,dx = (x/8) - (\sin 4x/32) + C$. Thus

$$\int_0^{2\pi}\sin^2x\cos^2x\,dx = \left(\frac{x}{8} - \frac{\sin 4x}{32}\right)\Bigg|_0^{2\pi} = \frac{\pi}{4},$$

so the average value is $(1/2\pi)\cdot\pi/4 = 1/8$. ▲

Exercises for Section 10.1

Evaluate the integrals in Exercises 1–12.

1. $\int\sin^3x\cos^3x\,dx$
2. $\int\sin^2x\cos^5x\,dx$
3. $\int_0^{2\pi}\sin^4t\,dt$
4. $\int_0^{\pi/2}\cos^4x\sin^2x\,dx$
5. $\int(\cos 2x - \cos^2x)\,dx$
6. $\int\cos 2x\sin x\,dx$
7. $\int_0^{\pi/4}\sin^2x\cos 2x\,dx$
8. $\int_0^{\pi/4}\left(\frac{\sin^2\theta}{\cos^2\theta}\right)d\theta$
9. $\int\sin 4x\sin 2x\,dx$
10. $\int\sin 2\theta\cos 5\theta\,d\theta$
11. $\int_0^{2\pi}\sin 5x\sin 2x\,dx$
12. $\int_{-\pi}^{\pi}\cos 2u\sin\frac{1}{2}u\,du$
13. Evaluate $\int\tan^3x\sec^3x\,dx$. [*Hint*: Convert to sines and cosines.]
14. Show that $\int\sin^6x\,dx = \frac{1}{192}(60x - 48\sin 2x + 4\sin^3 2x + 9\sin 4x) + C$.
15. Evaluate $\int[1/(1+x^2)]\,dx$ (a) as $\tan^{-1}x$ and (b) by the substitution $x = \tan u$. Compare your answers.
16. Evaluate the integral $\int[1/(4+9x^2)]\,dx$ by using the substitutions (a) $x = \frac{2}{3}u$ and (b) $x = \frac{2}{3}\tan\theta$. Compare your answers.

Evaluate the integrals in Exercises 17–28.

17. $\int\frac{\sqrt{x^2-4}}{x}\,dx$
18. $\int\frac{\sqrt{x^2-9}}{x}\,dx$
19. $\int\sqrt{1-u^2}\,du$
20. $\int\sqrt{9-16t^2}\,dt$
21. $\int\frac{s}{\sqrt{4+s^2}}\,ds$
22. $\int\frac{x}{\sqrt{x^2-1}}\,dx$
23. $\int\frac{x^3}{\sqrt{4-x^2}}\,dx$
24. $\int\frac{x^2}{(1+x^2)^{3/2}}\,dx$
25. $\int\frac{dx}{\sqrt{4x^2+x+1}}$
26. $\int\frac{dx}{\sqrt{5-4x-x^2}}$
27. $\int\frac{x\,dx}{\sqrt{3x^2+x-1}}$
28. $\int\sqrt{3+2x-x^2}\,dx$
29. Find the average value of $\cos^nx$ on the interval $[0, 2\pi]$ for $n = 0, 1, 2, 3, 4, 5, 6$.
30. Find the volume of the solid obtained by revolving the region under the graph $y = \sin^2x$ on $[0, 2\pi]$ about the x axis.
31. Find the center of mass of the region under the graph of $1/\sqrt{x^2+2x+2}$ on $[0, 1]$.
32. A plating company wishes to prepare the bill for a silver plate job of 200 parts. Each part has the shape of the region bounded by $y = \sqrt{x^2 - 9/x^2}$, $y = 0$, $x = 5$.
    (a) Find the area enclosed.
    (b) Assume that all units are centimeters. Only one side of the part is to receive the silver plate. The customer was charged \$25 for 1460 square centimeters previously. How much should the 200 parts cost?
33. The *average power* P for a resistance R and associated current i of period T is

$$P = \frac{1}{T}\int_0^T Ri^2\,dt.$$

That is, P is the average value of the instantaneous power Ri^2 on $[0, T]$. Compute the power for $R = 2.5$, $i = 10\sin(377t)$, $T = 2\pi/377$.

34. The current I in a certain RLC circuit is given by $I(t) = Me^{-\alpha t}[\sin^2(\omega t) + 2\cos(2\omega t)]$. Find the charge Q in coulombs, given by

$$Q(t) = Q_0 + \int_0^t I(s)\,ds.$$

35. The *root mean square current* and *voltage* are

$$I_{\text{rms}} = \left(\frac{1}{T}\int_0^T i^2\,dt\right)^{1/2}, \quad\text{and}$$

$$E_{\text{rms}} = \left(\frac{1}{T}\int_0^T e^2\,dt\right)^{1/2}$$

where $i(t)$ and $e(t)$ are the current through and voltage across a pure resistance R. (The current flowing through R is assumed to be periodic with period T.) Compute these numbers, given that $e(t) = 3 + (1.5)\cos(100t)$ volts, and $i(t) = 1 - 2\sin(100t - \pi/6)$ amperes, which corresponds to period $T = 2\pi/100$.

36. The *average power* $P = (1/T)\int_0^T Ri^2\,dt$ for periodic waveshapes does not in general obey a superposition principle. Two voltage sources e_1 and e_2 may individually supply 5 watts (when the other is dead), but when both sources are present the power can be zero (not 10). Compare $\int_0^T R(i_1 + i_2)^2\,dt$ with $\int_0^T Ri_1^2\,dt + \int_0^T Ri_2^2\,dt$ when $i_1 = I_1\cos(m\omega t + \phi_1)$, $i_2 = I_2\cos(n\omega t + \phi_2)$, $m \neq n$ (m, n positive integers), $T = 2\pi/\omega$, and $R, I_1, I_2, \omega, \phi_1, \phi_2$ are constants.

37. A charged particle is constrained by magnetic fields to move along a straight line, oscillating back and forth from the origin with higher and higher amplitude.

    Let $S(t)$ be the directed distance from the origin, and assume that $S(t)$ satisfies the equation

    $$[S(t)]^2 S'(t) = t\sin t + \sin^2 t\cos^2 t.$$

    (a) Prove $[S(t)]^3 = 3\int_0^t (x\sin x + \sin^2 x\cos^2 x)\,dx$.
    (b) Find $S(t)$.
    (c) Find all zeros of $S'(t)$ for $t > 1$. Which zeros correspond to times of maximum excursion from the origin?

★38. Show that the integral in Example 5 is a continuous function of b for fixed a and x.

10.2 Partial Fractions

By the method of partial fractions, one can evaluate any integral of the form

$$\int \frac{P(x)}{Q(x)}\,dx, \text{ where } P \text{ and } Q \text{ are polynomials.}$$

The integral of a polynomial can be expressed simply by the formula

$$\int (a_n x^n + a_{n-1}x^{n-1} + \cdots + a_1 x + a_0)\,dx = \frac{a_n x^{n+1}}{n+1} + \frac{a_{n-1}x^n}{n} + \cdots + a_0 x + C,$$

but there is no simple general formula for integrals of *quotients* of polynomials, i.e., for rational functions. There is, however, a general *method* for integrating rational functions, which we shall learn in this section. This method demonstrates clearly the need for evaluating integrals by hand or by a computer program such as MACSYMA, which automatically carries out the procedures to be described in this section, since tables cannot include the infinitely many possible integrals of this type.

One class of rational functions which we can integrate simply are the reciprocal powers. Using the substitution $u = ax + b$, we find that $\int [dx/(ax+b)^n] = \int (du/au^n)$, which is evaluated by the power rule. Thus, we get

$$\int \frac{dx}{(ax+b)^n} = \begin{cases} \dfrac{-1}{a(n-1)(ax+b)^{n-1}} + C, & \text{if } n \neq 1, \\[2ex] \dfrac{1}{a}\ln|ax+b| + C, & \text{if } n = 1. \end{cases}$$

More generally, we can integrate any rational function whose denominator can be factored into linear factors. We shall give several examples before presenting the general method.

Example 1 Evaluate $\displaystyle\int \frac{x+1}{(x-1)(x-3)}\,dx$.

Solution We shall try to write

$$\frac{x+1}{(x-1)(x-3)} = \frac{A}{x-1} + \frac{B}{x-3},$$

for constants A and B. To determine them, note that

$$\frac{A}{x-1}+\frac{B}{x-3}=\frac{(A+B)x-3A-B}{(x-1)(x-3)}.$$

Thus, we should choose

$$A+B=1 \quad \text{and} \quad -3A-B=1.$$

Solving, $A=-1$ and $B=2$. Thus,

$$\frac{x+1}{(x-1)(x-3)}=-\frac{1}{x-1}+\frac{2}{x-3},$$

so

$$\int \frac{x+1}{(x-1)(x-3)}\,dx=-\ln|x-1|+2\ln|x-3|+C$$

$$=\ln\left(\frac{|x-3|^2}{|x-1|}\right)+C. \blacktriangle$$

Example 2 Evaluate

(a) $\displaystyle\int \frac{4x^2+2x+3}{(x-2)^2(x+3)}\,dx$; (b) $\displaystyle\int_{-1}^{1} \frac{4x^2+2x+3}{(x-2)^2(x+3)}\,dx.$

Solution (a) As in Example 1, we might expect to decompose the quotient in terms of $1/(x-2)$ and $1/(x+3)$. In fact, we shall see that we can write

$$\frac{4x^2+2x+3}{(x-2)^2(x+3)}=\frac{A}{x-2}+\frac{B}{(x-2)^2}+\frac{C}{x+3} \tag{1}$$

if we choose the constants A, B, and C suitably. Adding the terms on the right-hand side of equation (1) over the common denominator, we get

$$\frac{A(x-2)(x+3)+B(x+3)+C(x-2)^2}{(x-2)^2(x+3)}.$$

The numerator, when multiplied out, would be a polynomial $a_2x^2+a_1x+a_0$, where the coefficients a_2, a_1, and a_0 depend on A, B, and C. The idea is to choose A, B, and C so that we get the numerator $4x^2+2x+3$ of our integration problem. (Notice that we have exactly three unknowns A, B, and C at our disposal to match the three coefficients in the numerator.)

To choose A, B, and C, it is easiest not to multiply out but simply to write

$$4x^2+2x+3=A(x-2)(x+3)+B(x+3)+C(x-2)^2 \tag{2}$$

and make judicious substitutions for x. For instance, $x=-3$ gives

$$4\cdot 9-2\cdot 3+3=C(-3-2)^2,$$
$$33=25C,$$
$$C=\tfrac{33}{25}.$$

Next, $x=2$ gives

$$4\cdot 4+2\cdot 2+3=B(2+3),$$
$$23=5B,$$
$$B=\tfrac{23}{5}.$$

To solve for A, we may use either of two methods.

Method 1. Let $x = 0$ in equation (2):

$$3 = -6A + 3B + 4C$$
$$= -6A + 3 \cdot \tfrac{23}{5} + 4 \cdot \tfrac{33}{25},$$
$$0 = -6A + 3 \cdot \tfrac{18}{5} + 4 \cdot \tfrac{33}{25},$$
$$6A = 3 \cdot \tfrac{134}{25} \quad \text{so} \quad A = \tfrac{67}{25}.$$

Method 2. Differentiate equation (2) to give

$$8x + 2 = A\big[(x-2) + (x+3)\big] + B + 2C(x-2)$$

and then substitute $x = 2$ again:

$$8 \cdot 2 + 2 = A(2+3) + B,$$
$$18 = 5A + B = 5A + \tfrac{23}{5},$$
$$5A = 18 - \tfrac{23}{5} = \tfrac{67}{5},$$
$$A = \tfrac{67}{25}.$$

This gives

$$\frac{4x^2 + 2x + 3}{(x-2)^2(x+3)} = \frac{67}{25}\frac{1}{x-2} + \frac{23}{5}\frac{1}{(x-2)^2} + \frac{33}{25}\frac{1}{x+3}.$$

(At this point, it is a good idea to check your answer, either by adding up the right-hand side or by substituting a few values of x, using a calculator.)

We can now integrate:

$$\int \frac{4x^2 + 2x + 3}{(x-2)^2(x+3)}\,dx = \frac{67}{25}\int \frac{dx}{x-2} + \frac{23}{5}\int \frac{dx}{(x-2)^2} + \frac{33}{25}\int \frac{dx}{x+3}$$
$$= \frac{67}{25}\ln|x-2| - \frac{23}{5}\frac{1}{x-2} + \frac{33}{25}\ln|x+3| + C.$$

(b) Since the integrand "blows up" at $x = -3$ and $x = 2$, it only makes sense to evaluate definite integrals over intervals which do not contain these points; $[-1, 1]$ is such an interval. Thus, by (a), the definite integral is

$$\left(\frac{67}{25}\ln|x-2| - \frac{23}{5}\frac{1}{x-2} + \frac{33}{25}\ln|x+3|\right)\Bigg|_{-1}^{1}$$
$$= \frac{67}{25}(\ln 1 - \ln 3) - \frac{23}{5}\left(\frac{1}{-1} - \frac{1}{-3}\right) + \frac{33}{25}(\ln 4 - \ln 2)$$
$$\approxeq -2.944 + 3.067 + 0.915 \approxeq 1.037. \blacktriangle$$

Not every polynomial can be written as a product of linear factors. For instance, $x^2 + 1$ cannot be factored further (unless we use complex numbers) nor can any other quadratic function $ax^2 + bx + c$ for which $b^2 - 4ac < 0$; but any polynomial can, in principle, be factored into linear and quadratic factors. (This is proved in more advanced algebra texts.) This factorization is not always so easy to carry out in practice, but whenever we manage to factor the denominator of a rational function, we can integrate that function by the method of partial fractions.

Example 3 Integrate $\displaystyle\int \frac{1}{x^3 - 1}\,dx$.

Solution The denominator factors as $(x-1)(x^2 + x + 1)$, and $x^2 + x + 1$ cannot be further factored (since $b^2 - 4ac = 1 - 4 = -3 < 0$). Now write

$$\frac{1}{x^3-1} = \frac{a}{x-1} + \frac{Ax+B}{x^2+x+1}.$$

Thus $1 = a(x^2+x+1) + (x-1)(Ax+B)$. We substitute values for x:

$$x = 1\colon 1 = 3a \qquad \text{so} \quad a = \tfrac{1}{3};$$

$$x = 0\colon 1 = \tfrac{1}{3} - B \qquad \text{so} \quad B = -\tfrac{2}{3}.$$

Comparing the x^2 terms, we get $0 = a + A$, so $A = -\frac{1}{3}$. Hence

$$\frac{1}{x^3-1} = \frac{1}{3}\left(\frac{1}{x-1} - \frac{x+2}{x^2+x+1}\right).$$

(This is a good point to check your work.)

Now

$$\int \frac{1}{x-1}\,dx = \ln|x-1| + C$$

and, writing $x + 2 = \frac{1}{2}(2x+1) + \frac{3}{2}$,

$$\int \frac{x+2}{x^2+x+1}\,dx = \frac{1}{2}\int \frac{2x+1}{x^2+x+1}\,dx + \frac{3}{2}\int \frac{dx}{(x+1/2)^2+3/4}$$

$$= \frac{1}{2}\ln|x^2+x+1| + \frac{3}{2}\cdot\sqrt{\frac{4}{3}}\tan^{-1}\left(\frac{x+1/2}{\sqrt{3/4}}\right) + C$$

$$= \frac{1}{2}\ln|x^2+x+1| + \sqrt{3}\tan^{-1}\left(\frac{2x+1}{\sqrt{3}}\right) + C.$$

Thus

$$\int \frac{dx}{x^3-1} = \frac{1}{3}\ln|x-1| - \frac{1}{6}\ln|x^2+x+1| - \frac{1}{\sqrt{3}}\tan^{-1}\left(\frac{2x+1}{\sqrt{3}}\right) + C$$

$$= \frac{1}{3}\left[\frac{1}{2}\ln\left|\frac{(x-1)^2}{x^2+x+1}\right| - \sqrt{3}\tan^{-1}\left(\frac{2x+1}{\sqrt{3}}\right)\right] + C.$$

Observe that the innocuous-looking integrand $1/(x^3-1)$ has brought forth both logarithmic and trigonometric functions. ▲

Now we are ready to set out a systematic method for the integration of $P(x)/Q(x)$ by partial fractions. (See the box on p. 469.) A few remarks may clarify the procedures given in the box. In case the denominator Q factors into n distinct linear factors, which we denote $Q = (x - r_1)(x - r_2) \dots (x - r_n)$, we write

$$\frac{P}{Q} = \frac{\alpha_1}{x-r_1} + \frac{\alpha_2}{x-r_2} + \dots + \frac{\alpha_n}{x-r_n}$$

and determine the n coefficients $\alpha_1, \dots, \alpha_n$ by multiplying by Q and matching P to the resulting polynomial. The division in step 1 has guaranteed that P has degree at most $n-1$, containing n coefficients. This is consistent with the number of constants $\alpha_1, \dots, \alpha_n$ we have at our disposal. Similarly, if the denominator has repeated roots, or if there are quadratic factors in the denominator, it can be checked that the number of constants at our disposal is equal to the number of coefficients in the numerator to be matched. A system of n equations in n unknowns is likely to have a unique solution, and in this case, one can prove that it does.[2]

[2] See Review Exercise 88, Chapter 13 for a special case, or H. B. Fine, *College Algebra*, Dover, New York (1961), p. 241 for the general case.

Partial Fractions

To integrate $P(x)/Q(x)$, where P and Q are polynomials containing no common factor:

1. If the degree of P is larger than or equal to the degree of Q, divide Q into P by long division, obtaining a polynomial plus $R(x)/Q(x)$, where the degree of R is less than that of Q. Thus we need only investigate the case where the degree of P is less than that of Q.
2. Factor the denominator Q into linear and quadratic factors—that is, factors of the form $(x - r)$ and $ax^2 + bx + c$. (Factor the quadratic expressions if $b^2 - 4ac > 0$.)
3. If $(x - r)^m$ occurs in the factorization of Q, write down a sum of the form

$$\frac{a_1}{(x-r)} + \frac{a_2}{(x-r)^2} + \cdots + \frac{a_m}{(x-r)^m},$$

where $a_1, a_2, \ldots$ are constants. Do so for each factor of this form (using constants $b_1, b_2, \ldots, c_1, c_2, \ldots,$ and so on) and add the expressions you get. The constants $a_1, a_2, \ldots, b_1, b_2, \ldots,$ and so on will be determined in step 5.

4. If $(ax^2 + bx + c)^p$ occurs in the factorization of Q with $b^2 - 4ac < 0$, write down a sum of the form

$$\frac{A_1 x + B_1}{ax^2 + bx + c} + \frac{A_2 x + B_2}{(ax^2 + bx + c)^2} + \cdots + \frac{A_p x + B_p}{(ax^2 + bx + c)^p}.$$

Do so for each factor of this form and add the expressions you get. The constants $A_1, A_2, \ldots, B_1, B_2, \ldots$ are determined in step 5. Add this expression to the one obtained in step 3.

5. Equate the expression obtained in steps 3 and 4 to $P(x)/Q(x)$. Multiply through by $Q(x)$ to obtain an equation between two polynomials. Comparing coefficients of these polynomials, determine equations for the constants $a_1, a_2, \ldots, A_1, A_2, \ldots, B_1, B_2, \ldots$ and solve these equations. Sometimes the constants can be determined by substituting convenient values of x in the equality or by differentiation of the equality.
6. Check your work by adding up the partial fractions or substituting a few values of x.
7. Integrate the expression obtained in step 5 by using

$$\int \frac{dx}{(x-r)^j} = -\left[\frac{1}{(j-1)(x-r)^{j-1}}\right] + C, \qquad j > 1$$

and $\displaystyle\int \frac{dx}{x-r} = \ln|x - r| + C.$

The terms with a quadratic denominator may be integrated by a manipulation which makes the derivative of the denominator appear in the numerator, together with completing the square (see Examples 3 and 6).

Example 4 Integrate $\int \frac{x^5 - x^4 + 1}{x^3 - x^2}\,dx$.

Solution First we divide out the fraction to get

$$\int \frac{x^5 - x^4 + 1}{x^3 - x^2}\,dx = \int \left(x^2 + \frac{1}{x^3 - x^2}\right)dx = \frac{1}{3}x^3 + \int \frac{dx}{x^3 - x^2}.$$

The denominator $x^3 - x^2$ is of degree 3, and the numerator is of degree zero. Thus we proceed to step 2 and factor:

$$x^3 - x^2 = x^2(x - 1).$$

Here $x = x - 0$ occurs to the power 2, so by step 3, we write down

$$\frac{a_1}{x} + \frac{a_2}{x^2}.$$

We also add the term $b_1/(x - 1)$ for the second factor.

$$\frac{a_1}{x} + \frac{a_2}{x^2} + \frac{b_1}{x - 1}.$$

Since there are no quadratic factors, we omit step 4. By step 5, we equate the preceding expression to $1/(x^3 - x^2)$:

$$\frac{a_1}{x} + \frac{a_2}{x^2} + \frac{b_1}{x - 1} = \frac{1}{x^2(x - 1)}.$$

Then we multiply by $x^2(x - 1)$:

$$a_1x(x - 1) + a_2(x - 1) + b_1x^2 = 1.$$

Setting $x = 0$, we get $a_2 = -1$. Setting $x = 1$, we get $b_1 = 1$. Comparing the coefficients of x^2 on both sides of the equation gives $a_1 + b_1 = 0$, so $a_1 = -b_1 = -1$. Thus $a_2 = -1$, $a_1 = -1$, and $b_1 = 1$. (We can check by substitution into the preceding equation: the left side is $(-1)x(x - 1) - (x - 1) + x^2$, which is just 1.)

Thus

$$\frac{1}{x^3 - x^2} = -\frac{1}{x^2} - \frac{1}{x} + \frac{1}{x - 1},$$

and so

$$\int \frac{dx}{x^3 - x^2} = \frac{1}{x} - \ln|x| + \ln|x - 1| + C = \frac{1}{x} + \ln\left|\frac{x - 1}{x}\right| + C.$$

Finally,

$$\int \frac{x^5 - x^4 + 1}{x^3 - x^2}\,dx = \frac{1}{3}x^3 + \frac{1}{x} + \ln\left|\frac{x - 1}{x}\right| + C. \blacktriangle$$

Example 5 Integrate $\int \frac{x^2}{(x^2 - 2)^2}\,dx$.

Solution The denominator factors as $(x - \sqrt{2})^2(x + \sqrt{2})^2$, so we write

$$\frac{x^2}{(x^2 - 2)^2} = \frac{a_1}{x - \sqrt{2}} + \frac{a_2}{(x - \sqrt{2})^2} + \frac{b_1}{x + \sqrt{2}} + \frac{b_2}{(x + \sqrt{2})^2}.$$

Thus

$$x^2 = a_1(x-\sqrt{2})(x+\sqrt{2})^2 + a_2(x+\sqrt{2})^2 + b_1(x+\sqrt{2})(x-\sqrt{2})^2 + b_2(x-\sqrt{2})^2.$$

We substitute values for x:

$$x=\sqrt{2}: 2 = 8a_2 \qquad \text{so} \quad a_2 = \tfrac{1}{4};$$

$$x=-\sqrt{2}: 2 = 8b_2 \qquad \text{so} \quad b_2 = \tfrac{1}{4}.$$

Therefore

$$\begin{aligned} x^2 &= a_1(x^2-2)(x+\sqrt{2}) + \tfrac{1}{4}(x^2+2\sqrt{2x}+2) \\ &\quad + b_1(x^2-2)(x-\sqrt{2}) + \tfrac{1}{4}(x^2-2\sqrt{2}\,x+2) \\ &= (a_1+b_1)x^3 + \left(\sqrt{2}\,a_1 + \tfrac{1}{2} - \sqrt{2}\,b_1\right)x^2 \\ &\quad + (-2a_1-2b_1)x - 2\sqrt{2}\,a_1 + 1 + 2\sqrt{2}\,b_1, \end{aligned}$$

and so

$$a_1 + b_1 = 0 \quad \text{and} \quad \sqrt{2}\,a_1 + \tfrac{1}{2} - \sqrt{2}\,b_1 = 1.$$

Thus

$$a_1 = \frac{1}{4\sqrt{2}}, \qquad b_1 = -\frac{1}{4\sqrt{2}}.$$

Hence

$$\begin{aligned} \frac{x^2}{(x^2-2)^2} &= \frac{1}{4\sqrt{2}\,(x-\sqrt{2})} + \frac{1}{4(x-\sqrt{2})^2} \\ &\quad - \frac{1}{4\sqrt{2}\,(x+\sqrt{2})} + \frac{1}{4(x+\sqrt{2})^2}, \end{aligned}$$

and so

$$\begin{aligned} \int \frac{x^2}{(x^2-2)^2}\,dx &= \frac{1}{4\sqrt{2}} \ln\left|\frac{x-\sqrt{2}}{x+\sqrt{2}}\right| - \frac{1}{4(x-\sqrt{2})} - \frac{1}{4(x+\sqrt{2})} + C \\ &= \frac{1}{4\sqrt{2}} \ln\left|\frac{x-\sqrt{2}}{x+\sqrt{2}}\right| - \frac{x}{2(x^2-2)} + C. \ \blacktriangle \end{aligned}$$

Example 6 Integrate $\displaystyle\int \frac{x^3}{(x-1)(x^2+2x+2)^2}\,dx.$

Solution For the factor $x-1$ we write

$$\frac{a_1}{x-1},$$

and for $(x^2+2x+2)^2$ (which does not factor further since x^2+2x+2 does not have real roots since $b^2-4ac = 4-4\cdot1\cdot2 = -4 < 0$) we write

$$\frac{A_1x+B_1}{x^2+2x+2} + \frac{A_2x+B_2}{(x^2+2x+2)^2}.$$

We then set

$$\frac{a_1}{x-1} + \frac{A_1x + B_1}{x^2+2x+2} + \frac{A_2x+B_2}{(x^2+2x+2)^2} = \frac{x^3}{(x-1)(x^2+2x+2)^2}$$

and multiply by $(x-1)(x^2+2x+2)^2$:

$$a_1(x^2+2x+2)^2 + (A_1x+B_1)(x-1)(x^2+2x+2)$$
$$+(A_2x+B_2)(x-1) = x^3.$$

Setting $x=1$ gives $a_1(25)=1$ or $a_1 = \frac{1}{25}$. Expanding the left-hand side, we get:

$$\tfrac{1}{25}(x^4+4x^3+8x^2+8x+4) + A_1x^4 + (A_1+B_1)x^3 + B_1x^2$$
$$-2A_1x - 2B_1 + A_2x^2 + (B_2-A_2)x - B_2 = x^3.$$

Comparing coefficients:

$$x^4\colon\ \tfrac{1}{25} + A_1 = 0; \tag{3}$$
$$x^3\colon\ \tfrac{4}{25} + (A_1+B_1) = 1; \tag{4}$$
$$x^2\colon\ \tfrac{8}{25} + B_1 + A_2 = 0; \tag{5}$$
$$x\colon\ \tfrac{8}{25} - 2A_1 + (B_2 - A_2) = 0; \tag{6}$$
$$x^0(=1)\colon\ \tfrac{4}{25} - 2B_1 - B_2 = 0. \tag{7}$$

Thus

$A_1 = -\frac{1}{25}$ (from equation (3));

$B_1 = \frac{22}{25}$ (from equation (4));

$A_2 = -\frac{30}{25}$ (from equation (5));

$B_2 = -\frac{40}{25}$ (from equation (6)).

At this stage you may check the algebra by substitution into equation (7). Algebraic errors are easy to make in integration by partial fractions.

We have thus far established

$$\frac{x^3}{(x-1)(x^2+2x+2)^2} = \frac{1}{25}\left[\frac{1}{x-1} + \frac{-x+22}{x^2+2x+2} + \frac{-30x-40}{(x^2+2x+2)^2}\right].$$

We compute the integrals of the first two terms as follows:

$$\int \frac{1}{x-1}\,dx = \ln|x-1| + C$$

$$\int \frac{-x+22}{x^2+2x+2}\,dx = \int \frac{-x-1+23}{x^2+2x+2}\,dx$$
$$= -\frac{1}{2}\int \frac{2x+2}{x^2+2x+2}\,dx + 23\int \frac{dx}{x^2+2x+2}$$
$$= -\frac{1}{2}\ln|x^2+2x+2| + 23\int \frac{dx}{(x+1)^2+1}$$
$$= -\frac{1}{2}\ln|x^2+2x+2| + 23\tan^{-1}(x+1) + C.$$

Finally, for the last term, we rearrange the numerator to make the derivative of the quadratic polynomial in the denominator appear:

$$\int \frac{-30x-40}{(x^2+2x+2)^2}\,dx = \int \frac{-15(2x+2)-10}{(x^2+2x+2)^2}\,dx$$

$$= 15\cdot\frac{1}{(x^2+2x+2)} - 10\int \frac{1}{\left[(x+1)^2+1\right]^2}\,dx.$$

Let $x+1=\tan\theta$, so $dx=\sec^2\theta\,d\theta$ and $(x+1)^2+1=\sec^2\theta$. Then

$$\int \frac{1}{\left[(x+1)^2+1\right]^2}\,dx = \int \frac{\sec^2\theta\,d\theta}{\sec^4\theta} = \int \cos^2\theta\,d\theta$$

$$= \int \frac{1+\cos 2\theta}{2}\,d\theta = \frac{\theta}{2} + \frac{\sin 2\theta}{4} + C$$

$$= \frac{1}{2}\tan^{-1}(x+1) + \frac{1}{2}\sin\theta\cos\theta + C$$

$$= \frac{1}{2}\tan^{-1}(x+1) + \frac{1}{2}\cdot\frac{x+1}{(x+1)^2+1} + C$$

(see Fig. 10.2.1).

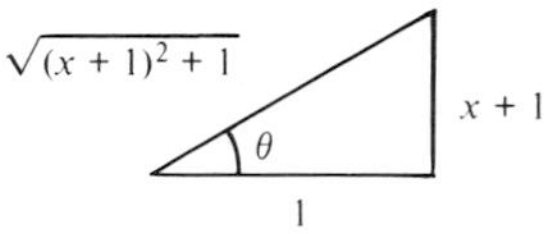

$\sqrt{(x+1)^2+1}$ $\quad x+1 \quad \theta \quad 1$

Figure 10.2.1. Geometry of the substitution $x+1=\tan\theta$.

Adding the results obtained above, we find

$$\int \frac{x^3}{(x-1)(x^2+2x+2)}\,dx$$

$$= \frac{1}{25}\left[\ln|x-1| - \frac{1}{2}\ln(x^2+2x+2) + 23\tan^{-1}(x+1) + 15\frac{1}{x^2+2x+2} - 5\tan^{-1}(x+1) - 5\frac{x+1}{x^2+2x+2}\right] + C$$

$$= \frac{1}{25}\left[\ln\left(\frac{|x-1|}{\sqrt{x^2+2x+2}}\right) + 18\tan^{-1}(x+1) + \frac{10-5x}{x^2+2x+2}\right] + C. \blacktriangle$$

Integrands with a single power $(x-a)^r$ in the denominator may appear to require partial fractions but are actually easiest to evaluate using a simple substitution.

Example 7 Integrate $\displaystyle\int \frac{x^3+2x+1}{(x-1)^5}\,dx$.

Solution Let $u=x-1$ so $du=dx$ and $x=u+1$. Then

$$\int \frac{x^3+2x+1}{(x-1)^5}\,dx = \int \frac{(u+1)^3+2(u+1)+1}{u^5}\,du$$

$$= \int \frac{u^3+3u^2+5u+4}{u^5}\,du = \int\left(\frac{1}{u^2}+\frac{3}{u^3}+\frac{5}{u^4}+\frac{4}{u^5}\right)du$$

$$= -\frac{1}{u} - \frac{3}{2u^2} - \frac{5}{3u^3} - \frac{4}{4u^4} + C$$

$$= -\left[\frac{1}{x-1} + \frac{3}{2(x-1)^2} + \frac{5}{3(x-1)^3} + \frac{1}{(x-1)^4}\right] + C. \blacktriangle$$

To conclude this section, we present a couple of techniques in which an integrand is converted by a substitution into a rational function which can then be integrated by partial fractions. The first such technique, called the method of *rationalizing substitutions*, applies when an integrand involves a fractional power. The idea is to express the fractional power as an integer power of a new variable.

Example 8 Eliminate the fractional power from $\int \frac{(1+x)^{2/3}}{1+2x}\,dx$.

Solution To get rid of the fractional power, substitute $u=(1+x)^{1/3}$. Then $u^3=1+x$ and $3u^2\,du=dx$, so the integral becomes

$$\int \frac{u^2}{1+2(u^3-1)}\cdot 3u^2\,du=\int \frac{3u^4\,du}{2u^3-1}. \blacktriangle$$

After the rationalizing substitution as made, the method of partial functions can be used to evaluate the integral. (Evaluating the integral above is left as Exercise 24).

Example 9 Try the substitution $u=\sqrt[3]{x^2+4}$ in the integrals:

(a) $\int \frac{x^4\,dx}{\sqrt[3]{x^2+4}}$ and (b) $\int \frac{2x^7\,dx}{\sqrt[3]{x^2+4}}$.

Solution We have $u^3=x^2+4$ and $3u^2\,du=2x\,dx$, so integral (a) becomes

$$\int \frac{x^4}{u}\cdot\frac{3u^2}{2x}\,du=\int \frac{3}{2}ux^3\,du,$$

from which we cannot eliminate x without introducing a new fractional power. However, (b) is

$$\int \frac{2x^7}{u}\cdot\frac{3u^2}{2x}\,du=\int 3ux^6\,du=\int 3u(u^3-4)^3\,du$$

(which can be evaluated as the integral of a polynomial). The reason the method works in case (b) lies in the special relation between the exponents of x inside and outside the radical (see Exercise 27). ▲

The second general technique applies when the integrand is built up by rational operations from $\sin x$ and $\cos x$ (and hence from the other trigonometric functions as well). The substitution $u=\tan(x/2)$ turns such an integrand into a rational function of u by virtue of the following trigonometric identities:

$$\sin x=\frac{2u}{1+u^2}, \tag{8}$$

$$\cos x=\frac{1-u^2}{1+u^2}, \tag{9}$$

and

$$dx=\frac{2\,du}{1+u^2}. \tag{10}$$

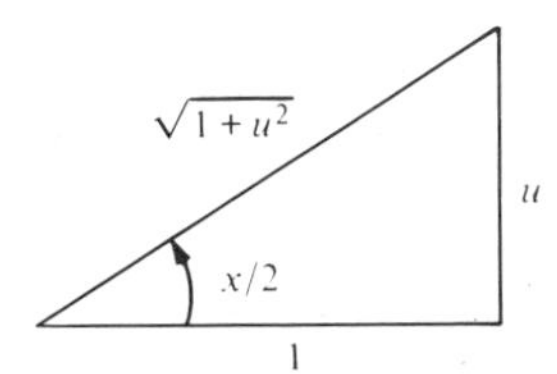

Figure 10.2.2. With the substitution $\tan(x/2)=u$, $\sin(x/2)=u/\sqrt{1+u^2}$.

To prove equation (8), use the addition formula

$$\sin x=\sin\left(\frac{x}{2}+\frac{x}{2}\right)=2\sin\left(\frac{x}{2}\right)\cos\left(\frac{x}{2}\right)$$

$$=2\left(\frac{u}{\sqrt{1+u^2}}\right)\left(\frac{1}{\sqrt{1+u^2}}\right)=\frac{2u}{1+u^2} \quad \text{(see Fig. 10.2.2).}$$

Similarly we derive equation (9). Equation (10) holds since

$$\frac{du}{dx} = \frac{1}{2}\sec^2\frac{x}{2} = \frac{1}{2}\frac{1}{1+u^2}.$$

Example 10 Evaluate $\int \frac{dx}{2+\cos x}$.

Solution Using equations (8), (9), and (10), we convert the integral to

$$\int \frac{2\,du}{1+u^2}\cdot\frac{1}{2+\left[(1-u^2)/(1+u^2)\right]} = \int \frac{2\,du}{2+2u^2+1-u^2} = \int \frac{2\,du}{3+u^2},$$

which is rational in u. No partial fraction decomposition is necessary; the substitution $u = \sqrt{3}\tan\theta$ converts the integral to

$$\int \frac{2\cdot\sqrt{3}\sec^2\theta\,d\theta}{3+3\tan^2\theta} = \frac{2}{\sqrt{3}}\int d\theta = \frac{2\theta}{\sqrt{3}} + C$$

(using the identity $1+\tan^2\theta = \sec^2\theta$). Writing the answer in terms of x, we get

$$\frac{2}{\sqrt{3}}\tan^{-1}\left(\frac{u}{\sqrt{3}}\right) + C = \frac{2}{\sqrt{3}}\tan^{-1}\left(\frac{1}{\sqrt{3}}\tan\frac{x}{2}\right) + C$$

for our final answer. ▲

Rational Expressions in $\sin x$ and $\cos x$

If $f(x)$ is a rational expression in $\sin x$ and $\cos x$, then substitute $u = \tan(x/2)$. Using equations (8), (9), and (10), transform $\int f(x)\,dx$ into the integral of a rational function of u to which the method of partial fractions can be applied.

Example 11 Find $\int_0^{\pi/4}\sec\theta\,d\theta$.

Solution First we find $\int\sec\theta\,d\theta = \int[d\theta/\cos\theta]$. We use equations (9) and (10) (with x replaced by θ) to get

$$\int\frac{d\theta}{\cos\theta} = \int\frac{1+u^2}{1-u^2}\,\frac{2\,du}{1+u^2} = 2\int\frac{du}{1-u^2}$$

$$= \int\left(\frac{1}{1+u} + \frac{1}{1-u}\right)du = \ln|1+u| - \ln|1-u| + C$$

$$= \ln\left|\frac{1+u}{1-u}\right| + C = \ln\left|\frac{1+\tan(\theta/2)}{1-\tan(\theta/2)}\right| + C.$$

(Compare this procedure with the method we used to find $\int\sec\theta\,d\theta$ in Example 6(b), Section 10.1.) Finally,

$$\int_0^{\pi/4}\sec\theta\,d\theta = \left(\ln\left|\frac{1+\tan(\theta/2)}{1-\tan(\theta/2)}\right|\right)\Bigg|_0^{\pi/4} = \ln\left(\frac{1+\tan(\pi/8)}{1-\tan(\pi/8)}\right) - \ln 1$$

$$= \ln\left(\frac{1+\tan(\pi/8)}{1-\tan(\pi/8)}\right) \approx 0.881. \ ▲$$

Exercises for Section 10.2

Evaluate the integrals in Exercises 1–12 using the method of partial fractions.

1. $\int \frac{1}{(x-2)^2(x^2+1)}\,dx.$
2. $\int \frac{x^4+2x^3+3}{(x-4)^6}\,dx.$
3. $\int_0^1 \frac{x^4}{(x^2+1)^2}\,dx$
4. $\int_0^1 \frac{2x^3-1}{x^2+1}\,dx$
5. $\int \frac{x^2}{(x-2)(x^2+2x+2)}\,dx.$
6. $\int \frac{dx}{(x-2)(x^2+3x+1)}$
7. $\int_2^4 \frac{x^3+1}{x^3-1}\,dx$
8. $\int_0^1 \frac{dx}{8x^3+1}$
9. $\int \frac{x}{x^4+2x^2-3}\,dx.$
10. $\int \frac{2x^2-x+2}{x^5+2x^3+x}\,dx$
11. $\int_{\pi/6}^{\pi/2} \frac{\cos x\,dx}{\sin x+\sin^3 x}$
12. $\int_0^{\pi/4} \frac{(\sec^2 x+1)\sec^2 x\,dx}{1+\tan^3 x}$

Evaluate the integrals in Exercises 13–16 using a rationalizing substitution.

13. $\int \frac{\sqrt{x}}{1+x}\,dx$
14. $\int \frac{x}{\sqrt{x+1}}\,dx$
15. $\int x\sqrt[3]{x^2+1}\,dx$
16. $\int x^3\sqrt[3]{x^2+1}\,dx$

Evaluate the integrals in Exercises 17–20.

17. $\int \frac{dx}{1+\sin x}$
18. $\int \frac{dx}{1+2\cos x}$
19. $\int_0^{\pi/4} \frac{d\theta}{1+\tan\theta}$
20. $\int_{\pi/4}^{\pi/2} \frac{d\theta}{1-\cos\theta}$
21. Find the volume of the solid obtained by revolving the region under the graph of the function $y = 1/[(1-x)(1-2x)]$ on $[5,6]$ about the y axis.
22. Find the center of mass of the region under the graph of $1/(x^2+4)$ on $[1,3]$.
23. Evaluate $\int \frac{(1+x)^{3/2}}{x}\,dx.$

★24. Evaluate the integral in Example 8.

25. A chemical reaction problem leads to the following equation:

$$\int \frac{dx}{(80-x)(60-x)} = k\int dt, \qquad k = \text{constant}.$$

In this formula, $x(t)$ is the number of kilograms of reaction product present after t minutes, starting with 80 kilograms and 60 kilograms of two reacting substances which obey the *law of mass action*.

(a) Integrate to get a logarithmic formula involving x and t ($x = 0$ when $t = 0$).

(b) Convert the answer to an exponential formula for x (assume $x < 60$).

(c) How much reaction product is present after 15 minutes, assuming $x = 20$ when $t = 10$?

26. Partial fractions appear in electrical engineering as a convenient means of analyzing and describing circuit responses to applied voltages. By means of the Laplace transform, circuit responses are associated with rational funcitons. Partial fraction methods are used to decompose these rational functions to elementary quotients, which are recognizable to engineers as arising from standard kinds of circuit responses. For example, from

$$\frac{s+1}{(s+2)(s^2+1)(s^2+4)} = \frac{A}{s+2} + \frac{Bs+C}{s^2+1} + \frac{Ds+2E}{s^2+4},$$

an engineer can easily see that this rational function represents the response

$$Ae^{-2t} + B\cos t + C\sin t + D\cos 2t + E\sin 2t.$$

Find the constants A, B, C, D, E.

★27. (a) Try evaluating $\int (x^m+b)^{p/q}x^r\,dx$, where m, p, q, and r are integers and b is a constant by the substitution $u = (x^m+b)^{1/q}$.

(b) Show that the integral in (a) becomes the integral of a rational function of u when the number $r - m + 1$ is evenly divisible by m.

★28. Any rational function which has the form $p(x)/(x-a)^m(x-b)^n$, where $\deg p < m+n$, can be integrated in the following way:

(i) Write

$$\frac{p(x)}{(x-a)^m(x-b)^n} = \frac{q(x)}{(x-a)^m} + \frac{r(x)}{(x-b)^n},$$

where $\deg q < m$ and $\deg r < n$.

(ii) Integrate each term, using the substitutions $u = x - a$ and $v = x - b$.

(a) Use this procedure to find

$$\int \frac{dx}{(x-2)^2(x-3)^3}.$$

(b) Find the same integral by the ordinary partial fraction method.

(c) Compare answers and the efficiency of the two methods.

10.3 Arc Length and Surface Area

Integration can be used to find the length of graphs in the plane and the area of surfaces of revolution.

In Sections 4.6, 9.1, and 9.2, we developed formulas for areas under and between graphs and for volumes of solids of revolution. In this section we continue applying integration to geometry and obtain formulas for lengths and areas.

The length of a piece of curve in the plane is sometimes called the *arc length* of the curve. As we did with areas and volumes, we assume that the length exists and will try to express it as an integral. For now, we confine our attention to curves which are graphs of functions; general curves are considered in the next section.

We shall begin with an argument involving infinitesimals to derive the formula for arc length. Following this, a different derivation will be given using step functions. The second method is the "honest" one, but it is also more technical.

We consider a curve that is a graph $y = f(x)$ from $x = a$ to $x = b$, as in Fig. 10.3.1. The curve may be thought of as being composed of infinitely

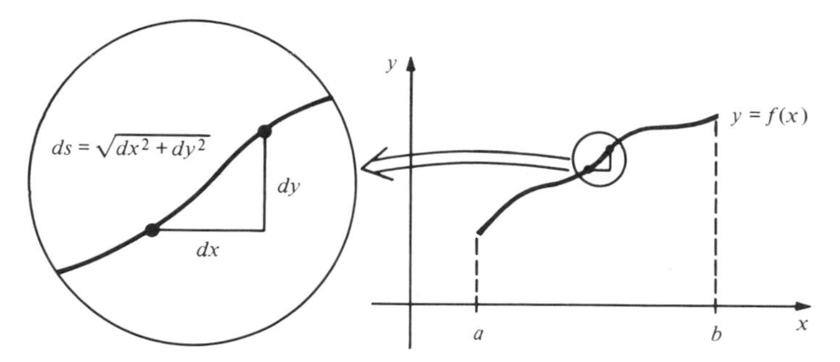

Figure 10.3.1. An "infinitesimal segment" of the graph of f.

many infinitesimally short segments. By the theorem of Pythagoras, the length ds of each segment is equal to $\sqrt{dx^2 + dy^2}$. But $dy/dx = f'(x)$, so $dy = f'(x)\,dx$ and $ds = \sqrt{dx^2 + [f'(x)]^2 dx^2} = \sqrt{1 + [f'(x)]^2}\,dx$. To get the total length, we add up all the infinitesimal lengths: $\int_a^b ds = \int_a^b \sqrt{1 + [f'(x)]^2}\,dx$.

Length of Curves

Suppose that the function f is continuous on $[a, b]$, and that the derivative f' exists and is continuous (except possibly at finitely many points) on $[a, b]$. Then the length of the graph of f on $[a, b]$ is:

$$L = \int_a^b \sqrt{1 + [f'(x)]^2}\,dx. \tag{1}$$

Let us check that formula (1) gives the right result for the length of an arc of a circle.

Example 1 Use integration to find the length of the graph of $f(x)=\sqrt{1-x^2}$ on $[0,b]$, where $0<b<1$. Then find the length geometrically and compare the results.

Solution By formula (1), the length is $\int_0^b \sqrt{1+[f'(x)]^2}\,dx$, where $f(x)=\sqrt{1-x^2}$. We have

$$f'(x)=\frac{-x}{\sqrt{1-x^2}}, \qquad [f'(x)]^2=\frac{x^2}{1-x^2}, \qquad 1+[f'(x)]^2=\frac{1}{1-x^2}.$$

Hence

$$L=\int_0^b \frac{dx}{\sqrt{1-x^2}}=\sin^{-1}(b)-\sin^{-1}(0)=\sin^{-1}(b).$$

Examining Fig. 10.3.2, we see that $\sin^{-1}(b)$ is equal to θ, the angle intercepted

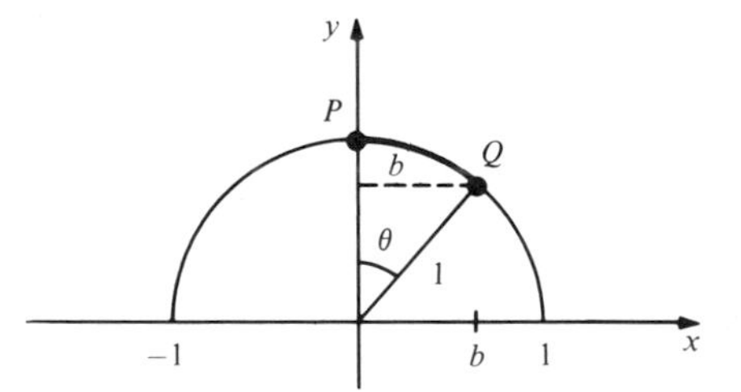

Figure 10.3.2. The length of the arc PQ is $\theta=\sin^{-1}b$.

by the arc whose length we are computing. By the definition of radian measure, the length of the arc is equal to the angle $\theta=\sin^{-1}(b)$, which agrees with our calculation by means of the integral. ▲

Example 2 Find the length of the graph of $f(x)=(x-1)^{3/2}+2$ on $[1,2]$.

Solution We are given $f(x)=(x-1)^{3/2}+2$ on $[1,2]$. Since $f'(x)=\frac{3}{2}(x-1)^{1/2}$, the length of the graph is

$$\begin{aligned}\int_a^b \sqrt{1+[f'(x)]^2}\,dx&=\int_1^2 \sqrt{1+\frac{9}{4}(x-1)}\,dx=\frac{1}{2}\int_1^2 \sqrt{9x-5}\,dx\\&=\frac{1}{18}\int_4^{13} u^{1/2}\,du \qquad (\text{where } u=9x-5)\\&=\tfrac{1}{27}(13^{3/2}-8)\approx 1.44.\ ▲\end{aligned}$$

Due to the square root, the integral in formula (1) is often difficult or even impossible to evaluate by elementary means. Of course, we can always approximate the result numerically (see Section 11.5 for specific examples). The following example shows how a simple-looking function can lead to a complicated integral for arc length.

Example 3 Find the length of the parabola $y=x^2$ from $x=0$ to $x=1$.

Solution We substitute $f(x)=x^2$ and $f'(x)=2x$ into formula (1):

$$L=\int_0^1 \sqrt{1+(2x)^2}\,dx=2\int_0^1 \sqrt{\left(\tfrac{1}{2}\right)^2+x^2}\,dx.$$

Now substitute $x = \frac{1}{2}\tan\theta$ and $\sqrt{(1/2)^2 + x^2} = \frac{1}{2}\sec\theta$:

$$\int\sqrt{\left(\tfrac{1}{2}\right)^2 + x^2}\,dx = \int\left(\frac{1}{2}\sec\theta\right)\left(\frac{1}{2}\sec^2\theta\,d\theta\right) = \frac{1}{4}\int\sec^3\theta\,d\theta.$$

We evaluate the integral of $\sec^3\theta$ using the following trickery[3]:

$$\begin{aligned}\int\sec^3\theta\,d\theta &= \int\sec\theta\sec^2\theta\,d\theta = \int\sec\theta(\tan^2\theta + 1)\,d\theta\\ &= \int\sec\theta\tan^2\theta\,d\theta + \int\sec\theta\,d\theta\\ &= \int(\sec\theta\tan\theta)\tan\theta\,d\theta + \ln|\sec\theta + \tan\theta|.\end{aligned}$$

(see Example 10, Section 10.2.) Now integrate by parts:

$$\begin{aligned}\int(\sec\theta\tan\theta)\tan\theta\,d\theta &= \int\frac{d}{d\theta}(\sec\theta)\tan\theta\,d\theta\\ &= \sec\theta\tan\theta - \int\sec\theta\sec^2\theta\,d\theta\\ &= \sec\theta\tan\theta - \int\sec^3\theta\,d\theta.\end{aligned}$$

Substituting this formula into the last expression for $\int\sec^3\theta\,d\theta$ gives

$$\int\sec^3\theta\,d\theta = \sec\theta\tan\theta - \int\sec^3\theta\,d\theta + \ln|\sec\theta + \tan\theta|;$$

so

$$\int\sec^3\theta\,d\theta = \frac{1}{2}(\sec\theta\tan\theta + \ln|\sec\theta + \tan\theta|) + C.$$

Since $2x = \tan\theta$ and $\sec\theta = 2\cdot\sqrt{(1/2)^2 + x^2} = \sqrt{1 + 4x^2}$, we can express the integral $\int\sec^3\theta\,d\theta$ in terms of x as

$$x\sqrt{1 + 4x^2} + \tfrac{1}{2}\ln|2x + \sqrt{1 + 4x^2}| + C.$$

[One may also evaluate the integral $\int\sqrt{\left(\frac{1}{2}\right)^2 + x^2}\,dx$ using integral formula (43) from the endpapers.] Substitution into the formula for L gives

$$\begin{aligned}L &= \tfrac{1}{2}\left(x\sqrt{1 + 4x^2} + \tfrac{1}{2}\ln|2x + \sqrt{1 + 4x^2}|\right)\Big|_0^1\\ &= \tfrac{1}{2}\left[\sqrt{5} + \tfrac{1}{2}\ln(2 + \sqrt{5})\right] \approx 1.479. \blacktriangle\end{aligned}$$

Example 4 Express the length of the graph of $f(x) = \sqrt{1 - k^2x^2}$ on $[0, b]$ as an integral.

Solution We get

$$f'(x) = -\frac{k^2x}{\sqrt{1 - k^2x^2}}, \qquad \text{so} \qquad \sqrt{1 + [f'(x)]^2} = \sqrt{\frac{1 + (k^4 - k^2)x^2}{1 - k^2x^2}},$$

[3] We can also write

$$\int\sec^3\theta\,d\theta = \int\frac{\cos\theta}{\cos^4\theta}\,d\theta = \int\frac{\cos\theta}{(1 - \sin^2\theta)^2}\,d\theta = \int\frac{du}{(1 - u^2)^2} \qquad (u = \sin\theta).$$

The last integral may now be evaluated by partial fractions.

Thus,

$$L=\int_0^b\sqrt{\frac{1+(k^4-k^2)x^2}{1-k^2x^2}}\;dx.$$

It turns out that the antiderivative

$$\int\sqrt{\frac{1+(k^4-k^2)x^2}{1-k^2x^2}}\;dx$$

cannot be expressed (unless $k^2=0$ or 1) in terms of algebraic, trigonometric, or exponential functions. It is a new kind of function called an *elliptic* function. (See Review Exercises 85 and 92 for more examples of such functions.) ▲

We now turn to the derivation of formula (1) using step functions.

Our first principle for arc length is that the length of a straight line segment is equal to the distance between its endpoints. Thus, if $f(x)=mx+q$ on $[a,b]$, the endpoints (see Fig. 10.3.3) of the graph are $(a, ma+q)$ and

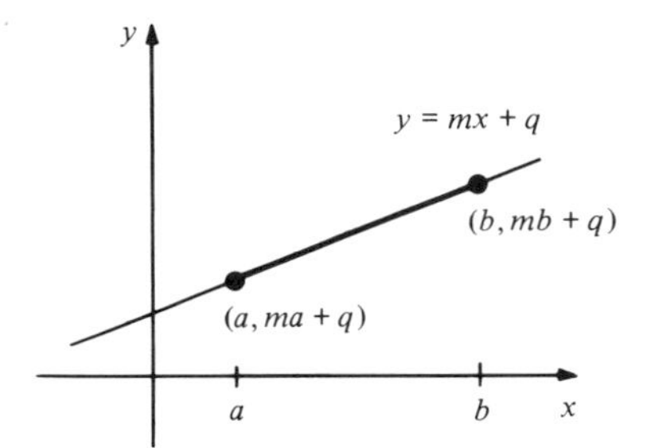

Figure 10.3.3. The length of the dark segment is $(b-a)\sqrt{1+m^2}$, where m is the slope.

$(b, mb+q)$, and the distance between them is

$$\sqrt{(a-b)^2+\left[(ma+q)-(mb+q)\right]^2}=\sqrt{(a-b)^2+m^2(a-b)^2}$$
$$=(b-a)\sqrt{1+m^2}\,.$$

(Since $a<b$, the square root of $(a-b)^2$ is $b-a$.)

Our strategy, as in Chapter 9, will be to interpret the arc length for a simple curve as an integral and then use the same formula for general curves. In the case of the straight line segment, $f(x)=mx+q$, whose length between $x=a$ and $x\doteq b$ is $(b-a)\sqrt{1+m^2}$, we can interpret m as the derivative $f'(x)$, so that

$$\text{Length}=(b-a)\sqrt{1+m^2}=\int_a^b\sqrt{1+\left[f'(x)\right]^2}\;dx.$$

Since the formula for the length is an integral of f', rather than of f, it is natural to look next at the functions for which f' is a step function. If f' is constant on an interval, f is linear on that interval; thus the functions with which we will be dealing are the *piecewise linear* (also called *ramp*, or *polygonal*) functions.

To obtain a piecewise linear function, we choose a partition of the interval $[a,b]$, say, $(x_0, x_1, \ldots, x_n)$ and specify the values $(y_0, y_1, \ldots, y_n)$ of the function f at these points. For each $i = 1, 2, \ldots, n$, we then connect the point (x_{i-1}, y_{i-1}) to the point (x_i, y_i) by a straight line segment (see Fig. 10.3.4).

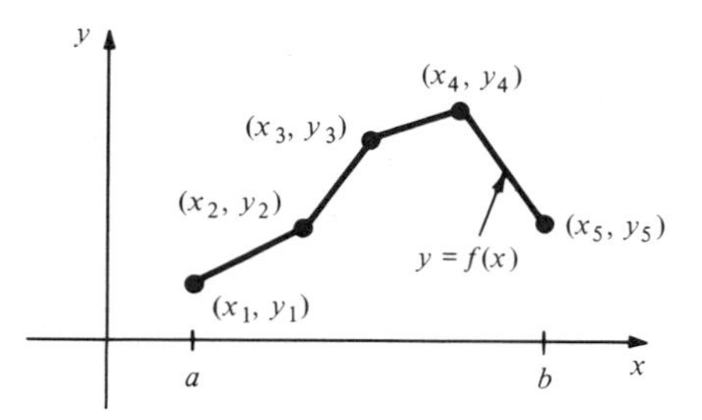

Figure 10.3.4. The graph of a piecewise linear function.

The function $f(x)$ is differentiable on each of the intervals (x_{i-1}, x_i), where its derivative is constant and equal to the slope $(y_i - y_{i-1})/(x_i - x_{i-1})$. Thus the function $\sqrt{1 + [f'(x)]^2}$ is a step function $[a, b]$, with value

$$k_i = \sqrt{1 + \left(\frac{y_i - y_{i-1}}{x_i - x_{i-1}}\right)^2}$$

on (x_{i-1}, x_i).[4] Therefore,

$$\begin{aligned}\int_a^b \sqrt{1 + [f'(x)]^2}\, dx &= \sum_{i=1}^n k_i \Delta x_i \\ &= \sum_{i=1}^n \sqrt{1 + \left(\frac{y_i - y_{i-1}}{x_i - x_{i-1}}\right)^2}\,(x_i - x_{i-1}) \\ &= \sum_{i=1}^n \sqrt{(x_i - x_{i-1})^2 + (y_i - y_{i-1})^2}\,.\end{aligned}$$

Note that the ith term in this sum, $\sqrt{(x_i - x_{i-1})^2 + (y_i - y_{i-1})^2}$, is just the length of the segment of the graph of f between (x_{i-1}, y_{i-1}) and (x_i, y_i).

Now we invoke a second principle of arc length: if n curves are placed end to end, the length of the total curve is the sum of the lengths of the pieces. Using this principle, we see that the preceding sum is just the length of the graph of f on $[a, b]$. So we have now shown, for piecewise linear functions, that the length of the graph of f on $[a, b]$ equals the integral $\int_a^b \sqrt{1 + [f'(x)]^2}\, dx$.

Example 5 Let the graph of f consist of straight line segments joining $(1, 0)$ to $(2, 1)$ to $(3, 3)$ to $(4, 1)$. Verify that the length of the graph, as computed directly, is given by the formula $\int_a^b \sqrt{1 + [f'(x)]^2}\, dx$.

Solution The graph is sketched in Fig. 10.3.5. The length is

$$d_1 + d_2 + d_3 = \sqrt{1 + 1^2} + \sqrt{1 + 2^2} + \sqrt{1 + (-2)^2} = \sqrt{2} + 2\sqrt{5}\,.$$

[4] Actually, $\sqrt{1 + [f'(x)]^2}$ is not defined at the points $x_0, x_1, \ldots, x_n$, but this does not matter when we take its integral, since the integral is not affected by changing the value of the integrand at isolated points.

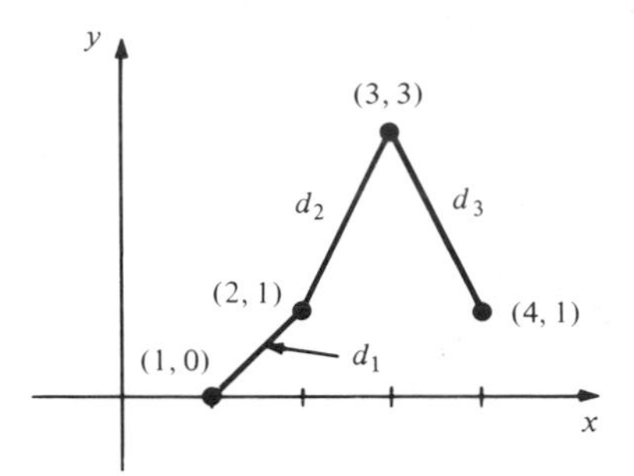

Figure 10.3.5. The length of this graph is $d_1 + d_2 + d_3$.

On the other hand,

$$f'(x) = \begin{cases} 1 & \text{on } (1,2) \\ 2 & \text{on } (2,3) \\ -2 & \text{on } (3,4) \end{cases}$$

(and is not defined at $x = 1,2,3,4$). Thus, by the definition of the integral of a step function (see Section 4.3),

$$\int_1^4 \sqrt{1 + [f'(x)]^2}\, dx = \left(\sqrt{1+1^2}\right)\cdot 1 + \left(\sqrt{1+2^2}\right)\cdot 1 + \left[\sqrt{1+(-2)^2}\right]\cdot 1$$
$$= \sqrt{2} + 2\sqrt{5}$$

which agrees with the preceding answer. ▲

Justifying the passage from step functions to general functions is more complicated than in the case of area, since we cannot, in any straightforward way, squeeze a general curve between polygons as far as length is concerned. Nevertheless, it is plausible that any reasonable graph can be approximated by a piecewise linear function, so formula (1) should carry over. These considerations lead to the technical conditions stated in conjunction with formula (1).

If we revolve the region R under the graph of $f(x)$ (assumed nonnegative) on $[a,b]$, about the x axis, we obtain a solid of revolution S. In Section 9.1 we saw how to express the volume of such a solid as an integral. Suppose now that instead of revolving the region, we revolve the graph $y = f(x)$ itself. We obtain a curved surface Σ, called a *surface of revolution*, which forms part of the boundary of S. (The remainder of the boundary consists of the disks at the ends of the solid, which have radii $f(a)$ and $f(b)$; see Fig. 10.3.6.) Our next goal is to obtain a formula for the *area* of the surface Σ. Again we give the argument using infinitesimals first.

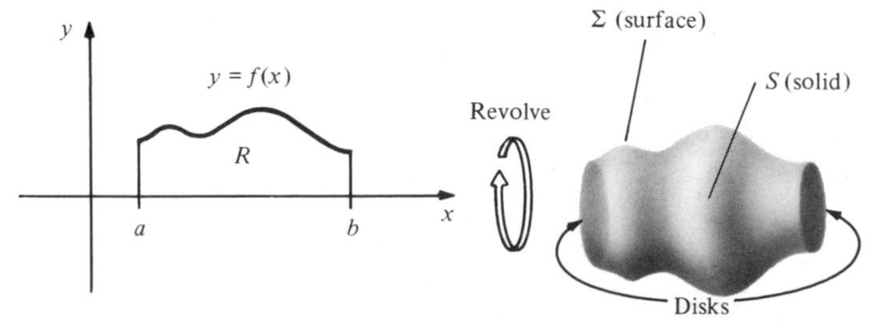

Figure 10.3.6. The boundary of the solid of revolution S consists of the surface of revolution Σ obtained by revolving the graph, together with two disks.

Referring to Figure 10.3.7, we may think of a smooth surface of revolution as being composed of infinitely many infinitesimal bands, as in Fig.

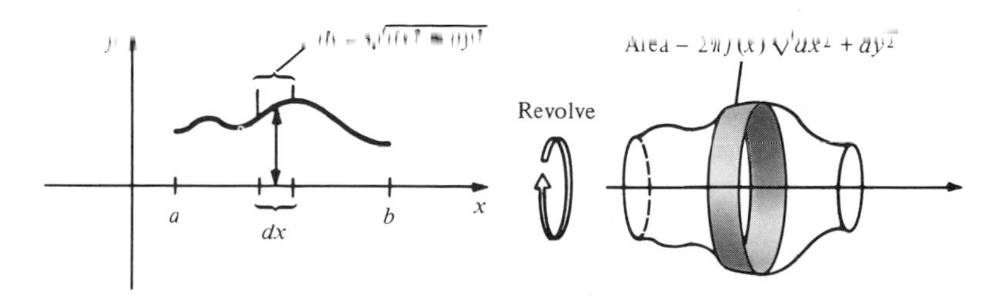

Figure 10.3.7. The surface of revolution may be considered as composed of infinitely many infinitesimal frustums.

10.3.7. The area of each band is equal to its circumference $2\pi f(x)$ times its width $ds = \sqrt{dx^2 + dy^2}$, so the total area is

$$\int_a^b 2\pi f(x)\sqrt{dx^2 + dy^2} = \int_a^b 2\pi f(x)\sqrt{1 + [f'(x)]^2}\, dx$$

since $dy = f'(x)\,dx$.

Area of a Surface of Revolution about the x Axis

The area of the surface obtained by revolving the graph of $f(x)$ ($\geqslant 0$) on $[a,b]$ about the x axis is

$$A = 2\pi\int_a^b f(x)\sqrt{1 + [f'(x)]^2}\, dx = 2\pi\int_a^b y\sqrt{1 + \left(\frac{dy}{dx}\right)^2}\, dx. \quad (2)$$

We now check that formula (2) gives the correct area for a sphere.

Example 6 Find the area of the spherical surface of radius r obtained by revolving the graph of $y = \sqrt{r^2 - x^2}$ on $[-r, r]$ about the x axis.

Solution As in Example 1, we have $\sqrt{1 + [f'(x)]^2} = r/\sqrt{r^2 - x^2}$, so the area is

$$\int_{-r}^{r} 2\pi\sqrt{r^2 - x^2}\,\frac{r}{\sqrt{r^2 - x^2}}\, dx = 2\pi r\int_{-r}^{r} dx = 2\pi r \cdot 2r = 4\pi r^2$$

which is the usual value for the area of a sphere. ▲

If, instead of the entire sphere, we take the band obtained by restricting x to $[a,b]$ $(-r \leqslant a \leqslant b \leqslant r)$, the area is $2\pi\int_a^b r\,dx = 2\pi r(b - a)$. Thus the area obtained by slicing a sphere by two parallel planes and taking the middle piece is equal to $2\pi r$ times the distance between the planes, regardless of where the two planes are located (see Fig. 10.3.8). Why doesn't the "longer" band around the middle have more area?

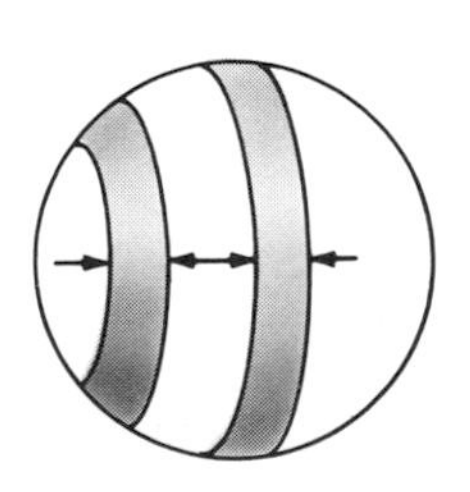

Figure 10.3.8. Bands of equal width have equal area.

As with arc length, the factor $\sqrt{1 + [f'(x)]^2}$ in the integrand sometimes makes it impossible to evaluate the surface area integrals by any means other than numerical methods (see Section 11.5). To get a problem which can be solved, we must choose f carefully.

Example 7 Find the area of the surface obtained by revolving the graph of x^3 on $[0,1]$ about the x axis.

Solution We find that $f'(x)=3x^2$ and $\sqrt{1+[f'(x)]^2}=\sqrt{1+9x^4}$, so

$$A=2\pi\int_0^1\sqrt{1+9x^4}\,x^3\,dx=\frac{\pi}{2}\int_0^1\sqrt{1+9u}\,du \qquad (u=x^4,\ du=4x^3\,dx)$$

$$=\frac{\pi}{18}\int_0^9(1+v)^{1/2}\,dv \qquad \left(u=\frac{1}{9}v,\ du=\frac{1}{9}\,dv\right)$$

$$=\frac{\pi}{18}\left[\frac{2}{3}(1+v)^{3/2}\right]\Bigg|_0^9=\frac{\pi}{27}(10^{3/2}-1)\approx 3.56. \ \blacktriangle$$

By a method similar to that for deriving equation (2), we can derive a formula for the area obtained by revolving the graph $y=f(x)$ about the y axis for $a\leqslant x\leqslant b$. Referring to Figure 10.3.9, the area of the shaded band is

$$\text{Width}\times\text{Circumference}=ds\cdot 2\pi x=2\pi x\sqrt{dx^2+dy^2}$$

Thus the surface area is $\int_a^b 2\pi x\sqrt{1+[f'(x)]^2}\,dx$.

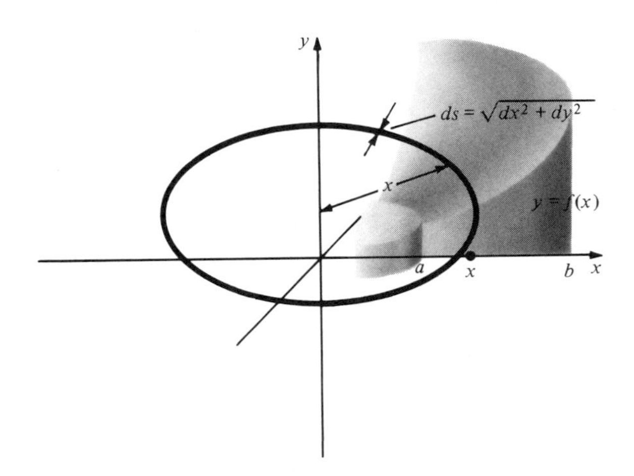

Figure 10.3.9. Rotating $y=f(x)$ about the y axis.

Area of a Surface of Revolution about the y Axis

The area of the surface obtained by revolving the graph of $f(x)$ ($\geqslant 0$) on $[a,b]$ about the y axis is

$$A=2\pi\int_a^b x\sqrt{1+[f'(x)]^2}\,dx. \tag{3}$$

Example 8 Find the area of the surface obtained by revolving the graph $y=x^2$ about the y axis for $1\leqslant x\leqslant 2$.

Solution If $f(x)=x^2$, $f'(x)=2x$ and $\sqrt{1+[f'(x)]^2}=\sqrt{1+4x^2}$. Then

$$A=2\pi\int_1^2 x\sqrt{1+4x^2}\,dx=\frac{\pi}{4}\int_5^{17}u^{1/2}\,du \qquad (u=1+4x^2,\ du=8x\,dx)$$

$$=\frac{\pi}{4}\left(\frac{2}{3}u^{3/2}\right)\Bigg|_5^{17}=\frac{\pi}{6}(17^{3/2}-5^{3/2})\approx 30.85. \ \blacktriangle$$

Finally we sketch how one derives formula (2) using step functions. The derivation of formula (3) is similar (see Exercise 41).

If $f(x) = k$, a constant, the surface is a cylinder of radius k and height $b - a$. Unrolling the cylinder, we obtain a rectangle with dimensions $2\pi k$ and $b - a$ (see Fig. 10.3.10), whose area is $2\pi k(b - a)$, so we can say that the area of the cylinder is $2\pi k(b - a)$.

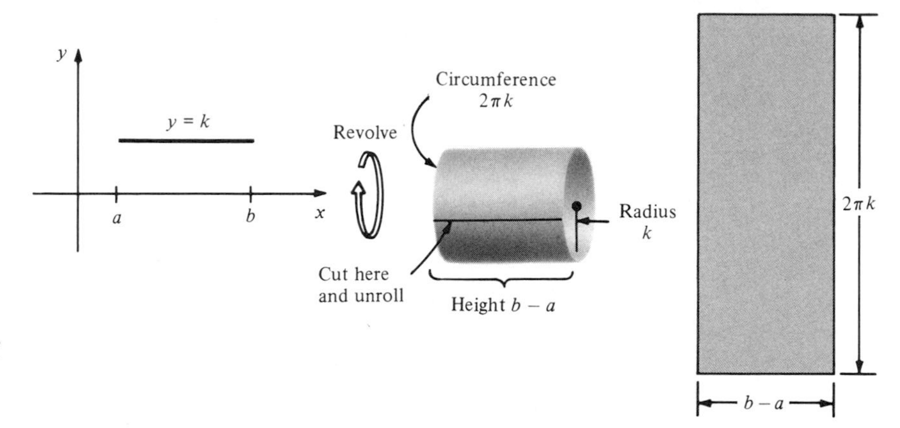

Figure 10.3.10. The area of the shaded cylinder is $2\pi k(b - a)$.

Next we look at the case where $f(x) = mx + q$, a linear function. The surface of revolution, as shown in Fig. 10.3.11, is a frustum of a cone—that is, the surface obtained from a right circular cone by cutting it with two planes perpendicular to the axis. To find the area of this surface, we may slit the frustum along a line and unroll it into the plane, as in Fig. 10.3.12, obtaining a circular sector of radius r and angle θ with a concentric sector of radius $r - s$ removed. By the definition of radian measure, we have $\theta r = 2\pi r_2$ and $\theta(r - s) = 2\pi r_1$, so $\theta s = 2\pi(r_2 - r_1)$, or $\theta = 2\pi[(r_2 - r_1)/s]$; from this we find that $r = r_2 s/(r_2 - r_1)$. The area of the figure is

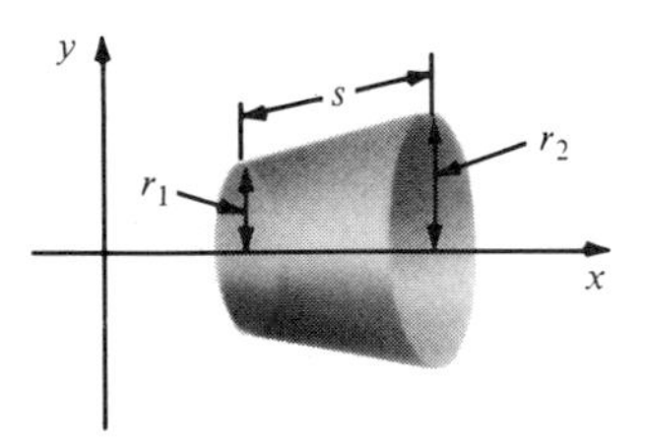

Figure 10.3.11. A frustum of a cone.

$$\begin{aligned}\frac{\theta}{2\pi}\left[\pi r^2 - \pi(r - s)^2\right] &= \frac{\theta}{2}\left[r^2 - (r^2 - 2rs + s^2)\right] = \frac{\theta}{2}(2rs - s^2)\\ &= \theta s\left(r - \frac{s}{2}\right) = 2\pi(r_2 - r_1)\left(\frac{r_2 s}{r_2 - r_1} - \frac{s}{2}\right)\\ &= 2\pi s\left(r_2 - \frac{r_2 - r_1}{2}\right) = \pi s(r_1 + r_2).\end{aligned}$$

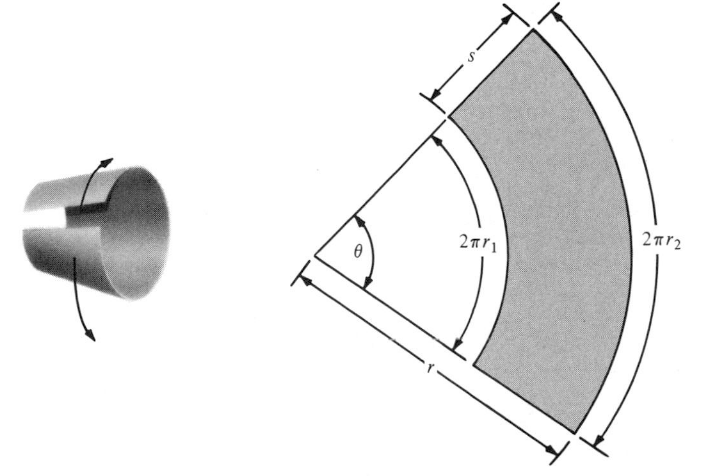

Figure 10.3.12. The area of the frustum, found by cutting and unrolling it, is $\pi s(r_1 + r_2)$.

(Notice that the proof breaks down in the case $r_1 = r_2 = k$, a cylinder, since r is then "infinite." Nevertheless, the resulting formula $2\pi ks$ for the area is still correct!)

We now wish to express $\pi s(r_1 + r_2)$ as an integral involving the function $f(x) = mx + q$. We have $r_1 = ma + q$, $r_2 = mb + q$, and $s = \sqrt{1+m^2}\,(b-a)$ (see Fig. 10.3.3), so the area is

$$\begin{aligned}
&\pi\sqrt{1+m^2}\,(b-a)\big[m(b+a)+2q\big] \\
&\quad = \pi\sqrt{1+m^2}\,\big[m(b^2-a^2)+2q(b-a)\big] \\
&\quad = 2\pi\sqrt{1+m^2}\left[m\frac{b^2-a^2}{2}+q(b-a)\right] \\
&\quad = 2\pi\sqrt{1+m^2}\left(m\frac{x^2}{2}+qx\right)\Bigg|_a^b .
\end{aligned}$$

Since $m(x^2/2) + qx$ is the antiderivative of $mx + q$, we have

$$\begin{aligned}
&2\pi\sqrt{1+m^2}\left(m\frac{x^2}{2}+qx\right)\Bigg|_a^b \\
&\quad = 2\pi\int_a^b \sqrt{1+m^2}\,(mx+q)\,dx \\
&\quad = 2\pi\int_a^b\left[\sqrt{1+f'(x)^2}\,\right]f(x)\,dx,
\end{aligned}$$

so we have succeeded in expressing the surface area as an integral.

Now we are ready to work with general surfaces. If $f(x)$ is *piecewise* linear on $[a,b]$, the surface obtained is a "conoid," produced by pasting together a finite sequence of frustums of cones, as in Fig. 10.3.13.

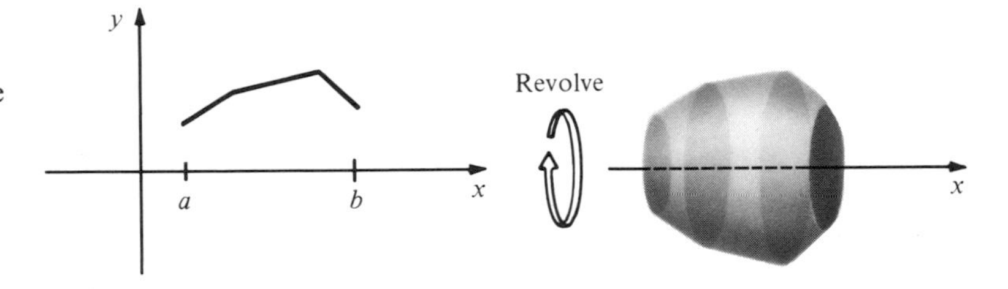

Figure 10.3.13. The surface obtained by revolving the graph of a piecewise linear function is a "conoid" consisting of several frustums pasted together.

The area of the conoid is the sum of the areas of the component frustums. Since the area of each frustum is given by the integral of the function $2\pi\sqrt{1+[f'(x)]^2}\,f(x)$ over the appropriate interval, the additivity of the integral implies that the area of the conoid is given by the same formula:

$$A = 2\pi\int_a^b f(x)\sqrt{1+[f'(x)]^2}\,dx.$$

We now assert, as we did for arc length, that this formula is true for general functions f. [To do this rigorously, we would need a precise definition of surface area, which is rather complicated to give (much more complicated, even, than for arc length).]

Example 9 The polygon joining the points $(2,0)$, $(4,4)$, $(7,5)$, and $(8,3)$ is revolved about the x axis. Find the area of the resulting surface of revolution.

Solution The function f whose graph is the given polygon is

$$f(x)=\begin{cases}2(x-2), & 2\leqslant x\leqslant 4,\\ \frac{1}{3}(x-4)+4, & 4\leqslant x\leqslant 7,\\ -2(x-7)+5, & 7\leqslant x\leqslant 8.\end{cases}$$

Then we have

$$f'(x)=\begin{cases}2, & 2<x<4,\\ \frac{1}{3}, & 4<x<7,\\ -2, & 7<x<8.\end{cases}$$

Thus, $A=2\pi\int_2^8 f(x)\sqrt{1+f'(x)^2}\,dx$

$$=2\pi\left[\int_2^4 f(x)\sqrt{1+f'(x)^2}\,dx+\int_4^7 f(x)\sqrt{1+f'(x)^2}\,dx\right.$$
$$\left.+\int_7^8 f(x)\sqrt{1+f'(x)^2}\,dx\right]$$
$$=2\pi\left\{\int_2^4[2(x-2)]\sqrt{1+4}\,dx+\int_4^7\left[\frac{1}{3}(x-4)+4\right]\sqrt{1+\frac{1}{9}}\,dx\right.$$
$$\left.+\int_7^8[-2(x-7)+5]\sqrt{1+4}\,dx\right\}.$$

Using $\int(x-a)\,dx=\frac{1}{2}(x-a)^2+C$, we find

$$A=2\pi\left\{\sqrt{5}\left[(x-2)^2\right]\Big|_2^4+\frac{\sqrt{10}}{3}\left[\frac{1}{6}(x-4)^2+4x\right]\Big|_4^7+\sqrt{5}\left[-(x-7)^2+5x\right]\Big|_7^8\right\}$$
$$=2\pi\left(4\sqrt{5}+\frac{9}{2}\sqrt{10}+4\sqrt{5}\right)\approx 201.8. \ \blacktriangle$$

Exercises for Section 10.3

1. Find the length of the graph of the function $f(x)=x^4/8+1/4x^2$ on $[1,3]$.
2. Find the length of the graph of the function $f(x)=(x^4-12x+3)/6x$ on $[2,4]$.
3. Find the length of the graph of the function $y=[x^3+(3/x)]/6$ on $1\leqslant x\leqslant 3$.
4. Find the length of the graph of the function $y=\sqrt{x}\,(4x-3)/6$ on $1\leqslant x\leqslant 9$.
5. Express the length of the graph of x^n on $[a,b]$ as an integral. (Do not evaluate.)
6. Express the length of the graph of $f(x)=\sin x$ on $[0,2\pi]$ as an integral. (Do not evaluate.)
7. Express the length of the graph of $f(x)=x\cos x$ on $[0,1]$ as an integral. (Do not evaluate.)
8. Express the length of the graph $y=e^{-x}$ on $[-1,1]$ as an integral. (Do not evaluate.)

In Exercises 9–12, let the graph of f consist of straight line segments joining the given points. Verify that the length of the graph as computed directly is equal to that given by the arc length formula.

9. $(0,0)$ to $(1,2)$ to $(2,1)$ to $(5,0)$.
10. $(1,1)$ to $(2,2)$ to $(3,0)$.
11. $(-1,-1)$ to $(0,1)$ to $(1,2)$ to $(2,-2)$.
12. $(-2,2)$ to $(-1,-3)$ to $(3,1)$.

Find the area of the surfaces obtained by revolving the curves in Exercises 13–20.

13. The graph of $\sqrt{x+1}$ on $[0,2]$ about the x axis.
14. The graph of $y=[x^3+(3/x)]/6$, $1\leqslant x\leqslant 3$ about the x axis.
15. The graph of $y=\sqrt{x}\,(4x-3)/6$, $1\leqslant x\leqslant 9$ about the x axis.
16. The graph of e^x on $[0,1]$ about the x axis.
17. The graph of $y=\cos x$ on $[-\pi/2,\pi/2]$ about the x axis.
18. The parabola $y=\sqrt{x}$ on $[4,5]$ about the x axis.
19. The graph of $x^{1/3}$ on $[1,3]$ about the y axis.
20. The graph of $y=\ln x$ on $[2,3]$ about the y axis.

In Exercises 21–24, the polygon joining the given points is revolved about the x axis. Find the area of the resulting surface of revolution.

21. $(0,0)$ to $(1,1)$ to $(2,0)$.
22. $(1,0)$ to $(3,2)$ to $(4,0)$.
23. $(2,1)$ to $(3,2)$ to $(4,1)$ to $(5,3)$.
24. $(4,0)$ to $(5,2)$ to $(6,1)$ to $(8,0)$.

25. Find the length of the graph of $a(x+b)^{3/2}+c$ on $[0,1]$, where a, b, and c are constants. What is the effect of changing the value of c?
26. Find the length of the graph of $y=x^2$ on $[0,b]$.
27. Express the length of the graph of $f(x)=2x^3$ on $[-1,2]$ as an integral. Evaluate numerically to within 1.0 by finding upper and lower sums. Compare your results with a string-and-ruler measurement.
28. Find the length, accurate to within 1 centimeter, of the curve in Fig. 10.3.14.

Figure 10.3.14. Find the length of this curve.

For each of the functions and intervals in Exercises 29–32, express as an integral: (a) the length of the curve; (b) the area of the surface obtained by revolving the curve about the x axis. (Do not evaluate the integrals.)

29. $\tan x + 2x$ on $[0,\pi/2]$
30. x^3+2x-1 on $[1,3]$
31. $1/x + x$ on $[1,2]$
32. $e^x + x^3$ on $[0,1]$

33. Find the area, accurate to within 5 square centimeters, of the surface obtained by revolving the curve in Fig. 10.3.15 around the x axis.

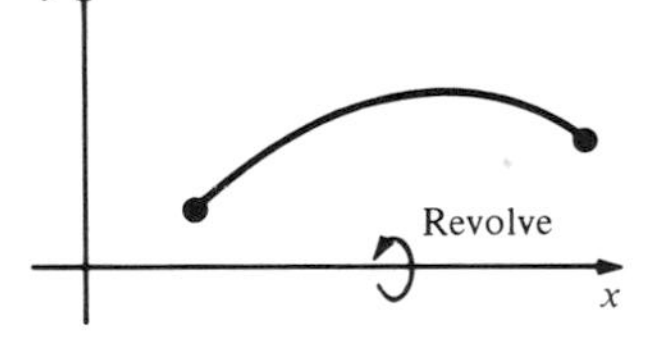

Figure 10.3.15. Find the area of the surface obtained by revolving this curve.

34. Use upper and lower sums to find the area, accurate to within 1 unit, of the surface obtained by revolving the graph of x^4 on $[0,1]$ about the x axis.
35. Prove that the length of the graph of $f(x)=\cos(\sqrt{3}\,x)$ on $[0,2\pi]$ is less than or equal to 4π.
36. Suppose that $f(x)\geqslant g(x)$ for all x in $[a,b]$. Does this imply that the length of the graph of f on $[a,b]$ is greater than or equal to that for g? Justify your answer by a proof or an example.

★37. Show that the length of the graph of $\sin x$ on $[0.1,1]$ is less than the length of the graph of $1+x^4$ on $[0.1,1]$.

★38. Suppose that the function f on $[a,b]$ has an inverse function g defined on $[\alpha,\beta]$. Assume that $0<a<b$ and $0<\alpha<\beta$.
   (a) Find a formula, in terms of f, for the area of the surface obtained by revolving the graph of g on $[\alpha,\beta]$ about the x axis.
   (b) Show that this formula is consistent with the one in formula (3) for the area of the surface obtained by revolving the graph of f on $[a,b]$ about the y axis.

★39. Write an integral representing the area of the surface obtained by revolving the graph of $1/(1+x^2)$ about the x axis. Do not evaluate the integral, but show that it is less than $2\sqrt{5}\,\pi^2$ no matter how long an interval is taken.

★40. Craftsman Cabinet Company was preparing a bid on a job that required epoxy coating of several tank interiors. The tanks were constructed from steel cylinders C feet in circumference and height H feet, with spherical steel caps welded to each end (see Fig. 10.3.16). Specifications required a $\frac{1}{8}$-inch coating. The 20-year-old estimator quickly figured the cylindrical part as HC square feet. For the spherical cap he stretched a tape measure over the cap to obtain S ft.

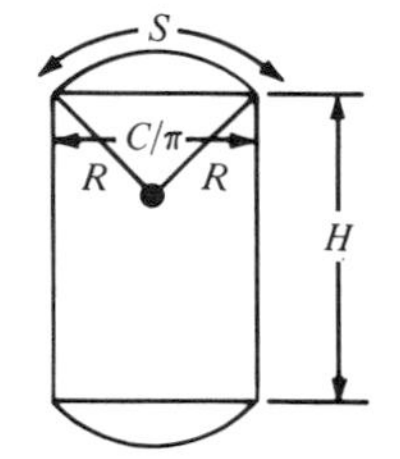

Figure 10.3.16. A cross section of a tank requiring an epoxy coating on its interior.

   (a) Find an equation relating the surface area of the steel cap and the tape measurements S and C. [*Hint*: Revolve $y=\sqrt{R^2-x^2}$ about the y axis, $0\leqslant x\leqslant c/2\pi$.]
   (b) Find an equation relating the surface area of the tank, S, C, and H.
   (c) Determine the cost for six tanks with $H=16$ feet, $C=37.7$ feet, $S=13.2$ feet, given that the coating costs \$2.10 per square foot.

★41. (a) Calculate the area of the frustum shown in Fig. 10.3.17 using geometry alone. (b) Derive formula (3) using step functions.

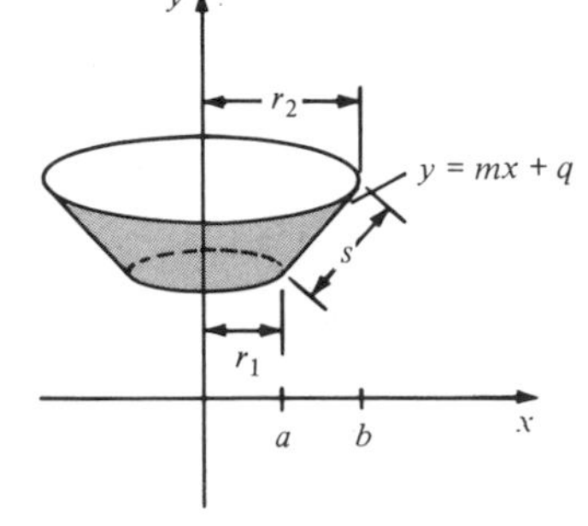

Figure 10.3.17. A line segment revolved around the y axis becomes a frustum of a cone.

10.4 Parametric Curves

Arc lengths may be found by integral calculus for curves which are not graphs of functions.

We begin this section with a study of the differential calculus of parametric curves, a topic which was introduced in Section 2.4. The arc length of a parametric curve is then expressed as an integral.

Recall from Section 2.4 that a *parametric curve* in the xy plane is specified by a pair of functions: $x = f(t)$, $y = g(t)$. The variable t, called the *parameter* of the curve, may be thought of as time; the pair $(f(t), g(t))$ then describes the path in the plane of a moving point. Many physical situations, such as the motion of the Earth about the sun and a car moving on a twisting highway, can be conveniently idealized as parametric curves.

Example 1 (a) Describe the motion of the point (x, y) if $x = \cos t$ and $y = \sin t$, for t in $[0, 2\pi]$. (b) Describe the motion of the point (t, t^3) for t in $(-\infty, \infty)$.

Solution (a) At $t = 0$, the point is at $(1,0)$. Since $\cos^2 t + \sin^2 t = 1$, the point (x, y) satisfies $x^2 + y^2 = 1$, so it moves on the unit circle. As t increases from zero, $x = \cos t$ decreases and $y = \sin t$ increases, so the point moves in a counterclockwise direction. Finally, since $(\cos(2\pi), \sin(2\pi)) = (1,0)$, the point makes a full rotation after 2π units of time (see Fig. 10.4.1).
(b) We have $x^3 = t^3 = y$, so the point is on the curve $y = x^3$. As t increases so does x, and the point moves from left to right (see Fig. 10.4.2). ▲

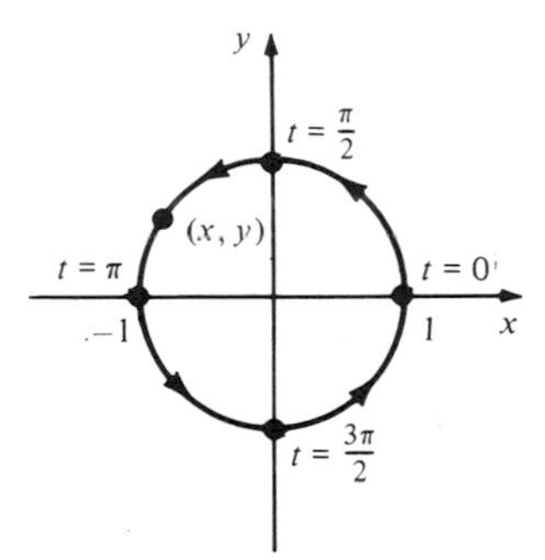

Figure 10.4.1. The point $(\cos t, \sin t)$ moves in a circle.

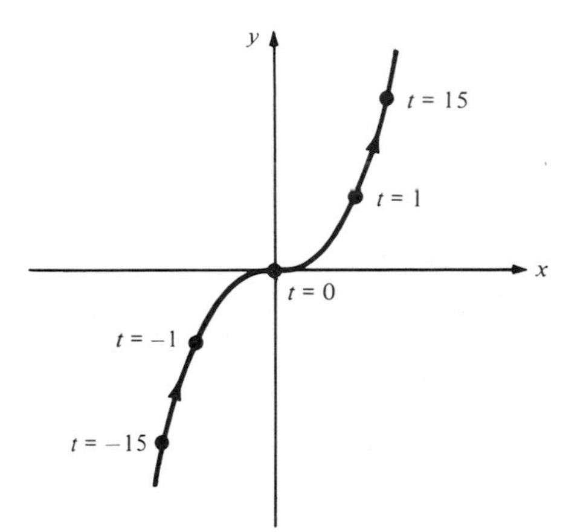

Figure 10.4.2. The motion of the point (t, t^3).

Example 1(b) illustrates a general fact: Any curve $y = f(x)$ which is the graph of a function can be described parametrically: we set $x = t$ and $y = f(t)$. However, parametric equations can describe curves which are not the graphs of functions, like the circle in Example 1(a).

The equations

$$x = at + b, \qquad y = ct + d$$

describe a straight line. To show this, we *eliminate the parameter* t in the following way. If $a \neq 0$, solve the first equation for t, getting $t = (x - b)/a$. Substituting this into the second equation gives $y = c[(x - b)/a] + d$; that is, $y = (c/a)x + (ad - bc)/a$, which is a straight line with slope c/a. If $a = 0$, we have $x = b$ and $y = ct + d$. If $c \neq 0$, then y takes all values as t varies and b is fixed, so we have the vertical line $x = b$ (which is not the graph of a function). If $c = 0$ as well as $a = 0$, then $x = b$ and $y = d$, so the graph is a "stationary" point (b, d).

Similarly, we can see that

$$x = r\cos t + x_0, \qquad y = r\sin t + y_0$$

describes a circle by writing

$$\frac{x - x_0}{r} = \cos t, \qquad \frac{y - y_0}{r} = \sin t.$$

Therefore,

$$\left(\frac{x - x_0}{r}\right)^2 + \left(\frac{y - y_0}{r}\right)^2 = \cos^2 t + \sin^2 t = 1$$

or $(x - x_0)^2 + (y - y_0)^2 = r^2$, which is the equation of a circle with radius r and center (x_0, y_0). As t varies from 0 to 2π, the point (x, y) moves once around the circle.

Parametric Equations of Lines and Circles

Straight line

$x = at + b,$ $\quad -\infty < t < \infty;$
$y = ct + d$ $\quad a$ and c not both zero; the line passes through (b, d) with slope c/a.

Circle

$x = r\cos t + x_0,$ $\quad 0 \leqslant t \leqslant 2\pi;$
$y = r\sin t + y_0,$ $\quad r > 0, r = \text{radius}, (x_0, y_0) = \text{center}.$

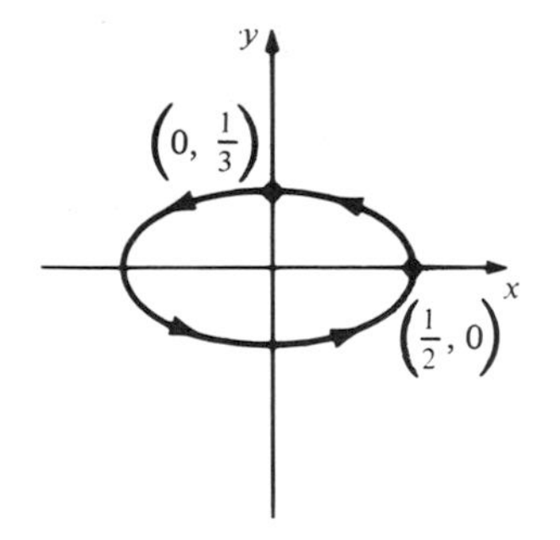

Figure 10.4.3. The parametric curve $x = \frac{1}{2}\cos t, y = \frac{1}{3}\sin t$ is an ellipse.

Other curves can be written conveniently in parametric form as well. For example, $4x^2 + 9y^2 = 1$ (an ellipse) can be written as $x = \frac{1}{2}\cos t, y = \frac{1}{3}\sin t$. As t goes from 0 to 2π, the point moves once around the ellipse (see Fig. 10.4.3). General properties of ellipses are studied in Section 14.1.

The same geometric curve can often be represented parametrically in more than one way. For example, the line $x = at + b, y = ct + d$ can also be represented by

$$x = t, \qquad y = \frac{ct}{a} + \frac{ad - bc}{a}$$

or by

$$x = t^3, \qquad y = \frac{ct^3}{a} + \frac{ad - bc}{a}.$$

(If we used t^2, we would get only half of the line since $t^2 \geqslant 0$ for all t.)

In Section 2.4 we saw that the tangent line to a parametric curve $(x, y) = (f(t), g(t))$ at the point $(f(t_0), g(t_0))$ has slope

$$\frac{dy}{dx} = \frac{dy/dt}{dx/dt} = \frac{g'(t_0)}{f'(t_0)}$$

If $f'(t_0) = 0$ and $g'(t_0) \neq 0$, the tangent line is vertical; if $f'(t_0)$ and $g'(t_0)$ are both zero, the tangent line is not defined. Since the tangent line passes through $(f(t_0), g(t_0))$, we may write its equation in point-slope form:

$$y = \frac{g'(t_0)}{f'(t_0)}\left[x - f(t)\right] + g(t_0). \tag{1}$$

Example 2 Find the equation of the tangent line when $t = 1$ for the curve $x = t^4 + 2\sqrt{t}$, $y = \sin(t\pi)$.

Solution When $t = 1$, $x = 3$ and $y = \sin\pi = 0$. Furthermore, $dx/dt = 4t^3 + 1/\sqrt{t}$, which equals 5 when $t = 1$; $dy/dt = \pi\cos(t\pi)$, which equals $-\pi$ when $t = 1$. Thus the equation of the tangent line is, by formula (1),

$$y = -\frac{\pi}{5}(x - 3) + 0 \quad \text{or} \quad y = -\frac{\pi}{5}x + \frac{3\pi}{5}. \ \blacktriangle$$

If a curve is given parametrically, it is natural to express its tangent line parametrically as well. To do this, we transform equation (1) to the form

$$\frac{y - g(t_0)}{g'(t_0)} = \frac{x - f(t_0)}{f'(t_0)}.$$

We can set both sides of this equation equal to t, obtaining

$$x = tf'(t_0) + f(t_0), \qquad y = tg'(t_0) + g(t_0). \tag{2}$$

Equation (2) is the parametric equation for a line with slope $g'(t_0)/f'(t_0)$ if $f'(t_0) \neq 0$. If $f'(t_0) = 0$ but $g'(t_0) \neq 0$, equations (2) describe a vertical line. If $f'(t_0)$ and $g'(t_0)$ are both zero, equations (2) describe a stationary point.

It is convenient to make one more transformation of equations (2), so that the tangent line passes through (x_0, y_0) at the same time t_0 as the curve, rather than at $t = 0$. Substituting $t - t_0$ for t, we obtain the formulas

$$x = f'(t_0)(t - t_0) + f(t_0), \qquad y = g'(t_0)(t - t_0) + g(t_0). \tag{3}$$

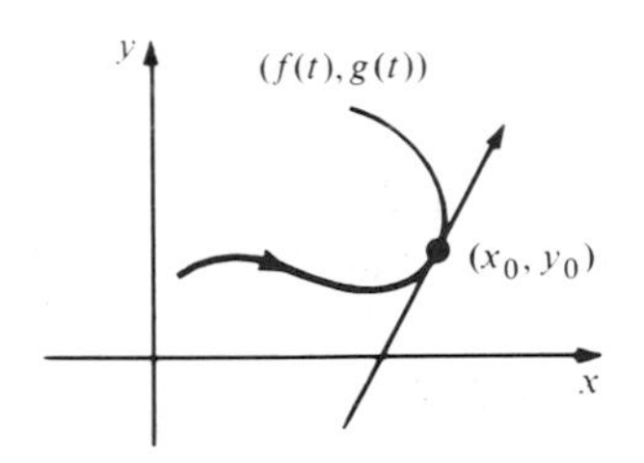

Figure 10.4.4. If the forces constraining a particle to the curve $(f(t), g(t))$ are removed at t_0, then the particle will follow the tangent line at t_0.

Notice that the functions in formulas (3) which define the tangent line to a curve are exactly the *linear approximations* to the functions defining the curve itself. If we think of $(x, y) = (f(t), g(t))$ as the position of a moving particle, then the tangent line at t_0 is the path which the particle would follow if, at time t_0, all constraining forces were suddenly removed and the particle were allowed to move freely in a straight line. (See Fig. 10.4.4.)

Example 3 A child is whirling an object on a string, letting out string at a constant rate, so that the object follows the path $x = (1 + t)\cos t$, $y = (1 + t)\sin t$.

(a) Sketch the path for $0 \leqslant t \leqslant 4\pi$.
(b) At $t = 4\pi$ the string breaks, so that the object follows its tangent line. Where is the object at $t = 5\pi$?

Solution (a) By plotting some points and thinking of (x, y) as moving in an ever enlarging circle, we obtain the sketch in Fig. 10.4.5.

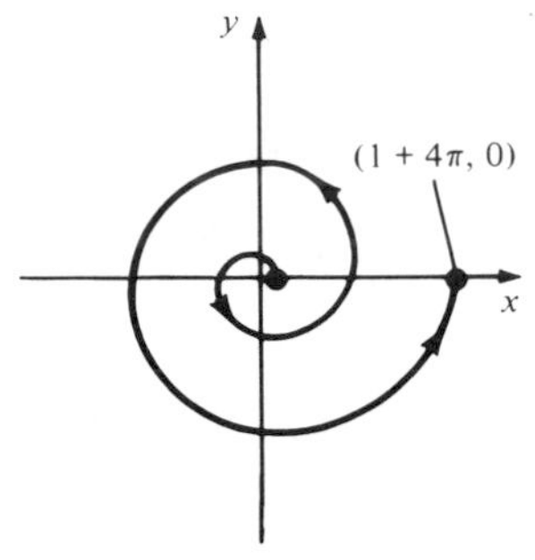

Figure 10.4.5. The curve $((1+t)\cos t, (1+t)\sin t)$ for t in $[0, 4\pi]$.

(b) We differentiate:

$$f'(t) = \frac{dx}{dt} = (1+t)(-\sin t) + \cos t, \quad \text{and}$$

$$g'(t) = \frac{dy}{dt} = (1+t)\cos t + \sin t.$$

When $t_0 = 4\pi$, we have

$$f(t_0) = (1+4\pi)\cos 4\pi = 1 + 4\pi, \quad \text{and} \quad g(t_0) = (1+4\pi)\sin 4\pi = 0,$$

$$f'(t_0) = (1+4\pi)\cdot 0 + 1 = 1, \quad \text{and} \quad g'(t_0) = (1+4\pi)\cdot 1 + 0 = 1 + 4\pi.$$

By formulas (3), the equations of the tangent line are

$$x = t - 4\pi + (1+4\pi), \qquad y = (1+4\pi)(t-4\pi) + 0.$$

When $t = 5\pi$, the object, which is now following the tangent line, is at $x = 1 + 5\pi \approx 16.71$, $y = (1+4\pi)\pi \approx 42.62$. ▲

Tangents to Parametric Curves

Let $x = f(t)$ and $y = g(t)$ be the parametric equations of a curve C. If f and g are differentiable at t_0, and $f'(t_0)$ and $g'(t_0)$ are not both zero, then the tangent line to C at t_0 is defined by the parametric equations:

$$x = f'(t_0)(t - t_0) + f(t_0), \qquad y = g'(t_0)(t - t_0) + g(t_0).$$

If $f'(t_0) \neq 0$, this line has slope $g'(t_0)/f'(t_0)$, and its equation can be written as

$$y = \frac{g'(t_0)}{f'(t_0)}\left[x - f(t_0)\right] + g(t_0).$$

If $f'(t_0) = 0$ and $g'(t_0) \neq 0$, the line is vertical; its equation is

$$x = f(t_0).$$

Example 4 Consider the curve $x = t^3 - t$, $y = t^2$.

(a) Plot the points corresponding to $t = -2, -1, -\frac{1}{2}, 0, \frac{1}{2}, 1, 2$.

(b) Using these points, together with the behavior of the functions t^2 and $t^3 - t$, sketch the entire curve.

(c) Find the slope of the tangent line at the points corresponding to $t = 1$ and $t = -1$.

(d) Eliminate the parameter t to obtain an equation in x and y for the curve.

Solution (a) We begin by making a table:

t	-2	-1	$-\frac{1}{2}$	0	$\frac{1}{2}$	1	2
$x = t^3 - t$	-6	0	$\frac{3}{8}$	0	$-\frac{3}{8}$	0	6
$y = t^2$	4	1	$\frac{1}{4}$	0	$\frac{1}{4}$	1	4

These points are plotted in Fig. 10.4.6. The number next to each point is the corresponding value of t. Notice that the point $(0, 1)$ occurs for $t = -1$ and $t = 1$.

(b) We plot x and y against t in Fig. 10.4.7. From the graph of x against t, we conclude that as t goes from $-\infty$ to ∞, the point comes in from the left, reverses direction for a while, and then goes out to the right. From the graph

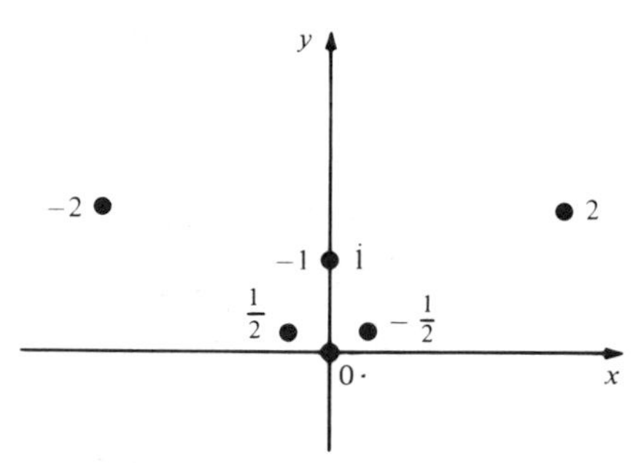

Figure 10.4.6. Some points on the curve $(t^3 - t, t^2)$.

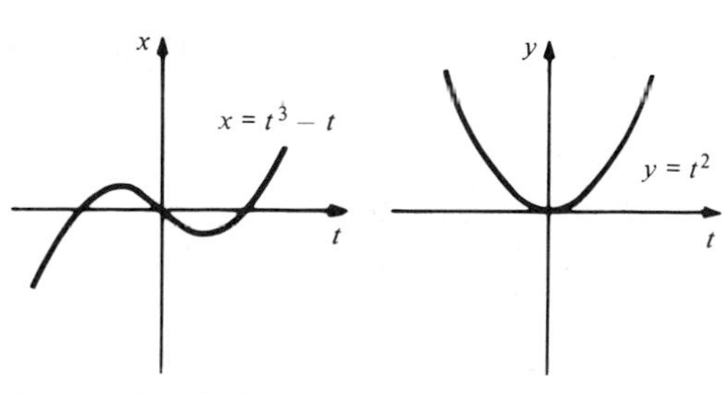

Figure 10.4.7. The graphs of x and y plotted separately against t.

of y against t, we see that the point descends for $t < 0$, reaches the bottom at $y = 0$ when $t = 0$, and then ascends for $t > 0$. Putting this information together with the points we have plotted, we sketch the curve in Fig. 10.4.8.

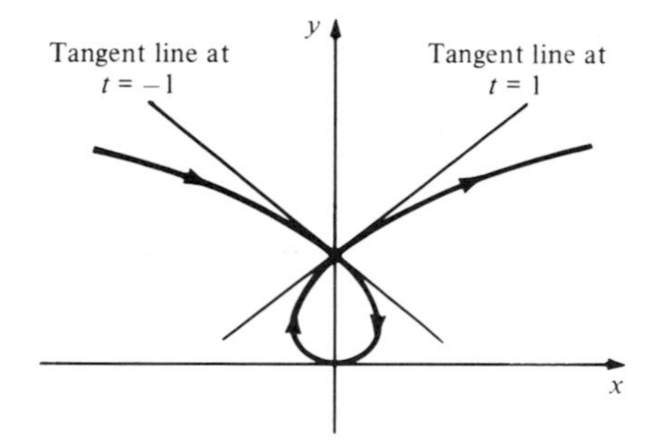

Figure 10.4.8. The parametric curve $(t^3 - t, t^2)$.

(c) The slope of the tangent line at time t is

$$\frac{dy}{dx} = \frac{dy/dt}{dx/dt} = \frac{2t}{3t^2 - 1}.$$

When $t = -1$, the slope is -1; when $t = 1$, the slope is 1. (See Fig. 10.4.8.)

(d) We can eliminate t by solving the second equation for t to get $t = \pm\sqrt{y}$ and substituting in the first to get $x = \pm(y^{3/2} - y^{1/2})$. To obtain an equation without fractional powers, we square both sides. The result is $x^2 = y(y-1)^2$, or $x^2 = y^3 - 2y^2 + y$. In this form, it is not so easy to predict the behavior of the curve, particularly at the "double point" (0, 1). ▲

Example 5 (a) Sketch the curve $x = t^3, y = t^2$. (b) Find the equation of the tangent line at $t = 1$. (c) What happens at $t = 0$?

Solution (a) Eliminating the parameter t, we have $y = x^{2/3}$. The graph has a cusp at the origin, as in Fig. 10.4.9. (Cusps were discussed in Section 3.4.)

(b) When $t = 1$, we have $x = t^3 = 1$, $y = t^2 = 1$, $dx/dt = 3t^2 = 3$, and $dy/dt = 2t = 2$, so the tangent line is given by

$$x = 3(t-1) + 1, \qquad y = 2(t-1) + 1.$$

It has slope $\frac{2}{3}$. (You can also see this by differentiating $y = x^{2/3}$ and setting $x = 1$.)

(c) When $t = 0$, we have $dx/dt = 0$ and $dy/dt = 0$, so the tangent line is not defined. ▲

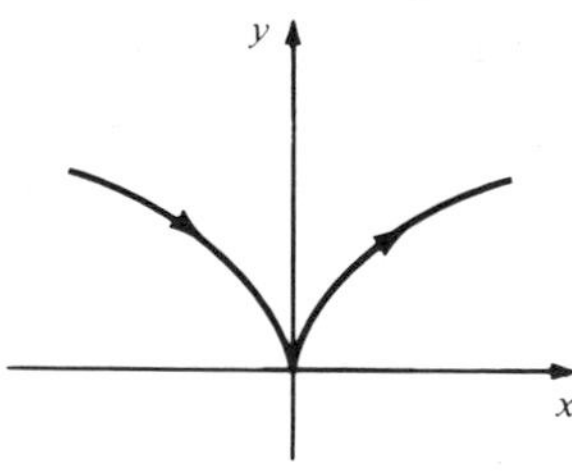

Figure 10.4.9. The curve (t^3, t^2) has a cusp at the origin.

Example 6 Consider the curve $x = \cos 3t$, $y = \sin t$. Find the points where the tangent is horizontal and those where it is vertical. Use this information to sketch the curve.

Solution The tangent line is vertical when $dx/dt = 0$ and horizontal when $dy/dt = 0$. (If both are zero, there is no tangent line.)

We have $dx/dt = -3\sin 3t$, which is zero when $t = 0, \pi/3, 2\pi/3, \pi, 4\pi/3, 5\pi/3$ (the curve repeats itself when t reaches 2π); $dy/dt = \cos t$, which is zero when $t = \pi/2$ or $3\pi/2$. We make a table:

t	0	$\frac{\pi}{3}$	$\frac{\pi}{2}$	$\frac{2\pi}{3}$	π	$\frac{4\pi}{3}$	$\frac{3\pi}{2}$	$\frac{5\pi}{3}$
$x = \cos 3t$	1	-1	0	1	-1	1	0	-1
$y = \sin t$	0	$\frac{\sqrt{3}}{2}$	1	$\frac{\sqrt{3}}{2}$	0	$-\frac{\sqrt{3}}{2}$	-1	$-\frac{\sqrt{3}}{2}$
Tangent	vert	vert	hor	vert	vert	vert	hor	vert

Using the fact that $\sqrt{3}/2 \approx 0.866$, we sketch this information in Fig. 10.4.10. Connecting these points in the proper order with a smooth curve, we obtain Fig. 10.4.11. This curve is an example of a *Lissajous figure* (see Review Exercise 93 and 94 at the end of this chapter). ▲

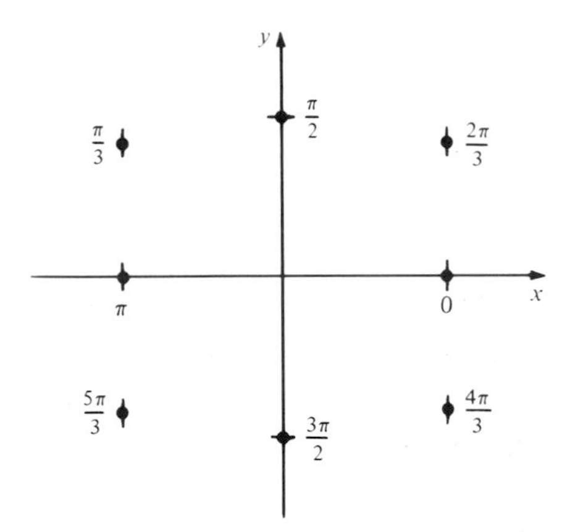

Figure 10.4.10. Points on the curve $(\cos 3t, \sin t)$ with horizontal and vertical tangent.

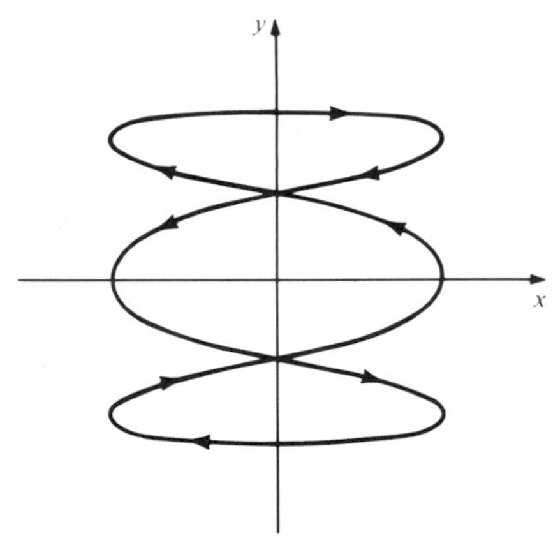

Figure 10.4.11. The curve $(\cos 3t, \sin t)$ is an example of a Lissajous figure.

What is the length of the curve given by $(x, y) = (f(t), g(t))$ for $a \leqslant t \leqslant b$? To get a formula in terms of f and g, we begin by considering the case in which the point $(f(t), g(t))$ moves along the graph of a function $y = h(x)$; that is, $g(t) = h(f(t))$.

If $f(a) = \alpha$ and $f(b) = \beta$, the length of the curve is $\int_\alpha^\beta \sqrt{1 + [h'(x)]^2}\, dx$ by formula (1) of Section 10.3. If we change variables from x to t in this integral, we have $dx = f'(t)\,dt$, so the length is

$$\int_a^b \sqrt{1 + [h'(f(t))]^2}\, f'(t)\, dt.$$

To eliminate the function h from this formula, we may apply the chain rule to

$g(t) = h(f(t))$, getting $g'(t) = h'(f(t)) \cdot f'(t)$. Solving for $h'(f(t))$ and substituting in the integral gives

$$\int_a^b \sqrt{1 + \left[\frac{g'(t)}{f'(t)}\right]^2} f'(t)\,dt$$

or

$$L = \int_a^b \sqrt{[f'(t)]^2 + [g'(t)]^2}\,dt = \int_a^b \sqrt{\left(\frac{dx}{dt}\right)^2 + \left(\frac{dy}{dt}\right)^2}\,dt. \tag{4}$$

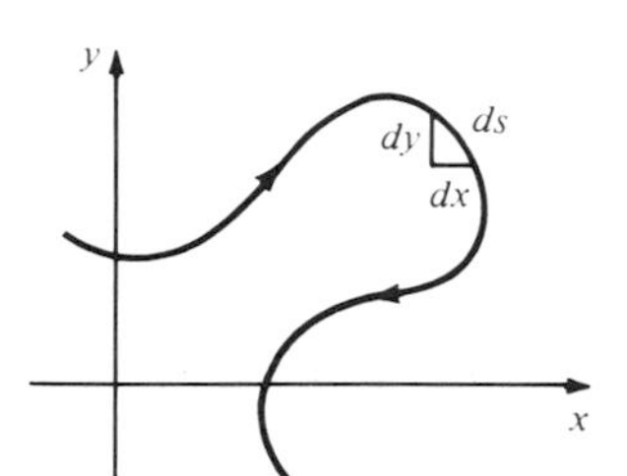

Figure 10.4.12. Finding the length of a parametric curve.

Formula (4) involves only the information contained in the parametrization. Since we can break up any reasonably behaved parametric curve into segments, each of which is the graph of a function or a vertical line (for which we see that equation (4) gives the correct length, since $f'(t) \equiv 0$), we conclude that equation (4) ought to be valid for any parametric curve.

Equation (4) may be derived using infinitesimals in the following way. Refer to Fig. 10.4.12 and note that $ds^2 = dx^2 + dy^2$. Thus

$$ds = \sqrt{dx^2 + dy^2} = \sqrt{\left(\frac{dx}{dt}\right)^2 + \left(\frac{dy}{dt}\right)^2}\,dt.$$

Integrating from $t = a$ to $t = b$ reproduces formula (4).

Length of a Parametric Curve

Suppose that a parametric curve C is given by continuous functions $x = f(t)$, $y = g(t)$, for $a \leqslant t \leqslant b$, and that $f'(t)$ and $g'(t)$ exist and are continuous, except possibly for finitely many points. Then the length of C is given by

$$L = \int_a^b \sqrt{[f'(t)]^2 + [g'(t)]^2}\,dt = \int_a^b \sqrt{\left(\frac{dx}{dt}\right)^2 + \left(\frac{dy}{dt}\right)^2}\,dt.$$

Example 7 Find the length of the circle of radius 2 which is given by the parametric equations $x = 2\cos t + 3$, $y = 2\sin t + 4$, $0 \leqslant t \leqslant 2\pi$.

Solution We find $f'(t) = dx/dt = -2\sin t$ and $g'(t) = dy/dt = 2\cos t$, so

$$L = \int_0^{2\pi} \sqrt{4\sin^2 t + 4\cos^2 t}\,dt$$

$$= \int_0^{2\pi} 2\sqrt{\sin^2 t + \cos^2 t}\,dt = \int_0^{2\pi} 2\,dt = 4\pi$$

(which equals 2π times the radius). ▲

Example 8 Find the length of (a) $x = t^8$, $y = t^4$ on $[1,3]$ and (b) $x = t\sin t$, $y = t\cos t$ on $[0, 4\pi]$.

Solution (a) We are given $x = f(t) = t^8$ and $y = g(t) = t^4$ on $[1,3]$. The length is

$$L = \int_1^3 \sqrt{(8t^7)^2 + (4t^3)^2}\,dt = \int_1^3 4t^3\sqrt{(2t^4)^2 + 1}\,dt.$$

Letting $u = 2t^4$, we have the length

$$L = \frac{1}{2}\int_2^{162}\sqrt{u^2+1}\,du.$$

Making the substitution $u = \tan\theta$, $du = \sec^2\theta\,d\theta$, we get

$$\int\sqrt{u^2+1}\,du = \int\sec^3\theta\,d\theta = I.$$

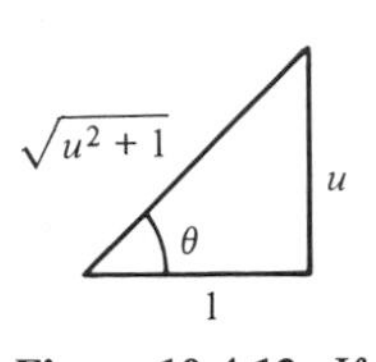

Figure 10.4.13. If $\tan\theta = u$, $\sqrt{u^2+1} = \sec\theta$.

(see Fig. 10.4.13). Integrating by parts,

$$I = \sec\theta\tan\theta - \int\sec\theta\tan^2\theta\,d\theta = \sec\theta\tan\theta - \int\sec\theta(\sec^2\theta - 1)\,d\theta.$$

Since

$$\int\sec\theta\,d\theta = \int\sec\theta\cdot\frac{\tan\theta+\sec\theta}{\tan\theta+\sec\theta}\,d\theta = \ln|\tan\theta+\sec\theta|,$$

we get $I = \sec\theta\tan\theta - I + \ln|\tan\theta + \sec\theta| + C$. Thus

$$I = \int\sec^3\theta\,d\theta = \frac{\sec\theta\tan\theta + \ln|\tan\theta+\sec\theta|}{2} + C.$$

(Compare Example 3, Section 10.3.) Putting everything together in terms of u,

$$L = \frac{1}{4}\left[\sqrt{u^2+1}\cdot u + \ln\left|u + \sqrt{u^2+1}\,\right|\right]\Big|_2^{162} \approx 6561.1.$$

(b) If $x = t\sin t$ and $y = t\cos t$, $dx/dt = \sin t + t\cos t$ and $dy/dt = \cos t - t\sin t$. Therefore,

$$\left(\frac{dx}{dt}\right)^2 + \left(\frac{dy}{dt}\right)^2 = \sin^2 t + 2t\sin t\cos t + t^2\cos^2 t + \cos^2 t - 2t\sin t\cos t + t^2\sin^2 t$$
$$= 1 + t^2.$$

Thus, using (a), the length is

$$\int_0^{4\pi}\sqrt{1+t^2}\,dt = \frac{1}{2}\left[t\sqrt{1+t^2} + \ln\left|t+\sqrt{1+t^2}\,\right|\right]\Big|_0^{4\pi}$$
$$= \frac{1}{2}\left[4\pi\sqrt{1+16\pi^2} + \ln\left(4\pi + \sqrt{1+16\pi^2}\right)\right] \approx 80.8\ \blacktriangle$$

Example 9 Show that if $x = f(t)$ and $y = g(t)$ is any curve with $(f(0), g(0)) = (0,0)$ and $(f(1), g(1)) = (0, a)$, then the length of the curve for $0 \leqslant t \leqslant 1$ is at least equal to a. What can you say if the length is exactly equal to a?

Solution It is evident that $[g'(t)]^2 \leqslant [f'(t)]^2 + [g'(t)]^2$, so

$$g'(t) \leqslant \sqrt{[f'(t)]^2 + [g'(t)]^2}\ .$$

Integrating from 0 to 1, we have

$$\int_0^1 g'(t)\,dt \leqslant \int_0^1\sqrt{[f'(t)]^2 + [g'(t)]^2}\,dt.$$

By the fundamental theorem of calculus, the left-hand side is equal to $g(1) - g(0) = a - 0 = a$; the right-hand side is the length L of the curve, so we have $a \leqslant L$. If $a = L$, the integrands must be equal; that is, $g'(t) = \sqrt{f'(t)^2 + g'(t)^2}$, which is possible only if $f'(t)$ is identically zero; that is, $f(t)$ is constant. Since $f(0) = f(1) = 0$, we must have $f(t)$ identically zero; that is, the point (x, y) stays on the y axis.

We have shown that the shortest curve between the points $(0,0)$ and $(0,a)$ is the straight line segment which joins them. $\blacktriangle$

Given a point moving according to $x = f(t)$, $y = g(t)$, the integral

$$D(t) = \int_a^t \sqrt{[f'(s)]^2 + [g'(s)]^2}\, ds$$

is the distance (along the curve) travelled by the point between time a and time t. The derivative $D'(t)$ should then represent the speed of the point along the curve. By the fundamental theorem of calculus (alternative version), we have

$$D'(t) = \sqrt{[f'(t)]^2 + [g'(t)]^2}\ .$$

Speed

Let a point move according to the equations $x = f(t)$, $y = g(t)$. Then the speed of the point at time t is

$$\sqrt{[f'(t)]^2 + [g'(t)]^2} = \sqrt{\left(\frac{dx}{dt}\right)^2 + \left(\frac{dy}{dt}\right)^2}\ .$$

Suppose that an object is constrained to move along the curve $x = f(t)$, $y = g(t)$ and that at time t_0 the constraining forces are removed, so the particle continues along the tangent line

$$x = f'(t_0)(t - t_0) + f(t_0), \qquad y = g'(t_0)(t - t_0) + g(t_0).$$

At time $t_0 + \Delta t$, the particle is at $(f'(t_0)\Delta t + f(t_0),\ g'(t_0)\Delta t + g(t_0))$, which is at distance $\sqrt{f'(t_0)^2 + g'(t_0)^2}\ \Delta t$ from $(f(t_0), g(t_0))$. Thus the distance travelled in time Δt after the force is removed is equal to Δt times the speed at t_0, so we have another justification of our formula for the speed.

Example 10 A particle moves around the elliptical track $4x^2 + y^2 = 4$ according to the equations $x = \cos t$, $y = 2\sin t$. When is the speed greatest? Where is it least?

Solution The speed is

$$\sqrt{\left[\frac{d(\cos t)}{dt}\right]^2 + \left[\frac{d(2\sin t)}{dt}\right]^2} = \sqrt{\sin^2 t + 4\cos^2 t} = \sqrt{1 + 3\cos^2 t}\ .$$

Without any further calculus, we observe that the speed is greatest when $\cos t = \pm 1$; that is, $t = 0, \pi, 2\pi$, and so forth. The speed is least when $\cos t = 0$; that is, $t = \pi/2, 3\pi/2, 5\pi/2$, and so on. ▲

Example 11 The position (x, y) of a bulge in a bicycle tire as it rolls down the street can be parametrized by the angle θ shown in Fig. 10.4.14. Let the radius of the tire be a. It can be verified by methods of plane trigonometry that $x = a\theta - a\sin\theta$, $y = a - a\cos\theta$. (This curve is called a *cycloid*.)

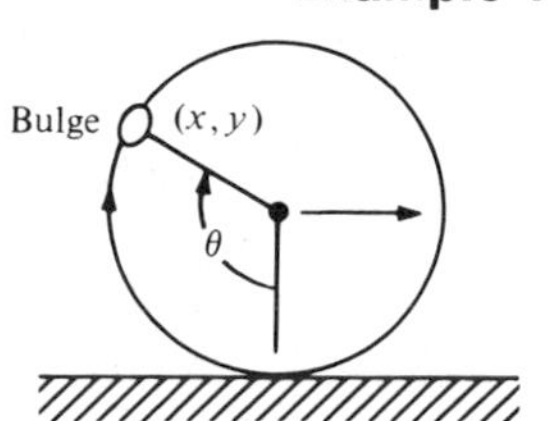

Figure 10.4.14. Investigate how a bulge on a tire moves.

(a) Find the distance travelled by the bulge for $0 \leqslant \theta \leqslant 2\pi$, using the identity $1 - \cos\theta = 2\sin^2(\theta/2)$. This distance is greater than $2\pi a$ (distance the tire rolls).

(b) Draw a figure for one arch of the cycloid, and superimpose the circle of radius a with center at $(\pi a, a)$, together with the line segment $0 \leqslant x \leqslant 2\pi a$ on the x axis. Show that the three enclosed areas are each πa^2.

Solution (a) The distance d is the arc length of the cycloid for $0 \leqslant \theta \leqslant 2\pi$. Thus,

$$d = \int_0^{2\pi} \sqrt{(a - a\cos\theta)^2 + (a\sin\theta)^2}\, d\theta$$

$$= a\int_0^{2\pi} \sqrt{1 - 2\cos\theta + \cos^2\theta + \sin^2\theta}\, d\theta = a\sqrt{2}\int_0^{2\pi}\sqrt{1 - \cos\theta}\, d\theta$$

$$= a\sqrt{2}\int_0^{2\pi}\sqrt{2}\sin\left(\frac{\theta}{2}\right) d\theta = -4a\cos(\theta/2)\Big|_0^{2\pi} = -4a(-1-1) = 8a.$$

(b) Refer to Fig. 10.4.15. The total area beneath the arch is

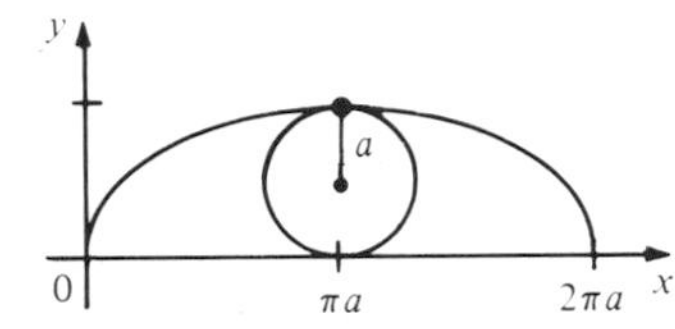

Figure 10.4.15. One arch of the cycloid.

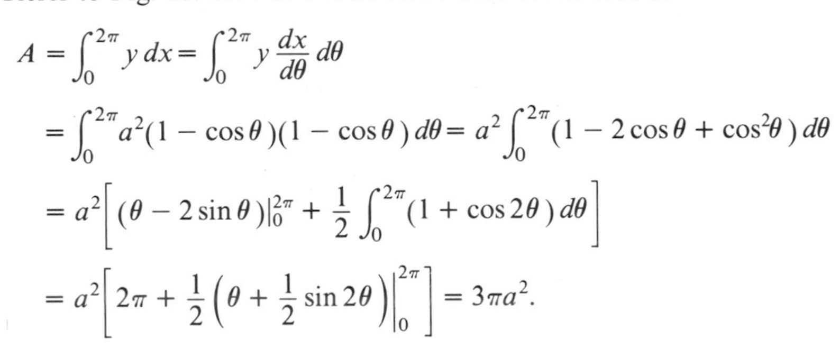

$$A = \int_0^{2\pi} y\, dx = \int_0^{2\pi} y\frac{dx}{d\theta}\, d\theta$$

$$= \int_0^{2\pi} a^2(1 - \cos\theta)(1 - \cos\theta)\, d\theta = a^2\int_0^{2\pi}(1 - 2\cos\theta + \cos^2\theta)\, d\theta$$

$$= a^2\left[(\theta - 2\sin\theta)\Big|_0^{2\pi} + \frac{1}{2}\int_0^{2\pi}(1 + \cos 2\theta)\, d\theta\right]$$

$$= a^2\left[2\pi + \frac{1}{2}\left(\theta + \frac{1}{2}\sin 2\theta\right)\Big|_0^{2\pi}\right] = 3\pi a^2.$$

The area of the circle is πa^2, so by symmetry each of the other two congruent regions also has area πa^2. ▲

Exercises for Section 10.4

For the parametric curves in Exercises 1–4, sketch the curve and find an equation in x and y by eliminating the parameter.

1. $x = 4t - 1, y = t + 2$.
2. $x = 2t + 1, y = t^2$.
3. $x = \cos\theta + 1, y = \sin\theta$.
4. $x = \sin\theta, y = \cos\theta - 3$.

Find a parametric representation for each of the curves in Exercises 5–12.

5. $2x^2 + y^2 = 1$.
6. $16x^2 + 9y^2 = 1$.
7. $4xy = 1$.
8. $y = 3x - 2$.
9. $y = x^3 + 1$.
10. $3x^2 - y^2 = 1$.
11. $y = \cos(2x)$.
12. $y^2 = x + x^2$.

Find the equation of the tangent line to each of the curves in Exercises 13–16 at the given point.

13. $x = \frac{1}{2}t^2 + t, y = t^{2/3}$; $t_0 = 1$.
14. $x = 1/t, y = \sqrt{t+1}$; $t_0 = 2$.
15. $x = \cos^2(t/2), y = \frac{1}{2}\sin t$; $t_0 = \pi/2$.
16. $x = \theta - \sin\theta, y = 1 - \cos\theta$; $\theta_0 = \pi/4$.

17. A bead is sliding on a wire, having position $x = (2 - 3t)^2$, $y = 2 - 3t$ at time t. If the bead flies off the wire at time $t = 1$, where is it when $t = 3$?
18. A piece of mud on a bicycle tire is following the cycloid $x = 6t - 3\sin 2t$, $y = 3 - 3\cos 2t$. At time $t = \pi/2$, the mud becomes detached from the tire. Along what line is it moving? (Ignore gravity.)

Sketch each of the parametric curves in Exercises 19–22, find an equation in x and y by eliminating the parameter, and find the points where the tangent line is horizontal or vertical.

19. $x = t^2$; $y = \cos t$. (What happens at $t = 0$?)
20. $x = \pi/2 - s$; $y = 2\sin 2s$.
21. $x = \cos 2t$; $y = \sin t$.
22. $(x, y) = (\cos t, \sin 2t)$.

23. Find the length of $x = t^2, y = t^3$ on $[0, 1]$.
24. Find the length of the curve given by $x = \frac{1}{2}\sin 2t$, $y = 3 + \cos^2 t$ on $[0, \pi]$.
25. Find the length of the curve (t^2, t^4) on $0 \leqslant t \leqslant 1$.
26. Find the length of the parametric curve $(e^t(\cos t)\sqrt{t^3+1}, 2e^t(\cos t)\sqrt{t^3+1})$ on $[0, 1]$.
27. Show that if $x = a\cos t + b$ and $y = a\sin t + d$: (a) the speed is constant; (b) the length of the curve on $[t_0, t_1]$ is equal to the speed times the elapsed time $(t_1 - t_0)$.
28. An object moves from left to right along the curve $y = x^{3/2}$ at constant speed. If the point is at $(0, 0)$ at noon and at $(1, 1)$ at 1:00 P.M., where is it at 1:30 P.M.?
29. Consider the parametrized curve $x = 2\cos\theta$, $y = \theta - \sin\theta$.
   (a) Find the equation of the tangent line at $\theta = \pi/2$.
   (b) Sketch the curve.
   (c) Express the length of the curve on $[0, \pi]$ as an integral.

30. Show that if

$$\frac{dx}{dt}\frac{d^2x}{dt^2} = -\frac{dy}{dt}\frac{d^2y}{dt^2},$$

then the speed of the curve $x = f(t)$, $y = g(t)$ is constant.

31. A particle travels a path in space with speed $s(t) = \sin^2(\pi t) + \tan^4(\pi t)\sec^2(\pi t)$. Find the distance $\int_0^{10} s(t)\,dt$ travelled in the first ten seconds.

32. A car loaded with skiers climbs a hill to a ski resort, constantly changing gears due to variations in the incline. Assume, for simplicity, that the motion of the auto is planar: $x = x(t)$, and $y = y(t)$, $0 \leqslant t \leqslant T$. Let $s(t)$ be the distance travelled along the road at time t (Fig. 10.4.16).

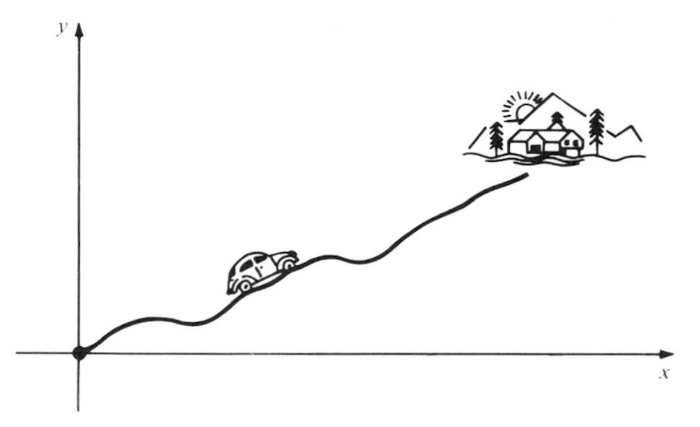

Figure 10.4.16. A car on its way to a ski cabin.

(a) The value $s(10)$ is the difference in the odometer readings from $t = 0$ to $t = 10$. Explain.
(b) The value $s'(t)$ is the speedometer reading at time t. Explain.
(c) The value $y'(t)$ is the rate of change in altitude, while $x'(t)$ is the rate of horizontal approach to the resort. Explain.
(d) What is the average rate of vertical ascent? What is the average speed for the trip?

33. A child walks with speed k from the center of a merry-go-round to its edge, while the equipment rotates counterclockwise with constant angular speed ω. The motion of the child relative to the ground is $x = kt\cos\omega t$, $y = kt\sin\omega t$.
(a) Find the *velocities* $\dot{x} = dx/dt$, $\dot{y} = dy/dt$.
(b) Determine the *speed*.
(c) The child experiences a *Coriolis* force opposite to the direction of rotation, tangent to the edge of the merry-go-round. The magnitude of this force is the mass m of the child times the factor $\sqrt{\ddot{x}(0)^2 + \ddot{y}(0)^2}$, where $\ddot{x} = d^2x/dt^2$. Find this force.

★34. (a) Find a parametric curve $x = f(t)$, $y = g(t)$ passing through the points $(1,1)$, $(2,2)$, $(4,2)$, $(5,1)$, $(3,0)$, and $(1,1)$ such that the functions f and g are both piecewise linear and the curve is a polygon whose vertices are the given points in the given order.
(b) Compute the length of this curve by formula (4) and then by elementary geometry. Compare the results.
(c) What is the area of the surface obtained by revolving the given curve about the y axis?

★35. At each point (x_0, y_0) of the parabola $y = x^2$, the tangent line is drawn and a point is marked on this line at a distance of 1 unit from (x_0, y_0) to the right of (x_0, y_0).
(a) Describe the collection of points thus obtained as a parametrized curve.
(b) Describe the collection of points thus obtained in terms of a relation between x and y.

★36. If $x = t$ and $y = g(t)$, show that the points where the speed is maximized are points of inflection of $y = g(x)$.

★37.[5] (a) Looking at a map of the United States, estimate the length of the coastline of Maine.
(b) Estimate the same length by looking at a map of Maine.
(c) Suppose that you used detailed local maps to compute the length of the coastline of Maine. How would the results compare with that obtained in part (b)?
(d) What is the "true" length of the coastline of Maine?
(e) What length for the coastline can you find given in an atlas or almanac?

★38. On a movie set, an auto races down a street. A follow-spot lights the action from 20 meters away, keeping a constant distance from the auto in order to maintain the same reflected light intensity for the camera. The follow-spot location (x, y) is the *pursuit curve*

$$x = t - 20\,\mathrm{sech}\left(\frac{t}{20}\right), \qquad y = 20\,\mathrm{sech}\left(\frac{t}{20}\right),$$

called a *tractrix*. Graph it.

[5] For further information on the ideas in this exercise, see B. Mandelbrot, *Fractals: Form, Chance and Dimension*, Freeman, New York (1977).

10.5 Length and Area in Polar Coordinates

Some length and area problems are most easily solved in polar coordinates.

The formula $L = \int_a^b \sqrt{(dx/dt)^2 + (dy/dt)^2}\, dt$ for the length of a parametric curve can be applied to the curve $r = f(\theta)$ in polar coordinates if we take the parameter to be θ in place of t. We write:

$$x = r\cos\theta = f(\theta)\cos\theta \quad \text{and} \quad y = r\sin\theta = f(\theta)\sin\theta.$$

Suppose that θ runs from α to β (see Fig. 10.5.1). By formula (4) of Section 10.4, the length is

$$\int_\alpha^\beta \sqrt{[f'(\theta)\cos\theta - f(\theta)\sin\theta]^2 + [f'(\theta)\sin\theta + f(\theta)\cos\theta]^2}\, d\theta$$

which simplifies to

$$\int_\alpha^\beta \sqrt{[f'(\theta)]^2 + [f(\theta)]^2}\, d\theta.$$

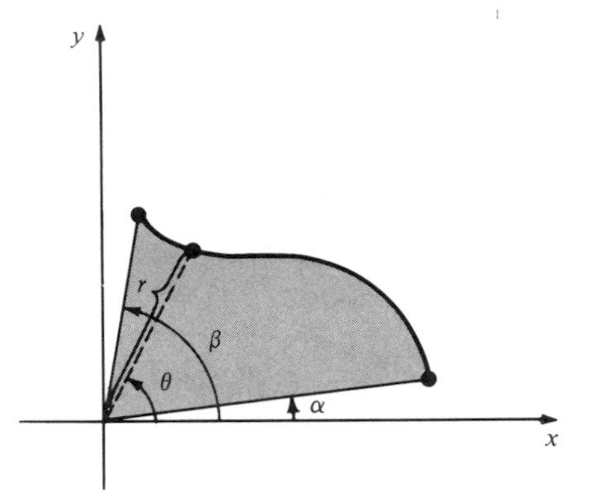

Figure 10.5.1. The length of the curve is $\int_\alpha^\beta \sqrt{(dr/d\theta)^2 + r^2}\, d\theta$.

Arc Length in Polar Coordinates

The length of the curve $r = f(\theta)$, $\alpha \leqslant \theta \leqslant \beta$, is given by

$$L = \int_\alpha^\beta \sqrt{f'(\theta)^2 + f(\theta)^2}\, d\theta = \int_\alpha^\beta \sqrt{\left(\frac{dr}{d\theta}\right)^2 + r^2}\, d\theta. \tag{1}$$

One can obtain the same formula by an infinitesmal argument, following Fig. 10.5.2. By Pythagoras' theorem, $ds^2 = dr^2 + (r\,d\theta)^2$, or $ds = \sqrt{dr^2 + r^2\,d\theta^2}$. If we use $dr = f'(\theta)\,d\theta$, this becomes

$$ds = \sqrt{\left(\frac{dr}{d\theta}\right)^2 d\theta^2 + r^2\,d\theta^2} = \sqrt{\left(\frac{dr}{d\theta}\right)^2 + r^2}\, d\theta,$$

so

$$L = \int_\alpha^\beta ds = \int_\alpha^\beta \sqrt{\left(\frac{dr}{d\theta}\right)^2 + r^2}\, d\theta.$$

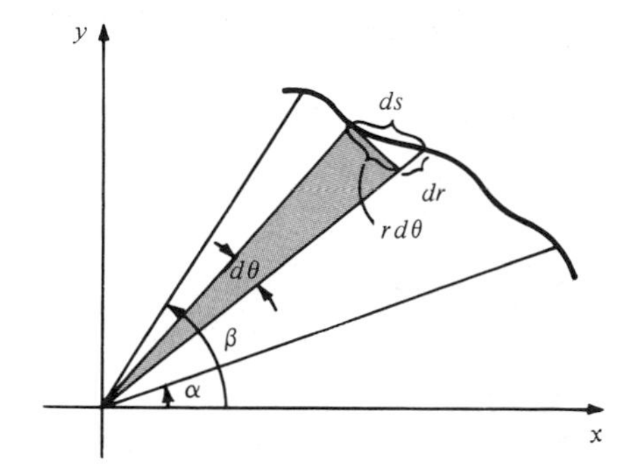

Figure 10.5.2. The infinitesimal element of arc length ds equals $\sqrt{dr^2 + r^2\,d\theta^2}$.

Example 1 Find the length of the curve $r = 1 - \cos\theta$, $0 \leqslant \theta \leqslant 2\pi$.

Solution We find $dr/d\theta = \sin\theta$, so by equation (1),

$$
\begin{aligned}
L &= \int_0^{2\pi} \sqrt{\sin^2\theta + (1-\cos\theta)^2}\, d\theta = \int_0^{2\pi} \sqrt{2 - 2\cos\theta}\, d\theta \\
&= \sqrt{2}\int_0^{2\pi} \sqrt{1-\cos\theta}\, d\theta = \sqrt{2}\int_0^{2\pi} \sqrt{2\sin^2\frac{\theta}{2}}\, d\theta \\
&= 2\int_0^{2\pi} \sin\frac{\theta}{2}\, d\theta = 4\int_0^{\pi} \sin u\, du \qquad \left(u = \frac{\theta}{2}\right) \\
&= 4(-\cos u)\Big|_0^{\pi} = 8. \blacktriangle
\end{aligned}
$$

Example 2 Find the length of the cardioid $r = 1 + \cos\theta$ $(0 \leqslant \theta \leqslant 2\pi)$.

Solution (This curve is sketched in Fig. 5.6.6.) The length is

$$
\begin{aligned}
L &= \int_0^{2\pi} \sqrt{r^2 + \left(\frac{dr}{d\theta}\right)^2}\, d\theta = \int_0^{2\pi} \sqrt{(1+\cos\theta)^2 + \sin^2\theta}\, d\theta \\
&= \int_0^{2\pi} \sqrt{2 + 2\cos\theta}\, d\theta.
\end{aligned}
$$

This can be simplified, by the half-angle formula $\cos^2(\theta/2) = (1+\cos\theta)/2$, to

$$L = \int_0^{2\pi} 2\cos\frac{\theta}{2}\, d\theta = 0.$$

Something is wrong here! We forgot that $\cos(\theta/2)$ can be negative, while the square root $\sqrt{2+2\cos\theta}$ must be positive; i.e.,

$$\sqrt{2+2\cos\theta} = \sqrt{4\cos^2\frac{\theta}{2}} = 2\left|\cos\frac{\theta}{2}\right|.$$

The correct evaluation of L is as follows:

$$L = \int_0^{2\pi} 2\left|\cos\frac{\theta}{2}\right| d\theta = \int_0^{\pi} 2\cos\frac{\theta}{2}\, d\theta - \int_{\pi}^{2\pi} 2\cos\frac{\theta}{2}\, d\theta$$

since $\cos(\theta/2) > 0$ on $(0,\pi)$ and $\cos(\theta/2) < 0$ on $(\pi, 2\pi)$. Thus

$$L = 4\sin\frac{\theta}{2}\Big|_0^{\pi} - 4\sin\frac{\theta}{2}\Big|_{\pi}^{2\pi} = 4(1-0) - 4(0-1) = 8. \blacktriangle$$

The curve expressed in polar coordinates by the equation $r = f(\theta)$, together with the rays $\theta = \alpha$ and $\theta = \beta$, encloses a region of the type shown (shaded) in Fig. 10.5.3. We call this the region *inside* the graph of f on $[\alpha, \beta]$.

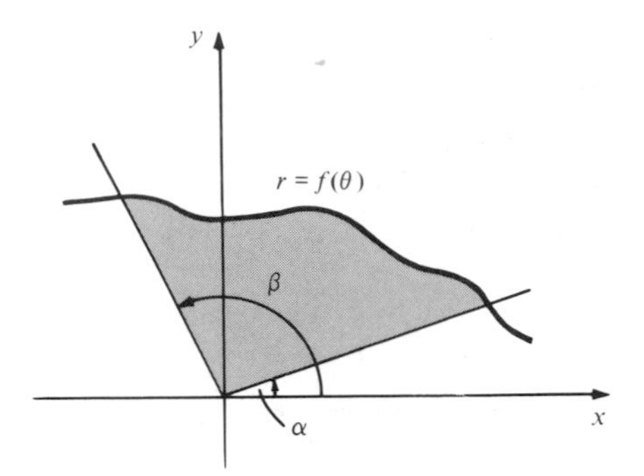

Figure 10.5.3. The region inside the graph $r = f(\theta)$ on $[\alpha, \beta]$ is shaded.

We wish to find a formula for the area of such a region as an integral involving the function f. We begin with the simplest case, in which f is a constant function $f(\theta) = k$. The region inside the curve $r = k$ on $[\alpha, \beta]$ is then a circular sector with radius k and angle $\beta - \alpha$ (see Fig. 10.5.4). The area is $(\beta - \alpha)/2\pi$ times the area πk^2 of a circle of radius k, or $\frac{1}{2}k^2(\beta - \alpha)$. We can express this as the integral $\int_\alpha^\beta \frac{1}{2} f(\theta)^2\, d\theta$.

If f is a step function, with $f(\theta) = k_i$, on (θ_{i-1}, θ_i), then the region inside the graph of f is of the type shown in Fig. 10.5.5. Its area is equal to the sum of the areas of the individual sectors, or

$$\sum_{i=1}^{n} \frac{1}{2} k_i^2 \Delta\theta_i = \int_\alpha^\beta \frac{1}{2} [f(\theta)]^2\, d\theta.$$

By approximating f with step functions, we conclude that the same formula holds for general f.

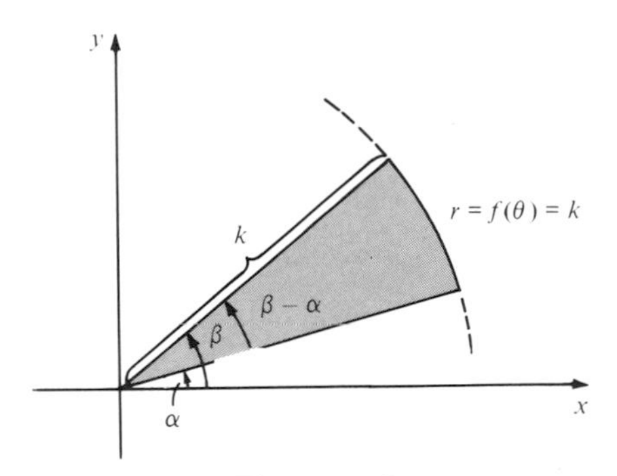

Figure 10.5.4. The area of the sector is $\frac{1}{2}k^2(\beta - \alpha)$.

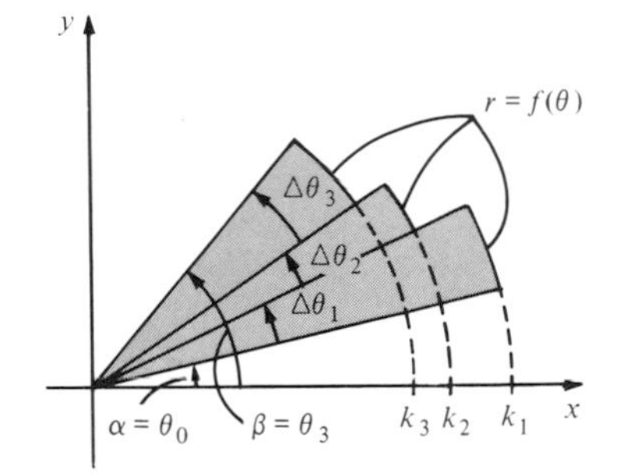

Figure 10.5.5. The area of the shaded region is $\sum \frac{1}{2} k_i^2 \Delta\theta_i$.

Area in Polar Coordinates

The area of the region enclosed by the curve $r = f(\theta)$ and the rays $\theta = \alpha$ and $\theta = \beta$ is given by

$$A = \frac{1}{2} \int_\alpha^\beta f(\theta)^2\, d\theta = \frac{1}{2} \int_\alpha^\beta r^2\, d\theta.$$

This formula can also be obtained by an infinitesimal argument. Indeed, the area dA of the shaded triangle in Fig. 10.5.2 is $\frac{1}{2}$(base) $\times$ (height)

$= \frac{1}{2}(r\,d\theta)r = \frac{1}{2}r^2\,d\theta$, so the area inside the curve is

$$\int_\alpha^\beta dA = \frac{1}{2}\int_\alpha^\beta r^2\,d\theta$$

which agrees with the formula in the preceding box.

Example 3 Find the area enclosed by one petal of the four-petaled rose $r = \cos 2\theta$ (see Fig. 5.6.3).

Solution The petal shown in Fig. 10.5.6 is enclosed by the arc $r = \cos 2\theta$ and the rays $\theta = -\pi/4$ and $\theta = \pi/4$. Notice that the rays do not actually appear in the boundary of the figure, since the radius $r = \cos(\pm\pi/2)$ is zero there. The area is given by $\frac{1}{2}\int_{-\pi/4}^{\pi/4} r^2\,d\theta = \frac{1}{2}\int_{-\pi/4}^{\pi/4}\cos^2 2\theta\,d\theta$. By the half-angle formula this is

$$\frac{1}{2}\int_{-\pi/4}^{\pi/4}\frac{1+\cos 4\theta}{2}\,d\theta = \frac{1}{4}\left(\theta + \frac{\sin 4\theta}{4}\right)\Bigg|_{-\pi/4}^{\pi/4} = \frac{\pi}{8}\,. \ \blacktriangle$$

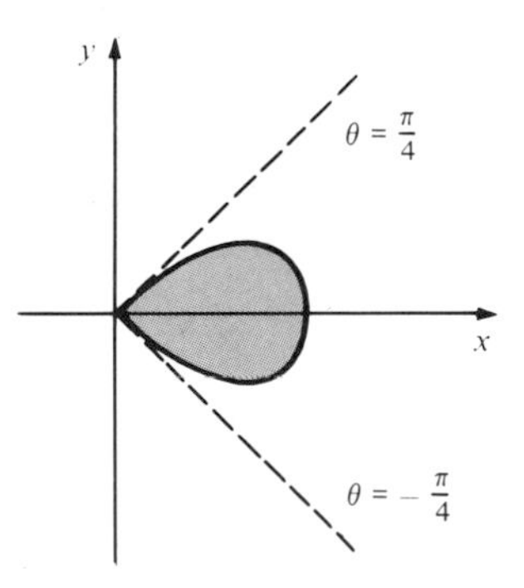

Figure 10.5.6. One leaf of the four-petaled rose $r = \cos 2\theta$.

Example 4 Find the area enclosed by the cardioid $r = 1 + \cos\theta$ (see Fig. 5.6.6).

Solution The area enclosed is defined by $r = 1 + \cos\theta$ and the full range $0 \leqslant \theta \leqslant 2\pi$, so

$$A = \frac{1}{2}\int_0^{2\pi}(1+\cos\theta)^2\,d\theta = \frac{1}{2}\int_0^{2\pi}(1 + 2\cos\theta + \cos^2\theta)\,d\theta.$$

Again using the half-angle formula,

$$A = \frac{1}{2}\int_0^{2\pi}\left(\frac{3}{2} + 2\cos\theta + \frac{\cos 2\theta}{2}\right)d\theta$$

$$= \frac{1}{2}\left[\frac{3\theta}{2} + 2\sin\theta + \frac{\sin 2\theta}{4}\right]\Bigg|_0^{2\pi} = \frac{3\pi}{2}\,. \ \blacktriangle$$

Example 5 Find a formula for the area between two curves in polar coordinates.

Solution Suppose $r = f(\theta)$ and $r = g(\theta)$ are the two curves with $f(\theta) \geqslant g(\theta) > 0$. We are required to find a formula for the shaded area in Fig. 10.5.7. The area is just the difference betweeen the areas for f and g; that is,

$$A = \frac{1}{2}\int_\alpha^\beta\left[f(\theta)^2 - g(\theta)^2\right]d\theta. \ \blacktriangle$$

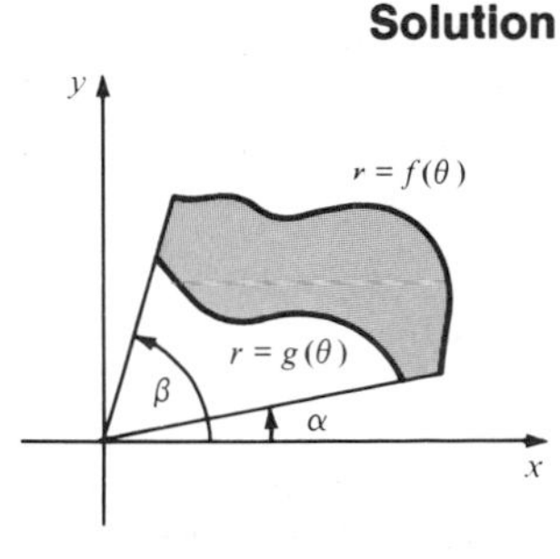

Figure 10.5.7. The area of the shaded region is $\frac{1}{2}\int_\alpha^\beta[f(\theta)^2 - g(\theta)^2]\,d\theta$.

Example 6 Sketch and find the area of the region between the curves $r = \cos 2\theta$ and $r = 2 + \sin\theta$, $0 \leqslant \theta \leqslant 2\pi$.

Solution The curves are sketched in Fig. 10.5.8. To do this, we plotted points for θ at multiples of $\pi/4$ and then noted whether r was increasing or decreasing on each of the intervals between these θ values. To find the shaded area, we must be careful because of the sign changes of $g(\theta) = \cos 2\theta$. The inner loops are described in the following way by positive functions:

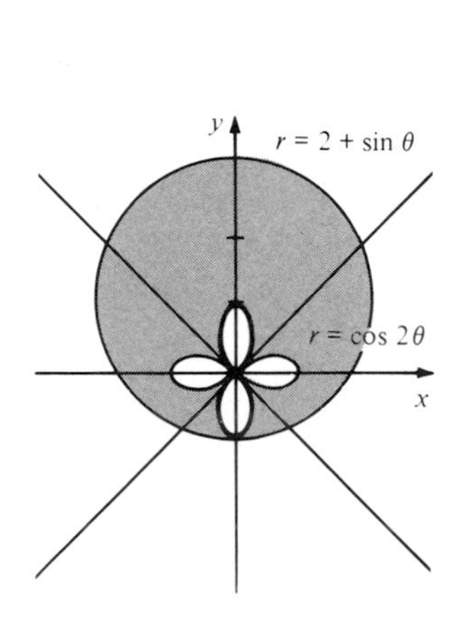

$$g(\theta) = \begin{cases} \cos 2\theta, & 0 \leqslant \theta \leqslant \frac{\pi}{4} \\ -\cos 2\theta, & \frac{\pi}{4} \leqslant \theta \leqslant \frac{3\pi}{4} \\ \cos 2\theta, & \frac{3\pi}{4} \leqslant \theta \leqslant \frac{5\pi}{4} \\ -\cos 2\theta, & \frac{5\pi}{4} \leqslant \theta \leqslant \frac{7\pi}{4} \\ \cos 2\theta, & \frac{7\pi}{4} \leqslant \theta \leqslant 2\pi \end{cases}.$$

In fact, we are lucky because in the formula in Example 5, $g(\theta)$ is squared anyway, so the shaded area is simply

$$\frac{1}{2}\int_0^{2\pi}\left[(2 + \sin\theta)^2 - \cos^2 2\theta\right] d\theta$$
$$= \frac{1}{2}\left[\int_0^{2\pi} 4\, d\theta + \int_0^{2\pi} 4\sin\theta\, d\theta + \int_0^{2\pi} \sin^2\theta\, d\theta - \int_0^{2\pi} \cos^2 2\theta\, d\theta\right]$$
$$= \frac{1}{2}\left[8\pi + 0 + \pi - \pi\right] = 4\pi. \blacktriangle$$

Exercises for Section 10.5

Find the length of the curves in Exercises 1–4.

1. $r = 3(1 + \sin\theta)$; $0 \leqslant \theta \leqslant 2\pi$.
2. $r = 1/(\cos\theta + \sin\theta)$; $0 \leqslant \theta \leqslant \pi/2$.
3. $r = 4\theta^2$; $0 \leqslant \theta \leqslant 3$.
4. $r = 8\theta^2$; $0 \leqslant \theta \leqslant 1$.

Sketch and find the area of the region bounded by the curves in Exercises 5–10.

5. $r = 3\sin\theta$; $0 \leqslant \theta \leqslant \pi$.
6. $r = 2(1 + \sin\theta)$; $0 \leqslant \theta \leqslant 2\pi$.
7. $r = \theta$; $0 \leqslant \theta \leqslant 3\pi/2$.
8. $r = \theta\cos(\theta^3)$; $0 \leqslant \theta \leqslant \pi/4$.
9. $r = 4 + \sin\theta$; $0 \leqslant \theta \leqslant 2\pi$.
10. $r = \theta + \sin 4\theta$; $\pi/4 \leqslant \theta \leqslant \pi$. [*Hint*: Find the critical points of r.]
11. Check the arc length formula in polar coordinates for a circle.
12. Check the area formula in polar coordinates for a segment of a circle and a whole circle.

In Exercises 13–16, sketch and find the length (as an integral) of the graph of $r = f(\theta)$, $\alpha \leqslant \theta \leqslant \beta$. (The answer may be in the form of an integral.) Then find the area of the region bounded by this graph and the rays $\theta = \alpha$ and $\theta = \beta$.

13. $r = \tan(\theta/2)$; $-\pi/2 \leqslant \theta \leqslant \pi/2$.
14. $r = \theta + \sin(\theta^2)$; $-\pi/4 \leqslant \theta \leqslant 3\pi/4$.
15. $r = \sec\theta + 2$; $0 \leqslant \theta \leqslant \pi/4$.
16. $r = 2e^{3\theta}$; $\ln 2 \leqslant \theta \leqslant \ln 3$.

In Exercises 17–20, find the length of and areas bounded by the following curves between the rays indicated. Express the areas as numbers but leave the length as integrals.

17. $r = \theta(1 + \cos\theta)$; $\theta = 0$, $\theta = \pi/2$.
18. $r = 1/\theta$; $\theta = 1$, $\theta = \pi$.
19. $f(\theta) = \sqrt{1 + 2\sin 2\theta}$; $\theta = 0$, $\theta = \pi/2$.
20. $f(\theta) = \theta^2 - (\pi/2)\theta + 4$; $\theta = 0$, $\theta = \pi/2$.

In Exercises 21–24, sketch and find the area of each of the regions between each of the following pairs of curves ($0 \leqslant \theta \leqslant 2\pi$). Then find the length of the curves which bound the regions.

21. $r = \cos\theta$, $r = \sqrt{3}\sin\theta$.
22. $r = 3$, $r = 2(1 + \cos\theta)$.
23. $r = 2\cos\theta$, $r = 1 + \cos\theta$.
24. $r = 1$, $r = 1 + \cos\theta$.
25. The curve $r = e^\theta$ is called a *logarithmic spiral*. Find the length of the loop of the logarithmic spiral for θ in $[2n\pi, 2(n + 1)\pi]$.

26. Suppose that the distance from the origin to $(x, y) = (f(t), g(t))$ attains its maximum value at $t = t_0$. Show that the tangent line at t_0 is perpendicular to the line from the origin to the point $(f(t_0), g(t_0))$.

★27. An elliptical orbit is parametrized by $x = a\cos\theta$, $y = b\sin\theta$, $0 \leqslant \theta \leqslant 2\pi$. This parametrization is 2π-periodic. In Chapter 18 we shall show that for any T-periodic parametrization of a continuously differentiable closed curve $x = x(t)$, $y = y(t)$ which is a *simple* (never crosses itself),

$$\text{area enclosed} = \int_0^T \frac{1}{2}[x(t)\dot{y}(t) - \dot{x}(t)y(t)]\,dt,$$

where $\dot{x}(t) = dx/dt$ and $\dot{y}(t) = dy/dt$. (See also Review Exercise 95 for this Chapter.)

(a) Use this formula to verify that the area enclosed by an ellipse of semiaxes a and b is πab.

(b) Apply the formula to the case of a curve $x(t) = r\cos t$, $y(t) = r\sin t$, where $r = r(t)$, showing that the area enclosed is $\frac{1}{2}\int_0^T r^2\,dt$.

Review Exercises for Chapter 10

Evaluate the integrals in Exercises 1–50.

1. $\int 3\sin^2 x\cos x\,dx$
2. $\int \sin^2 2x\cos^3 2x\,dx$
3. $\int \sin 3x\cos 5x\,dx$
4. $\int \cos 4x\sin 6x\,dx$
5. $\int \frac{x^3}{\sqrt{1-x^2}}\,dx \quad (|x| < 1)$
6. $\int \frac{dx}{(x^2+2)^2}$
7. $\int \frac{\sqrt{x^2-16}}{x}\,dx \quad (x > 4)$
8. $\int \frac{dx}{\sqrt{x^2-16}} \quad (x > 4)$
9. $\int \frac{dx}{x^2+x+2}$
10. $\int \frac{dx}{\sqrt{x^2+x+2}}$
11. $\int \frac{dx}{x^3+x^2}$
12. $\int \frac{dx}{x^3-27}$
13. $\int \frac{x^3}{(x^2+1)^2}\,dx$
14. $\int \frac{x^2}{(x+1)^3}\,dx$
15. $\int \frac{dx}{x^2+4x+5}$
16. $\int \sec^6\theta\,d\theta$
17. $\int \sin\sqrt{x}\,dx$
18. $\int \frac{dx}{1+\cos ax}$
19. $\int \frac{dx}{(1-\cos ax)^2}$
20. $\int \frac{dx}{1-x^4}$
21. $\int \frac{\sin^2 x}{\cos x}\,dx$
22. $\int \ln\left(\frac{x+a}{x-a}\right)dx$
23. $\int \frac{\tan^{-1}x}{1+x^2}\,dx$
24. $\int \frac{dx}{x^4+1}$
25. $\int \frac{x}{x^3-9}\,dx$
26. $\int (x+5)\ln x\,dx$
27. $\int e^{\sqrt{x}}\,dx$
28. $\int x^3\sqrt{1-x^2}\,dx$
29. $\int \frac{dx}{1+e^x}$
30. $\int \frac{x}{(x-3)^8}\,dx$
31. $\int \left(\frac{x}{x^2-1}\right)^3 dx$
32. $\int \sqrt{x^2+2x+3}\,dx$
33. $\int \sin 3x\cos 2x\,dx$
34. $\int \sin^2 3x\cos^4 3x\,dx$
35. $\int \frac{x}{x^2+1}\,dx$
36. $\int \frac{x}{(x^2+1)^2}\,dx$
37. $\int \frac{e^{\sqrt{x}}}{\sqrt{x}}\,dx$
38. $\int (e^x+1)^3 e^x\,dx$
39. $\int_2^3 \frac{x}{x^2+1}\,dx$
40. $\int_0^{\pi/2} \sin x\, e^{\cos x}\,dx$
41. $\int \frac{x}{x^2+3}\,dx$
42. $\int \frac{x}{\sqrt{x+1}}\,dx$
43. $\int x^3\ln x\,dx$
44. $\int \sqrt{1+\sin x}\cdot\cos x\,dx$
45. $\int_1^2 \frac{(\ln 3x+5)^3}{x}\,dx$
46. $\int_e^{e^4} \frac{\ln t^2}{t^2}\,dt$
47. $\int_0^1 \sinh^2 x\,dx$
48. $\int_{\pi/8}^{\pi/4} \frac{\sin\theta}{\sqrt{1-\cos^2\theta}}\,d\theta$
49. $\int_0^{2\pi} \frac{\sin\theta}{1+\cos\theta+\cos^2\theta}\,d\theta$
50. $\int_0^{\pi/100} \sec^2 100x\,dx$

In Exercises 51–54, find the length of the given graph.

51. $y = 3x^{3/2}$, $0 \leqslant x \leqslant 9$
52. $y = (x+1)^{3/2} + 1$, $0 \leqslant x \leqslant 2$.
53. $y = \frac{x^3}{3} + \frac{1}{4x}$, $1 \leqslant x \leqslant 2$.
54. $y = \frac{x^4}{4} + \frac{1}{8x^3}$, $1 \leqslant x \leqslant 2$.

In Exercises 55–58, find the area of the surface obtained by revolving the given graph about the given axis.

55. $y = x^2$, $0 \leqslant x \leqslant 1$, about the y axis.
56. $y = \sqrt{x}$, $0 \leqslant x \leqslant 1$, about the x axis.
57. $y = \log_{10}x$, $10 \leqslant x \leqslant 100$, about the y axis.
58. $y = 2^x$, $3 \leqslant x \leqslant 4$, about the x axis.

For each of the pairs of parametric equations in Exercises 59–64, sketch the curve and find an equation in x and y by eliminating the parameter.

59. $x = t^2$; $y = t - 1$
60. $x = 2t + 5$; $y = t^3$
61. $x = 3t$; $y = 2t + 1$
62. $x = t$; $y = t$
63. $x = 0$; $y = t^4$
64. $x = t^2\sqrt{t^2 - 1}$; $y = t^2$
65. Find the equation of the tangent line to the curve $x = t^4$, $y = 1 + t^3$ at $t = 1$.
66. Find the equation of the tangent line to the parametric curve $x = 3\cos t$, $y = \sin t$ at $t = \pi/4$.
67. Find the arc length of $x = t^2$, $y = 2t^4$ from $t = 0$ to $t = 2$.
68. Find the length of $x = e^t\sin t$, $y = e^t\cos t$ from $t = 0$ to $t = \pi/2$.

Find the arc length (as an integral if necessary) and area enclosed by each of the graphs given in polar coordinates in Exercises 69–74.

69. $r = \theta^2$; $0 \leqslant \theta \leqslant \frac{\pi}{2}$.
70. $r = \dfrac{1}{\cos\theta}$; $0 \leqslant \theta \leqslant \dfrac{\pi}{4}$.
71. $r = \frac{1}{2} + \cos 2\theta$; $0 \leqslant \theta \leqslant \pi$.
72. $r = 2|\cos\theta|$; $0 \leqslant \theta \leqslant 2\pi$.
73. $r = 3\cos^4\dfrac{\theta}{4}$; $0 \leqslant \theta \leqslant \pi$.
74. $r = \dfrac{1}{2}\sin^2\dfrac{\theta}{2}$; $\dfrac{\pi}{4} \leqslant \theta \leqslant \dfrac{3\pi}{4}$.

If f is a function on $[0, 2\pi]$, then the numbers

$$\left.\begin{aligned} a_m &= \frac{1}{\pi}\int_0^{2\pi} f(x)\cos mx\,dx \\ b_m &= \frac{1}{\pi}\int_0^{2\pi} f(x)\sin mx\,dx \end{aligned}\right\} \qquad (m = 0, 1, 2, \dots)$$

are called the *Fourier coefficients* of f. Find all the Fourier coefficients of each of the functions in Exercises 75–82.

75. $\sin 2x$
76. $\sin 5x$
77. $\cos 3x$
78. $\cos 8x$
79. $3\cos 4x$
80. $2\cos 8x + \sin 7x + \cos 9x$
81. $\sin^2 x$
82. $\cos^3 x$
83. The solution of the *logistic equation* of population biology, $dN/dt = (k_1N - k_2)N$, $N(0) = N_0$, requires the evaluation of the definite integral
$$\int_{N_0}^{N(t)} \frac{du}{(k_1u - k_2)u}.$$
 (a) Evaluate by means of partial fraction methods and compare your answer with Exercise 19, Section 8.5.
 (b) The integral is just the time t. Solve for $N(t)$ in terms of t, using exponentials.
 (c) Find $\lim_{t\to\infty} N(t)$ when it exists.
84. Kepler's second law of planetary motion says that *the radial segment drawn from the sun to a planet sweeps out equal areas in equal times*. Locate the origin $(0,0)$ at the sun and introduce polar coordinates (γ, θ) for the planet location. Assume the *angular momentum* of the planet (of mass m) about the sun is constant; $mr^2\dot{\theta} = mk$, $k =$ constant, and $\dot{\theta} = d\theta/dt$. Establish Kepler's second law by showing $\int_s^{s+h} r^2\dot{\theta}\,dt$ is the same for all times s; thus the area swept out is the same for all time intervals of length h.
85. An elliptical satellite circuits the earth in a circular orbit. The angle ϕ between its major axis and the direction to the earth's center oscillates between $+\phi_m$ and $-\phi_m$ (*librations* of the earth satellite). It is assumed that $0 < \phi_m < \pi/2$, so that the satellite does not tumble end over end. The time T for one complete cycle of this oscillation is given by
$$T = \frac{4}{\pi}\int_0^{\phi_m} \frac{d\phi}{\sqrt{\cos 2\phi - \cos 2\phi_m}}.$$
Change variables in the integral via the formulas
$$\sin\phi = \sin\phi_m \sin\beta \text{ (which defines } \beta\text{)},$$
$$\cos 2\phi = 1 - 2\sin^2\phi,$$
$$\cos 2\phi_m = 1 - 2\sin^2\phi_m,$$
to obtain the *elliptic integral representation*
$$T = \frac{4}{\pi\sqrt{2}}\int_0^{\pi/2} \frac{d\beta}{\sqrt{1 - k^2\sin^2\beta}}$$
for the period of libration T, where $k^2 = \sin^2\phi_m$.

★86. An engineer is studying the impact of an infinite bar by a short round-headed bar, making a maximum indentation α_1. Applying Hertz' theory of impact, she obtains the equation $\frac{1}{2}\rho c_0\Omega\alpha' = k(\alpha_1^{3/2} - \alpha^{3/2})$ for the indentation α at time t. The symbols ρ, c_0, Ω, k are constants. The equation is solved by an initial integration to get
$$t = \frac{\rho c_0\Omega}{2k\sqrt{\alpha_1}}\int_0^{\alpha/\alpha_1} \frac{du}{1 - u^{3/2}}.$$
 (a) Evaluate the integral by making the substitution $v = \sqrt{u}$, followed by the method of partial fractions.
 (b) Substitute $s = (4tk\sqrt{\alpha_1})/(3\rho c_0\Omega)$ to obtain
$$9s = \frac{2\pi}{\sqrt{3}} + 2\ln\left|\frac{1 + y + y^2}{(1 - y)^2}\right| - 4\sqrt{3}\tan^{-1}\left(\frac{2y + 1}{\sqrt{3}}\right),$$
 where $y = \sqrt{\alpha/\alpha_1}$.

★87. Find a general formula for $\int dx/\sqrt{ax^2 + bx + c}$; $a \neq 0$. There will be two cases, depending upon the sign of a.

★88. Let $f(x) = x^n$, $0 < a \leqslant x \leqslant b$. For which rational values of n can you evaluate the integral occurring in the formula for:
 (a) The area under the graph of f?
 (b) The length of the graph of f?
 (c) The volume of the surface obtained by revolving the region under the graph of f about the x axis? The y axis?

(d) The area of the surface of revolution obtained by revolving the graph of f about the x axis? The y axis?

Evaluate these integrals.

★89. Same as Exercise 88, but with $f(x) = 1 + x^n$.

★90. Same as Exercise 88, but with $f(x) = (1 + x^2)^n$.

★91. (a) Find the formula for the area of the surface obtained by revolving the graph of $r = f(\theta)$ about the x axis, $\alpha \leqslant \theta \leqslant \beta$.
(b) Find the area of the surface obtained by revolving $r = \cos 2\theta$, $-\pi/4 \leqslant \theta \leqslant \pi/4$ about the x axis (express as an integral if necessary).

★92. Consider the integral

$$\int \frac{dx}{\sqrt{(1 - x^2)(1 - k^2x^2)}}.$$

(a) Show that, for $k = 0$ and $k = 1$, this integral can be evaluated in terms of trigonometric and exponential functions and their inverses.

(b) Show that, for any k, the integral may be transformed to one of the form

$$\int \frac{d\theta}{\sqrt{1 - k^2\sin^2\theta}}.$$

(This integrand occurs in the sunshine formula—see the supplement to Section 9.5.)

(c) Show that the integral

$$\int \sqrt{1 - k^2\sin^2\theta}\, d\theta$$

(which also occurs in the sunshine formula) arises when one tries to find the arc length of an ellipse $x^2/a^2 + y^2/b^2 = 1$. Express k in terms of a and b.

Due to the result of part (c), the integrals in parts (a), (b), and (c) are called *elliptic integrals*.

★93. Consider the parametric curve given by $x = \cos mt$, $y = \sin nt$, when m and n are integrals. Such a curve is called a *Lissajous figure* (see Example 6, Section 10.4).

(a) Plot the curve for $m = 1$ and $n = 1, 2, 3, 4$.

(b) Describe the general behavior of the curve if $m = 1$, for any value of n. Does it matter whether n is even or odd?

(c) Plot the curve for $m = 2$ and $n = 1, 2, 3, 4, 5$.

(d) Plot the curve for $m = 3$ and $n = 4, 5$.

★94. (Lissajous figures continued). The path $x = x(t)$, $y = y(t)$ of movement of the *tri-suspension pendulum* of Fig. 10.R.1 produces a Lissajous figure of the general form $x = A_1\cos(\omega_1 t + \theta_1)$, $y = A_2\sin(\omega_2 t + \theta_0)$.

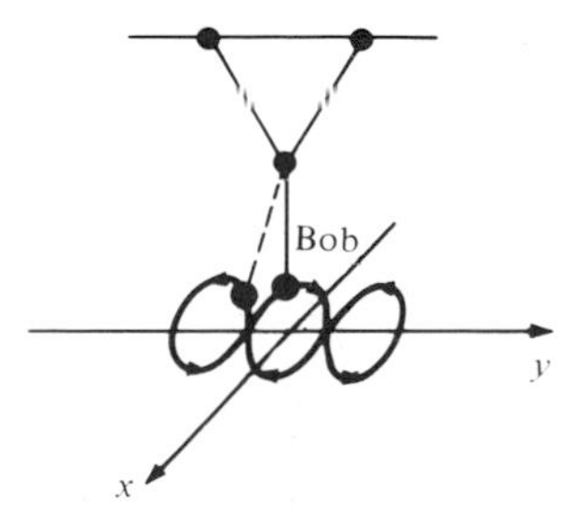

Figure 10.R.1. The bob on this pendulum traces out a Lissajous figure.

(a) Draw the Lissajous figures for $\omega_1 = \omega_2 = 1$, $A_1 = A_2$, for some sample values of θ_1, θ_2. The figures should come out to be straight lines, circles, ellipses.

(b) When $\omega_1 = 1$, $\omega_2 = 3$, $A_1 = A_2 = 1$, the bob retraces its path, but has two self-intersections. Verify this using the results of Exercise 93. Conjecture what happens when ω_1/ω_1 is the ratio of integers.

(c) When $\omega_2/\omega_1 = \pi$, $A_1 = A_2 = 1$, the bob does not retrace its path, and has infinitely many self-intersections. Verify this, graphically. Conjecture what happens when ω_2/ω_1 is irrational (not the quotient of integers).

★95. Consider the curve $r = f(\theta)$ for $0 \leqslant \theta \leqslant 2\pi$ as a parametric curve: $x = f(t)\cos t$, $y = f(t)\sin t$. Assuming that $f(\theta) > 0$ for all θ in $[0, 2\pi]$ and that $f(2\pi) = f(0)$, show that the area enclosed by the curve is given by

$$-\int_0^{2\pi} y\frac{dx}{dt}\,dt \qquad \text{(A)}$$

as well as by $\int_0^{2\pi} x(dy/dt)\,dt$ and by the more symmetric formula

$$\frac{1}{2}\int_0^{2\pi}\left[x\frac{dy}{dt} - y\frac{dx}{dt}\right]dt.$$

[*Hint*: Substitute the definitions of x and y into (A), integrate by parts, and use the formula for area in polar coordinates.] These formulas are in fact valid for any closed parametric curve. (See Section 18.4.)

★96. (a) If r is a non-repeated root of $Q(x)$, show that the portion of the partial fraction expansion of $P(x)/Q(x)$ corresponding to the factor $x - r$ is $A/(x - r)$ where $A = P(r)/Q'(r)$. (b) Use (a) to calculate $\int [(x^2 + 2)/(x^3 - 6x + 11x - 6)]\,dx$.

Chapter 11

Limits, L'Hôpital's Rule, and Numerical Methods

Limits are used in both the theory and applications of calculus.

Our treatment of limits up to this point has been rather casual. Now, having learned some differential and integral calculus, you should be prepared to appreciate a more detailed study of limits.

The chapter begins with formal definitions for limits and a review of computational techniques for limits of functions, including infinite and one-sided limits. The next topic is l'Hôpital's rule, which employs differentiation to compute limits. Infinite limits are used to study improper integrals. The chapter ends with some numerical methods involving limits of sequences.

11.1 Limits of Functions

There are many kinds of limits, but they all obey similar laws.

In Section 1.2, we discussed on an intuitive basis what $\lim_{x\to x_0} f(x)$ means and why the limit notion is important in understanding the derivative. Now we are ready to take a more careful look at limits.

Recall that the statement $\lim_{x\to x_0} f(x) = l$ means, roughly speaking, that $f(x)$ comes close to and remains arbitrarily close to l as x comes close to x_0. Thus we start with a positive "tolerance" ε and try to make $|f(x) - l|$ less than ε by requiring x to be close to x_0. The closeness of x to x_0 is to be measured by another positive number—mathematical tradition dictates the use of the Greek letter δ for this number. Here, then, is the famous ε-δ definition of a limit—it was first stated in this form by Karl Weierstrass around 1850.

The ε-δ Definition of $\lim\limits_{x\to x_0} f(x)$

Let f be a function defined at all points near x_0, except perhaps at x_0 itself, and let l be a real number. We say that *l is the limit of $f(x)$ as x approaches x_0* if, for every positive number ε, there is a positive number δ such that $|f(x) - l| < \varepsilon$ whenever $|x - x_0| < \delta$ and $x \neq x_0$. We write $\lim_{x\to x_0} f(x) = l$.

The purpose of giving the ε-δ definition is to enable us to be more precise in dealing with limits. Proofs of some of the basic theorems in this chapter and the next require this definition; however, practical computations can often be done without a full mastery of the theory. Your instructor should tell you how much theory you are expected to know.

The ε-δ definition of limit is illustrated in Figure 11.1.1. We shade the region consisting of those (x, y) for which:

1. $|x - x_0| > \delta$ (region I in Fig. 11.1.1(b));
2. $x = x_0$ (the vertical line II in Fig. 11.1.1(b));
3. $x \neq x_0$, $|x - x_0| < \delta$, and $|y - l| < \varepsilon$ (region III in Fig. 11.1.1(b)).

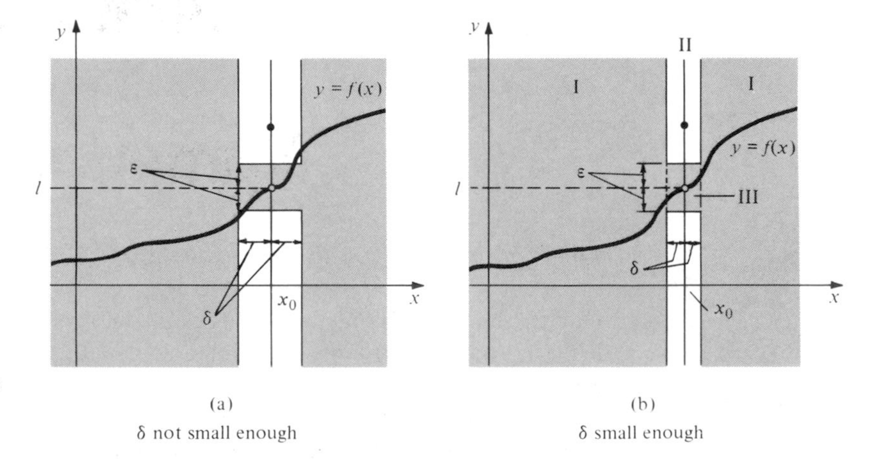

Figure 11.1.1. When $\lim_{x\to x_0} f(x) = l$, we can, for any $\varepsilon > 0$, catch the graph of f in the shaded region by making δ small enough. The value of f at x_0 is irrelevant, since the line $x = x_0$ is always "shaded."

If $\lim_{x\to x_0} f(x) = l$, then we can catch the graph of f in the shaded region by making δ small enough—that is, by making the unshaded strips sufficiently narrow.

Notice the statement $x \neq x_0$ in the definition. This means that the limit depends only upon the values of $f(x)$ for x *near* x_0, and not on $f(x_0)$ itself. (In fact, $f(x_0)$ might not even be defined.)

Here are two examples of how the ε-δ condition is verified.

Example 1 (a) Prove that $\lim_{x\to 2}(x^2 + 3x) = 10$ using the ε-δ definition. (b) Prove that $\lim_{x\to a}\sqrt{x} = \sqrt{a}$, where $a > 0$, using the ε-δ definition.

Solution (a) Here $f(x) = x^2 + 3x$, $x_0 = 2$, and $l = 10$. Given $\varepsilon > 0$ we must find $\delta > 0$ such that $|f(x) - l| < \varepsilon$ if $|x - x_0| < \delta$.

A useful general rule is to write down $f(x) = l$ and then to express it in terms of $x - x_0$ as much as possible, by writing $x = (x - x_0) + x_0$. In our case we replace x by $(x - 2) + 2$:

$$\begin{aligned} f(x) - l &= x^2 + 3x - 10 \\ &= (x - 2 + 2)^2 + 3(x - 2 + 2) - 10 \\ &= (x - 2)^2 + 4(x - 2) + 4 + 3(x - 2) + 6 - 10 \\ &= (x - 2)^2 + 7(x - 2). \end{aligned}$$

Now we use the properties $|a + b| \leqslant |a| + |b|$ and $|a^2| = |a|^2$ of the absolute value to note that

$$|f(x) - l| \leqslant |x - 2|^2 + 7|x - 2|.$$

If this is to be less than ε, we should choose δ so that $\delta^2 + 7\delta \leqslant \varepsilon$. We may require at the outset that $\delta \leqslant 1$. Then $\delta^2 \leqslant \delta$, so $\delta^2 + 7\delta \leqslant 8\delta$. Hence we pick δ so that $\delta \leqslant 1$ and $\delta \leqslant \varepsilon/8$.

With this choice of δ, we shall now verify that $|f(x) - l| < \varepsilon$ whenever

$|x - x_0| < \delta$. In our case $|x - x_0| < \delta$ means $|x - 2| < \delta$, so for such an x,

$$\begin{aligned} |f(x) - l| &\leqslant |x - 2|^2 + 7|x - 2| \\ &< \delta^2 + 7\delta \\ &\leqslant \delta + 7\delta \\ &= 8\delta \\ &\leqslant \varepsilon, \end{aligned}$$

and so $|f(x) - l| < \varepsilon$.

(b) Here $f(x) = \sqrt{x}$, $x_0 = a$, and $l = \sqrt{a}$. Given $\varepsilon > 0$ we must find a $\delta > 0$ such that $|\sqrt{x} - \sqrt{a}| < \varepsilon$ when $|x - a| < \delta$. To do this we write $\sqrt{x} - \sqrt{a} = (x - a)/(\sqrt{x} + \sqrt{a})$. Since f is only defined for $x \geqslant 0$, we confine our attention to these x's. Then

$$|\sqrt{x} - \sqrt{a}| = \frac{|x - a|}{\sqrt{x} + \sqrt{a}} \leqslant \frac{|x - a|}{\sqrt{a}} \quad \text{(decreasing the denominator increases the fraction).}$$

Thus, given $\varepsilon > 0$ we can choose $\delta = \sqrt{a}\,\varepsilon$; then $|x - a| < \delta$ implies $|\sqrt{x} - \sqrt{a}| < \varepsilon$, as required. ▲

In practice, it is usually more efficient to use the laws of limits than the ε-δ definition, to evaluate limits. These laws were presented in Section 1.3 and are recalled here for reference.

Basic Properties of Limits

Assume that $\lim_{x\to x_0} f(x)$ and $\lim_{x\to x_0} g(x)$ exist:

Sum rule:

$$\lim_{x\to x_0}[f(x) + g(x)] = \lim_{x\to x_0} f(x) + \lim_{x\to x_0} g(x).$$

Product rule:

$$\lim_{x\to x_0}[f(x)g(x)] = \lim_{x\to x_0} f(x) \lim_{x\to x_0} g(x).$$

Reciprocal rule:

$$\lim_{x\to x_0}[1/f(x)] = 1/\lim_{x\to x_0} f(x) \quad \text{if} \quad \lim_{x\to x_0} f(x) \neq 0.$$

Constant function rule:

$$\lim_{x\to x_0} c = c.$$

Identity function rule:

$$\lim_{x\to x_0} x = x_0.$$

Replacement rule: If the functions f and g agree for all x near x_0 (not necessarily including $x = x_0$), then

$$\lim_{x\to x_0} f(x) = \lim_{x\to x_0} g(x).$$

Rational functional rule: If P and Q are polynomials and $Q(x_0) \neq 0$, then P/Q is continuous at x_0; i.e.,

$$\lim_{x\to x_0}[P(x)/Q(x)] = P(x_0)/Q(x_0).$$

Composite function rule: If h is continuous at $\lim_{x\to x_0} f(x)$, then

$$\lim_{x\to x_0} h(f(x)) = h\left(\lim_{x\to x_0} f(x)\right).$$

The properties of limits can all be proved using the ε-δ definition. The theoretically inclined student is urged to do so by studying Exercises 75–77 at the end of this section.

Let us recall how to use the properties of limits in specific computations.

Example 2 Using the fact that $\lim_{\theta\to 0}\left(\frac{1-\cos\theta}{\theta}\right)=0$, find $\lim_{\theta\to 0}\cos\left(\frac{1-\cos\theta}{\theta}\right)$.

Solution The composite function rule says that $\lim_{x\to x_0}h(f(x)) = h(\lim_{x\to x_0}f(x))$ if h is continuous at $\lim_{x\to x_0}f(x)$. We let $f(\theta)=(1-\cos\theta)/\theta$, and $h(\theta)=\cos\theta$ so that $h(f(\theta))=\cos[(1-\cos\theta)/\theta]$. Hence the required limit is

$$\lim_{\theta\to 0}h(f(\theta)) = h\left(\lim_{\theta\to 0}\frac{1-\cos\theta}{\theta}\right)=\cos 0 = 1,$$

since cos is continuous at $\theta=0$. ▲

Example 3 Find (a) $\lim_{x\to 2}\left(\frac{x^2-5x+6}{x-2}\right)$ and (b) $\lim_{x\to 1}\left(\frac{x-1}{\sqrt{x}-1}\right)$.

Solution (a) Since the denominator vanishes at $x=2$, we cannot plug in this value. The numerator may be factored, however, and for any $x\neq 2$ our function is

$$\frac{x^2-5x+6}{x-2}=\frac{(x-3)(x-2)}{x-2}=x-3.$$

Thus, by the replacement rule,

$$\lim_{x\to 2}\frac{x^2-5x+6}{x-2}=\lim_{x\to 2}(x-3)=2-3=-1.$$

(b) Again we cannot plug in $x=1$. However, we can rationalize the denominator by multiplying numerator and denominator by $\sqrt{x}+1$. Thus (if $x\neq 1$):

$$\frac{x-1}{\sqrt{x}-1}=\frac{(x-1)(\sqrt{x}+1)}{(\sqrt{x}-1)(\sqrt{x}+1)}=\frac{(x-1)(\sqrt{x}+1)}{x-1}=\sqrt{x}+1.$$

As x approaches 1, this approaches 2, so $\lim_{x\to 1}[(x-1)/(\sqrt{x}-1)]=2$. ▲

Limits of the form $\lim_{x\to\pm\infty}f(x)$, called *limits at infinity*, are dealt with by a modified version of the ideas above. Let us motivate the ideas by a physical example.

Let $y=f(t)$ be the length, at time t, of a spring with a bobbing mass on the end. If no frictional forces act, the motion is sinusoidal, given by an equation of the form $f(t)=y_0+a\cos\omega t$.[1] In reality, a spring does not go on bobbing forever; frictional forces cause *damping*, and the actual motion has the form

$$y=f(t)=y_0+ae^{-bt}\cos\omega t, \tag{1}$$

where b is positive. A graph of this function is sketched in Fig. 11.1.2.

As time passes, we observe that the length becomes and remains arbitrarily near to the equilibrium length y_0. (Even though $y=y_0$ already for $t=\pi/2\omega$, this is not the same thing because the length does not yet *remain* near y_0.) We express this mathematical property of the function f by writing $\lim_{t\to\infty}f(t)=y_0$. The limiting behavior appears graphically as the fact that the

[1] This is derived in Section 8.1., but if you have not studied that section, you should simply take for granted the formulas given here.

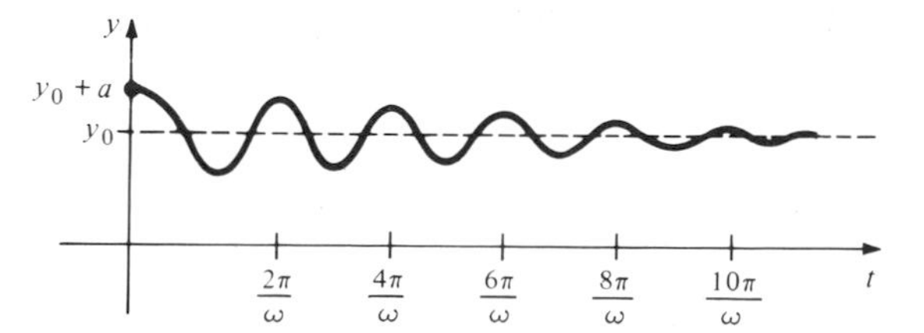

Figure 11.1.2. The motion of a damped spring has the form $y = f(t) = y_0 + ae^{-bt}\cos\omega t$.

graph of f remains closer and closer to the line $y = y_0$ as we look farther to the right.

The precise definition is analogous to that for $\lim_{x\to x_0} f(x)$. As is usual in our general definitions, we denote the independent variable by x rather than t.

The ε-A Definition of $\lim_{x\to+\infty} f(x)$

Let f be a function whose domain contains an interval of the form (a, ∞). We say that a real number l *is the limit of* $f(x)$ *as* x *approaches* ∞ if, for every positive number ε, there is a number $A > a$ such that $|f(x) - l| < \varepsilon$ whenever $x > A$. We write $\lim_{x\to\infty} f(x) = l$.

A similar definition is used for $\lim_{x\to-\infty} f(x) = l$.

When $\lim_{x\to\infty} f(x) = l$ or $\lim_{x\to-\infty} f(x) = l$, the line $y = l$ is called a *horizontal asymptote* of the graph $y = f(x)$.

We illustrate this definition in Figs. 11.1.3 and 11.1.4 by shading the region consisting of those points (x, y) for which $x \leqslant A$ or for which $x > A$ and $|y - l| < \varepsilon$. If $\lim_{x\to\infty} f(x) = l$, we should be able to "catch" the graph of f in this region by choosing A large enough—that is, by sliding the point A sufficiently far to the right.

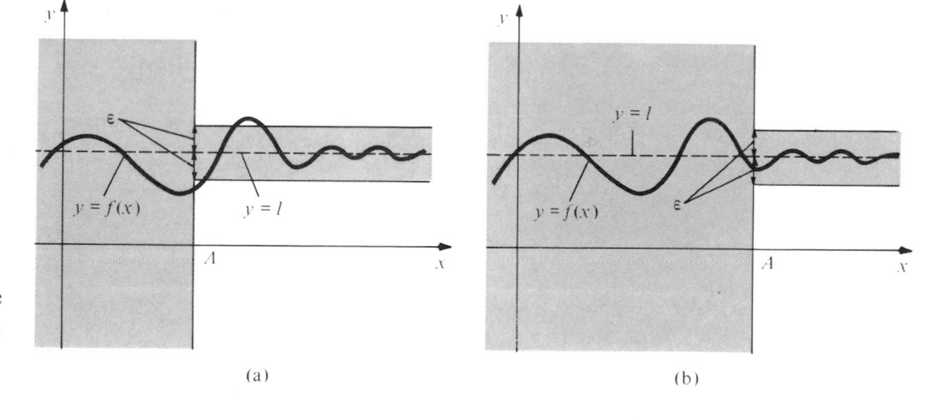

Figure 11.1.3. When $\lim_{x\to\infty} f(x) = l$, we can catch the graph in the shaded region by sliding the region sufficiently far to the right. This is true no matter how small ε may be.

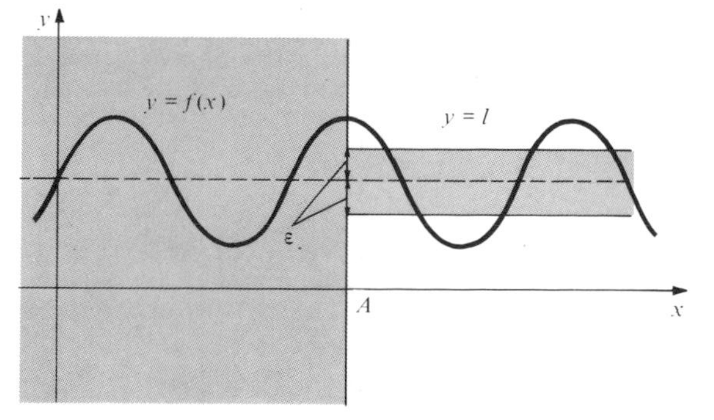

Figure 11.1.4. When it is not true that $\lim_{x\to\infty} f(x) = l$, then for some ε, we can never catch the graph of f in the shaded region, no matter how far to the right we slide the region.

There is an analogous definition for $\lim_{x\to-\infty} f(x)$ in which we require a number A (usually large and negative) such that $|f(x) - l| < \varepsilon$ if $x < A$.

Example 4 Prove that $\lim\limits_{x\to\infty} \dfrac{x^2}{1+x^2} = 1$ by using the ε-A definition.

Solution Given $\varepsilon > 0$, we must choose A such that $|x^2/(1+x^2) - 1| < \varepsilon$ for $x > A$. We have

$$\left|\frac{x^2}{1+x^2} - 1\right| = \left|\frac{x^2 - 1 - x^2}{1+x^2}\right| = \frac{1}{|1+x^2|} < \frac{1}{x^2}.$$

To make this less than ε, we observe that $1/x^2 < \varepsilon$ whenever $x > 1/\sqrt{\varepsilon}$, so we may choose $A = 1/\sqrt{\varepsilon}$. (See Fig. 11.1.5.) ▲

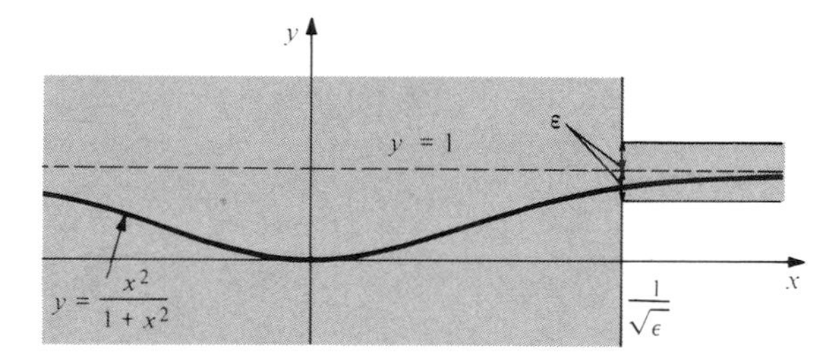

Figure 11.1.5. Illustrating the fact that $\lim_{x\to\infty}[x^2/(1+x^2)] = 1$.

At the beginning of Section 6.4. we stated several limit properties for e^x and $\ln x$. Some simple cases can be verified by the ε-A definition; others are best handled by l'Hôpital's rule, which is introduced in the next section.

Example 5 Use the ε-A definition to show that for $k < 0$, $\lim_{x\to\infty} e^{kx} = 0$.

Solution First of all, we note that $f(x) = e^{kx}$ is a decreasing positive function. Given $\varepsilon > 0$, we wish to find A such that $x > A$ implies $e^{kx} < \varepsilon$. Taking logarithms of the last inequality gives $kx < \ln\varepsilon$, or $x > (\ln\varepsilon)/k$. So we may let $A = (\ln\varepsilon)/k$. (If ε is small, $\ln\varepsilon$ is a large negative number.) ▲

The examples above illustrate the ε-A method, but limit computations are usually done using laws analogous to those for limits as $x \to x_0$, which are stated in the box on the facing page.

Example 6 Find (a) $\lim\limits_{x\to\infty}\left(\dfrac{1}{x} + \dfrac{3}{x^2} + 5\right)$ and (b) $\lim\limits_{x\to\infty} \dfrac{8x+2}{3x-1}$.

Solution (a) We have

$$\lim_{x\to\infty}\left(\frac{1}{x} + \frac{3}{x^2} + 5\right) = \lim_{x\to\infty}\frac{1}{x} + 3\left(\lim_{x\to\infty}\frac{1}{x}\right)^2 + \lim_{x\to\infty} 5 = 0 + 3\cdot 0^2 + 5 = 5.$$

(b) We cannot simply apply the quotient rule, since the limits of the numerator and denominator do not exist. Instead we use a trick: if $x \neq 0$, we can multiply the numerator and denominator by $1/x$ to obtain

$$\frac{8x+2}{3x-1} = \frac{8 + (2/x)}{3 - (1/x)} \quad \text{for} \quad x \neq 0.$$

By the replacement rule (with $A = 0$), we have

$$\lim_{x\to\infty}\frac{8x+2}{3x-1} = \lim_{x\to\infty}\frac{8+(2/x)}{3-(1/x)} = \frac{8+0}{3-0} = \frac{8}{3}.$$

(The values of $(8x+2)/(3x-1)$ for $x = 10^2, 10^4, 10^6, 10^8$ are $2.682\ldots$, $2.66682\ldots$, $2.6666682\ldots$, $2.666666682\ldots$.) ▲

Limits of Functions as x Approaches ∞

Constant function rule:

$$\lim_{x\to\infty} c = c.$$

$1/x$ rule:

$$\lim_{x\to\infty} \frac{1}{x} = 0.$$

Assuming that $\lim_{x\to\infty} f(x)$ and $\lim_{x\to\infty} g(x)$ exist, we have these additional rules:

Sum rule:

$$\lim_{x\to\infty} [f(x) + g(x)] = \lim_{x\to\infty} f(x) + \lim_{x\to\infty} g(x).$$

Product rule:

$$\lim_{x\to\infty} [f(x)g(x)] = \lim_{x\to\infty} f(x) \lim_{x\to\infty} g(x).$$

Quotient rule: If $\lim_{x\to\infty} g(x) \neq 0$, then

$$\lim_{x\to\infty}\left[\frac{f(x)}{g(x)}\right] = \frac{\lim_{x\to\infty} f(x)}{\lim_{x\to\infty} g(x)}.$$

Replacement rule: If for some real number A, the functions $f(x)$ and $g(x)$ agree for all $x > A$, then

$$\lim_{x\to\infty} f(x) = \lim_{x\to\infty} g(x).$$

Composite function rule: If h is continuous at $\lim_{x\to\infty} f(x)$, then

$$\lim_{x\to\infty} h(f(x)) = h\left(\lim_{x\to\infty} f(x)\right).$$

All these rules remain true if we replace ∞ by $-\infty$ (and "$> A$" by "$< A$" in the replacement rule).

The method used in Example 6 also shows that

$$\lim_{x\to\infty} \frac{a_n x^n + a_{n-1}x^{n-1} + \cdots + a_1 x + a_0}{b_n x^n + b_{n-1}x^{n-1} + \cdots + b_1 x + b_0} = \frac{a_n}{b_n}$$

as long as $b_n \neq 0$.

Example 7 Find $\lim_{x\to\infty}(\sqrt{x^2+1} - x)$. Interpret the result geometrically in terms of right triangles.

Solution Multiplying the numerator and denominator by $\sqrt{x^2+1} + x$ gives

$$\sqrt{x^2+1} - x = \left(\sqrt{x^2+1} - x\right)\frac{\sqrt{x^2+1}+x}{\sqrt{x^2+1}+x}$$

$$= \frac{x^2+1-x^2}{\sqrt{x^2+1}+x} = \frac{1}{\sqrt{x^2+1}+x}.$$

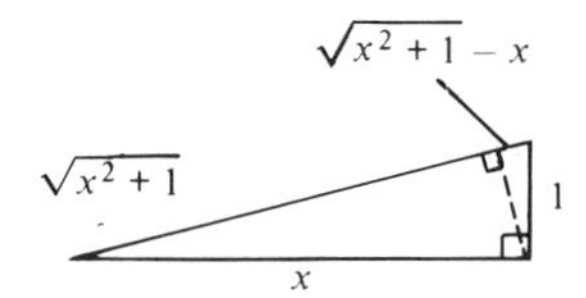

Figure 11.1.6. As the length x goes to ∞, the difference $\sqrt{x^2+1} - x$ between the lengths of the hypotenuse and the long leg goes to zero.

As $x \to \infty$, the denominator becomes arbitrarily large, so we find that $\lim_{x\to\infty}(\sqrt{x^2+1} - x) = 0$. For a geometric interpretation, see Fig. 11.1.6. ▲

Example 8 Find the horizontal asymptotes of $f(x) = \dfrac{x}{\sqrt{x^2+1}}$. Sketch.

Solution We find

$$\lim_{x\to+\infty} \frac{x}{\sqrt{x^2+1}} = \lim_{x\to+\infty} \frac{1}{\sqrt{1+1/x^2}} = 1$$

and

$$\lim_{x\to-\infty} \frac{x}{\sqrt{x^2+1}} = \lim_{x\to-\infty} \frac{-\sqrt{x^2}}{\sqrt{x^2+1}} = \lim_{x\to-\infty} \frac{-1}{\sqrt{1+1/x^2}} = -1$$

(in the second limit we may take $x<0$, so $x = -\sqrt{x^2}$). Hence the horizontal asymptotes are the lines $y = \pm 1$. See Fig. 11.1.7. ▲

Figure 11.1.7. The curve $y = x/\sqrt{x^2+1}$ has the lines $y=-1$ and $y=1$ as horizontal asymptotes.

Consider the limits $\lim_{x\to0}\sin(1/x)$ and $\lim_{x\to0}(1/x^2)$. Neither limit exists, but the functions $\sin(1/x)$ and $1/x^2$ behave quite differently as $x\to0$. (See Fig. 11.1.8.) In the first case, for x in the interval $(-\delta,\delta)$, the quantity $1/x$ ranges

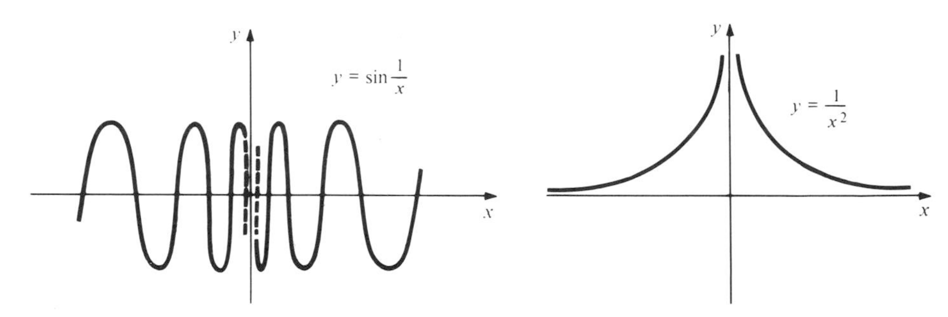

Figure 11.1.8. $\lim_{x\to0}f(x)$ does not exist for either of these functions.

over all numbers with absolute value greater than $1/\delta$, and $\sin(1/x)$ oscillates back and forth infinitely often. The function $\sin(1/x)$ takes each value between -1 and 1 infinitely often but remains close to no particular number. In the case of $1/x^2$, the value of the function is again near no particular number, but there is a definite "trend" to be seen; as x comes nearer to zero, $1/x^2$ becomes a larger positive number; we may say that $\lim_{x\to0}(1/x^2)=\infty$.

Here is a precise definition.

The B-δ Definition of $\lim_{x\to x_0} f(x) = \infty$

Let f be a function defined in an interval about x_0, except possibly at x_0 itself. We say that $f(x)$ *approaches* ∞ as x *approaches* x_0 if, given any real number B, there is a positive number δ such that for all x satisfying $|x-x_0|<\delta$ and $x\neq x_0$, we have $f(x)>B$. We write $\lim_{x\to x_0}f(x)=\infty$.

The definition of $\lim_{x\to x_0}f(x) = -\infty$ is similar: replace $f(x)>B$ in the B-δ definition by $f(x)<B$.

Remarks 1. In the preceding definition, we usually think of δ as being small, while B is large positive if the limit is ∞ and large negative if the limit is $-\infty$.

2. If $\lim_{x\to x_0} f(x)$ is equal to $\pm\infty$, we still may say that "$\lim_{x\to x_0} f(x)$ does not exist," since it does not approach any particular number.

3. One can define the statements $\lim_{x\to\infty} f(x) = \pm\infty$ in an analogous way.

The following test provides a useful technique for detecting "infinite limits."

Reciprocal Test for $\lim_{x\to x_0} f(x) = \infty$

Let f be defined in an open interval about x_0, except possibly at x_0 itself. Then $\lim_{x\to x_0} f(x) = \infty$ if:

1. For all $x \neq x_0$ in some interval about x_0, $f(x)$ is positive; and
2. $\lim_{x\to x_0}[1/f(x)] = 0$.

Similarly, if $f(x)$ is negative and $\lim_{x\to x_0}[1/f(x)] = 0$, then $\lim_{x\to x_0} f(x) = -\infty$.

The complete proof of the reciprocal test is left to the reader in Exercise 79. However, the basic idea is very simple: $f(x)$ is very large if and only if $1/f(x)$ is very small.

A similar result is true for limits of the form $\lim_{x\to\infty} f(x)$; namely, if $f(x)$ is positive for large x and $\lim_{x\to\infty}[1/f(x)] = 0$, then $\lim_{x\to\infty} f(x) = \infty$.

Example 9 Find the following limits: (a) $\lim\limits_{x\to 1} \dfrac{1}{(x-1)^2}$; (b) $\lim\limits_{x\to\infty} \dfrac{1-x^2}{x^{3/2}}$.

Solution (a) We note that $1/(x-1)^2$ is positive for all $x \neq 1$. We look at the reciprocal: $\lim_{x\to 1}(x-1)^2 = 0$; thus, by the reciprocal test, $\lim_{x\to 1}[1/(x-1)^2] = \infty$.
(b) For $x > 1$, $(1-x^2)/x^{3/2}$ is negative. Now we have

$$\lim_{x\to\infty} \frac{x^{3/2}}{1-x^2} = \lim_{x\to\infty} \frac{1}{x^{-3/2} - x^{1/2}} = \lim_{x\to\infty} \frac{1}{x^{1/2}} \frac{1}{1/x^2 - 1}$$

$$= \lim_{x\to\infty} \frac{1}{x^{1/2}} \lim_{x\to\infty} \frac{1}{1/x^2 - 1} = 0\cdot(-1) = 0,$$

so $\lim_{x\to\infty}[(1-x^2)/x^{3/2}] = -\infty$, by the reciprocal test. ▲

If we look at the function $f(x) = 1/(x-1)$ near $x_0 = 1$ we find that $\lim_{x\to 1}[1/f(x)] = 0$, but $f(x)$ has different signs on opposite sides of 1, so $\lim_{x\to 1}[1/(x-1)]$ is neither ∞ nor $-\infty$. This example suggests the introduction of the notion of a "one-sided limit." Here is the definition.

One-Sided Limits

Let f be defined for all x in an interval of the form (x_0, b). We say that *$f(x)$ approaches l as x approaches x_0 from the right* if, for any positive number ε, there is a positive number δ such that for all x such that $x_0 < x < x_0 + \delta$, we have $|f(x) - l| < \varepsilon$. We write $\lim_{x\to x_0+} f(x) = l$.

A similar definition holds for the limit of $f(x)$ as x approaches x_0 from the left; this limit is written as $\lim_{x\to x_0-} f(x) = l$.

In the definition of a one-sided limit, only the values of $f(x)$ for x on one side of x_0 are taken into account. Precise definitions of statements like $\lim_{x\to x_0+} f(x) = \infty$ are left to you. We remark that the reciprocal test extends to one-sided limits.

Example 10 Find (a) $\lim_{x\to 1+}\frac{1}{(1-x)}$, (b) $\lim_{x\to 1-}\frac{1}{(1-x)}$,

(c) $\lim_{x\to 0+}\frac{(x^2+2)|x|}{x}$, and (d) $\lim_{x\to 0-}\frac{(x^2+2)|x|}{x}$.

Solution (a) For $x>1$, we find that $1/(1-x)$ is negative, and we have $\lim_{x\to 1}(1-x) = 0$, so $\lim_{x\to 1+}[1/(1-x)] = -\infty$. Similarly, $\lim_{x\to 1-}[1/(1-x)] = +\infty$, so we get $+\infty$ for (b).
(c) For x positive, $|x|/x = 1$, so $(x^2+2)|x|/x = x^2+2$ for $x>0$. Thus the limit is $0^2+2=2$.
(d) For $x<0$, $|x|/x = -1$, so

$$\lim_{x\to 0-}\frac{(x^2+2)|x|}{x} = -\lim_{x\to 0-}\left[x^2+2\right] = -2. \blacktriangle$$

If a one-sided limit of $f(x)$ at x_0 is equal to ∞ or $-\infty$, then the graph of f lies closer and closer to the line $x = x_0$; we call this line a *vertical asymptote* of the graph.

Example 11 Find the vertical asymptotes and sketch the graph of

$$f(x) = \frac{1}{(x-1)(x-2)^2}.$$

Solution Vertical asymptotes occur where $\lim_{x\to x_0\pm}[1/f(x)] = 0$; in this case, they occur at $x_0 = 1$ and $x_0 = 2$. We observe that $f(x)$ is negative on $(-\infty, 1)$, positive on $(1,2)$, and positive on $(2,\infty)$. Thus we have $\lim_{x\to 1-} f(x) = -\infty$, $\lim_{x\to 1+} f(x) = \infty$, $\lim_{x\to 2-} f(x) = \infty$, and $\lim_{x\to 2+} f(x) = \infty$. The graph of f is sketched in Fig. 11.1.9. $\blacktriangle$

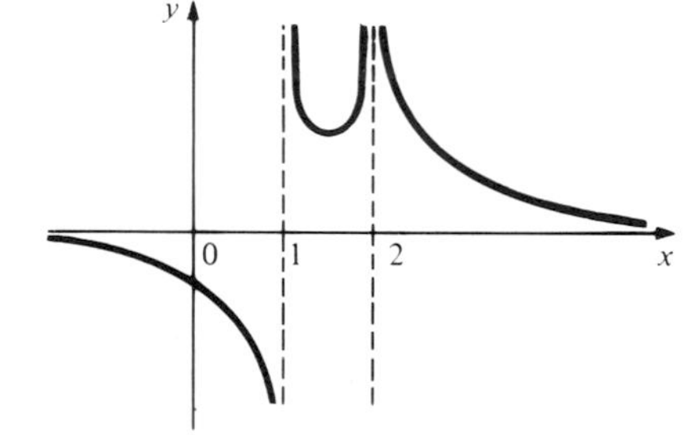

Figure 11.1.9. The graph $y = 1/(x-1)(x-2)^2$ has the lines $x=1$ and $x=2$ as vertical asymptotes.

We conclude this section with an additional law of limits. In the next sections we shall consider various additional techniques and principles for evaluating limits.

Comparison Test

1. If $\lim_{x\to x_0} f(x) = 0$ and $|g(x)| \leqslant |f(x)|$ for all x near x_0 with $x \neq x_0$, then $\lim_{x\to x_0} g(x) = 0$.
2. If $\lim_{x\to\infty} f(x) = 0$ and $|g(x)| \leqslant |f(x)|$ for all large x, then $\lim_{x\to\infty} g(x) = 0$.

Some like to call this the "sandwich principle" since $g(x)$ is sandwiched between $-|f(x)|$ and $|f(x)|$ which are squeezing down on zero as $x\to x_0$ (or $x\to\infty$ in case 2).

Example 12 (a) Establish comparison test 1 using the ε-δ definition of limit.
(b) Show that $\lim_{x\to 0}\left[x\sin\left(\frac{1}{x}\right)\right] = 0$.

Solution (a) Given $\varepsilon > 0$, there is a $\delta > 0$ such that $|f(x)| < \varepsilon$ if $|x - x_0| < \delta$, by the assumption that $\lim_{x\to x_0} f(x) = 0$. Given that $\varepsilon > 0$, this same δ also gives $|g(x)| < \varepsilon$ if $|x - x_0| < \delta$ since $|g(x)| \leqslant |f(x)|$. Hence g has limit zero as $x \to x_0$ as well.
(b) Let $g(x) = x\sin(1/x)$ and $f(x) = x$. Then $|g(x)| \leqslant |x|$ for all $x \neq 0$, since $|\sin(1/x)| \leqslant 1$, so the comparison test applies. Since x approaches 0 as $x \to 0$, so does $g(x)$, ▲

Exercises for Section 11.1

Verify the limit statements in Exercises 1–4 using the ε-δ definition.

1. $\lim_{x\to a} x^2 = a^2$
2. $\lim_{x\to 3}(x^2 - 2x + 4) = 7$
3. $\lim_{x\to 3}(x^3 + 2x^2 + 2) = 47$
4. $\lim_{x\to 3}(x^3 + 2x) = 33$
5. Using the fact that $\lim_{\theta\to 0}[(\tan\theta)/\theta] = 1$, find $\lim_{\theta\to 0}\exp[(3\tan\theta)/\theta]$.
6. Using the fact that $\lim_{\theta\to 0}[(\sin\theta)/\theta] = 1$, find $\lim_{\theta\to 0}\cos[(\pi\sin\theta)/(4\theta)]$.

Find the limits in Exercises 7–12.

7. $\lim\limits_{x\to 3}(x^2 - 2x + 2)$
8. $\lim\limits_{x\to -2}\dfrac{(x^2 - 4)}{x^2 + 4}$
9. $\lim\limits_{x\to 2}\dfrac{(x^2 - 4)}{(x^2 - 5x + 6)}$
10. $\lim\limits_{x\to 27}\dfrac{(\sqrt[3]{x} - 3)}{(x - 2)}$
11. $\lim\limits_{x\to 0}\dfrac{(3 + x)^2 - 9}{x}$
12. $\lim\limits_{x\to 2}\dfrac{x - 2}{x^2 - 3x + 2}$

Verify the limit statements in Exercises 13–16 using the ε-A definition.

13. $\lim\limits_{x\to\infty}\dfrac{1 + x^3}{x^3} = 1$
14. $\lim\limits_{x\to\infty}\dfrac{3x}{x^2 + 2} = 0$
15. $\lim\limits_{x\to\infty}(1 + e^{-3x}) = 1$
16. $\lim\limits_{x\to\infty}\dfrac{1}{\ln x} = 0$

Find the limits in Exercises 17–24.

17. $\lim\limits_{x\to\infty}\left(\dfrac{3}{x} + \dfrac{5}{x^2} - 2\right)$
18. $\lim\limits_{x\to\infty}\left(\dfrac{8}{x^2} - \dfrac{1}{x^3} + 5\right)$
19. $\lim\limits_{x\to\infty}\dfrac{10x^2 - 2}{15x^2 - 3}$
20. $\lim\limits_{x\to\infty}\dfrac{-4x + 3}{x + 2}$
21. $\lim\limits_{x\to\infty}\dfrac{3x^2 + 2x + 4}{5x^2 + x + 7}$
22. $\lim\limits_{x\to\infty}\dfrac{x^2 + xe^{-x}}{6x^2 + 2}$
23. $\lim\limits_{x\to\infty}\dfrac{x + 2 + 1/x}{2x + 3 + 2/x}$
24. $\lim\limits_{x\to\infty}\dfrac{x - 3 - 1/x^2}{2x + 5 + 1/x^2}$
25. Find $\lim_{x\to\infty}[\sqrt{x^2 + a^2} - x]$ and interpret your answer geometrically.
26. Find $\lim_{x\to\infty}[\sqrt{c^2x^2 + 1} - cx]$ and interpret your answer geometrically.
27. Find the horizontal asymptotes of the graph of $\sqrt{x^2 + 1} - (x + 1)$. Sketch.
28. Find the horizontal asymptotes of the graph $y = (x + 1)/\sqrt{x^2 + 2}$. Sketch.

Find the limits in Exercises 29–32 using the reciprocal test.

29. $\lim\limits_{x\to 2}\dfrac{1}{(x - 2)^2}$
30. $\lim\limits_{x\to 2}\dfrac{x^2}{(x - 2)^4}$
31. $\lim\limits_{x\to\infty}\dfrac{x^2 + 2}{\sqrt{x}}$
32. $\lim\limits_{x\to\infty}\dfrac{x^4 + 8}{x^{5/2}}$

Find the one-sided limits in Exercises 33–40.

33. $\lim\limits_{x\to 2+}\dfrac{x^2 - 4}{(x - 2)^2}$
34. $\lim\limits_{x\to 2-}\dfrac{x^2 - 4}{(x - 2)^2}$
35. $\lim\limits_{x\to 0-}\dfrac{(x - 1)(x - 2)}{x(x + 1)(x + 2)}$
36. $\lim\limits_{x\to 1+}\dfrac{x(x + 3)}{(x - 1)(x - 2)}$
37. $\lim\limits_{x\to 0+}\dfrac{(x^3 - 1)|x|}{x}$
38. $\lim\limits_{x\to 0-}\dfrac{(x^4 + 2)|x|}{x}$
39. $\lim\limits_{x\to \frac{1}{2}-}\dfrac{2x - 1}{\sqrt{(2x - 1)^2}}$
40. $\lim\limits_{x\to \frac{1}{2}+}\dfrac{\sqrt{(2x - 1)^2}}{x - 1/2}$

Find the vertical and horizontal asymptotes of the functions in Exercises 41–44 and sketch their graphs.

41. $f(x) = \dfrac{1}{x^2 - 5x + 6}$
42. $f(x) = \dfrac{1}{2x + 3}$
43. $f(x) = \dfrac{1}{x^2 - 1}$
44. $f(x) = \dfrac{x^2}{x^2 - 1}$
45. (a) Establish the comparison test 2 using the ε-A definition of limit. (b) Use (a) to find
$$\lim_{x\to\infty}\left[\frac{1}{x}\sin\left(\frac{1}{x}\right)\right].$$
46. (a) Use the B-δ definition of limit to show that if $\lim_{x\to x_0} f(x) = \infty$ and $g(x) \geqslant f(x)$ for x close to x_0, $x \neq x_0$, then $\lim_{x\to x_0} g(x) = \infty$. (b) Use (a) to show that $\lim_{x\to 1}[(1 + \cos^2 x)/(1 - x)^2] = \infty$.

Find the limits in Exercises 47–60.

47. $\lim_{x\to 1} \frac{3+4x}{4+5x}$
48. $\lim_{x\to 1} \frac{x^3-1}{x-1}$
49. $\lim_{x\to 1} \frac{x^3-1}{x^2-1}$
50. $\lim_{x\to 2} \frac{x-2}{x^2+3x+2}$
51. $\lim_{x\to -3} \frac{x^2+2x-3}{x^2+x-6}$
52. $\lim_{x\to 1} \frac{x^n-1}{x-1}$
53. $\lim_{x\to -1} \frac{x^{2n+1}+1}{x+1}$
54. $\lim_{x\to 2} \frac{x-2}{\sqrt{x}-\sqrt{2}}$
55. $\lim_{x\to 2} \frac{x^2+3x+6}{9x-1}$
56. $\lim_{x\to\infty} \sin(\frac{1}{x})$
57. $\lim_{x\to 1+} \frac{e^x-1}{x-1}$
58. $\lim_{x\to\infty} \sin\left(\frac{\pi x^2+4}{6x^2+9}\right)$
59. $\lim_{x\to 1-} \frac{\ln 2x}{x-1}$
60. $\lim_{x\to -\infty} \ln(x^2)$

Find the horizontal and vertical asymptotes of the functions in Exercises 61–64.

61. $y = \frac{x}{x^2-1}$
62. $y = \frac{(x+1)(x-1)}{(x-2)x(x+2)}$
63. $y = \frac{e^x+2x}{e^x-2x}$
64. $y = \frac{\ln x - 1}{\ln x + 1}$

65. Let $f(x)$ and $g(x)$ be polynomials such that $\lim_{x\to\infty}[f(x)/g(x)] = l$. Prove that the limit $\lim_{x\to-\infty} f(x)/g(x)$ is equal to l as well. What happens if $l = \infty$ or $-\infty$?
66. How close to 3 does x have to be to ensure that $|x^3 - 2x - 21)| < \frac{1}{1000}$?
67. Let $f(x) = |x|$.
    (a) Find $f'(x)$ and sketch its graph.
    (b) Find $\lim_{x\to 0-} f'(x)$ and $\lim_{x\to 0+} f'(x)$.
    (c) Does $\lim_{x\to 0} f'(x)$ exist?
68. (a) Give a precise definition of this statement: $\lim_{x\to\infty} f(x) = -\infty$. (b) Draw figures like Figs. 11.1.1, 11.1.3, and 11.1.4 to illustrate your definition.
69. Draw figures like Figs. 11.1.1, 11.1.3, and 11.1.4 to illustrate the definition of these statements: (a) $\lim_{x\to x_0+} f(x) = l$; (b) $\lim_{x\to x_0+} f(x) = \infty$. [*Hint*: The shaded region should include all points with $x \leqslant x_0$.]
70. (a) Graph $y = f(x)$, where
$$f(x) = \begin{cases} |x|/x, & x \neq 0, \\ 0, & x = 0. \end{cases}$$
    Does $\lim_{x\to 0} f(x)$ exist?
    (b) Graph $y = g(x)$, where
$$g(x) = \begin{cases} x+1, & x < 0, \\ 2x-1, & x \geqslant 0. \end{cases}$$
    Does $\lim_{x\to 0} g(x)$ exist?
    (c) Let $f(x)$ be as in part (a) and $g(x)$ as in part (b). Graph $y = f(x) + g(x)$. Does $\lim_{x\to 0}[f(x) + g(x)]$ exist? Conclude that the limit of a sum can exist even though the limits of the summands do not.
71. The number $N(t)$ of individuals in a population at time t is given by
$$N(t) = N_0 \frac{e^{3t}}{(3/2) + e^{3t}}.$$
    Find the value of $\lim_{t\to\infty} N(t)$ and discuss its biological meaning.
72. The current in a certain RLC circuit is given by $I(t) = \{[(1/3)\sin t + \cos t]e^{-t/2} + 4\}$ amperes. The value of $\lim_{t\to\infty} I(t)$ is called the *steady-state current*; it respresents the current present after a long period of time. Find it.
73. The temperature $T(x,t)$ at time t at position x of a rod located along $0 \leqslant x \leqslant l$ on the x axis is given approximately by the rule $T(x,t) = B_1 e^{-\mu_1 t}\sin\lambda_1 x + B_2 e^{-\mu_2 t}\sin\lambda_2 x + B_3 e^{-\mu_3 t}\sin\lambda_3 x$, where $\mu_1, \mu_2, \mu_3, \lambda_1, \lambda_2, \lambda_3$ are all positive. Show that $\lim_{t\to\infty} T(x,t) = 0$ for each fixed location x along the rod. The model applies to a rod without heat sources, with the heat allowed to radiate from the right end of the rod; zero limit means all heat eventually radiates out the right end.
74. A psychologist doing some manipulations with testing theory wishes to replace the reliability factor
$$R = \frac{nr}{1+(n-1)r} \qquad (\textit{Spearman–Brown formula})$$
    by unity, because someone told her that she could do this for large extension factors n. She formally replaces n by $1/x$, simplifies, and then sets $x = 0$, to obtain 1. What has she done, in the language of limits?

★75. Study this ε-δ proof of the sum rule: Let $\lim_{x\to x_0} f(x) = L$ and $\lim_{x\to x_0} g(x) = M$. Given $\varepsilon > 0$, choose $\delta_1 > 0$ such that $|x - x_0| < \delta_1$ and $x \neq x_0$ implies $|f(x) - L| < \varepsilon/2$; choose $\delta_2 > 0$ such that $|x - x_0| < \delta_2$, $x \neq x_0$, implies that $|g(x) - M| < \varepsilon/2$. Let δ be the smaller of δ_1 and δ_2. Then $|x - x_0| < \delta$, and $x \neq x_0$ implies $|(f(x) + g(x)) - (L + M)| \leqslant |f(x) - L| + |g(x) - M|$ (by the triangle inequality $|x + y| \leqslant |x| + |y|$). This is less than $\varepsilon/2 + \varepsilon/2 = \varepsilon$, and therefore $\lim_{x\to x_0}[f(x) + g(x)] = L + M$.

Now prove that $\lim_{x\to x_0}[af(x) + bg(x)] = a\lim_{x\to x_0} f(x) + b\lim_{x\to x_0} g(x)$.

★76. Study this ε-δ proof of the product rule: If $\lim_{x\to x_0} f(x) = L$ and $\lim_{x\to x_0} g(x) = M$, then $\lim_{x\to x_0} f(x)g(x) = LM$.

Proof: Let $\varepsilon > 0$ be given. We must find a number $\delta > 0$ such that $|f(x)g(x) - LM| < \varepsilon$ whenever $|x - x_0| < \delta$, $x \neq x_0$. Adding and subtracting $f(x)M$, we have
$$\begin{aligned} &|f(x)g(x) - LM| \\ &= |f(x)g(x) - f(x)M + f(x)M - LM| \\ &\leqslant |f(x)|\cdot|g(x) - M| + |f(x) - L||M|. \end{aligned}$$
The closeness of $g(x)$ to M and $f(x)$ to L must depend upon the size of $f(x)$ and $|M|$. Choose δ_1 such that $|f(x) - L| < (\varepsilon/2)M$ whenever $|x - x_0| < \delta_1$, $x \neq x_0$. Also, choose δ_2 such that $|x - x_0| < \delta_2$, $x \neq x_0$, implies that $|f(x) - L|$

< 1, which in turn implies that $|f(x)| < |L| + 1$ (since $|f(x)| = |f(x) - L + L| \leqslant |f(x) - L| + |L| < 1 + |L|$). Finally, choose $\delta_3 > 0$ such that $|g(x) - M| < (\varepsilon/2)(|L| + 1)$ whenever $|x - x_0| < \delta_3$, $x \neq x_0$. Let δ be the smallest of δ_1, δ_2, and δ_3. If $|x - x_0| < \delta$, $x \neq x_0$, then $|x - x_0| < \delta_1$, $|x - x_0| < \delta_2$, and $|x - x_0| < \delta_3$, so by the choice of $\delta_1, \delta_2, \delta_3$, we have

$$|f(x)g(x) - LM|$$
$$\leqslant |f(x)||g(x) - M| + |f(x) - L||M|$$
$$< (|L| + 1)\frac{\varepsilon}{2(|L| + 1)} + \frac{\varepsilon}{2|M|} \cdot |M|$$
$$= \varepsilon,$$

and so $|f(x)g(x) - LM| < \varepsilon$.

Now prove the quotient rule for limits.

★77. Study the following proof of the one-sided composite function rule: If $\lim_{x \to x_0+} f(x) = L$ and g is continuous at L, then $g(f(x))$ is defined for all x in some interval of the form (x_0, b), and $\lim_{x \to x_0+} g(f(x)) = g(L)$.

Proof: Let $\varepsilon > 0$. We must find a positive number δ such that whenever $x_0 < x < x_0 + \delta$, $g(f(x))$ is defined and $|g(f(x)) - g(L)| < \varepsilon$. Since g is continuous at L, there is a positive number ρ such that whenever $|y - L| < \rho$, $g(y)$ is defined and $|g(y) - g(L)| < \varepsilon$. Now since $\lim_{x \to x_0+} f(x) = L$, we can find a positive number δ such that whenever $x_0 < x < x_0 + \delta$, $|f(x) - L| < \rho$. For such x, we apply the previously obtained property of ρ, with $y = f(x)$, to conclude that $g(f(x))$ is defined and that $|g(f(x)) - g(L)| < \varepsilon$.

Now prove the composite function rule.

★78. Use the ε-A definition to prove the sum rule for limits at infinity.

★79. Use the B-δ definition to prove the reciprocal test for infinite limits.

★80. Suppose that a function f is defined on an open interval I containing x_0, and that there are numbers m and K such that we have the inequality $|f(x) - f(x_0) - m(x - x_0)| \leqslant K|x - x_0|^2$ for all x in I. Prove that f is differentiable at x_0 with derivative $f'(x_0) = m$.

★81. Show that $\lim_{x \to \infty} f(x) = l$ if and only if $\lim_{y \to 0+} f(1/y) = l$. (This reduces the computation of limits at infinity to one-sided limits at zero.)

11.2 L'Hôpital's Rule

Differentiation can be used to evaluate limits.

L'Hôpital's rule[2] is a very efficient way of using differential calculus to evaluate limits. It is not necessary to have mastered the theoretical portions of the previous sections to use l'Hôpital's rule, but you should review some of the computational aspects of limits from either Section 11.1 or Section 1.3.

L'Hôpital's rule deals with limits of the form $\lim_{x \to x_0}[f(x)/g(x)]$, where $\lim_{x \to x_0} f(x)$ and $\lim_{x \to x_0} g(x)$ are both equal to zero or infinity, so that the quotient rule cannot be applied. Such limits are called *indeterminate forms*. (One can also replace x_0 by ∞, x_0+, or x_0-.)

Our first objective is to calculate $\lim_{x \to x_0}[f(x)/g(x)]$ if $f(x_0) = 0$ and $g(x_0) = 0$. Substituting $x = x_0$ gives us $\frac{0}{0}$, so we say that we are dealing with an *indeterminate form of type* $\frac{0}{0}$. Such forms occurred when we considered the derivative as a limit of difference quotients; in Section 1.3 we used the limit rules to evaluate some simple derivatives. Now we can work the other way around, using our ability to calculate derivatives in order to evaluate quite complicated limits: l'Hôpital's rule provides the means for doing this.

The following box gives the simplest version of l'Hôpital's rule.

[2] In 1696, Guillaume F. A. l'Hôpital published in Paris the first calculus textbook: *Analyse des Infiniment Petits (Analysis of the infinitely small)*. Included was a proof of what is now referred to as l'Hôpital's rule; the idea, however, probably came from J. Bernoulli. This rule was the subject of some work by A. Cauchy, who clarified its proof in his *Cours d'Analyse (Course in analysis)* in 1823. The foundations were in debate until almost 1900. See, for instance, the very readable article, "The Law of the Mean and the Limits $\frac{0}{0}$, $\frac{\infty}{\infty}$," by W. F. Osgood, *Annals of Mathematics*, Volume 12 (1898–1899), pp. 65–78.

L'Hôpital's Rule: Preliminary Version

Let f and g be differentiable in an open interval containing x_0; assume that $f(x_0) = g(x_0) = 0$. If $g'(x_0) \neq 0$, then

$$\lim_{x \to x_0} \frac{f(x)}{g(x)} = \frac{f'(x_0)}{g'(x_0)}.$$

To prove this, we use the fact that $f(x_0) = 0$ and $g(x_0) = 0$ to write

$$\frac{f(x)}{g(x)} = \frac{f(x) - f(x_0)}{g(x) - g(x_0)} = \frac{[f(x) - f(x_0)]/(x - x_0)}{[g(x) - g(x_0)]/(x - x_0)}.$$

As x tends to x_0, the numerator tends to $f'(x_0)$, and the denominator tends to $g'(x_0) \neq 0$, so the result follows from the quotient rule for limits.

Let us verify this rule on a simple example.

Example 1 Find $\lim_{x \to 1}\left[\frac{x^3 - 1}{x - 1}\right]$.

Solution Here we take $x_0 = 1$, $f(x) = x^3 - 1$, and $g(x) = x - 1$. Since $g'(1) = 1$, the preliminary version of l'Hôpital's rule applies to give

$$\lim_{x \to 1} \frac{x^3 - 1}{x - 1} = \frac{f'(1)}{g'(1)} = \frac{3}{1} = 3.$$

We know two other ways (from Chapter 1) to calculate this limit. First, we can factor the numerator:

$$\frac{x^3 - 1}{x - 1} = \frac{(x - 1)(x^2 + x + 1)}{x - 1} = x^2 + x + 1 \qquad (x \neq 1).$$

Letting $x \to 1$, we again recover the limit 3. Second, we can recognize the function $(x^3 - 1)/(x - 1)$ as the different quotient $[h(x) - h(1)]/(x - 1)$ for $h(x) = x^3$. As $x \to 1$, this different quotient approaches the derivative of h at $x = 1$, namely 3. ▲

The next example begins to show the power of l'Hôpital's rule in a more difficult limit.

Example 2 Find $\lim_{x \to 0} \frac{\cos x - 1}{\sin x}$.

Solution We apply l'Hôpital's rule with $f(x) = \cos x - 1$ and $g(x) = \sin x$. We have $f(0) = 0$, $g(0) = 0$, and $g'(0) = 1 \neq 0$, so

$$\lim_{x \to 0} \frac{\cos x - 1}{\sin x} = \frac{f'(0)}{g'(0)} = \frac{-\sin(0)}{\cos(0)} = 0. \text{ ▲}$$

This method does not solve all $\frac{0}{0}$ problems. For example, suppose we wish to find

$$\lim_{x \to 0} \frac{\sin x - x}{x^3}.$$

If we differentiate the numerator and denominator, we get $(\cos x - 1)/3x^2$, which becomes $\frac{0}{0}$ when we set $x = 0$. This suggests that we use l'Hôpital's rule again, but to do so, we need to know that $\lim_{x \to x_0}[f'(x)/g(x)]$ is equal to $\lim_{x \to x_0}[f''(x)/g''(x)]$, even when $f'(x_0)/g'(x_0)$ is again indeterminate. The following strengthened version of l'Hôpital's rule is the result we need. Its proof is given later in the section.

L'Hôpital's Rule

Let f and g be differentiable on an open interval containing x_0, except perhaps at x_0 itself. Assume:

(i) $g(x) \neq 0$,
(ii) $g'(x) \neq 0$ for x in an interval about x_0, $x \neq x_0$,
(iii) f and g are continuous at x_0 with $f(x_0) = g(x_0) = 0$, and
(iv) $\displaystyle\lim_{x \to x_0} \frac{f'(x)}{g'(x)} = l.$

Then $\displaystyle\lim_{x \to x_0} \frac{f(x)}{g(x)} = l.$

Example 3 Calculate $\displaystyle\lim_{x \to 0} \frac{\cos x - 1}{x^2}$.

Solution This is in $\frac{0}{0}$ form, so by l'Hôpital's rule,

$$\lim_{x \to 0} \frac{\cos x - 1}{x^2} = \lim_{x \to 0} \frac{-\sin x}{2x}$$

if the latter limit can be shown to exist. However, we can use l'Hôpital's rule again to write

$$\lim_{x \to 0} \frac{-\sin x}{2x} = \lim_{x \to 0} \frac{-\cos x}{2}.$$

Now we may use the continuity of $\cos x$ to substitute $x = 0$ and find the last limit to be $-\frac{1}{2}$; thus

$$\lim_{x \to 0} \frac{\cos x - 1}{x^2} = -\frac{1}{2}.$$

To keep track of what is going on, some students like to make a table:

	form	type	limit
$\dfrac{f}{g}$	$\dfrac{\cos x - 1}{x^2}$	$\dfrac{0}{0}$ indeterminate	?
$\dfrac{f'}{g'}$	$\dfrac{-\sin x}{2x}$	$\dfrac{0}{0}$ indeterminate	?
$\dfrac{f''}{g''}$	$\dfrac{-\cos x}{2}$	determinate	$\boxed{-\dfrac{1}{2}}$

▲

Each time the numerator and denominator are differentiated, we must check the type of limit; if it is $\frac{0}{0}$, we proceed and are sure to stop when the limit becomes determinate, that is, when it can be evaluated by substitution of the limiting value.

Warning If l'Hôpital's rule is used when the limit is determinate, incorrect answers can result. For example, $\lim_{x\to 0}[(x^2+1)/x] = \infty$ but l'Hôpital's rule would lead to $\lim_{x\to 0}(2x/1)$ which is zero (and is incorrect).

Example 4 Find $\displaystyle\lim_{x\to 0}\frac{\sin x - x}{\tan x - x}$.

Solution This is in $\frac{0}{0}$ form, so we use l'Hôpital's rule:

	form	type	limit
$\frac{f}{g}$	$\frac{\sin x - x}{\tan x - x}$	$\frac{0}{0}$	?
$\frac{f'}{g'}$	$\frac{\cos x - 1}{\sec^2 x - 1}$	$\frac{0}{0}$	?
$\frac{f''}{g''}$	$\frac{-\sin x}{2\sec x(\sec x\tan x)}$	$\frac{0}{0}$	?
$\frac{f'''}{g'''}$	$\frac{-\cos x}{4\sec^2 x\tan^2 x + 2\sec^4 x}$	determinate	$\boxed{-\frac{1}{2}}$

Thus $\displaystyle\lim_{x\to 0}\frac{\sin x - x}{\tan x - x} = -\frac{1}{2}$. ▲

L'Hôpital's rule also holds for one-sided limits, limits as $x\to\infty$, or if we have indeterminates of the form $\frac{\infty}{\infty}$. To prove the rule for the form $\frac{0}{0}$ in case $x\to\infty$, we use a trick: set $t = 1/x$, so that $x = 1/t$ and $t\to 0+$ as $x\to+\infty$. Then

$$\begin{aligned}\lim_{x\to+\infty}\frac{f'(x)}{g'(x)} &= \lim_{t\to 0+}\frac{f'(1/t)}{g'(1/t)}\\ &= \lim_{t\to 0+}\frac{-t^2 f'(1/t)}{-t^2 g'(1/t)}\\ &= \lim_{t\to 0+}\frac{(d/dt)f(1/t)}{(d/dt)g(1/t)} \qquad \text{(by the chain rule)}\\ &= \lim_{t\to 0+}\frac{f(1/t)}{g(1/t)} \qquad \text{(by l'Hôpital's rule)}\\ &= \lim_{x\to+\infty}\frac{f(x)}{g(x)}.\end{aligned}$$

It is tempting to use a similar trick for the $\frac{\infty}{\infty}$ form as $x\to x_0$, but it does not work. If we write

$$\frac{f(x)}{g(x)} = \frac{1/g(x)}{1/f(x)},$$

which is in the $\frac{0}{0}$ form, we get

$$\lim_{x\to x_0}\frac{f(x)}{g(x)} = \lim_{x\to x_0}\frac{-g'(x)/[g(x)]^2}{-f'(x)/[f(x)]^2}$$

which is no easier to handle. For the correct proof, see Exercise 42.

The use of l'Hôpital's rule is summarized in the following display.

L'Hôpital's Rule

To find $\lim_{x\to x_0}[f(x)/g(x)]$ where $\lim_{x\to x_0} f(x)$ and $\lim_{x\to x_0} g(x)$ are both zero or both infinite, differentiate the numerator and denominator and take the limit of the new fraction; repeat the process as many times as necessary, checking each time that l'Hôpital's rule applies.

If $\lim_{x\to x_0} f(x) = \lim_{x\to x_0} g(x) = 0$ (or each is $\pm\infty$), then

$$\lim_{x\to x_0} \frac{f(x)}{g(x)} = \lim_{x\to x_0} \frac{f'(x)}{g'(x)}$$

(x_0 may be replaced by $\pm\infty$ or $x_0\pm$).

The result of the next example was stated at the beginning of Section 6.4. The solution by l'Hôpital's rule is much easier than the one given in Review Exercise 90 of Chapter 6.

Example 5 Find $\lim_{x\to\infty} \frac{\ln x}{x^p}$, where $p>0$.

Solution This is in the form $\frac{\infty}{\infty}$. Differentiating the numerator and denominator, we find

$$\lim_{x\to\infty} \frac{\ln x}{x^p} = \lim_{x\to\infty} \frac{1/x}{px^{p-1}} = \lim_{x\to\infty} \frac{1}{px^p} = 0,$$

since $p>0$. ▲

Certain expressions which do not appear to be in the form $f(x)/g(x)$ can be put in that form with some manipulation. For example, the indeterminate form $\infty\cdot 0$ appears when we wish to evaluate $\lim_{x\to x_0} f(x)g(x)$ where $\lim_{x\to x_0} f(x) = \infty$ and $\lim_{x\to x_0} g(x) = 0$. This can be converted to $\frac{0}{0}$ or $\frac{\infty}{\infty}$ form by writing

$$f(x)g(x) = \frac{g(x)}{1/f(x)} \quad \text{or} \quad f(x)g(x) = \frac{f(x)}{1/g(x)}.$$

Example 6 Find $\lim_{x\to 0+} x\ln x$.

Solution We write $x\ln x$ as $(\ln x)/(1/x)$, which is now in $\frac{\infty}{\infty}$ form. Thus

$$\lim_{x\to 0+} x\ln x = \lim_{x\to 0+} \frac{\ln x}{1/x} = \lim_{x\to 0+} \frac{1/x}{-1/x^2} = \lim_{x\to 0+} (-x) = 0. \ ▲$$

Indeterminate forms of the type 0^0 and 1^∞ can be handled by using logarithms:

Example 7 Find (a) $\lim_{x\to 0+} x^x$ and (b) $\lim_{x\to 1} x^{1/(1-x)}$.

Solution (a) This is of the form 0^0, which is indeterminate because zero to any power is zero, while any number to the zeroth power is 1. To obtain a form to which l'Hôpital's rule is applicable, we write x^x as $\exp(x\ln x)$. By Example 6, we have $\lim_{x\to 0+} x\ln x = 0$. Since $g(x) = \exp(x)$ is continuous, the continuous function rule applies, giving $\lim_{x\to 0+}\exp(x\ln x) = \exp(\lim_{x\to 0+} x\ln x) = e^0 = 1$, so $\lim_{x\to 0+} x^x = 1$. (Numerically, $0.01^{0.1} = 0.79$, $0.001^{0.001} = 0.993$, and $0.00001^{0.00001} = 0.99988$.)

(b) This has the indeterminate form 1^∞. We have $x^{1/(x-1)} = e^{(\ln x)/(x-1)}$; applying l'Hôpital's rule gives

$$\lim_{x\to 1} \frac{\ln x}{x-1} = \lim_{x\to 1} \frac{1/x}{1} = 1,$$

so

$$\lim_{x\to 1} x^{1/(x-1)} = \lim_{x\to 1} e^{(\ln x)/(x-1)} = e^{\lim_{x\to 1}[(\ln x)/(x-1)]} = e^1 = e.$$

If we set $x = 1 + (1/n)$, then $x \to 1$ when $n \to \infty$; we have $1/(x-1) = n$, so the limit we just calculated is $\lim_{n\to\infty}(1 + 1/n)^n$. Thus l'Hôpital's rule gives another proof of the limit formula $\lim_{n\to\infty}(1 + 1/n)^n = e$. ▲

The next example is a limit of the form $\infty - \infty$.

Example 8 Find $\lim_{x\to 0}\left(\frac{1}{x \sin x} - \frac{1}{x^2}\right)$.

Solution We can convert this limit to $\frac{0}{0}$ form by bringing the expression to a common denominator:

	form	type	limit
	$\frac{1}{x\sin x} - \frac{1}{x^2}$	$\infty - \infty$	?
$\frac{f}{g}$	$\frac{x - \sin x}{x^2 \sin x}$	$\frac{0}{0}$	?
$\frac{f'}{g'}$	$\frac{1 - \cos x}{2x \sin x + x^2 \cos x}$	$\frac{0}{0}$	?
$\frac{f''}{g''}$	$\frac{\sin x}{2 \sin x + 4x \cos x - x^2 \sin x}$	$\frac{0}{0}$	?
$\frac{f'''}{g'''}$	$\frac{\cos x}{6 \cos x - 6x \sin x - x^2 \cos x}$	determinate	$\boxed{\frac{1}{6}}$

Thus $\lim_{x\to 0}\left(\frac{1}{x \sin x} - \frac{1}{x^2}\right) = \frac{1}{6}$. ▲

Finally, we shall prove l'Hôpital's rule. The proof relies on a generalization of the mean value theorem.

Cauchy's mean value theorem *Suppose that f and g are continuous on $[a,b]$ and differentiable on (a,b) and that $g(a) \neq g(b)$. Then there is a number c in (a,b) such that*

$$\frac{f(b) - f(a)}{g(b) - g(a)} = \frac{f'(c)}{g'(c)}.$$

Proof First note that if $g(x) = x$, we recover the mean value theorem in its usual form. The proof of the mean value theorem in Section 3.6 used the function

$$l(x) = f(a) + (x - a)\frac{f(b) - f(a)}{b - a}.$$

For the Cauchy mean value theorem, we replace $x - a$ by $g(x) - g(a)$ and look at

$$h(x) = f(a) + [g(x) - g(a)]\frac{f(b) - f(a)}{g(b) - g(a)}.$$

Notice that $f(a) = h(a)$ and $f(b) = h(b)$. By the horserace theorem (see Section 3.6), there is a point c such that $f'(c) = h'(c)$; that is,

$$f'(c) = g'(c)\left[\frac{f(b) - f(a)}{g(b) - g(a)}\right],$$

which is what we wanted to prove. ■

We now prove the final version of l'Hôpital's rule. Since $f(x_0) = g(x_0) = 0$, we have

$$\frac{f(x)}{g(x)} = \frac{f(x) - f(x_0)}{g(x) - g(x_0)} = \frac{f'(c_x)}{g'(c_x)}, \tag{1}$$

where c_x (which depends on x) lies between x and x_0. Note that $c_x \to x_0$ as $x \to x_0$. Since, by hypothesis, $\lim_{x\to x_0}[f'(x)/g'(x)] = l$, it follows that we also have $\lim_{x\to x_0}[f'(c_x)/g'(c_x)] = l$, and so by equation (1), $\lim_{x\to x_0}[f(x)/g(x)] = l$. ■

Exercises for Section 11.2

Use the preliminary version of l'Hôpital's rule to evaluate the limits in Exercises 1–4.

1. $\lim_{x\to 3} \frac{x^4 - 81}{x - 3}$
2. $\lim_{x\to 2} \frac{3x^2 - 12}{x - 2}$
3. $\lim_{x\to 0} \frac{x^2 + 2x}{\sin x}$
4. $\lim_{x\to 1} \frac{x^3 + 3x - 4}{\sin(x - 1)}$

Use the final version of l'Hôpital's rule to evaluate the limits in Exercises 5–8.

5. $\lim_{x\to 0} \frac{\cos 3x - 1}{5x^2}$
6. $\lim_{x\to 0} \frac{\cos 10x - 1}{8x^2}$
7. $\lim_{x\to 0} \frac{\sin 2x - 2x}{x^3}$
8. $\lim_{x\to 0} \frac{\sin 3x - 3x}{x^3}$

Evaluate the $\frac{\infty}{\infty}$ forms in Exercises 9–12.

9. $\lim_{x\to\infty} \frac{e^x}{x^{375}}$
10. $\lim_{x\to\infty} \frac{x^4 + \ln x}{3x^4 + 2x^2 + 1}$
11. $\lim_{x\to 0} \frac{\ln x}{x^{-2}}$
12. $\lim_{x\to 0+} \frac{e^{1/x}}{1/x}$

Evaluate the $0 \cdot \infty$ forms in Exercises 13–16.

13. $\lim_{x\to 0} [x^4 \ln x]$
14. $\lim_{x\to 1} [\tan \frac{\pi x}{2} \ln x]$
15. $\lim_{x\to 0} [x^{\pi} e^{-\pi x}]$
16. $\lim_{x\to \pi} [(x^2 - 2\pi x + \pi^2)\csc^2 x]$

Evaluate the limits in Exercises 17–36.

17. $\lim_{x\to 0} [(\tan x)^x]$
18. $\lim_{x\to\infty} \left[\left(1 + \frac{1}{x}\right)^x\right]$
19. $\lim_{x\to 0} (\csc x - \cot x)$
20. $\lim_{x\to\infty} [\ln x - \ln(x - 1)]$
21. $\lim_{x\to 0} \frac{\sqrt{1 + x^2} - 1}{\sin 2x}$
22. $\lim_{x\to 1} \frac{\sqrt{x^2 - 1}}{\cos(\pi x/2)}$
23. $\lim_{x\to 1} \frac{1 - x^2}{1 + x^2}$
24. $\lim_{x\to 0} \frac{x + \sin 2x}{2x + \sin 3x}$
25. $\lim_{x\to\infty} \frac{x}{x^2 + 1}$
26. $\lim_{x\to\infty} \frac{(2x + 1)^3}{x^3 + 2}$
27. $\lim_{x\to 5+} \frac{\sqrt{x^2 - 25}}{x - 5}$
28. $\lim_{x\to 5+} \frac{\sqrt{x^2 - 25}}{x + 5}$
29. $\lim_{x\to -1} \frac{x^2 + 2x + 1}{x^2 - 1}$
30. $\lim_{x\to 1-} x^{1/(1 - x^2)}$
31. $\lim_{x\to 0} \frac{\cos x - 1 + x^2/2}{x^4}$
32. $\lim_{x\to 1} \frac{\ln x}{e^x - 1}$
33. $\lim_{x\to \pi} \frac{1 + \cos x}{x - \pi}$
34. $\lim_{x\to \pi/2} (x - \frac{\pi}{2})\tan x$
35. $\lim_{x\to 0} \frac{\sin x - x + (1/6)x^3}{x^5}$
36. $\lim_{x\to\infty} \frac{x^3 + \ln x + 5}{5x^3 + e^{-x} + \sin x}$
37. Find $\lim_{x\to 0+} x^p \ln x$, where p is positive.
38. Use l'Hôpital's rule to show that as $x \to \infty$, $x^n/e^x \to 0$ for any integer n; that is, e^x goes to infinity faster than any power of x. (This was proved by another method in Section 6.4.)

★39. Give a geometric interpretation of the Cauchy mean value theorem. [*Hint*: Consider the curve given in parametric form by $y = f(t)$, $x = g(t)$.]

★40. Suppose that f is continuous at $x = x_0$, that $f'(x)$ exists for x in an interval about x_0, $x \neq x_0$, and that $\lim_{x\to x_0} f'(x) = m$. Prove that $f'(x_0)$ exists and equals m. [*Hint*: Use the mean value theorem.]

★41. Graph the function $f(x) = x^x$, $x > 0$.

★42. Prove the $\frac{\infty}{\infty}$ version of l'Hôpital's rule. Do this as follows:

(i) Use Cauchy's mean value theorem to prove that for every $\varepsilon > 0$, there is an $M > a$ such that for $x > M$,

$$\left| \frac{f(x) - f(M)}{g(x) - g(M)} - l \right| < \varepsilon$$

(ii) Write

$$\frac{f(x)}{g(x)} = \frac{f(x) - f(M)}{g(x) - g(M)} \cdot \frac{f(x)}{f(x) - f(M)} \cdot \frac{g(x) - g(M)}{g(x)}$$

to conclude that

$$\left| \frac{f(x)}{g(x)} - l \right| < 2\varepsilon$$

for large x.

(iii) Complete the proof using (ii).

★43. (a) Find $\lim_{a \to 0} \frac{1}{a} \ln\left(\frac{e^a - 1}{a} \right)$

(b) Find $\lim_{a \to \infty} \frac{1}{a} \ln\left(\frac{e^a - 1}{a} \right)$

(c) Are your results consistent with the computations of Exercise 30, Section 9.4?

11.3 Improper Integrals

The area of an unbounded region is defined by a limiting process.

The definite integral $\int_a^b f(x)\,dx$ of a function f which is non-negative on the interval $[a, b]$ equals the area of the region under the graph of f between a and b. If we let b go to infinity, the region becomes unbounded, as in Fig. 11.3.1. One's first inclination upon seeing such unbounded regions may be to assert that their areas are infinite; however, examples suggest otherwise.

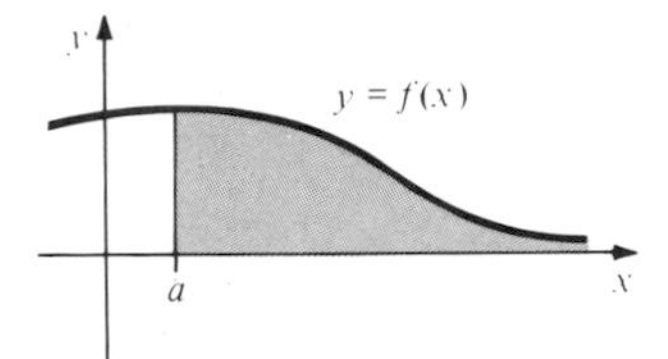

Figure 11.3.1. The region under the graph of f on $[a, \infty)$ is unbounded.

Example 1 Find $\int_1^b \frac{1}{x^4}\,dx$. What happens as b goes to infinity?

Solution We have

$$\int_1^b \frac{dx}{x^4} = \int_1^b x^{-4}\,dx = \left. \frac{x^{-3}}{-3} \right|_1^b = \frac{1/b^3 - 1}{-3} = \frac{1 - 1/b^3}{3}.$$

As b becomes larger and larger, this integral always remains less than $\frac{1}{3}$; furthermore, we have

$$\lim_{b \to \infty} \int_1^b \frac{dx}{x^4} = \lim_{b \to \infty} \frac{1 - 1/b^3}{3} = \frac{1}{3}. \ \blacktriangle$$

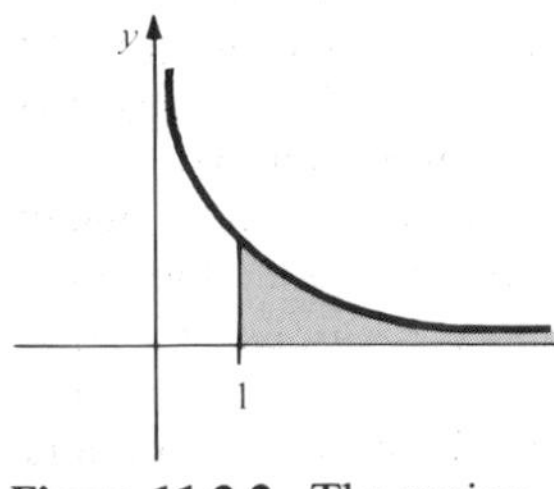

Figure 11.3.2. The region under the graph of $1/x^4$ on $[1, \infty)$ has finite area. It is $\int_1^\infty (dx/x^4) = \frac{1}{3}$.

Example 1 suggests that $\frac{1}{3}$ is the area of the unbounded region consisting of those points (x, y) such that $1 \leqslant x$ and $0 \leqslant y \leqslant 1/x^4$. (See Fig. 11.3.2.) In accordance with our notation for finite intervals, we denote this area by $\int_1^\infty (dx/x^4)$. Guided by this example, *we define integrals over unbounded intervals as limits of integrals over finite intervals*. The general definition follows.

Integrals over Unbounded Intervals

Suppose that for a fixed, f is integrable on $[a,b]$ for all $b>a$. If the limit $\lim_{b\to\infty}\int_a^b f(x)\,dx$ exists, we say that the *improper integral* $\int_a^\infty f(x)\,dx$ *is convergent*, and we define its value by

$$\int_a^\infty f(x)\,dx=\lim_{b\to\infty}\int_a^b f(x)\,dx.$$

Similarly, if, for fixed b, f is integrable on $[a,b]$ for all $a<b$, we define

$$\int_{-\infty}^b f(x)\,dx=\lim_{a\to-\infty}\int_a^b f(x)\,dx,$$

if the limit exists.

Finally, if f is integrable on $[a,b]$ for all $a<b$, we define

$$\int_{-\infty}^\infty f(x)\,dx=\int_{-\infty}^0 f(x)\,dx+\int_0^\infty f(x)\,dx,$$

if the improper integrals on the right-hand side are *both* convergent.

If an improper integral is not convergent, it is called *divergent*.

Example 2 For which values of the exponent r is $\int_1^\infty x^r\,dx$ convergent?

Solution We have

$$\lim_{b\to\infty}\int_1^b x^r\,dx=\lim_{b\to\infty}\frac{x^{r+1}}{r+1}\bigg|_1^b=\lim_{b\to\infty}\frac{b^{r+1}-1}{r+1}\qquad(r\neq-1).$$

If $r+1>0$ (that is, $r>-1$), the limit $\lim_{b\to\infty}b^{r+1}$ does not exist and the integral is divergent. If $r+1<0$ (that is, $r<-1$), we have $\lim_{b\to\infty}b^{r+1}=0$ and the integral is convergent—its value is $-1/(r+1)$. Finally, if $r=-1$ we have $\int_1^b x^{-1}\,dx=\ln b$, which does not converge as $b\to\infty$. We conclude that $\int_1^\infty x^r\,dx$ is convergent just for $r<-1$. ▲

Example 3 Find $\displaystyle\int_{-\infty}^\infty \frac{dx}{1+x^2}$.

Solution We write $\int_{-\infty}^\infty (dx/(1+x^2))=\int_{-\infty}^0 (dx/(1+x^2))+\int_0^\infty (dx/(1+x^2))$. To evaluate these integrals, we use the formula $\int (dx/(1+x^2))=\tan^{-1}x$. Then

$$\int_{-\infty}^0\frac{dx}{1+x^2}=\lim_{a\to-\infty}(\tan^{-1}0-\tan^{-1}a)$$
$$=0-\lim_{a\to-\infty}\tan^{-1}a=0-\left(\frac{-\pi}{2}\right)=\frac{\pi}{2}.$$

(See Fig. 5.4.5 for the horizontal asymptotes of $y=\tan^{-1}x$.) Similarly, we have

$$\int_0^\infty\frac{dx}{1+x^2}=\lim_{b\to\infty}(\tan^{-1}b-\tan^{-1}0)=\frac{\pi}{2},$$

so $\displaystyle\int_{-\infty}^\infty\frac{dx}{1+x^2}=\frac{\pi}{2}+\frac{\pi}{2}=\pi.$ ▲

Sometimes we wish to know that an improper integral converges, even though we cannot find its value explicitly. The following test is quite effective for this situation.

Comparison Test

Suppose that f and g are functions such that

(i) $|f(x)| \leqslant g(x)$ for all $x \geqslant a$ and

(ii) $\int_a^b f(x)\,dx$ and $\int_a^b g(x)\,dx$ exist for every $b > a$.

Then

(1) If $\int_a^\infty g(x)\,dx$ is convergent, so is $\int_a^\infty f(x)\,dx$, and

(2) if $\int_a^\infty f(x)\,dx$ is divergent, so is $\int_a^\infty g(x)\,dx$.

Similar statements hold for integrals of the type

$$\int_{-\infty}^{b} f(x)\,dx \quad \text{and} \quad \int_{-\infty}^{\infty} f(x)\,dx.$$

Here we shall explain the idea behind the comparison test. A detailed proof is given at the end of this section.

If $f(x)$ and $g(x)$ are both positive functions (Fig. 11.3.3(a)), then the region under the graph of f is contained in the region under the graph of g, so

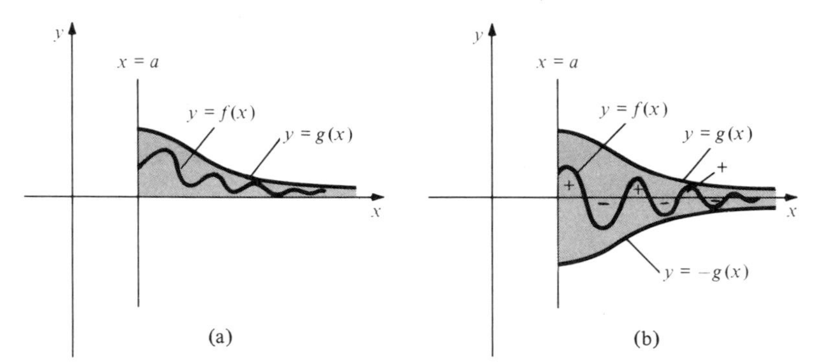

Figure 11.3.3. Illustrating the comparison test.

the integral $\int_a^b f(x)\,dx$ increases and remains bounded as $b \to \infty$. We expect, therefore, that it should converge to some limit. In the general case (Fig. 11.3.3(b)), the sums of the plus areas and the minus areas are both bounded by $\int_a^\infty g(x)\,dx$, and the cancellations can only help the integral to converge.

Note that in the event of convergence, the comparison test only gives the inequality $-\int_a^\infty g(x)\,dx \leqslant \int_a^\infty f(x)\,dx \leqslant \int_a^\infty g(x)\,dx$, but it does not give us the value of $\int_a^\infty f(x)\,dx$.

Example 4 Show that $\displaystyle\int_0^\infty \frac{dx}{\sqrt{1+x^8}}$ is convergent, by comparison with $1/x^4$.

Solution We have $1/\sqrt{1+x^8} < 1/\sqrt{x^8} = 1/x^4$, so it is tempting to compare with $\int_0^\infty (dx/x^4)$. Unfortunately, the latter integral is not defined because $1/x^4$ is unbounded near zero. However, we can break the original integral in two parts:

$$\int_0^\infty \frac{dx}{\sqrt{1+x^8}} = \int_0^1 \frac{dx}{\sqrt{1+x^8}} + \int_1^\infty \frac{dx}{\sqrt{1+x^8}}$$

The first integral on the right-hand side exists because $1/\sqrt{1+x^8}$ is continuous on [0, 1]. The second integral is convergent by the comparison test, taking $g(x) = 1/x^4$ and $f(x) = 1/\sqrt{1+x^8}$. Thus $\int_0^\infty (dx/\sqrt{1+x^8})$ is convergent. ▲

Example 5 Show that $\int_0^\infty \frac{\sin x}{(1+x)^2}\,dx$ converges (without attempting to evaluate).

Solution We may apply the comparison test by choosing $g(x) = 1/(1+x)^2$ and $f(x) = (\sin x)/(1+x)^2$, since $|\sin x| \leqslant 1$. To show that $\int_0^\infty (dx/(1+x)^2)$ is convergent, we can compare $1/(1+x)^2$ with $1/x^2$ on $[1, \infty)$, as in Example 4, or we can evaluate the integral explicitly:

$$\int_0^\infty \frac{dx}{(1+x)^2} = \lim_{b\to\infty} \int_0^a \frac{dx}{(1+x)^2} = \lim_{b\to\infty}\left[\frac{-1}{(1+x)}\right]\Bigg|_0^b$$

$$= \lim_{b\to\infty}\left[1 - \frac{1}{1+b}\right] = 1. \blacktriangle$$

Example 6 Show that $\int_1^\infty \frac{dx}{\sqrt{1+x^2}}$ is divergent.

Solution We use the comparison test in the reverse direction, comparing $1/\sqrt{1+x^2}$ with $1/x$. In fact, for $x \geqslant 1$, we have $1/\sqrt{1+x^2} \geqslant 1/\sqrt{x^2+x^2} = 1/\sqrt{2}\,x$. But $\int_1^b (dx/\sqrt{2}\,x) = (1/\sqrt{2})\ln b$, and this diverges as $b \to \infty$. Therefore, by statement (2) in the comparison test, the given integral diverges. ▲

We shall now discuss the second type of improper integral. If the graph of a function f has a vertical asymptote at one endpoint of the interval $[a, b]$, then the integral $\int_a^b f(x)\,dx$ is not defined in the usual sense, since the function f is not bounded on the interval $[a, b]$. As with integrals of the form $\int_a^\infty f(x)\,dx$, we are dealing with areas of unbounded regions in the plane—this time the unboundedness is in the vertical rather than the horizontal direction. Following our earlier procedure, we can define the integrals of unbounded functions as limits, which are again called improper integrals.

Integrals of Unbounded Functions

Suppose that the graph of f has $x_0 = b$ as a vertical asymptote and that for a fixed, f is integrable on $[a, q]$ for all q in $[a, b)$. If the limit $\lim_{q\to b-} \int_a^q f(x)\,dx$ exists, we shall say that the *improper integral* $\int_a^b f(x)\,dx$ *is convergent*, and we define

$$\int_a^b f(x)\,dx = \lim_{q\to b-} \int_a^q f(x)\,dx.$$

Similarly, if $x = a$ is a vertical asymptote, we define

$$\int_a^b f(x)\,dx = \lim_{p\to a+} \int_p^b f(x)\,dx,$$

if the limit exists. (See Fig. 11.3.4.)

If both $x = a$ and $a = b$ are vertical asymptotes, or if there are vertical asymptotes in the interior (a, b), we may break up $[a, b]$ into subintervals such that the integral of f on each subinterval is of the type considered in the preceding definition. If each part is convergent, we may add the results to get $\int_a^b f(x)\,dx$. The comparison test may be used to test each for convergence. (See Example 9 below.)

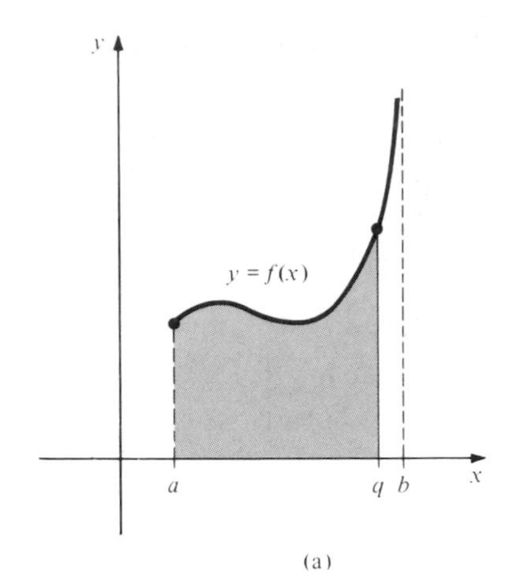

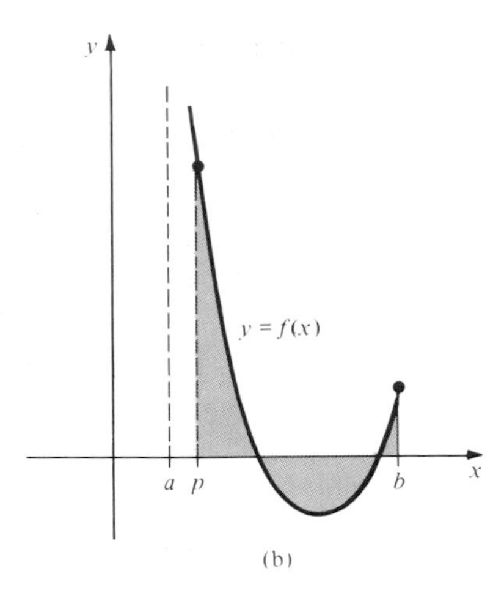

Figure 11.3.4. Improper integrals defined by (a) the limit $\lim_{q\to b-}\int_a^q f(x)\,dx$ and (b) the limit $\lim_{p\to a+}\int_p^b f(x)\,dx$.

Example 7 For which values of r is $\int_0^1 x^r\,dx$ convergent?

Solution If $r \geqslant 0$, x^r is continuous on $[0,1]$ and the integral exists in the ordinary sense. If $r<0$, we have $\lim_{x\to0+}x^r=\infty$, so we must take a limit. We have

$$\lim_{p\to0+}\int_p^1 x^r\,dx=\lim_{p\to0+}\left.\frac{x^{r+1}}{r+1}\right|_p^1=\frac{1}{r+1}\left(1-\lim_{p\to0+}p^{r+1}\right),$$

provided $r\neq-1$. If $r+1>0$ (that is, $r>-1$), we have $\lim_{p\to0+}p^{r+1}=0$, so the integral is convergent and equals $1/(r+1)$. If $r+1<0$ (that is, $r<-1$), $\lim_{p\to0+}p^{r+1}=\infty$, so the integral is divergent. Finally, if $r+1=0$, we have $\lim_{p\to0+}\int_p^1 x^r\,dx=\lim_{p\to0+}(0-\ln p)=\infty$. Thus the integral $\int_0^1 x^r\,dx$ converges just for $r>-1$. (Compare with Example 2.) ▲

Example 8 Find $\int_0^1 \ln x\,dx$.

Solution We know that $\int \ln x\,dx = x\ln x - x + C$, so

$$\int_0^1 \ln x\,dx=\lim_{p\to0+}(1\ln 1-1-p\ln p+p)$$

$$=0-1-0+0=-1$$

($\lim_{p\to0+}p\ln p=0$ by Example 6, Section 11.2). ▲

Example 9 Show that the improper integral $\int_0^\infty \frac{e^{-x}}{\sqrt{x}}\,dx$ is convergent.

Solution This integral is improper at both ends; we may write it as I_1+I_2, where

$$I_1=\int_0^1\left(e^{-x}/\sqrt{x}\right)dx \quad\text{and}\quad I_2=\int_1^\infty\left(e^{-x}/\sqrt{x}\right)dx$$

and then we apply the comparison test to each term. On $[0,1]$, we have $e^{-x}\leqslant 1$, so $e^{-x}/\sqrt{x}\leqslant 1/\sqrt{x}$. Since $\int_0^1(dx/\sqrt{x})$ is convergent (Example 7), so is I_1. On $[1,\infty)$, we have $1/\sqrt{x}\leqslant 1$, so $e^{-x}/\sqrt{x}\leqslant e^{-x}$; but $\int_1^\infty e^{-x}\,dx$ is convergent because

$$\int_1^\infty e^{-x}\,dx=\lim_{b\to\infty}\int_1^b e^{-x}\,dx=\lim_{b\to\infty}(e^{-1}-e^{-b})=e^{-1}$$

Thus I_2 is also convergent and so $\int_0^\infty(e^{-x}/\sqrt{x})\,dx$ is convergent. ▲

Improper integrals arise in arc length problems for graphs with vertical tangents.

Example 10 Find the length of the curve $y=\sqrt{1-x^2}$ for x in $[-1,1]$. Interpret your result geometrically.

Solution By formula (1), Section 10.3, the arc length is

$$\int_{-1}^{1}\sqrt{1+(dy/dx)^2}\,dx=\int_{-1}^{1}\sqrt{1+\left(-x/\sqrt{1-x^2}\right)^2}\,dx$$
$$=\int_{-1}^{1}\frac{dx}{\sqrt{1-x^2}}.$$

The integral is improper at both ends, since

$$\lim_{x\to-1+}\frac{1}{\sqrt{1-x^2}}=\lim_{x\to1-}\frac{1}{\sqrt{1-x^2}}=\infty.$$

We break it up as

$$\int_{-1}^{0}\frac{dx}{\sqrt{1-x^2}}+\int_{0}^{1}\frac{dx}{\sqrt{1-x^2}}$$
$$=\lim_{p\to-1+}\int_{p}^{0}\frac{dx}{\sqrt{1-x^2}}+\lim_{q\to1-}\int_{0}^{q}\frac{dx}{\sqrt{1-x^2}}$$
$$=\lim_{p\to-1+}(\sin^{-1}0-\sin^{-1}p)+\lim_{q\to1-}(\sin^{-1}q-\sin^{-1}0)$$
$$=0-\left(-\frac{\pi}{2}\right)+\frac{\pi}{2}-0=\pi.$$

Geometrically, the curve whose arc length we have just found is a semicircle of radius 1, so we recover the fact that the circumference of a circle of radius 1 is 2π. ▲

Example 11 Luke Skyrunner has just been knocked out in his spaceship by his archenemy, Captain Tralfamadore. The evil captain has set the controls to send the spaceship into the sun! His perverted mind insists on a slow death, so he sets the controls so that the ship makes a constant angle of 30° with the sun (Fig. 11.3.5). What path will Luke's ship follow? How long does Luke have to wake up if he is 10 million miles from the sun and his ship travels at a constant velocity of a million miles per hour?

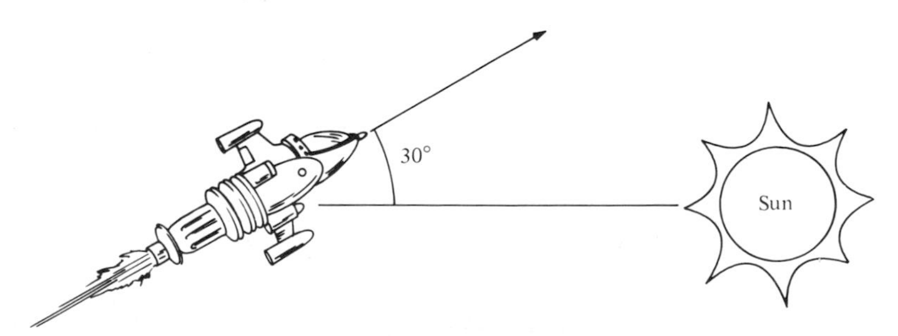

Figure 11.3.5. Luke Skyrunner's ill-fated ship.

Solution We use polar coordinates to describe a curve $(r(t),\theta(t))$ such that the radius makes a constant *angle* α with the tangent ($\alpha=30°$ in the problem). To find $dr/d\theta$, we observe, from Fig. 11.3.6(a), that

$$\Delta r\approx\frac{r\,\Delta\theta}{\tan\alpha}\quad\text{so}\quad\frac{dr}{d\theta}=\frac{r}{\tan\alpha}. \tag{1}$$

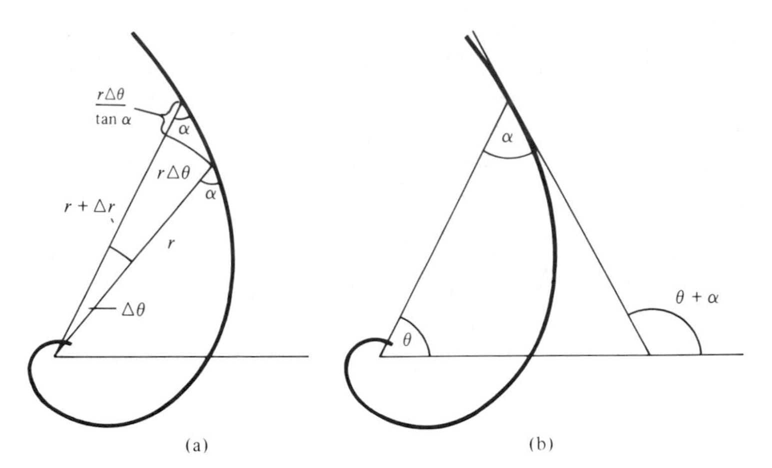

Figure 11.3.6. The geometry of Luke's path.

We can derive formula (1) rigorously, but also more laboriously, by calculating the slope of the tangent line in polar coordinates and setting it equal to $\tan(\theta + \alpha)$ as in Fig. 11.3.6(b). This approach gives

$$\frac{\tan\theta(dr/d\theta) + r}{dr/d\theta - r\tan\theta} = \tan(\theta + \alpha) = \frac{\tan\theta + \tan\alpha}{1 - \tan\theta\tan\alpha},$$

so that again

$$\frac{dr}{d\theta} = \frac{r}{\tan\alpha}.$$

The solution of equation (1) is[3]

$$r(\theta) = r(0)e^{\theta/\tan\alpha}. \tag{2}$$

For this solution to be valid, we must regard θ as a continuous variable ranging from $-\infty$ to ∞, not as being between zero and 2π. As $\theta \to \infty$, $r(\theta) \to \infty$ and as $\theta \to -\infty$, $r(\theta) \to 0$, so the curve spirals outward as θ increases and inward as θ decreases (if $0 < \alpha < \pi/2$). This answers the first question: Luke follows the logarithmic spiral given by equation (2), where $\theta = 0$ is chosen as the starting point.

From Section 10.6, the distance Luke has to travel is the arc length of equation (2) from $\theta = 0$ to $\theta = -\infty$, namely, the improper integral

$$\int_{-\infty}^{0} \sqrt{\left(\frac{dr}{d\theta}\right)^2 + r}\, d\theta = \int_{-\infty}^{0} \sqrt{\frac{r^2}{\tan^2\alpha} + r^2}\, d\theta = \int_{-\infty}^{0} r\sqrt{\cot^2\alpha + 1}\, d\theta$$

$$= \int_{-\infty}^{0} r(0)e^{\theta/\tan\alpha} \frac{1}{\sin\alpha}\, d\theta$$

$$= \frac{r(0)}{\cos\alpha} e^{\theta/\tan\alpha}\Big|_{-\infty}^{0} = \frac{r(0)}{\cos\alpha}.$$

With velocity $= 10^6$, $r(0) = 10^7$, and $\cos\alpha = \cos 30° = \sqrt{3}/2$, the time needed to travel the distance is

$$\text{time} = \frac{\text{distance}}{\text{velocity}} = \frac{10^7}{\sqrt{3}/2} \times \frac{1}{10^6} = \frac{20}{\sqrt{3}} \approx 11.547 \text{ hours}$$

Thus Luke has less than 11.547 hours to wake up. ▲

[3] See Section 8.2. If you have not read Chapter 8, you may simply check directly that equation (2) is a solution of (1).

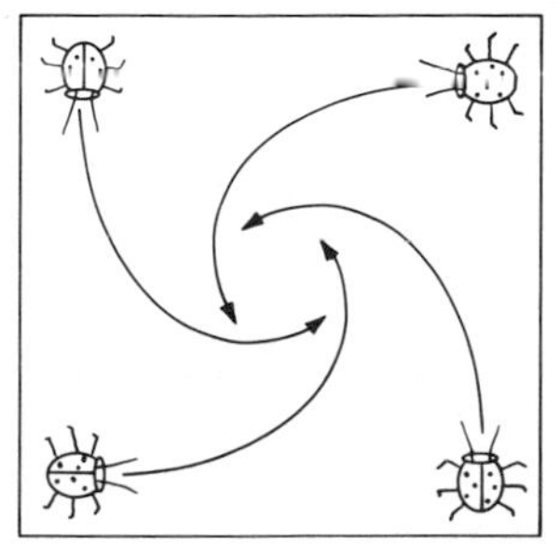

Figure 11.3.7. These love bugs follow logarithmic spirals.

The logarithmic spiral turns up in another interesting situation. Place four love bugs at the corners of a square (Fig. 11.3.7). Each bug, being in love, walks directly toward the bug in front of it, at constant top bug speed. The result is that the bugs all spiral in to the center of the square following logarithmic spirals. The time required for the bugs to reach the center can be calculated as in Example 11 (see Exercise 46).

We conclude this section with a proof of the comparison test. The proof is based on the following principle:

Let F be a function defined on $[a, \infty)$ such that

(i) *F is nondecreasing; i.e., $F(x_1) \leqslant F(x_2)$ whenever $x_1 < x_2$;*

(ii) *F is bounded above: there is a number M such that $F(x) \leqslant M$ for all x.*

Then $\lim_{x\to\infty} F(x)$ exists and is at most M.

The principle is quite plausible, since the graph of F never descends and never crosses the line $y = M$, so that we expect it to have a horizontal asymptote as $x \to \infty$. (See Fig. 11.3.8).

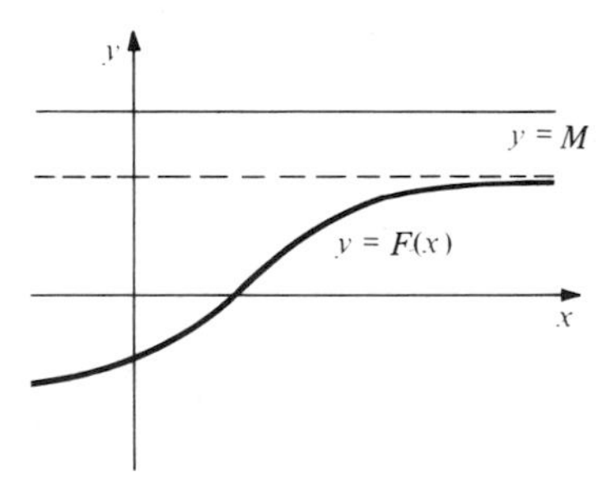

Figure 11.3.8. The graph of a nondecreasing function lying below the line $y = M$ has a horizontal asymptote.

A rigorous proof of the principle requires a careful study of the real numbers,[4] so we shall simply take the principle for granted, just as we did for some basic facts in Chapter 3. A similar principle holds for nonincreasing functions which are bounded below.

Now we are ready to prove statement (1) in the comparison test as stated in the box on p. 530. (Statement (2) follows from (1), for if $\int_a^\infty g(x)\,dx$ converged, so would $\int_a^\infty f(x)\,dx$.)

Let

$$f_1(x) = \begin{cases} f(x) & \text{if } f(x) \geqslant 0 \\ 0 & \text{if } f(x) < 0 \end{cases}$$

and

$$f_2(x) = \begin{cases} f(x) & \text{if } f(x) < 0 \\ 0 & \text{if } f(x) \geqslant 0 \end{cases}$$

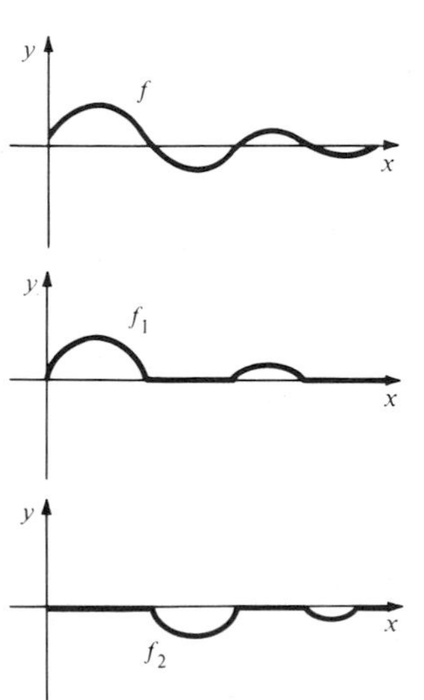

Figure 11.3.9. f_1 and f_2 are the positive and negative parts of f.

be the positive and negative parts of f, respectively. (See Fig. 11.3.9.)

Notice that $f = f_1 + f_2$. Let $F_1(x) = \int_a^x f_1(t)\,dt$ and $F_2(x) = \int_a^x f_2(t)\,dt$. Since f_1 is always non-negative, $F_1(x)$ is increasing. Moreover, by the assumptions of the comparison test,

$$F_1(x) \leqslant \int_a^x |f(t)|\,dt \leqslant \int_a^x g(t)\,dt \leqslant \int_a^\infty g(t)\,dt,$$

so F_1 is bounded above by $\int_a^\infty g(t)\,dt$. Thus, F_1 has a limit as $x \to \infty$. Likewise,

[4] See the theoretical references listed in the preface.

F_2 has a limit since F_2 is decreasing and bounded below. Since

$$\int_a^x f(t)\,dt = F_1(x) + F_2(x),$$

it, too, has a limit as $x \to \infty$. ■

Exercises for Section 11.3

Evaluate the improper integrals in Exercises 1–8.

1. $\int_1^{\infty} \frac{3}{x^2}\,dx$
2. $\int_2^{\infty} \frac{dx}{x^2}$
3. $\int_1^{\infty} e^{-5x}\,dx$
4. $\int_0^{\infty} e^{-3x}\,dx$
5. $\int_2^{\infty} \frac{dx}{x^2-1}$
6. $\int_0^{\infty} \frac{1}{(1+x)^2}\,dx$
7. $\int_{-\infty}^{\infty} \frac{dx}{4+x^2}$
8. $\int_{-\infty}^{\infty} \frac{dx}{9+x^2}$

Show, using the comparison test, that the integrals in Exercises 9–12 are convergent.

9. $\int_0^{\infty} \frac{dx}{3+x^3}$
10. $\int_0^{\infty} \frac{\sin x\,dx}{\sqrt{1+x^4}}$
11. $\int_0^{\infty} \frac{e^{-x}}{1+x}\,dx$
12. $\int_1^{\infty} \frac{e^{-x}}{1+\ln x}\,dx$

Show, using the comparison test, that the integrals in Exercises 13–16 are divergent.

13. $\int_0^{\infty} \frac{dx}{\sqrt{2+x^2}}$
14. $\int_0^{\infty} \frac{dx}{8+x+1/x}$
15. $\int_1^{\infty} \frac{(2+\sin x)\,dx}{1+x}$
16. $\int_1^{\infty} \frac{(3-\cos x)}{\sqrt{1+x^2}}\,dx$

Evaluate the improper integrals in Exercises 17–20.

17. $\int_0^{10} \frac{dx}{x^{2/3}}$
18. $\int_0^{1} \frac{dx}{x^{3/4}}$
19. $\int_0^{1} \frac{dx}{\sqrt{1-x}}$
20. $\int_0^{1} \frac{dx}{(1-x)^{2/3}}$

Using the comparison test, determine the convergence of the improper integrals in Exercises 21–24.

21. $\int_{-1}^{1} \frac{dx}{x^2+x}$
22. $\int_{-1}^{1} \frac{dx}{(x^2+x)^{1/3}}$
23. $\int_{-\infty}^{\infty} e^{-|x|}\,dx$
24. $\int_0^{\infty} \frac{dx}{(1+x^3)^{1/3}}$

Determine the convergence or divergence of the integrals in Exercises 25–40.

25. $\int_{-1}^{\infty} \frac{\tan^{-1}x}{(2+x)^3}\,dx$
26. $\int_0^{\infty} \frac{\sin x}{1+x^2}\,dx$
27. $\int_{-\infty}^{\infty} \frac{x}{(x^2+1)^{3/2}}\,dx$
28. $\int_2^{\infty} \frac{1}{t^2-1}\,dt$
29. $\int_1^{\infty} \frac{1}{(5x^2+1)^{2/3}}\,dx$
30. $\int_1^{\infty} \frac{1}{x^2}\left(1-\frac{1}{x}\right)dx$
31. $\int_1^{\infty} \frac{1}{(4x-3)^{1/3}}\,dx$
32. $\int_0^{\infty} \left[\cos x + \frac{1}{(x+1)^2}\right] dx$
33. $\int_{-\infty}^{-2} \left(\frac{1}{x^{6/5}} - \frac{1}{x^{4/3}}\right) dx$
34. $\int_0^{1} \frac{e^{-t}}{\sqrt[3]{t^2}}\,dt$
35. $\int_1^{2} \frac{1}{\sqrt{t-1}}\,dt.$
36. $\int_{-\infty}^{\infty} \frac{\cos(x^2+1)}{x^2}\,dx$. [*Hint*: Use the comparison test on a small interval.]
37. $\int_{-\infty}^{2} \left(\frac{1}{x^{5/3}} - \frac{1}{x^{4/3}}\right) dx$
38. $\int_{-4}^{10} \left[\frac{1}{(x+4)^{2/3}} + \frac{1}{(x-10)^{2/3}}\right] dx$
39. $\int_2^{\infty} \frac{dx}{x\ln x}$
40. $\int_1^{\infty} e^{-x}\ln x\,dx$
41. Consider the spirals defined in polar coordinates by the parametric equations $\theta = t$, $r = t^{-k}$. For which values of k does the spiral have finite arc length for $\pi/2 \leqslant t < \infty$? (Use the comparison test.)
42. Does the spiral $\theta = t$, $r = e^{-\sqrt{t}}$ have finite arc length for $\pi \leqslant t < \infty$?
43. Find the area under the graph of the function $f(x) = (3x+5)/(x^3-1)$ from $x = 2$ to $x = \infty$.
44. Find the area between the graphs $y = x^{-4/3}$ and $y = x^{-5/3}$ on $[1, \infty)$.
45. In Example 11, suppose that Luke's airhoses melt down when he is 10^6 miles form the sun. Now how long does he have to wake up?
46. Let α in Fig. 11.3.7 be 60° (α is defined in Example 11). Find the time required for the bugs to reach the center in terms of their speed and their initial distance from the center.
47. The region under curve $y = e^{-x}$ is rotated about the x axis to form a solid of revolution. Find the volume obtained by discarding the portion on $-\infty < x \leqslant 10$ (after slicing the solid at $x = 10$).

48. Determine the lateral surface area of the surface of revolution obtained by revolving $y = e^{-x}$, $0 \leqslant x < \infty$, about the x axis.
49. Show that $\lim_{A\to 0^+}[\int_{-3}^{-A}(dx/x) + \int_A^2(dx/x)]$ exists and determine its value.
50. Discuss the following "calculations":

(a) $\displaystyle\int_{-1}^{1}\frac{dx}{x^2} = -\frac{1}{x}\Big|_{-1}^{1} = -1+(-1) = -2;$

(b) $\displaystyle\int_{\pi/2}^{5\pi/2}\frac{\cos x}{(1+\sin x)^3}dx$

$$= -\frac{1}{2}\cdot\frac{1}{(1+\sin x)^2}\Big|_{\pi/2}^{5\pi/2} = 0.$$

51. You can simulate the logarithmic spiral yourself as follows: Stand in an open field containing a lone tree and lock your neck muscles so that your head is pointed at a fixed angle α to your body. Walk forward in such a way that you are always looking at the tree. Prove that you will walk along a logarithmic spiral.
52. The probability P that a phonograph needle will last in excess of 150 hours is given by the formula $P = \int_{150}^{\infty}\frac{1}{100}e^{-t/100}dt$. Find the value of P.
53. The probability p that the score on a reading comprehension test is no greater than the value a is

$$p = \int_{-\infty}^{a}\frac{1}{\sqrt{2\pi\sigma^2}}e^{-(\tau-\mu)^2/2\sigma^2}d\tau;$$

σ, μ are constants.

(a) Let $x = (\tau-\mu)/\sigma$ and $x_1 = (a-\mu)/\sigma$. Show that

$$p = \int_{-\infty}^{x_1}\frac{1}{\sqrt{2\pi}}e^{-x^2/2}dx.$$

(b) Show that $\int_{-\infty}^{\infty}e^{-x^2/2}dx < \infty$.

★54. Pearson and Lee studied the inheritance of physical characteristics in families in 1903. One law that resulted from these studies is

$$P = \int_{-\infty}^{(\tau-\mu)/\sigma}\frac{1}{\sqrt{2\pi}}e^{-x^2/2}dx$$

for the probability P that a mother's height is not greater than τ inches. The estimated values of μ and σ are $\mu = 62.484$ inches, $\sigma^2 = 5.7140$ square inches.

(a) Determine the value of P by appeal to integral tables for

$$\int_{-\infty}^{u}\frac{1}{\sqrt{2\pi}}e^{-x^2/2}dx$$

using $\tau = 63$ inches. Look in a mathematical table under *probability functions* or *normal distribution*.

(b) According to the study, how many mothers out of 100 are likely to have height not exceeding 63 inches?

★55. (a) Evaluate $\displaystyle\int_0^{\infty}\frac{du}{u^{1/2}+u^{3/2}}$.

(b) For what p and q is

$$\int_0^{\infty}\frac{dx}{x^p+x^q}$$

convergent?

★56. Consider the surface of revolution obtained by revolving the graph of $f(x) = 1/x$ on the interval $[1,\infty)$ about the x axis.

(a) Show that the area of this surface is infinite.
(b) Show that the volume of the solid of revolution bounded by this surface is finite.
(c) The results of parts (a) and (b) suggest that one could fill the solid with a finite amount of paint, but it would take an infinite amount of paint to paint the surface. Explain this paradox.

Next consider the surface of revolution obtained by revolving the curve $y = 1/x^r$ for x in $[1,\infty)$ about the x axis.

(d) For which values of r does this surface have finite area?
(e) For which values of r does the solid surrounded by this surface have finite volume? Compute the volume for these values of r.

★57. Show that if $0 < f'(x) < 1/x^2$ for all x in $[0,\infty)$, then $\lim_{x\to\infty}f(x)$ exists.

11.4 Limits of Sequences and Newton's Method

Solutions of equations can often be found as the limits of sequences.

This section begins with a discussion of sequences and their limits. The topic will be taken up again in Section 12.1 when we study infinite series. A *sequence* is just an "infinite list" of numbers: $a_1, a_2, a_3, \ldots,$ with one a_n for each natural number n. A number l is called the *limit* of this sequence if, roughly speaking, a_n comes and remains arbitrarily close to l as n increases.

Perhaps the most familiar example of a sequence with a limit is that of an infinite decimal expansion. Consider, for instance, the equation

$$\tfrac{1}{3} = 0.333\ldots \tag{1}$$

in which the dots on the right-hand side are taken to stand for "infinitely many 3's." We can interpret equation (1) without recourse to any metaphysical notion of infinity: the finite decimals $0.3, 0.33, 0.333$, and so on are approximations to $\frac{1}{3}$, and we can make the approximation as good as we wish by taking enough 3's. Our sequence $a_1, a_2, \ldots$ is defined in this case by $a_n = 0.33\ldots3$, with n 3's (here the three dots stand for only finitely many 3's). In other words,

$$a_n = \frac{3}{10} + \frac{3}{100} + \cdots + \frac{3}{10^n}. \tag{2}$$

We can estimate the difference between a_n and $\frac{1}{3}$ by using some algebra. Multiplying equation (2) by 10 gives

$$10a_n = 3 + \frac{3}{10} + \cdots + \frac{3}{10^{n-1}}, \tag{3}$$

and subtracting equation (2) from equation (3) gives

$$9a_n = 3 - \frac{3}{10^n},$$

$$a_n = \frac{1}{3} - \frac{1}{3}\left(\frac{1}{10^n}\right).$$

Finally,

$$\frac{1}{3} - a_n = \frac{1}{3}\left(\frac{1}{10^n}\right). \tag{4}$$

As n is taken larger and larger, the denominator 10^n becomes larger and larger, and so the difference $\frac{1}{3} - a_n$ becomes smaller and smaller. In fact, if n is chosen large enough, we can make $\frac{1}{3} - a_n$ as small as we please. (See Fig. 11.4.1.)

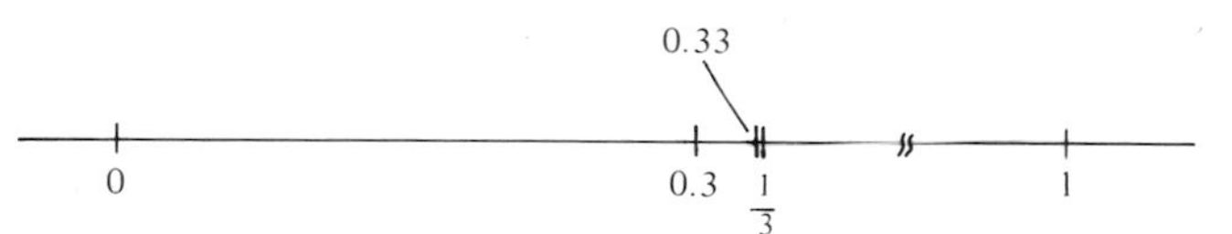

Figure 11.4.1. The decimal approximations to $\frac{1}{3}$ form a sequence converging to $\frac{1}{3}$.

Example 1 How large must n be for the error $\frac{1}{3} - a_n$ to be less than 1 part in 1 million?

Solution By equation (4), we must have

$$\frac{1}{3}\left(\frac{1}{10^n}\right) < 10^{-6}$$

or $10^{-n} < 3 \cdot 10^{-6}$. It suffices to have $n \geqslant 6$, so the finite decimal 0.333333 approximates $\frac{1}{3}$ to within 1 part in a million. So do the longer decimals 0.3333333, 0.33333333, and so on. ▲

There is nothing special about the number 10^{-6} in Example 1. Given *any* positive number ε, we will always be able to make $\frac{1}{3} - a_n = \frac{1}{3}(1/10^n)$ less than ε by letting n be sufficiently large. We express this fact by saying that $\frac{1}{3}$ is the *limit* of the numbers

$$a_n = \frac{3}{10} + \frac{3}{100} + \cdots + \frac{3}{10^n}$$

as n becomes arbitrarily large, or

$$\lim_{n\to\infty}\left(\frac{3}{10} + \frac{3}{100} + \cdots + \frac{3}{10^n}\right) = \frac{1}{3}.$$

We may think of a sequence $a_1, a_2, a_3, \ldots$ as a function whose domain consists of the natural numbers $1, 2, 3, \ldots$ (Occasionally, we allow the domain to start at zero or some other integer.) Thus we may represent a sequence graphically in two ways—either by plotting the points $a_1, a_2, \ldots$ on a number line or by plotting the pairs (n, a_n) in the plane.

Example 2 (a) Write the first six terms of the sequence $a_n = n/(n+1)$, $n = 1, 2, 3, \ldots$. Represent the sequence graphically in two ways. Find the value of $\lim_{n\to\infty}[n/(n+1)]$. (b) Repeat for $a_n = (-1)^n/n$. (c) Repeat for $a_n = (-1)^n n/(n+1)$.

Solution (a) We obtain the terms a_1 through a_6 by substituting $n = 1, 2, 3, \ldots, 6$ into the formula for a_n, giving $\frac{1}{2}, \frac{2}{3}, \frac{3}{4}, \frac{4}{5}, \frac{5}{6}, \frac{6}{7}$. These values are plotted in Fig. 11.4.2. As n gets larger, the fraction $n/(n+1)$ gets larger and larger but never exceeds 1; we may guess that the limit is equal to 1.

To verify this guess, we look at the difference $1 - n/(n+1)$. We have

$$1 - \frac{n}{n+1} = \frac{n+1-n}{n+1} = \frac{1}{n+1},$$

which does indeed become arbitrarily small as n increases, so

$$\lim_{n\to\infty}\frac{n}{(n+1)} = 1.$$

(b) The terms a_1 through a_6 are $-1, \frac{1}{2}, -\frac{1}{3}, \frac{1}{4}, -\frac{1}{5}, \frac{1}{6}$. They are plotted in Fig. 11.4.3. As n gets larger, the number $(-1)^n/n$ seems to get closer to zero. Therefore we guess that $\lim_{n\to\infty}[(-1)^n/n] = 0$.

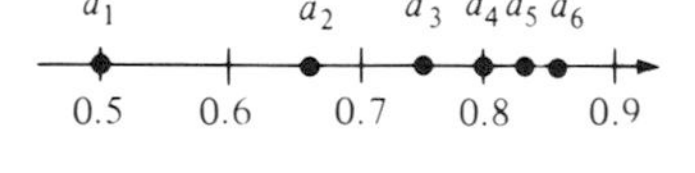

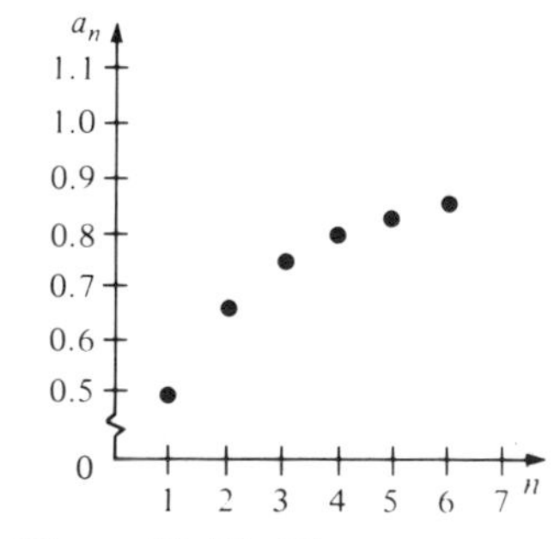

Figure 11.4.2. The sequence $a_n = n/(n+1)$ represented graphically in two different ways.

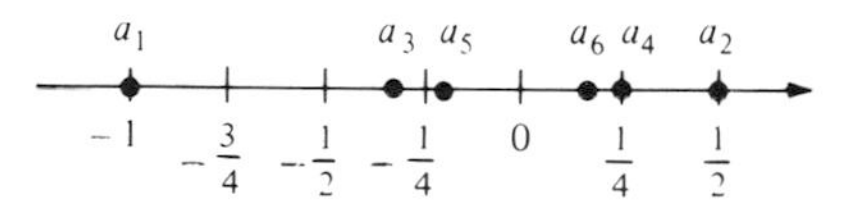

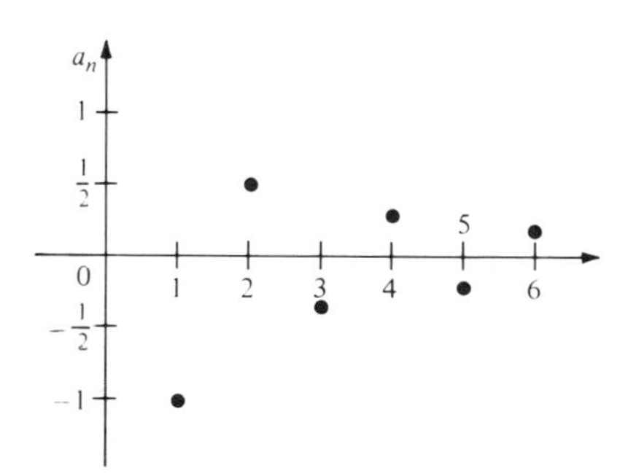

Figure 11.4.3. The sequence $a_n = (-1)^n/n$ plotted in two ways.

(c) We have, for a_1 through a_6, $-\frac{1}{2}, \frac{2}{3}, -\frac{3}{4}, \frac{4}{5}, -\frac{5}{6}, \frac{6}{7}$. They are plotted in Fig. 11.4.4. In this case, the numbers a_n do not approach any particular number. (Some of them are approaching 1, others -1.) We guess that the sequence does not have a limit. ▲

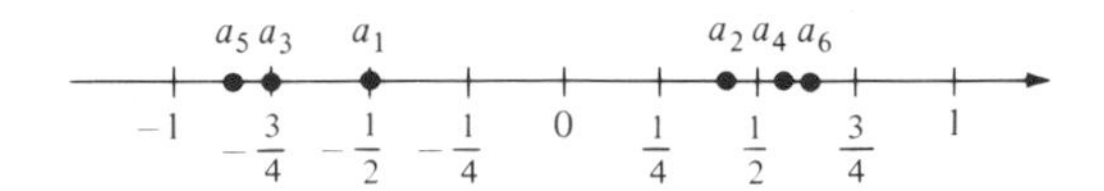

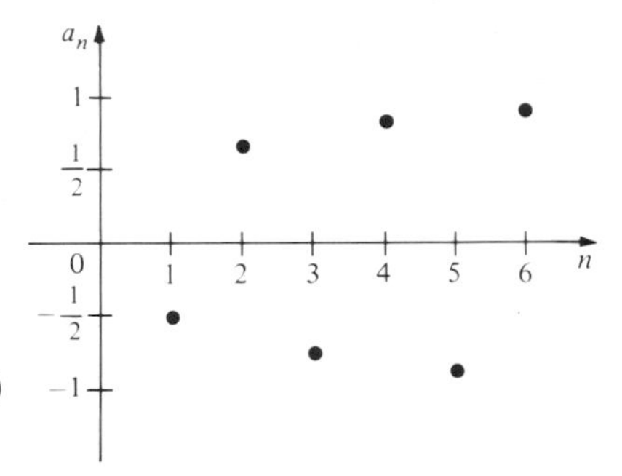

Figure 11.4.4. The sequence $a_n = (-1)^n n/(n+1)$ plotted in two ways.

Just as with the ε-δ definition for limits of functions, there is an ε-N definition for limits of sequences which makes the preceding ideas precise.

Limits of Sequences

The sequence $a_1, a_2, a_3, \ldots, a_n, \ldots$ approaches l as a limit if a_n gets close to and remains arbitrarily close to l as n becomes large. In this case, we write

$$\lim_{n\to\infty} a_n = l.$$

In precise terms, $\lim_{n\to\infty} a_n = l$ if, for every $\varepsilon > 0$, there is an N such that $|a_n - l| < \varepsilon$ for all $n \geqslant N$.

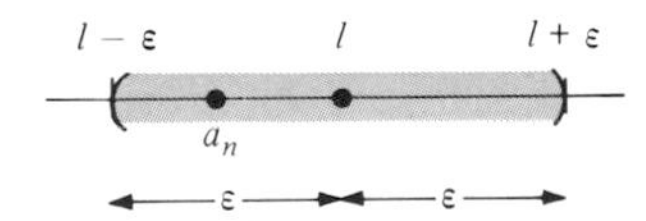

Figure 11.4.5. The relationship between a_n, l, and ε in the definition of the limit of a sequence.

It is useful to think of the number ε in this definition as a *tolerance*, or allowable error. The definition specifies that if l is to be the limit of the sequence a_n, then, given any tolerance, all the terms of the sequence beyond a certain point should be within that tolerance of l. Of course, as the tolerance is made smaller, it will usually be necessary to go farther out in the sequence to bring the terms within tolerance of the limit. (See Fig. 11.4.5.)

The purpose of the ε-N definition is to lay a framework for a precise discussion of limits of sequences and their properties—just as the definitions in Section 11.1.

Let us check the limit of a simple sequence using the ε-N definition.

Example 3 Prove that $\lim_{n\to\infty}(1/n) = 0$, using the ε-N definition.

Solution To show that the definition is satisfied, we must show that for any $\varepsilon > 0$ there is a number N such that $|1/n - 0| < \varepsilon$ if $n > N$. If we choose $N \geqslant 1/\varepsilon$, we get, for $n > N$,

$$\left|\frac{1}{n} - 0\right| = \frac{1}{n} < \frac{1}{N} \leqslant \varepsilon.$$

Thus the assertion is proved. ▲

Calculator Discussion

Limits of sequences can sometimes be visualized on a calculator. Consider the sequence obtained by taking successive square roots of a given positive number a:

$$a_0 = a, \qquad a_1 = \sqrt{a}\,, \qquad a_2 = \sqrt{\sqrt{a}}\,, \qquad a_3 = \sqrt{\sqrt{\sqrt{a}}}\,,$$

and so forth. (See Fig. 11.4.6.)

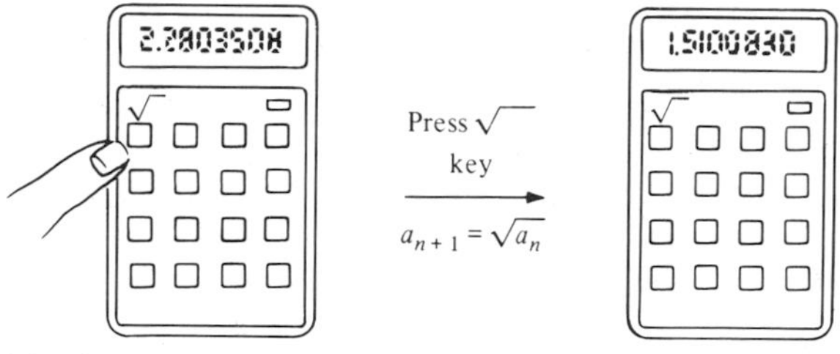

Figure 11.4.6. For a recursively defined sequence $a_{n+1} = f(a_n)$, the next member in the sequence is obtained by depressing the "f" key. Here $f = \sqrt{}$.

For instance, if we start by entering $a = 5.2$, we get

$$a_0 = 5.2,$$

$$a_1 = \sqrt{5.2} = 2.2803508,$$

$$a_2 = \sqrt{2.2803508} = 1.5100830,$$

$$a_3 = \sqrt{1.5100830} = 1.2288544,$$

and so on. After pressing the $\sqrt{}$ repeatedly you will see the numbers getting closer and closer to 1 until roundoff error causes the number 1 to appear and then stay forever. This sequence has 1 as a limit. (Of course, the calculation does not *prove* this fact, but does suggest it.) Observe that the sequence is defined *recursively*—that is, each member of the sequence is obtained from the previous one by some specific process. The sequence $1, 2, 4, 8, 16, 32, \ldots$ is another example; each term is twice the previous one: $a_{n+1} = 2a_n$. ▲

Limits of sequences are closely related to limits of functions. For example, if $f(x)$ is defined for $x \geqslant 0$, then $a_n = f(n)$ is a sequence. If $\lim_{x\to\infty} f(x)$ exists, then $\lim_{n\to\infty} a_n$ exists as well and these limits are equal. This fact can sometimes be used to evaluate some limits. For instance,

$$\lim_{x\to\infty} \frac{x}{x+1} = \lim_{x\to\infty} \frac{1}{1+1/x} = \frac{1}{1+\lim_{x\to\infty}(1/x)} = \frac{1}{1+0} = 1,$$

and so $\lim_{n\to\infty}[n/(n+1)] = 1$, confirming our calculations in Example 2(a).

Limits of sequences also obey rules similar to those for functions.[5] We illustrate:

Example 4 Find (a) $\displaystyle\lim_{n\to\infty}\left(\frac{n^2+1}{3n^2+n}\right)$

and (b) $\displaystyle\lim_{n\to\infty}\left(1 - \frac{3}{n} + \frac{n}{n+1}\right)$.

Solution (a) Write

$$\lim_{n\to\infty}\left(\frac{n^2+1}{3n^2+n}\right) = \lim_{n\to\infty}\left(\frac{1+1/n^2}{3+1/n}\right) \qquad \text{(dividing numerator and denominator by } n^2\text{)}$$

[5] These are written out formally in Section 12.1.

$$= \frac{1 + \lim_{n\to\infty}(1/n^2)}{3 + \lim_{n\to\infty}(1/n)} \qquad \text{(quotient and sum rules)}$$

$$= \frac{1+0}{3+0} = \frac{1}{3}.$$

(b) $$\lim_{n\to\infty}\left(1 - \frac{3}{n} + \frac{n}{n+1}\right) = 1 - 3\left(\lim_{n\to\infty}\frac{1}{n}\right) + \lim_{n\to\infty}\left(\frac{1}{1+1/n}\right)$$

$$= 1 - 3\cdot 0 + 1 = 2. \blacktriangle$$

The connection with limits of functions allows us to use l'Hôpital's rule to find limits of sequences.

Example 5 (a) Using numerical calculations, guess the value of $\lim_{n\to\infty}\sqrt[n]{n}$. (b) Use l'Hôpital's rule to verify the result in (a).

Solution (a) Using a calculator we find:

n	$\sqrt[n]{n}$
1	1
5	1.37973
10	1.25893
50	1.08138
100	1.04713
500	1.01251
1000	1.00693
5000	1.00170
10,000	1.00092

Thus it appears that $\lim_{n\to\infty}\sqrt[n]{n} = 1$.

(b) To verify this, we use l'Hôpital's rule to show that $\lim_{x\to\infty}x^{1/x} = 1$. The limit is in ∞^0 form, so we use logarithms:

$$x^{1/x} = e^{(\ln x)/x}.$$

Now $\lim_{x\to\infty}(\ln x/x)$ is in $\frac{\infty}{\infty}$ form, and l'Hôpital's rule gives

$$\lim_{x\to\infty}\frac{\ln x}{x} = \lim_{x\to\infty}\frac{1/x}{1} = 0.$$

Hence

$$\lim_{x\to\infty}x^{1/x} = \exp\left(\lim_{x\to\infty}\frac{\ln x}{x}\right)$$

$$= \exp(0) = 1,$$

confirming our numerical calculations. ▲

When we introduced limits of sequences in Example 1, we implicitly used the fact that $\lim_{n\to\infty}(1/10^n) = 0$. The following general fact is useful.

Limits of Powers

$$\lim_{n\to\infty} r^n = \begin{cases} \infty, & \text{if } r > 1, \\ 1, & \text{if } r = 1, \\ 0, & \text{if } 0 \leqslant r < 1. \end{cases}$$

To see this, first consider the case $r > 1$. We write r as $1 + s$ where $s > 0$. If we expand $r^n = (1+s)^n$, we get $r^n = 1 + ns +$ (other positive terms.) Therefore, $r^n \geqslant 1 + ns$, which goes to ∞ as $n \to \infty$. Second, if $r = 1$, then $r^n = 1$ for all n, so $\lim_{n\to\infty} r^n = 1$. Finally, if $0 \leqslant r < 1$, then excluding the easy case $r = 0$, we let $\rho = 1/r$ so $\rho > 1$, and so $\lim_{n\to\infty}\rho^n = \infty$. Therefore, $\lim_{n\to\infty} r^n = \lim_{n\to\infty}(1/\rho^n) = 0$ (compare the reciprocal test for limits of functions in Section 11.1).

Example 6 Evaluate (a) $\lim_{n\to\infty} 3^n$, (b) $\lim_{n\to\infty} e^{-n}$, (c) $\lim_{n\to\infty}(e + (\frac{2}{3})^n)^4$.

Solution (a) Here $r = 3 > 1$, so $\lim_{n\to\infty} 3^n = \infty$.
(b) $e^{-n} = (1/e)^n$, and $1/e < 1$, so $\lim_{n\to\infty} e^{-n} = 0$.
(c) $\lim_{n\to\infty}[e + (\frac{2}{3})^n]^4 = [e + \lim_{n\to\infty}(\frac{2}{3})^n]^4 = [e+0]^4 = e^4$. ▲

Another useful test is the comparison test: it says that if $\lim_{n\to\infty} a_n = 0$ and if $|b_n| \leqslant |a_n|$, then $\lim_{n\to\infty} b_n = 0$ as well. This is plausible since b_n is squeezed between $-|a_n|$ and $|a_n|$ which are tending to zero. We ask the reader to supply the proof in Exercise 56.

Comparison Test

If $\lim_{n\to\infty} a_n = 0$ and $|b_n| \leqslant |a_n|$ then $\lim_{n\to\infty} b_n = 0$.

Example 7 Find (a) $\displaystyle\lim_{n\to\infty}\frac{\sin n}{n}$ and (b) $\displaystyle\lim_{n\to\infty}\frac{(-1)^n + n}{n}$.

Solution (a) If $a_n = 1/n$ and $b_n = (\sin n)/n$, then $a_n \to 0$ and $|b_n| \leqslant |a_n|$, so by the comparison test, $\lim_{n\to\infty}(\sin n)/n = 0$.
(b) $|(-1)^n/n| \leqslant 1/n \to 0$, so $(-1)^n/n \to 0$ by the comparison test. Thus

$$\lim_{n\to\infty}\left(\frac{(-1)^n + n}{n}\right) = \lim_{n\to\infty}\left(\frac{(-1)^n}{n}\right) + \lim_{n\to\infty}\frac{n}{n} = 0 + 1 = 1. \quad ▲$$

Many questions in mathematics and its applications lead to the problem of solving an equation of the form

$$f(x) = 0, \tag{5}$$

where f is some function. The solutions of equation (5) are called the *roots* or *zeros* of f. If f is a polynomial of degree at most 4, one can find the roots of f by substituting the coefficients of f into a general formula (see pp. 17 and 173). On the other hand, if f is a polynomial of degree 5 or greater, or a function involving the trigonometric or exponential functions, there may be no explicit formula for the roots of f, and one may have to search for the solution numerically.

Newton's method uses linear approximations to produce a sequence $x_0, x_1, x_2, \ldots$ which converges to a solution of $f(x) = 0$. Let x_0 be a first guess. We seek to correct this guess by an amount Δx so that $f(x_0 + \Delta x) = 0$. Solving this equation for Δx is no easier than solving the original equation (5), so we manufacture an easier problem, replacing f by its first-order approximation at x_0; that is, we replace $f(x_0 + \Delta x)$ by $f(x_0) + f'(x_0)\Delta x$. If $f(x_0)$ is not equal to zero, we can solve the equation $f(x_0) + f'(x_0)\Delta x = 0$ to obtain $\Delta x = -f(x_0)/f'(x_0)$, so that our new guess is

$$x_1 = x_0 + \Delta x = x_0 - f(x_0)/f'(x_0).$$

Geometrically, we have found x_1 by following the tangent line to the graph of f at $(x_0, f(x_0))$ until it meets the x axis; the point where it meets is $(x_1, 0)$ (see Fig. 11.4.7).

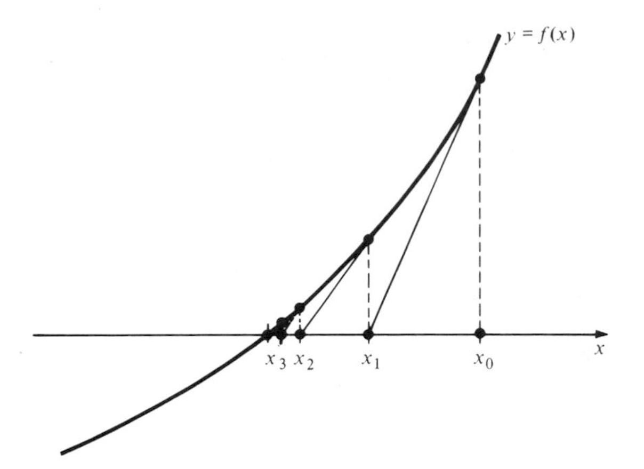

Figure 11.4.7. The geometry of Newton's method.

Now we find a new guess x_2 by repeating the procedure with x_1 in place of x_0; that is,

$$x_2 = x_1 - \frac{f(x_1)}{f'(x_1)}.$$

In general, once we have found x_n, we define x_{n+1} by

$$x_{n+1} = x_n - \frac{f(x_n)}{f'(x_n)}. \tag{6}$$

Let us see how the method works in a case where we know the answer in advance. (This iteration procedure is particularly easy to use on a programmable calculator.)

Example 8 Use Newton's method to find the first few approximations to a solution of the equation $x^2 = 4$, taking $x_0 = 1$.

Solution To put the equation $x^2 = 4$ in the form $f(x) = 0$, we let $f(x) = x^2 - 4$. Then $f'(x) = 2x$, so the iteration rule (6) becomes $x_{n+1} = x_n - (x_n^2 - 4)/2x_n$, which may be simplified to $x_{n+1} = \frac{1}{2}(x_n + 4/x_n)$. Applying this formula repeatedly, with $x_0 = 1$, we get (to the limits of our calculator's accuracy)

$$\begin{aligned} x_1 &= 2.5 \\ x_2 &= 2.05 \\ x_3 &= 2.000609756 \\ x_4 &= 2.000000093 \\ x_5 &= 2 \\ x_6 &= 2 \\ &\vdots \end{aligned}$$

and so on forever. The number 2 is, of course, precisely the positive root of our equation $x^2 = 4$. ▲

Example 9 Use Newton's method to locate a root of $x^5 - x^4 - x + 2 = 0$. Compare what happens with various starting values of x_0 and attempt to explain the phenomenon.

Solution The iteration formula is

$$x_{n+1} = x_n - \frac{x_n^5 - x_n^4 - x_n + 2}{5x_n^4 - 4x_n^3 - 1} = \frac{4x_n^5 - 3x_n^4 - 2}{5x_n^4 - 4x_n^3 - 1}.$$

For the purpose of convenient calculation, we may write this as

$$x_{n+1} = \frac{(4x_n - 3)x_n^4 - 2}{(5x_n - 4)x_n^3 - 1}.$$

Starting at $x_0 = 1$, we find that the denominator is undefined, so we can go no further. (Can you interpret this difficulty geometrically?)

Starting at $x_0 = 2$, we get

$$x_1 = 1.659574468,$$
$$x_2 = 1.372968569,$$
$$x_3 = 1.068606737,$$
$$x_4 = -0.5293374382,$$
$$x_5 = 169.5250382.$$

The iteration process seems to have sent us out on a wild goose chase. To see what has gone wrong, we look at the graph of $f(x) = x^5 - x^4 - x + 2$. (See Fig. 11.4.8.) There is a "bowl" near $x_0 = 2$; Newton's method attempts to take us down to a nonexistent root. (Only after many iterations does one converge to the root—see Exercise 60 and Example 10.)

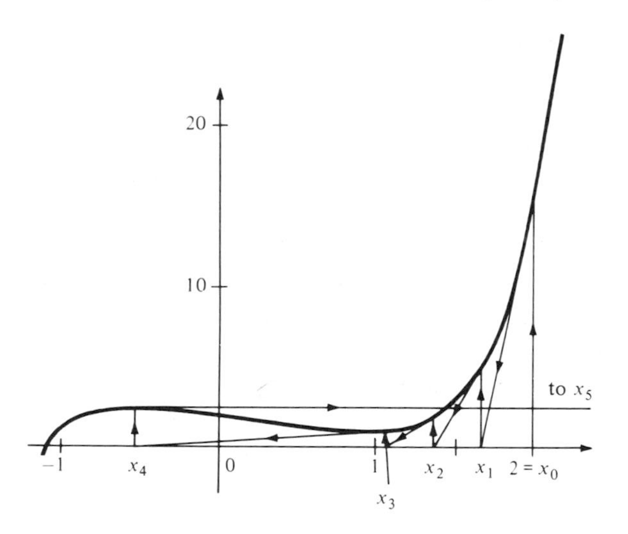

Figure 11.4.8. Newton's method does not always work.

Finally, we start with $x_0 = -2$. The iteration gives

$$x_0 = -2, \qquad f(x_0) = -44;$$
$$x_1 = -1.603603604, \qquad f(x_1) = -13.61361361;$$
$$x_2 = -1.323252501, \qquad f(x_2) = -3.799819057;$$
$$x_3 = -1.162229582, \qquad f(x_3) = -0.782974790;$$
$$x_4 = -1.107866357, \qquad f(x_4) = -0.067490713;$$
$$x_5 = -1.102228599, \qquad f(x_5) = -0.000663267;$$
$$x_6 = -1.102172085, \qquad f(x_6) = -0.000000061;$$
$$x_7 = -1.102172080, \qquad f(x_7) = -0.000000003.$$

Since the numbers in the $f(x)$ column appear to be converging to zero and those in the x column are converging, we obtain a root to be (approximately) -1.10217208. Since $f(x)$ is negative at this value (where $f(x) = -0.000000003$) and positive at -1.10217207 (where $f(x) = 0.000000115$), we can conclude, by the intermediate value theorem, that the root is between these two values. ▲

Example 9 illustrates several important features of Newton's method. First of all, it is important to start with an initial guess which is reasonably close to a root—graphing is a help in making such a guess. Second, we notice that once we get near a root, then convergence becomes very rapid—in fact, the number of correct decimal places is approximately *doubled* with each iteration. Finally, we notice that the process for passing from x_n to x_{n+1} is the same for each value of n; this feature makes Newton's method particularly attractive for use with a programmable calculator or a computer. Human intelligence still comes into play in the choice of the first guess, however.

Newton's Method

To find a root of the equation $f(x) = 0$, where f is a differentiable function such that f' is continuous, start with a guess x_0 which is reasonably close to a root. Then produce the sequence $x_0, x_1, x_2, \ldots$ by the iterative formula:

$$x_{n+1} = x_n - \frac{f(x_n)}{f'(x_n)}.$$

If $\lim_{n\to\infty} x_n = \bar{x}$, then $f(\bar{x}) = 0$.

To justify the last statement in the box above, we suppose that $\lim_{n\to\infty} x_n = \bar{x}$. Taking limits on both sides of the equation $x_{n+1} = x_n - f(x_n)/f'(x_n)$, we obtain $\bar{x} = \bar{x} - \lim_{n\to\infty}[f(x_n)/f'(x_n)]$, or $\lim_{n\to\infty}[f(x_n)/f'(x_n)] = 0$. Now let $a_n = f(x_n)/f'(x_n)$. Then we have $\lim_{n\to\infty} a_n = 0$, while $f(x_n) = a_n f'(x_n)$. Taking limits as $n \to \infty$ and using the continuity of f and f', we find

$$\lim_{n\to\infty} f(x_n) = \lim_{n\to\infty} a_n \lim_{n\to\infty} f'(x_n), \quad \text{so}$$

$$f(\bar{x}) = 0 \cdot f'(\bar{x}) = 0.$$

Newton's method, applied with care, can also be used to solve equations involving trigonometric or exponential functions.

Example 10 Use Newton's method to find a positive number x such that $\sin x = x/2$.

Solution With $f(x) = \sin x - x/2$, the iteration formula becomes

$$x_{n+1} = x_n - \frac{\sin x_n - x_n/2}{\cos x_n - 1/2} = \frac{2(x_n \cos x_n - \sin x_n)}{2\cos x_n - 1}.$$

Taking $x_0 = 0$ as our first guess, we get $x_1 = 0$, $x_2 = 0$, and so forth, since zero is already a root of our equation. To find a positive root, we try another guess, say $x_0 = 6$. We get

$x_1 = 13.12652598$ $\qquad x_5 = 266.0803351$

$x_2 = 30.50101246$ $\qquad x_6 = 143.3754278$

$x_3 = 176.5342378$ $\qquad \vdots$

$x_4 = 448.4888306$ $\qquad x_9 = -759.1194553$

$\qquad x_{10} = 3{,}572.554623$

We do not seem to be getting anywhere. To see what might be wrong, we draw a sketch (Fig. 11.4.9). The many places where the graph of $\sin x - x/2$

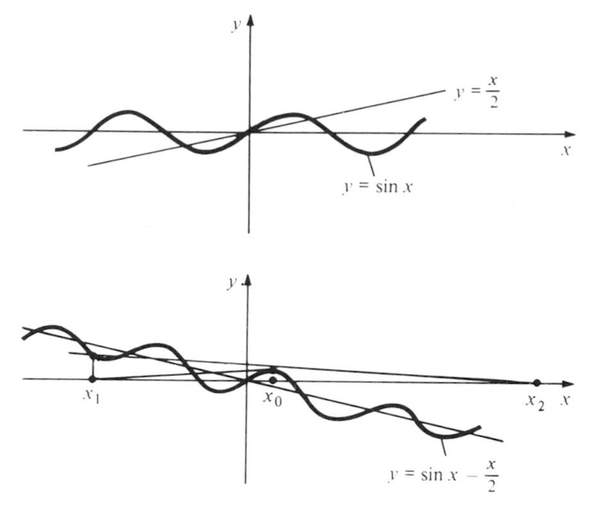

Figure 11.4.9. Newton's method goes awry.

has a horizontal or nearly horizontal tangent causes the Newton sequence to make wild excursions.[6] We need to make a better first guess; we try $x_0 = 3$. This gives

$$
\begin{aligned}
x_1 &= 2.08799541, & x_5 &= 1.89549427,\\
x_2 &= 1.91222926, & x_6 &= 1.89549427,\\
x_3 &= 1.89565263, & x_7 &= 1.89549427,\\
x_4 &= 1.89549428, & x_8 &= 1.89549427.
\end{aligned}
$$

We conclude that our root is somewhere near 1.89549427. Substituting this value for x in $\sin x - x/2$ gives 1.0×10^{-11}. There may be further doubt about the last figure, due to internal roundoff errors in the calculator; we are probably safe to announce our result as 1.8954943. ▲

You may find it amusing to try other starting values for x_0 in Example 10. For instance, the values 6.99, 7, and 7.01 seem to lead to totally different results. (This was on a HP 15C hand calculator. Numerical errors may be crucial in a calculation such as this.) Recently, the study of sequences defined by iteration has become important as a model for the long-time behavior of dynamical systems. For instance, sequences defined by simple rules of the form $x_{n+1} = ax_n(1 - x_n)$ display very different behavior according to the value of the constant a. (See the supplement to this section and Exercise 60.)

Supplement to Section 11.4
Newton's Method and Chaos

The sequences generated by Newton's method may exhibit several types of strange behavior if the starting guess is not close to a root:

(a) the sequence $x_0, x_1, x_2, \ldots$ may wander back and forth over the real line for some time before converging to a root;

[6] Try these calculations and those in Example 9 on your calculator and see if you converge to the root after many iterations. You will probably get different numbers from ours, probably due to roundoff errors, computer inaccuracies and the extreme sensitivity of the calculations. We got four different sets of numbers with four calculators. (The ones here were found on an HP 15C which also has a SOLVE algorithm which cleverly avoids many difficulties.)

(b) slightly different choices of x_0 or the use of different calculators may lead to very different sequences;
(c) the sequence $x_0, x_1, x_2, \ldots$ may eventually cycle between two or more values, none of which is a root of the equation we are trying to solve;
(d) the sequence $x_0, x_1, x_2, \ldots$ may wander "forever" without ever settling into a regular pattern.

Recent research in pure and applied mathematics has shown that the type of erratic behavior just described is the rule rather than the exception for many mathematical operations and the physical processes which they model (see Exercise 60 for a simple example). Indeed, "chaotic" behavior is observed in fluid flow, chemical reactions, and biological systems, and is responsible for the inherent unpredictability of the weather.

Some references on this work on "chaos", aimed at the nonexpert reader, are:

M. J. Feigenbaum, "Universal behavior in nonlinear systems," Los Alamos Science **1** (Summer 1980), 4–27.
D. R. Hofstadter, "Metamagical themas," Scientific American **245** (November 1981), 22–43.
L. P. Kadanoff, "Roads to chaos," Physics Today **36** (December 1983), 46–53.
D. Ruelle, "Strange attractors," Math. Intelligencer **2** (1980), 126–137.
D. G. Saari and J. B. Urenko, "Newton's method, circle maps, and chaotic motion," American Mathematical Monthly **91** (1984), 3–17.

Exercises for Section 11.4

1. If $a_n = 1/10 + 1/100 + \cdots + 1/10^n$, how large must n be for $\frac{1}{9} - a_n$ to be less than 10^{-6}?
2. If $a_n = 7/10 + 7/100 + \cdots + 7/10^n$, how large must n be for $\frac{7}{9} - a_n$ to be less than 10^{-8}?

Find the limits of the sequences in Exercises 3 and 4.

3. $a_n = 1 + 1/2 + 1/4 + \cdots + 1/2^n$.
4. $a_n = \sin(n\pi/2)$.

Write down the first six terms of the sequences in Exercises 5–10.

5. $k_n = n^2 - 2\sqrt{n}\,;\ n = 0, 1, 2, \ldots$.
6. $a_n = (-1)^{n+1}[(n-1)/n!];\ n = 0, 1, 2, \ldots$ $(0! = 1,\ n! = n(n-1)\cdots 3\cdot 2\cdot 1.)$
7. $b_n = nb_{n-1}/(1+n);\ b_0 = \frac{1}{7}$.
8. $c_{n+1} = -c_n/[2n(4n+1)];\ c_1 = 2$.
9. $a_{n+1} = [1/(n+1)]\sum_{i=0}^{n} a_i;\ a_0 = \frac{1}{2}$.
10. $k_n = \sqrt{3n^2 + 2n}\,;\ n = 1, 2, 3, \ldots$.

Establish the limits in Exercises 11–14 using the ε-N definition.

11. $\lim_{n\to\infty} \dfrac{3}{n} = 0$
12. $\lim_{n\to\infty} \left(1 - \dfrac{1}{n}\right) = 1$
13. $\lim_{n\to\infty} \dfrac{3}{2n+1} = 0$
14. $\lim_{n\to\infty} \dfrac{2}{2n+5} = 0$

Evaluate the limits in Exercises 15–24.

15. $\lim_{n\to\infty} \dfrac{3n}{n+1}$
16. $\lim_{n\to\infty} \dfrac{2n}{8n-1}$
17. $\lim_{n\to\infty} \dfrac{n-3n^2}{n^2+1}$
18. $\lim_{n\to\infty} \dfrac{n^3+3n^2+1}{n^4+8n^2+2}$
19. $\lim_{n\to\infty} \left[\dfrac{3n^2-2n+1}{n(n+1)} - \dfrac{n(n+2)}{(n+1)(n+3)}\right]^2$
20. $\lim_{n\to\infty} \dfrac{2+1/n}{(n^2-2)/(n^2+1)}$
21. $\lim_{n\to\infty} \dfrac{(\sin n)^2}{n+2}$
22. $\lim_{n\to\infty} \dfrac{(1+n)\cos(n+1)}{n^2+1}$
23. $\lim_{n\to\infty} \dfrac{(-1)^n\cdot 2}{n+1}$
24. $\lim_{n\to\infty} \dfrac{\cos(\pi\sqrt{n}\,)}{n^2}$

Using numerical calculations, guess the limit as $n\to\infty$ of the sequences in Exercises 25–28. Verify your answers using l'Hôpital's rule.

25. $\sqrt[n]{n/2}$
26. $\sqrt[n]{n/3}$
27. $\sqrt[n]{n(n+1)/4}$
28. $\sqrt[n]{n(n+3)/7}$

Find the limits in Exercises 29–34.

29. $\lim_{n\to\infty} \dfrac{1}{8^n}$
30. $\lim_{n\to\infty} \pi^n$
31. $\lim_{n\to\infty} \dfrac{n+(3/4)^n}{n^2+2}$
32. Find $\lim_{n\to\infty} (\pi + (\frac{2}{3})^n)^3$
33. $\lim_{n\to\infty} \left[\dfrac{3b+(1/2)^{2n}}{n^2-1}\right]^3$; b constant
34. $\lim_{n\to\infty} \left(\dfrac{2a+e^{-2n}}{n-1}\right)^2$; a constant

35. (a) Use Newton's method to find a solution of $x^3 - 8x^2 + 2x + 1 = 0$. (b) Use division and the quadratic formula to find the other two roots.[7]

36. Use Newton's method to find all real roots of $x^3 - x + \frac{1}{10}$.

37. Use Newton's method to locate a root of $f(x) = x^5 + x^2 - 3$ with starting values $x_0 = 0$, $x_0 = 2$.

38. Use Newton's method to locate a zero for $f(x) = x^4 - 2x^3 - 1$. Use $x_0 = 2$, 3, and -1 as starting values and compare the results.

39. Use Newton's method to locate a root of $\tan x = x$ in $[\pi/2, 3\pi/2]$.

40. Use Newton's method to find the following numbers: (a) $\sqrt{2}$; (b) $\sqrt[3]{2}$

41. The equation $\tan x = \alpha x$ appears in heat conduction problems to determine values $\lambda_1, \lambda_2, \lambda_3, \ldots$ that appear in the expression for the temperature distribution. The numbers $\lambda_1, \lambda_2, \ldots$ are the positive solutions of $\tan x = \alpha x$, listed in increasing order. Find the numbers $\lambda_1, \lambda_2, \lambda_3$ for $\alpha = 2$, 3, 5, by Newton's method. Display your answers in a table.

42. (a) Use Newton's method to solve the equation $x^2 - 2 = 0$ to 8 decimal places of accuracy, using the initial guess $x_0 = 2$.

★(b) Find a constant C such that $|x_n - \sqrt{2}| \leqslant C|x_{n-1} - \sqrt{2}|^2$ for $n = 1$, 2, 3, and 4. (See Review Exercise 101 for the theory of the rapid convergence of Newton's method.)

43. Experiment with Newton's method for evaluation of the root $1/e$ of the equation $e^{-ex} = 1/e$.

44. Enter the display value 1.0000000 on your calculator and repeatedly press the "sin" key using the "radian mode". This process generates display numbers $a_1 = 1.0000000$, $a_2 = 0.84147$, $a_3 = 0.74562, \ldots$.
(a) Write a formula for a_n, using function notation.
(b) Conjecture the value of $\lim_{n\to\infty} a_n$. Explain with a graph.

45. Display the number 2 on your calculator. Repeatedly press the "x^2" key. You should get the numbers 2, 4, 16, 256, 65536, ... Express the display value a_n after n repetitions by a formula.

46. Let $f(x) = 1 + 1/x$. Equipped with a calculator with a reciprocal function, complete the following:
(a) Write out $f(f(f(f(f(f(2))))))$ as a division problem, and calculate the value. We abbreviate this as $f^{(6)}(2)$, meaning to display the value 2, press the "$1/x$" key and add 1, successively six times.
(b) Experiment to determine $\lim_{n\to\infty}[1/f^{(n)}(2)]$ to five decimal places.

47. Suppose that $\lim_{n\to\infty} a_n = a$ and that $a > 0$. Prove that there is a positive integer N such that $a_n > 0$ for all $n > N$.

48. Let $a_n = 1$ if n is even and -1 if n is odd. Does $\lim_{n\to\infty} a_n$ exist?

49. If a radioactive substance has a half-life of T, so that half of it decays after time T, write a sequence a_n showing the fraction remaining after time nT. What is $\lim_{n\to\infty} a_n$?

50. Evaluate:

(a) $\displaystyle \lim_{n\to\infty}\left[\frac{(-1)^n}{3} + (-1)^{n+1}\left(\frac{1}{3} + \frac{2}{n} + \frac{1}{n^2}\right)\right]$;

(b) $\displaystyle \lim_{n\to\infty}\left\{\frac{(-1)^n}{2} + n\left[\frac{1 + (-1)^{n+1}(3n+1)}{6n^2 - 5n + 2}\right]\right\}$.

Find the limit or prove that the limit does not exist in Exercises 51–54.

51. $\displaystyle \lim_{n\to\infty}\left(\frac{1}{2} + (-1)^n\left(\frac{1}{2} - \frac{1}{n}\right)\right)$.

52. $\displaystyle \lim_{n\to\infty}\left(\frac{(-1)^n \sin(n\pi/2)}{n}\right)\left(\frac{n^2+1}{n+1}\right)$.

53. $\displaystyle \lim_{n\to\infty}\left(\frac{3n}{4n+1} + \frac{(-1)^n \sin n}{n+1}\right)$.

54. $\displaystyle \lim_{n\to\infty}\frac{(1+n)\cos n}{n}$.

★55. (a) Give an A-N definition of what $\lim_{n\to\infty} a_n = \infty$ means. (b) Prove, using your definition in part (a), that $\lim_{n\to\infty}[(1+n^2)/(1+8n)] = \infty$.

★56. If $a_n \to 0$ and $|b_n| \leqslant |a_n|$, show that $b_n \to 0$.

★57. Suppose that a_n, b_n, and c_n, $n = 1, 2, 3, \ldots$, are sequences of numbers such that for each n, we have $a_n < b_n < c_n$.
(a) If $\lim_{n\to\infty} a_n = L$ and if $\lim_{n\to\infty} b_n$ exists, show that $\lim_{n\to\infty} b_n \geqslant L$. [*Hint:* Suppose not!]
(b) If $\lim_{n\to\infty} a_n = L = \lim_{n\to\infty} c_n$, prove that $\lim_{n\to\infty} b_n = L$.

★58. A rubber ball is released from a height h. Each time it strikes the floor, it rebounds with two-thirds of its previous velocity.
(a) How far does the ball rise on each bounce? (Use the fact that the height y of the ball at time t from the beginning of each bounce is of the form $y = vt - \frac{1}{2}gt^2$ during the bounce. The constant g is the acceleration of gravity.)
(b) How long does each bounce take?
(c) Show that the ball stops bouncing after a finite time has passed.
(d) How far has the ball travelled when it stops bouncing?
(e) How would the results differ if this experiment were done on the moon?

[7] For a computer, this method is preferable to using the formula for the roots of a cubic!

★59. (Research Problems)

(a) Experiment with the recursion relation $x_{n+1} = ax_n(1 - x_n)$ for various values of the parameter a where $0 < a \leqslant 4$ and x_0 is in $[0, 1]$. How does the behavior of the sequences change when a varies?

(b) Study the bizarre behavior of Newton's method in Example 9 for various starting values x_0. Can you see a pattern? Does x_n always converge?

(c) Study the bizarre behavior of Newton's method in Example 10.

11.5 Numerical Integration

Integrals can be approximated by sequences which can be computed numerically.

The fundamental theorem of calculus does not solve all our integration problems. The antiderivative of a given integrand may not be easy or even possible to find. The integrand might be given, not by a formula, but by a table of values; for example, we can imagine being given power readings from an energy cell and asked to find the energy stored. In either case, it is necessary to use a method of numerical integration to find an approximate value for the integral.

In using a numerical method, it is important to estimate errors so that the final answer can be said, with confidence, to be correct to so many significant figures. The possible errors include errors in the method, roundoff errors, and roundoff errors in arithmetic operations. The task of keeping careful track of possible errors is a complicated and fascinating one, of which we can give only some simple examples.[8]

The simplest method of numerical integration is based upon the fact that the integral is a limit of Riemann sums (see Section 4.3). Suppose we are given $f(x)$ on $[a,b]$, and divide $[a,b]$ into subintervals $a = x_0 < x_1 < \cdots < x_n = b$. Then $\int_a^b f(x)\,dx$ is approximated by $\sum_{i=1}^{n} f(c_i)\Delta x_i$, where c_i lies in $[x_{i-1}, x_i]$. Usually, the points x_i are taken to be equally spaced, so $\Delta x_i = (b-a)/n$ and $x_i = a + i(b-a)/n$. Choosing $c_i = x_i$ or x_{i+1} gives the method in the following box.

Riemann Sums

To calculate an approximation to $\int_a^b f(x)\,dx$, let $x_i = a + i(b-a)/n$ and form the sum

$$\frac{b-a}{n}\left[f(x_0) + f(x_1) + \cdots + f(x_{n-1})\right] \tag{1a}$$

or

$$\frac{b-a}{n}\left[f(x_1) + f(x_2) + \cdots + f(x_n)\right]. \tag{1b}$$

[8] For a further discussion of error analysis in numerical integration, see, for example, P. J. Davis, *Interpolation and Approximation*, Wiley, New York (1963).

Example 1 Let $f(x) = \cos x$. Evaluate $\int_0^{\pi/2}\cos x\,dx$ by the method of Riemann sums, taking 10 equally spaced points: $x_0 = 0$, $x_1 = \pi/20$, $x_2 = 2\pi/20, \dots, x_{10} = 10\pi/20 = \pi/2$, and $c_i = x_i$. Compare the answer with the actual value.

Solution Formula (1a) gives

$$\begin{aligned}\int_0^{\pi/2}\cos x\,dx &\approx \frac{\pi}{20}\left(1+\cos\frac{\pi}{20}+\cos\frac{2\pi}{20}+\cdots+\cos\frac{9\pi}{20}\right)\\ &= \frac{\pi}{20}(1+0.98769+0.95106+\cdots+0.15643)\\ &= \frac{\pi}{20}(6.85310) = 1.07648.\end{aligned}$$

The actual value is $\sin(\pi/2) - \sin(0) = 1$, so our estimate is about 7.6% off. ▲

Unfortunately, this method is inefficient, because many points x_i are needed to get an accurate estimate of the integral. For this reason we will seek alternatives to the method of Riemann sums.

To get a better method, we estimate the area in each interval $[x_{i-1}, x_i]$ more accurately by replacing the rectangular approximation by a trapezoidal one. (See Fig. 11.5.1.) We join the points $(x_i, f(x_i))$ by straight line segments to obtain a set of approximating trapezoids. The area of the trapezoid between x_{i-1} and x_i is

$$A_i = \tfrac{1}{2}\left[f(x_{i-1}) + f(x_i)\right]\Delta x_i$$

since the area of a trapezoid is its average height times its width.

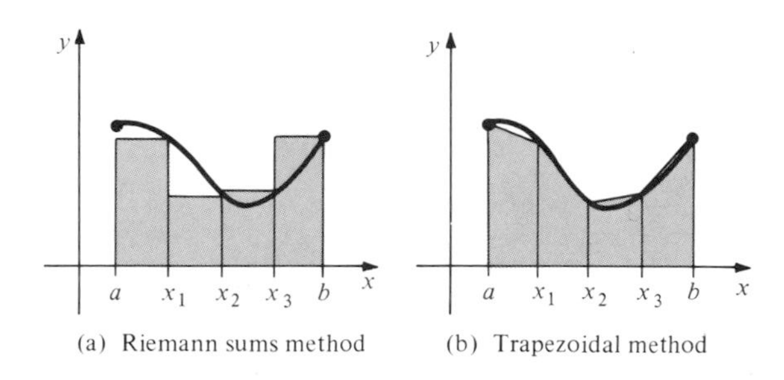

Figure 11.5.1. Comparing two methods of numerical integration.

The approximation to $\int_a^b f(x)\,dx$ given by the trapezoidal rule is $\sum_{i=1}^n \frac{1}{2}[f(x_{i-1}) + f(x_i)]\Delta x_i$. This becomes simpler if the points x_i are equally spaced. Then $\Delta x_i = (b-a)/n$, $x_i = a + i(b-a)/n$, and the sum is

$$\left(\frac{b-a}{n}\right)\sum_{i=1}^{n}\frac{1}{2}\left[f(x_{i-1}) + f(x_i)\right]$$

which can be rewritten as

$$\frac{b-a}{2n}\left[f(x_0) + 2f(x_1) + \cdots + 2f(x_{n-1}) + f(x_n)\right]$$

since every term occurs twice except those from the endpoints. Although we used areas to obtain this formula, we may apply it even if $f(x)$ takes negative values.

Trapezoidal Rule

To calculate an approximation to $\int_a^b f(x)\,dx$, let $x_i = a + i(b-a)/n$ and form the sum

$$\frac{b-a}{2n}\left[f(x_0) + 2f(x_1) + \cdots + 2f(x_{n-1}) + f(x_n)\right]. \tag{2}$$

Formula (2) turns out to be much more accurate than the method of Riemann sums, even though it is just the average of the Riemann sums (1a) and (1b). Using results of Section 12.5, one can show that the error in the *method* (apart from other roundoff or cumulative errors) is $\leqslant [(b-a)/12]M_2(\Delta x)^2$, where M_2 is the maximum of $|f''(x)|$ on $[a,b]$. Of course, if we are given only numerical data, we have no way of estimating M_2, but if a formula for f is given, M_2 can be determined. Note, however, that the error depends on $(\Delta x)^2$, so if we divide $[a,b]$ into k times as many divisions, the error goes down by a factor of $1/k^2$. The error in the Riemann sums method, on the other hand, is $\leqslant (b-a)M_1(\Delta x)$, where M_1 is the maximum of $|f'(x)|$ on $[a,b]$. Here Δx occurs only to the first power. Thus even if we do not know how large M_1 and M_2 are, if n is taken large enough, the error in the trapezoidal rule will eventually be much smaller than that in the Riemann sums method.

Example 2 Repeat Example 1 by using the trapezoidal rule. Compare the answer with the true value.

Solution Now formula (2) becomes

$$\frac{\pi/2}{2\cdot 10}\left(\cos 0 + 2\cos\frac{\pi}{20} + \cdots + 2\cos\frac{9\pi}{20} + \cos\frac{\pi}{2}\right)$$

$$\approx \frac{\pi}{40}\left[1 + 2(0.9877 + 0.9511 + \cdots + 0.1564) + 0\right] \approx 0.9979.$$

The answer is correct to within about 0.2%, much better than the accuracy in Example 1. ▲

Example 3 Use the trapezoidal rule with $n = 10$ to estimate numerically the area of the surface obtained by revolving the graph of $y = x/(1+x^2)$ about the x axis, $0 \leqslant x \leqslant 1$.

Solution The area is given by formula (2) on p. 483:

$$A = 2\pi\int_0^1\left(\frac{x}{1+x^2}\right)\sqrt{1 + \left[\frac{d}{dx}\left(\frac{x}{1+x^2}\right)\right]^2}\,dx$$

$$= 2\pi\int_0^1 \frac{x\sqrt{(1+x^2)^4 + (1-x^2)^2}}{(1+x^2)^3}\,dx.$$

There is little hope of carrying out this integration, so a numerical approach seems appropriate. We use the trapezoidal rule with the following values:

x_i	0	0.1	0.2	0.3	0.4	0.5	0.6	0.7	0.8	0.9	1.0
$y_i = f(x_i)$	0	0.13797	0.25713	0.34668	0.40650	0.44369	0.46684	0.48204	0.49216	0.49807	0.50000

where $f(x) = x\sqrt{(1+x^2)^4 + (1-x^2)^2} / (1+x^2)^3$. Inserting these data in the formula

$$\int_a^b f(x)\,dx \approx \left(\frac{b-a}{2n}\right)\left[f(x_0) + 2f(x_1) + \cdots + 2f(x_{n-1}) + f(x_n)\right]$$

with $x_i = a + [i(b-a)/n]$, $a = 0$ and $b = 1$, gives

$$\int_0^1 \frac{x\sqrt{(1+x^2)^4 + (1-x^2)^2}}{(1+x^2)^3}\,dx \approx 0.37811,$$

so the area is $A \approx (2\pi)(0.37811) = 2.3757$. Of course, we cannot be sure how many decimal places in this result are correct without an error analysis (see Exercise 17).[9] ▲

There is a yet more powerful method of numerical integration called Simpson's rule,[10] which is based on approximating the graph by parabolas rather than straight lines. To determine a parabola we need to specify three points through which it passes; we will choose the adjacent points

$$(x_{i-1}, f(x_{i-1})), \qquad (x_i, f(x_i)), \qquad (x_{i+1}, f(x_{i+1})).$$

It is easily proved (see Exercise 11) that the integral from x_{i-1} to x_{i+1} of the quadratic function whose graph passes through these three points is

$$A_i = \frac{\Delta x}{3}\left[f(x_{i-1}) + 4f(x_i) + f(x_{i+1})\right],$$

where $\Delta x = x_i - x_{i-1} = x_{i+1} - x_i$ (equally spaced points). See Fig. 11.5.2.

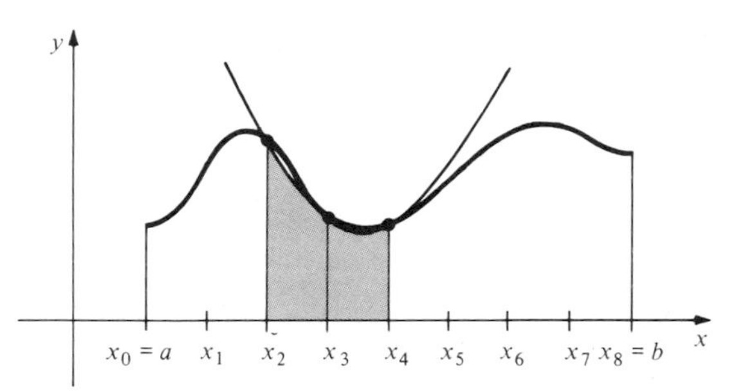

Figure 11.5.2. Illustrating Simpson's rule.

If we do this for every set of three adjacent points, starting at the left endpoint a—that is, for $\{x_0, x_1, x_2\}$, then $\{x_2, x_3, x_4\}$, then $\{x_4, x_5, x_6\}$, and so on—we will get an approximate formula for the area. In order for the points to fill the interval exactly, n should be even, say $n = 2m$.

As in the trapezoidal rule, the contributions from endpoints a and b are counted only once, as are those from the center points of triples $\{x_{i-1}, x_i, x_{i+1}\}$ (that is, x_i for i odd), while the others are counted twice. Thus we are led to Simpson's rule, stated in the box on the next page.

This method is very accurate; the error in using formula (3) does not exceed $[(b-a)/180]M_4(\Delta x)^4$, where M_4 is the maximum of the fourth derivative of $f(x)$ on $[a,b]$. As Δx is taken smaller and smaller, this error decreases much faster than in the other two methods. It is remarkable that juggling the

[9] The HP 15C has a clever integration program that is careful about errors. It gives 2.3832 for this integral in a few minutes.

[10] It was discussed by Thomas Simpson in his book, *Mathematical Dissertations on Physical and Analytical Subjects* (1743).

coefficients to give formula (3) in place of formula (1) or formula (2) can increase the accuracy so much.

Simpson's Rule

To calculate an approximation to $\int_a^b f(x)\,dx$, let $n = 2m$ be even and $x_i = a + i(b - a)/n$. Form the sum

$$\frac{b-a}{3n}\left[f(x_0) + 4f(x_1) + 2f(x_2) + 4f(x_3) + 2f(x_4) + \cdots + 2f(x_{n-2}) + 4f(x_{n-1}) + f(x_n)\right]. \tag{3}$$

Example 4 Repeat Example 1 using Simpson's rule. Compare the answer with the true value.

Solution Using a calculator, we can evaluate formula (3) by

$$\begin{aligned}&\frac{\pi/2}{3\cdot 10}\Big(\cos 0 + 4\cos\frac{\pi}{20} + 2\cos\frac{2\pi}{20} + 4\cos\frac{3\pi}{20} + 2\cos\frac{4\pi}{20}\\ &\qquad + 4\cos\frac{5\pi}{20} + 2\cos\frac{6\pi}{20} + 4\cos\frac{7\pi}{20} + 2\cos\frac{8\pi}{20} + 4\cos\frac{9\pi}{20} + \cos\frac{\pi}{2}\Big)\\ &\approx \frac{\pi}{60}(1 + 3.9507534 + \cdots + 0) = \frac{\pi}{60}\cdot 19.098658\\ &\approx 1.0000034.\end{aligned}$$

The error is less than four parts in a million. ▲

Example 5 Suppose that you are given the following table of data:

$f(0) = 0.846$	$f(0.4) = 1.121$	$f(0.8) = 2.321$
$f(0.1) = 0.928$	$f(0.5) = 1.221$	$f(0.9) = 3.101$
$f(0.2) = 0.882$	$f(0.6) = 1.661$	$f(1.0) = 3.010$
$f(0.3) = 0.953$	$f(0.7) = 2.101$	

Evaluate $\int_0^1 f(x)\,dx$ by Simpson's rule.

Solution By formula (3),

$$\int_0^1 f(x)\,dx \approx \frac{1}{30}\left[f(0) + 4f(0.1) + 2f(0.2) + 4f(0.3) + 2f(0.4) + 4f(0.5) + 2f(0.6) + 4f(0.7) + 2f(0.8) + 4f(0.9) + f(1.0)\right].$$

Inserting the given values of f and evaluating on a calculator, we get

$$\int_0^1 f(x)\,dx = \frac{1}{30}(49.042) = 1.635.$$

This should be quite accurate unless the fourth derivative of f is very large. ▲

Example 6 How small must we take Δx in the trapezoidal rule to evaluate $\int_2^4 e^{-x^2}\,dx$ to within 10^{-6}? For Simpson's rule?

Solution Let $f(x) = e^{-x^2}$, $a = 2$, and $b = 4$. The error in the trapezoidal rule is no more than $[(b - a)/12]M_2(\Delta x)^2$, where M_2 is the maximum of $|f''(x)|$ on $[a, b]$. We find

$$f'(x) = -2xe^{-x^2}, \quad \text{and} \quad f''(x) = -2e^{-x^2} + 4x^2e^{-x^2} = 2(2x^2 - 1)e^{-x^2}.$$

Now $f'''(x) = (12x - 8x^3)e^{-x^2} = 4x(3 - 2x^2)e^{-x^2} < 0$ on $[2,4]$, so $f''(x)$ is decreasing. Also, $f''(x) > 0$ on $[2,4]$, so $|f''(x)| = f''(x) \leqslant f''(2) = 14e^{-4} = M_2$, so the error is at most

$$\frac{b-a}{12} M_2(\Delta x)^2 = \frac{1}{6} \cdot 14e^{-4}(\Delta x)^2.$$

To make this less than 10^{-6}, we should choose Δx so that

$$\tfrac{1}{6} \cdot 14e^{-4}(\Delta x)^2 < 10^{-6},$$

$$(\Delta x)^2 < e^4 10^{-6} \cdot \tfrac{3}{7} = 0.0000234,$$

$$\Delta x < 0.0048.$$

That is, we should take at least $n = (b-a)/\Delta x = 416$ divisions.

For Simpson's rule, the error is at most $[(b-a)/180]M_4(\Delta x)^4$. Here

$$f''''(x) = 4(4x^4 - 12x^2 + 3)e^{-x^2}.$$

On $[2,4]$, we find that $4x^4 - 12x^2 + 3$ is increasing and e^{-x^2} is decreasing, so

$$|f''''(x)| \leqslant 4(4 \cdot 4^4 - 12 \cdot 4^2 + 3)e^{-4}$$

$$= 61.17 = M_4.$$

Thus $[(b-a)/180]M_4(\Delta x)^4 = \frac{1}{90} \cdot 61.17(\Delta x)^4 = 0.68(\Delta x)^4$. Hence if we are to have error less than 10^{-6}, it suffices to have

$$0.68(\Delta x)^4 \leqslant 10^{-6},$$

$$\Delta x \leqslant 0.035.$$

Thus we should take at least $n = (b-a)/\Delta x = 57$ divisions. ▲

Exercises for Section 11.5

Use the indicated numerical method(s) to approximate the integrals in Exercises 1–4.

1. $\int_{-1}^{1}(x^2+1)\,dx$. Use Riemann sums with $n = 10$ (that is, divide $[-1, 1]$ into 10 subintervals of equal length). Compare with the actual value.
2. $\int_0^{\pi/2}(x + \sin x)\,dx$. Use Riemann sums and the trapezoidal rule with $n = 8$. Compare these two approximate values with the actual value.
3. $\int_1^3[(\sin \pi x/2)/(x^2 + 2x - 1)]\,dx$. Use the trapezoidal rule and Simpson's rule with $n = 12$.
4. $\int_0^2(1/\sqrt{x^3+1}\,)\,dx$. Use the trapezoidal rule and Simpson's rule with $n = 20$.
5. Use Simpson's rule with $n = 10$ to find an approximate value for $\int_0^1(x/\sqrt{x^3+2}\,)\,dx$.
6. Estimate the value of $\int_1^3 e^{\sqrt{x}}\,dx$, using Simpson's rule with $n = 4$. Check your answer using $x = u^2$, $dx = 2u\,du$.
7. Suppose you are given the following table of data:

$f(0) = 1.384$	$f(0.4) = 0.915$	$f(0.8) = 0.935$
$f(0.1) = 1.179$	$f(0.5) = 0.768$	$f(0.9) = 1.262$
$f(0.2) = 0.973$	$f(0.6) = 0.511$	$f(1.0) = 1.425$
$f(0.3) = 1.000$	$f(0.7) = 0.693$	

Numerically evaluate $\int_0^1(x + f(x))\,dx$ by the trapezoidal rule.

8. Numerically evaluate $\int_0^1 2f(x)\,dx$ by Simpson's rule, where $f(x)$ is the function in Exercise 7.
9. Suppose that you are given the following table of data:

$f(0.0) = 2.037$	$f(1.3) = 0.819$
$f(0.2) = 1.980$	$f(1.4) = 1.026$
$f(0.4) = 1.843$	$f(1.5) = 0.799$
$f(0.6) = 1.372$	$f(1.6) = 0.662$
$f(0.8) = 1.196$	$f(1.7) = 0.538$
$f(1.0) = 0.977$	$f(1.8) = 0.555$
$f(1.2) = 0.685$	

Numerically evaluate $\int_0^{1.8} f(x)\,dx$ by using Simpson's rule. [*Hint*: Watch out for the spacing of the points.]

10. Numerically evaluate $\int_0^{1.8} f(x)\,dx$ by using the trapezoidal rule, where $f(x)$ is the function in Exercise 9.

11. Evaluate $\int_a^b (px^2 + qx + r)\,dx$. Verify that Simpson's rule with $n = 2$ gives the exact answer. What happens if you use the trapezoidal rule? Discuss.

12. Evaluate $\int_a^b (px^3 + qx^2 + rx + s)\,dx$ by Simpson's rule with $n = 2$ and compare the result with the exact integral.

13. How large must n be taken in the trapezoidal rule to guarantee an accuracy of 10^{-5} in the evaluation of the integral in Exercise 2? Answer the same question for Simpson's rule.

14. *Gaussian quadrature* is an approximation method based on interpolation. The formula for integration on the interval $[-1, 1]$ is $\int_{-1}^{1} f(x)\,dx = f(1/\sqrt{3}) + f(-1/\sqrt{3}) + R$, where the remainder R satisfies $|R| \leqslant M/135$, M being the largest value of $f^{(4)}(x)$ on $-1 \leqslant x \leqslant 1$.
    (a) The remainder R is zero for cubic polynomials. Check it for $x^3, x^3 - 1, x^3 + x + 1$.
    (b) Find $\int_{-1}^{1} [x^2/(1 + x^4)]\,dx$ to two places.
    (c) What is R for $\int_{-1}^{1} x^6\,dx$? Why is it so large?

15. A tank 15 meters by 60 meters is filled to a depth of 3.2 meters above the bottom. The time T it takes to empty half the tank through an orifice 0.5 meters wide by 0.2 meters high placed 0.1 meters from the bottom is given by

$$\frac{(2.2)T}{10^4} = \frac{1}{\sqrt{19.6}} \int_{130}^{190} \frac{dx}{(x + 20)^{3/2} - x^{3/2}}.$$

Compute T from Simpson's rule with $n = 6$.

★16. A metropolitan sports and special events complex is circular in shape with an irregular roof that appears from a distance to be almost hemispherical (Fig. 11.5.3).

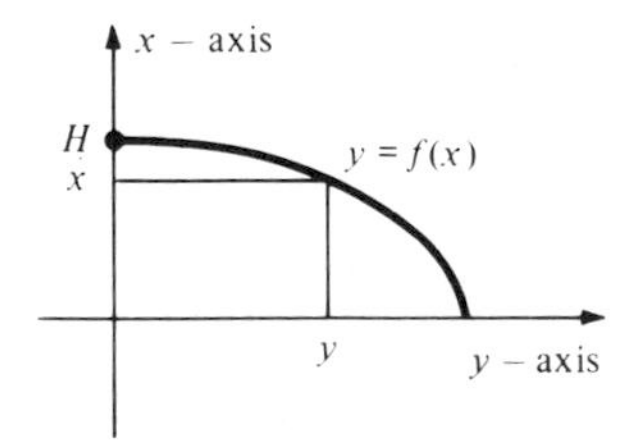

Figure 11.5.3. The profile of the roof of a sports complex.

A summer storm severely damaged the roof, requiring a roof replacement to go out for bid. Responding contractors were supplied with plans of the complex from which to determine an estimate. Estimators had to find the roof profile $y = f(x)$, $0 \leqslant x \leqslant H$, which generates the roof by revolution about the x axis (x and y in feet, x vertical, y horizontal).
    (a) Find the square footage of the roof via a surface area formula. This number determines the amount of roofing material required.
    (b) To check against construction errors, a tape measure is tossed over the roof and the measurement recorded. Give a formula for this measurement using the arc length formula.
    (c) Suppose the curve f is not given explicitly in the plans, but instead $f(0), f(4), f(8), f(12), \ldots, f(H)$ are given (complex center-to-ceiling distances every 4 feet). Discuss how to use this information to numerically evaluate the integrals in (a), (b) above, using Example 3 as a guide.
    (d) Find an expression which approximates the surface area of the roof by assuming it is a *conoid* produced by a piecewise linear function constructed from the numbers $f(0), f(4), f(8), \ldots, f(H)$.

★17. How many digits in the approximate value $A \approx 2.3757$ in Example 3 can be justified by an error analysis?

★18. (*Another numerical integration method*)
    (a) Let $(x_1, y_1), (x_2, y_2), \ldots, (x_n, y_n)$ be n points in the plane such that all the x_i's are different. Show that the polynomial of degree no more than $n - 1$ whose graph passes through the given points is

$$P(x) = y_1 L_1(x) + y_2 L_2(x) + \cdots + y_n L_n(x),$$

where $L_i(x) = A_i(x)/A'(x_i)$,

$$A(x) = (x - x_1)(x - x_2) \ldots (x - x_n),$$
$$A_i(x) = A(x)/(x - x_i), \quad i = 1, 2, \ldots, n.$$

(P is called the *Lagrange interpolation polynomial*.)
    (b) Suppose that you are given the following data for an unknown function $f(x)$:

$$f(0) = 0.01, \quad f(0.3) = 1.18,$$
$$f(0.1) = 0.12, \quad f(0.4) = 0.91.$$
$$f(0.2) = 0.82,$$

Estimate the value of $f(0.16)$ by using the Lagrange interpolation formula.
    (c) Estimate $\int_0^{0.4} f(x)\,dx$ (1) by using the trapezoidal rule, (2) by using Simpson's rule, and (3) by integrating the Lagrange interpolation polynomial.
    (d) Estimate $\int_0^{\pi/2} \cos x\,dx$ by using a Lagrange interpolation polynomial with $n = 4$. Compare your result with those obtained by the trapezoidal and Simpson's rules in Examples 2 and 4.

Review Exercises for Chapter 11

Verify the limits in Exercises 1–4 using the ε-δ definition.

1. $\lim_{x\to 1}(x^2+x-1)=1$
2. $\lim_{x\to 0}(x^3+3x+2)=2$
3. $\lim_{x\to 2}(x^2-8x+8)=-4$
4. $\lim_{x\to 5}(x^2-25)=0$

Calculate the limits in Exercises 5–16.

5. $\lim_{x\to 0}\tan\left(\frac{x+1}{x-1}\right)$
6. $\lim_{x\to 1}\cos\left[\left(\frac{x+1}{x+2}\right)\frac{3\pi}{2}\right]$
7. $\lim_{x\to\infty}\left(\frac{x+1}{x-1}\right)$
8. $\lim_{x\to\infty}\left(\frac{x^2+2}{3x^2+2x+1}\right)$
9. $\lim_{x\to\infty}(\sqrt{2x^2+1}-\sqrt{2}\,x)$
10. $\lim_{x\to\infty}(\sqrt{4x^2+1}-2x)$
11. $\lim_{x\to 1-}\frac{1}{\sqrt{1-x}}$
12. $\lim_{x\to 2+}\frac{\sin\sqrt{x-2}}{\sqrt{x-2}}$
13. $\lim_{x\to 0}[x\sin(3/x^2)]$
14. $\lim_{x\to 0}\sin(\sqrt{x^2+2}-x)$
15. $\lim_{x\to 0}\frac{x^4+8x}{3x^4+2}$
16. $\lim_{x\to 0}\frac{x+3}{3x+8}$
17. Find the horizontal asymptotes of the graph $y=\tan^{-1}(3x+2)$. Sketch.
18. Find the vertical asymptotes for the graph of $y=1/(x^2-3x-10)^2$. Sketch.

Find the horizontal and vertical asymptotes of the functions in Exercises 19 and 20, and sketch.

19. $f(x)=\frac{x-1}{x^2+1}$
20. $f(x)=\frac{2x+3}{3x+5}$

Find the limits if they exist, using l'Hôpital's rule, in Exercises 21–44.

21. $\lim_{x\to\infty}\frac{x^3+8x+9}{4x^3-9x^2+10}$
22. $\lim_{x\to\infty}\frac{x}{x+2}$
23. $\lim_{x\to 0}\frac{1-\cos x}{3^x-2^x}$
24. $\lim_{x\to 1}\frac{x^x-1}{x-1}$
25. $\lim_{x\to 0}\frac{(\sqrt{x^2+9}-3)}{\sin x}$
26. $\lim_{x\to 0}\frac{\sqrt[3]{x^3+27}-3}{x}$
27. $\lim_{x\to 0}\frac{\sin 5x}{x}$
28. $\lim_{x\to 0}\frac{\tan^2 x}{x^2}$
29. $\lim_{x\to 2}\frac{\sin(x-2)-x+2}{(x-2)^3}$
30. $\lim_{x\to 0}\frac{24\cos x-24+12x^2-x^4}{x^5}$
31. $\lim_{x\to 0}\frac{\tan(x+3)-\tan 3}{x}$
32. $\lim_{x\to\pi^2}\frac{\cos\sqrt{x}+1}{x-\pi^2}$
33. $\lim_{x\to 0}x\cot x$
34. $\lim_{x\to 0+}\frac{\cot x}{\ln x}$
35. $\lim_{x\to 0+}\left(\frac{1}{\sin x}-\frac{1}{x}\right)$
36. $\lim_{x\to 1}\left(\frac{1}{\ln x}-\frac{1}{x-1}-\frac{1}{2}\right)$
37. $\lim_{x\to\infty}x^2e^{-x}$
38. $\lim_{x\to 0+}x^3(\ln x)^2$
39. $\lim_{x\to 0+}x^{\sin x}$
40. $\lim_{x\to\infty}(\sin e^{-x})^{1/\sqrt{x}}$
41. $\lim_{x\to 0+}(1+\sin 2x)^{1/x}$
42. $\lim_{x\to 0+}(\cos 2x)^{1/x^2}$
43. $\lim_{x\to\infty}\frac{(\ln x)^2}{x}$
44. $\lim_{x\to 2}\frac{x^2+x-6}{x^2+2x-8}$

Decide which improper integrals in Exercises 45–54 are convergent. Evaluate when possible.

45. $\int_1^\infty\frac{1}{x^2}\,dx.$
46. $\int_{-\infty}^\infty\frac{\sin x}{x^2+3}\,dx.$
47. $\int_2^\infty\frac{dx}{\ln x}$. [*Hint*: Prove $\ln x\leqslant x$ for $x\geqslant 2$.]
48. $\int_1^\infty\frac{dx}{\sqrt{x^2+8x+12}}.$
49. $\int_1^2\frac{1}{\sqrt{x-1}}\,dx.$
50. $\int_{-1}^0\frac{x+1}{\sqrt{1-x^2}}\,dx.$
51. $\int_0^1\frac{dx}{(1-x)^{2/5}}.$
52. $\int_1^\infty x^2e^{-x}\,dx.$
53. $\int_0^1 x\ln x\,dx.$
54. $\int_0^\infty(x+2)e^{-(x^2+4x)}\,dx.$

Evaluate the limits in Exercises 55 and 56.

55. $\lim_{x\to\infty}\int_0^x\frac{dt}{t^2+t+1}$
56. $\lim_{x\to 0+}\int_x^1\frac{dt}{\sqrt{t}}$
57. The region under the curve $y=xe^{-x}$ on $[0,\infty)$ is revolved about the x axis. Find the volume of the resulting solid.
58. The curve $y=\sin x/x^2$ on $[1,\infty)$ is revolved around the x axis. Determine whether the resulting surface has finite area.

Evaluate the limits of the sequences in Exercises 59–72.

59. $\lim_{n\to\infty}\left(8+\left(\frac{2}{3}\right)^n\right)^5$
60. $\lim_{n\to\infty}\sqrt[3]{\frac{8n^3-2n+1}{n^3+1}}$
61. $\lim_{n\to\infty}\left(1+\frac{8}{n}\right)^n$
62. $\lim_{n\to\infty}\left(\frac{n-3}{n}\right)^{-2n}$
63. $\lim_{n\to\infty}(n^2+3n+1)e^{-n}$
64. $\lim_{n\to\infty}\frac{2^n}{n^2}$
65. $\lim_{n\to\infty}\frac{n}{n+2}$
66. $\lim_{n\to\infty}\frac{n^2+2n}{3n^2+1}$
67. $\lim_{n\to\infty}\tan\left[\frac{3n}{n+8}\right]$
68. $\lim_{n\to\infty}[\ln(n^2+1)-\ln(3n^2+5)]$

69. $\lim_{n\to\infty} \dfrac{\sin(\pi n/2)}{n^{-3}}$

70. $\lim_{n\to\infty} \dfrac{n\cos 4n\pi}{2n+1}$

71. $\lim_{n\to\infty} \dfrac{8-2n}{5n}$

72. $\lim_{n\to\infty} \left(1 - \dfrac{2+n}{3n+1}\right)$

Using l'Hôpital's rule if necessary, evaluate the limits of the sequences in Exercises 73–76.

73. $\lim_{n\to\infty} \sqrt[2n]{3n}$

74. $\lim_{m\to\infty} m\log_{10}(2^{-m})$

75. $\lim_{n\to\infty} \left[\dfrac{1}{n^2} - \dfrac{1}{(n+1)^2}\right]$

76. $\lim_{j\to\infty} \dfrac{2e^{3j}}{e^{3j}+j^5}$

Use Newton's method for Exercises 77–80.

77. Locate the roots of $x^3 - 3x^2 + 8 = 0$.

78. Find the cube root of 21.

79. Solve the equation $e^x = 2 + x$.

80. Find two numbers, each of whose square is one-tenth its natural logarithm.

81. Evaluate $\int_2^3 (x^2\,dx/\sqrt{x^2+1})$ by the trapezoidal rule with $n = 10$.

82. Evaluate the integral in Exercise 81 by Simpson's rule with $n = 10$.

83. Use Simpson's rule with $n = 10$ to calculate the volume obtained by revolving the curve $y = f(x)$ on $[1,3]$ about the x axis, given the data:

$f(1) = 2.03$	$f(2.2) = 3.16$
$f(1.2) = 2.08$	$f(2.4) = 3.01$
$f(1.4) = 2.16$	$f(2.6) = 2.87$
$f(1.6) = 2.34$	$f(2.8) = 2.15$
$f(1.8) = 2.82$	$f(3) = 1.96$
$f(2) = 3.01$	

84. (a) Evaluate $(2/\sqrt{\pi})\int_0^1 e^{-t^2}dt$ by using Simpson's rule with 10 subdivisions.
(b) Given an upper bound for the error in part (a). (See Example 6 of Section 11.5.)
(c) What does Simpson's rule with 10 subdivisions give for $(2/\sqrt{\pi})\int_0^{10} e^{-t^2}dt$?
(d) The function $(2/\sqrt{\pi})\int_0^x e^{-t^2}dt$ is denoted erf(x) and is called the error function. Its values are tabulated. (For example: *Handbook of Mathematical Functions*, National Bureau of Standards, Applied Mathematics Series 55, June 1964, pp. 310–311.) Compare your results with the tabulated results. *Note*: $\lim_{x\to\infty}\text{erf}(x) = 1$, and erf(10) is so close to 1 that it probably won't be listed in the tables. Explain your result in part (c).

85. Let $f(x) = \cos x$ for $x \geqslant 0$ and $f(x) = 1$ for $x < 0$. Decide whether or not f is continuous or differentiable or both.

86. Let $f(x) = x^{1/\sin(x-1)}$. How should $f(1)$ be defined in order to make f continuous?

87. Find a function on $[0,1]$ which is integrable (as an improper integral) but whose square is not.

88. Show that $\int_0^\infty [(\sin x)/(1+x)]\,dx$ is convergent. [*Hint*: Integrate by parts.]

89. (a) Show that
$$f''(x_0) = \lim_{h\to 0} \frac{f(x_0+h) - 2f(x_0) + f(x_0-h)}{h^2}$$
if f'' is continuous at x_0. [*Hint*: Use l'Hôpital's rule.]
★(b) Find a similar formula for $f'''(x_0)$.

90. Show that
$$f''(x_0) = \lim_{\Delta x\to 0} 2\left[\frac{f(x_0+\Delta x) - f(x_0) - f'(x_0)\Delta x}{\Delta x}\right]$$
if f'' is continuous at x_0.

91. Use Riemann sums to evaluate
$$\lim_{n\to\infty} \sum_{i=1}^{n} (\ln n - \ln i)/n.$$

92. Let
$$S_n = \sum_{i=1}^{2n}\left[2 - \cos\left(\frac{i}{n}\pi\right)\right]\frac{1}{n}.$$
Prove that $\lim_{n\to\infty} S_n = 4$ using Riemann sums.

93. Let
$$S_n = \sum_{i=1}^{n}\left(\frac{i}{n} + \frac{i^2}{n^2}\right)\frac{1}{n}.$$
Prove that $S_n \to \frac{5}{6}$ as $n\to\infty$ by using Riemann sums.

94. Expressing the following sums as Riemann sums, show that:
(a) $\displaystyle\lim_{n\to\infty}\sum_{i=1}^{n}\left[\sqrt{\frac{i}{n}} - \left(\frac{i}{n}\right)^{3/2}\right]\frac{1}{n} = \frac{4}{15}$;
(b) $\displaystyle\lim_{n\to\infty}\sum_{i=1}^{n}\frac{3n}{(2n+i)^2} = \frac{1}{2}$.

95. P dollars is deposited in an account each day for a year. The account earns interest at an annual rate r (e.g., $r = 0.05$ means 5%) compounded continuously. Use Riemann sums to show that the amount in the account at the end of the year is approximately
$$(365P/r)\,(e^r - 1).$$

★96. Evaluate:
$$\lim_{x\to\pi^2}\left[\frac{\sin\sqrt{x}}{(\sqrt{x}-\pi)(\sqrt{x}+\pi)} + \tan\sqrt{x}\right].$$

★97. Limits can sometimes be evaluated by geometric techniques. An important instance occurs when the curve $y = f(x)$ is trapped between the two intersecting lines through (a, L) with slopes m and $-m$, $0 < |x-a| \leqslant h$. Then $\lim_{x\to a} f(x) = L$, because points approaching $y = f(x)$ from the left or right are forced into a vertex, and therefore to the point (a, L).

(a) The equations of the two lines are $y = L + m(x - a)$, $y = L - m(x - a)$. Draw these on a figure and insert a representative graph for f which stays between the lines.

(b) Show that the algebraic condition that f stay between the two straight lines is

$$\left|\frac{f(x) - L}{x - a}\right| \leqslant m, \qquad 0 < |x - a| \leqslant h.$$

This is called a *Lipschitz condition*.

(c) Argue that a Lipschitz condition implies $\lim_{x \to a} f(x) = L$, by appeal to the definition of limit.

★98. Another geometric technique for evaluation of limits is obtained by requiring that $y = f(x)$ be trapped on $0 \leqslant |x - a| \leqslant h$ between two power curves

$$y = L + m(x - a)^{\alpha}, \qquad y = L - m(x - a)^{\alpha},$$

where $\alpha > 0$, $m > 0$. The resulting algebraic condition is called a *Hölder condition*:

$$\frac{|f(x) - L|}{|x - a|^{\alpha}} \leqslant m, \qquad 0 < |x - a| \leqslant h.$$

(a) Verify that the described geometry leads to the Hölder condition.

(b) Argue geometrically that, in the presence of a Hölder condition, $\lim_{x \to a} f(x) = L$.

(c) Prove the contention in (b) by appeal to the definition of limit.

★99. Prove the chain rule for differentiable functions, $(f \circ g)'(x_0) = f'(g(x_0)) \cdot g'(x_0)$, as follows:

(a) Let $y = g(x)$ and $z = f(y)$, and write $\Delta y = g'(x_0)\,\Delta x + \rho(x)$. Show that

$$\lim_{\Delta x \to 0} \frac{\rho(x)}{\Delta x} = 0$$

Also write $\Delta z = f'(y_0)\,\Delta y + \sigma(y)$, where $y_0 = g(x_0)$ and show that

$$\lim_{\Delta y \to 0} \frac{\sigma(y)}{\Delta y} = 0.$$

(b) Show that

$$\Delta z = f'(y_0)\, g'(x_0)\,\Delta x + f'(y_0)\rho(x) + \sigma(g(x)).$$

(c) Note that $\sigma(g(x)) = 0$ if $\Delta y = 0$. Thus show that

$$\frac{\sigma(g(x))}{\Delta x} = \begin{Bmatrix} \dfrac{\sigma(g(x))}{\Delta y}\dfrac{\Delta y}{\Delta x} & \text{if } \Delta y \neq 0 \\ 0 & \text{if } \Delta y = 0 \end{Bmatrix} \to 0$$

as $\Delta x \to 0$.

(d) Use parts (a), (b), and (c) above to show that $\lim_{\Delta x \to 0}[\Delta z/\Delta x] = f'(y_0)g'(x_0)$. (This proof avoids the problem of division by zero mentioned on p. 113.)

★100. An alternative to Newton's method for finding solutions of the equation $f(x) = 0$ is the iteration scheme

$$x_{n+1} = x_n - \frac{f(x_n)}{f'(x_0)},$$

sometimes known as *Picard's method*. Notice that this method requires evaluating f' only at the initial guess x_0 and so requires less computation at each step.

(a) Show that, if the sequence $x_0, x_1, x_2, \dots$ converges, then $\lim_{n \to \infty} x_n$ is a solution of $f(x) = 0$.

(b) Compare Picard's method and Newton's method on the problem $x^5 = x + 1$, using the initial guess $x_0 = 1$ in each case and iterating until the solution is found to six decimal places of accuracy.

(c) Suppose that $f(q) = 0$ and in addition that $0 < \frac{1}{2}f'(x_0) < f'(x) < \frac{3}{2}f'(x_0)$ for all x in the interval $I = (q - a, q + a)$. Prove that if x_0 is any initial guess in I, then $|x_{n+1} - q| < \frac{1}{2}|x_n - q|$, and so $\lim_{n \to \infty} x_n = q$. [*Hint*: $x_{n+1} = P(x_n)$, where $P(x) = x - f'(x)/f(x_0)$. Differentiate $P(x)$ and apply the mean value theorem.] (A similar analysis for Newton's method is presented in the following Review Exercise.)

★101. Newton's method for solving $f(x) = 0$ can be described by saying that $x_{n+1} = N(x_n)$, where the *Newton iteration* function N is defined by $N(x) = x - f(x)/f'(x)$ for all x such that $f'(x) \neq 0$.

(a) Show that $N(x) = x$ if and only if $f(x) = 0$

(b) Show that $N'(x) = f(x)f''(x)/[f'(x)]^2$

(c) Suppose that $\bar{x}$ is a root of f, that $[a, b]$ is an interval containing $\bar{x}$, and that there are numbers p, q and M such that

$$0 < p \leqslant f'(x) \leqslant q \quad \text{and} \quad |f''(x)| \leqslant M$$

for all x in $[a, b]$. Show that there is a constant C such that

$$|N(x) - \bar{x}| \leqslant C|x - \bar{x}|^2$$

for all x in $[a, b]$. Express C in terms of p, q and M.

This establishes the "quadratic convergence" of $x_0, x_1, x_2, \dots$ to $\bar{x}$ as soon as some x_i is in $[a, b]$. [*Hint*: Apply the mean value theorem to N to conclude $N(x) - \bar{x} = N'(\xi)(x - \bar{x})$ for some ξ between x and $\bar{x}$. Use the mean value theorem again to show that $|N'(\xi)| \leqslant D|\xi|$ for a constant $D > 0$.]

(d) How many iterations are needed to solve $x^2 - 2 = 0$ to within 20 decimal places, assuming an initial guess in the interval $[1.4, 1.5]$?

Chapter 12

Infinite Series

Infinite sums can be used to represent numbers and functions.

The decimal expansion $\frac{1}{3} = 0.3333\ldots$ is a representation of $\frac{1}{3}$ as an infinite sum $\frac{3}{10} + \frac{3}{100} + \frac{3}{1000} + \frac{3}{10,000} + \cdots$. In this chapter, we will see how to represent numbers as infinite sums and to represent functions of x by infinite sums whose terms are monomials in x. For example, we will see that

$$\ln 2 = 1 - \frac{1}{2} + \frac{1}{3} - \frac{1}{4} + \cdots$$

and

$$\sin x = x - \frac{x^3}{1\cdot 2\cdot 3} + \frac{x^5}{1\cdot 2\cdot 3\cdot 4\cdot 5} - \cdots.$$

Later in the chapter we shall use our knowledge of infinite series to study complex numbers and some differential equations. There are other important uses of series that are encountered in later courses. One of these is the topic of Fourier series; this enables one, for example, to decompose a complex sound into an infinite series of pure tones.

12.1 The Sum of an Infinite Series

The sum of infinitely many numbers may be finite.

An *infinite series* is a sequence of numbers whose terms are to be added up. If the resulting sum is finite, the series is said to be *convergent*. In this section, we define convergence in terms of limits, give the simplest examples, and present some basic tests. Along the way we discuss some further properties of the limits of sequences, but the reader should also review the basic facts about sequences from Section 11.4.

Our first example of the limit of a sequence was an expression for the number $\frac{1}{3}$:

$$\frac{1}{3} = \lim_{n\to\infty}\left(\frac{3}{10} + \frac{3}{100} + \cdots + \frac{3}{10^n}\right).$$

This expression suggests that we may consider $\frac{1}{3}$ as the sum

$$\frac{3}{10} + \frac{3}{100} + \cdots + \frac{3}{10^n} + \cdots$$

of infinitely many terms. Of course, not every sum of infinitely many terms gives rise to a number (consider $1 + 1 + 1 + \cdots$), so we must be precise about what we mean by adding together infinitely many numbers. Following the idea used in the theory of improper integrals (in Section 11.3), we will define the sum of an infinite series by taking finite sums and then passing to the limit as the sum includes more and more terms.

Convergence of Series

Let $a_1, a_2, \ldots$ be a sequence of numbers. The number $S_n = \sum_{i=1}^{n} a_i = a_1 + a_2 + \cdots + a_n$ is called the *nth partial sum* of the a_i's. If the sequence $S_1, S_2, \ldots$ of partial sums approaches a limit S as $n \to \infty$, we say that *the series* $a_1 + a_2 + \cdots = \sum_{i=1}^{\infty} a_i$ *converges*, and we write

$$\sum_{i=1}^{\infty} a_i = S;$$

that is,

$$\sum_{i=1}^{\infty} a_i \quad \text{is defined as} \quad \lim_{n\to\infty} \sum_{i=1}^{n} a_i$$

and is called the *sum* of the series.

If the series $\sum_{i=1}^{\infty} a_i$ does not converge, we say that it *diverges*. In this case, the series has no sum.

In summary:

$$\text{a series } \sum_{i=1}^{\infty} a_i \text{ converges if } \lim_{n\to\infty} \sum_{i=1}^{n} a_i \text{ exists (and is finite);}$$

$$\text{a series } \sum_{i=1}^{\infty} a_i \text{ diverges if } \lim_{n\to\infty} \sum_{i=1}^{n} a_i \text{ does not exist (or is infinite).}$$

Example 1 Write down the first four partial sums for each of the following series:

(a) $1 + \frac{1}{2} + \frac{1}{4} + \frac{1}{8} + \frac{1}{16} + \cdots;$

(b) $1 - \frac{1}{2} + \frac{1}{3} - \frac{1}{4} + \frac{1}{5} - \frac{1}{6} + \cdots;$

(c) $1 + \frac{1}{5} + \frac{1}{5^2} + \frac{1}{5^3} + \frac{1}{5^4} + \cdots;$

(d) $\sum_{i=0}^{\infty} \frac{3}{2^{i+1}}.$

Solution (a) $S_1 = 1$, $S_2 = 1 + 1/2 = 3/2$, $S_3 = 1 + \frac{1}{2} + \frac{1}{4} = \frac{7}{4}$, and $S_4 = 1 + \frac{1}{2} + \frac{1}{4} + \frac{1}{8} = \frac{15}{8}$.

(b) $S_1 = 1$, $S_2 = 1 - \frac{1}{2} = \frac{1}{2}$, $S_3 = 1 - \frac{1}{2} + \frac{1}{3} = \frac{5}{6}$, and $S_4 = 1 - \frac{1}{2} + \frac{1}{3} - \frac{1}{4} = \frac{7}{12}$.

(c) $S_1 = 1$, $S_2 = 1 + \frac{1}{5} = \frac{6}{5}$, $S_3 = 1 + \frac{1}{5} + \frac{1}{5^2} = \frac{31}{25}$, and $S_4 = 1 + \frac{1}{5} + \frac{1}{25} + \frac{1}{125} = \frac{156}{125}$.

(d) $\sum_{i=0}^{\infty} \frac{3}{2^{i+1}} = \frac{3}{2} + \frac{3}{2^2} + \frac{3}{2^3} + \frac{3}{2^4} + \cdots$, so $S_1 = \frac{3}{2}$, $S_2 = \frac{3}{2} + \frac{3}{4} = \frac{9}{4}$, $S_3 = 3(\frac{1}{2} + \frac{1}{4} + \frac{1}{8}) = \frac{21}{8}$, and $S_4 = 3(\frac{1}{2} + \frac{1}{4} + \frac{1}{8} + \frac{1}{16}) = \frac{45}{16}$. ▲

Do not confuse a *sequence* with a *series*. A sequence is simply an infinite list of numbers (separated by commas): $a_1, a_2, a_3, \ldots$. A series is an infinite list of numbers (separated by plus signs) which are meant to be added together: $a_1 + a_2 + a_3 + \cdots$. Of course, the terms in an infinite series may themselves be considered as a sequence, but the most important sequence associated with the series $a_1 + a_2 + \cdots$ is its sequence of partial sums: $S_1, S_2, S_3, \ldots$ —that is, the sequence

$$a_1, a_1 + a_2, a_1 + a_2 + a_3, \ldots .$$

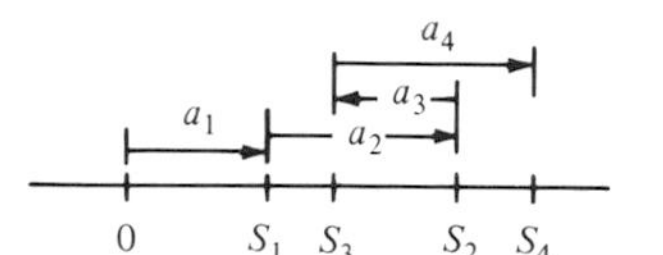

Figure 12.1.1. The term a_i of a series represents the "move" from the partial sum S_{i-1} to S_i. S_n is the cumulative result of the first n moves.

We may illustrate the difference between the a_i's and the S_n's pictorially. Think of $a_1, a_2, a_3, \ldots$ as describing a sequence of "moves" on the real number line, starting at 0. Then $S_n = a_1 + \cdots + a_n$ is the position reached after the nth move. (See Fig. 12.1.1.) Note that the term a_i can be recovered as the difference $S_i - S_{i-1}$.

To study the limits of partial sums, we will need to use some general properties of limits of sequences. The definition of convergence of a sequence was given in Section 11.4. The basic properties we need are proved and used in a manner similar to those for limits of functions (Section 11.1) and are summarized in the following display.

Properties of Limits of Sequences

Suppose that the sequences $a_1, a_2, \ldots$ and $b_1, b_2, \ldots$ are convergent, and that c is a constant. Then:

1. $\lim_{n\to\infty}(a_n + b_n) = \lim_{n\to\infty} a_n + \lim_{n\to\infty} b_n$.
2. $\lim_{n\to\infty}(ca_n) = c \lim_{n\to\infty} a_n$.
3. $\lim_{n\to\infty}(a_n b_n) = (\lim_{n\to\infty} a_n) \cdot (\lim_{n\to\infty} b_n)$.
4. If $\lim_{n\to\infty} b_n \neq 0$ and $b_n \neq 0$ for all n, then
$$\lim_{n\to\infty} \frac{a_n}{b_n} = \frac{\lim_{n\to\infty} a_n}{\lim_{n\to\infty} b_n}.$$
5. If f is continuous at $\lim_{n\to\infty} a_n$, then
$$\lim_{n\to\infty} f(a_n) = f\Big(\lim_{n\to\infty} a_n\Big).$$
6. $\lim_{n\to\infty} c = c$.
7. $\lim_{n\to\infty}(1/n) = 0$.
8. If $\lim_{x\to\infty} f(x) = l$, then $\lim_{n\to\infty} f(n) = l$.
9. If $|r| < 1$, then $\lim_{n\to\infty} r^n = 0$, and if $|r| > 1$ or $r = -1$, $\lim_{n\to\infty} r^n$ does not exist.

Here are a couple of examples of how the limit properties are used. We will see many more examples as we work with series.

Example 2 Find (a) $\lim_{n\to\infty} \frac{3+n}{2n+1}$ and (b) $\lim_{n\to\infty} \sin\Big(\frac{\pi n}{2n+1}\Big)$.

Solution (a) $$\lim_{n\to\infty} \frac{3+n}{2n+1} = \lim_{n\to\infty} \frac{3/n+1}{2+1/n} = \frac{3\lim_{n\to\infty} 1/n + \lim_{n\to\infty} 1}{\lim_{n\to\infty} 2 + \lim_{n\to\infty} 1/n} = \frac{3\cdot 0+1}{2+0} = \frac{1}{2}.$$

This solution used properties 1, 2, and 4 above, together with the facts that $\lim_{n\to\infty} 1/n = 0$ (property 7) and $\lim_{n\to\infty} c = c$ (property 6).
(b) Since $\sin x$ is a continuous function, we can use property 5 to get

$$\lim_{n\to\infty} \sin\left(\frac{\pi n}{2n+1}\right) = \sin\left[\lim_{n\to\infty}\left(\frac{\pi n}{2n+1}\right)\right]$$
$$= \sin\left[\lim_{n\to\infty}\left(\frac{\pi}{2+1/n}\right)\right]$$
$$= \sin\left(\frac{\pi}{2}\right) = 1. \blacktriangle$$

We return now to infinite series. A simple but basic example is the geometric series

$$a + ar + ar^2 + \cdots$$

in which the ratio between each two successive terms is the same. To write a geometric series in summation notation, it is convenient to allow the index i to start at zero, so that $a_0 = a$, $a_1 = ar$, $a_2 = ar^2$, and so on. The general term is then $a_i = ar^i$, and the series is compactly expressed as $\sum_{i=0}^{\infty} ar^i$. In our notation $\sum_{i=1}^{\infty} a_i$ for a general series, the index i will start at 1, but in special examples we may start it wherever we wish. Also, we may replace the index i by any other letter; $\sum_{i=1}^{\infty} a_i = \sum_{j=1}^{\infty} a_j = \sum_{n=1}^{\infty} a_n$, and so forth.

To find the sum of a geometric series, we must first evaluate the partial sums $S_n = \sum_{i=0}^{n} ar^i$. We write

$$S_n = a + ar + ar^2 + \cdots + ar^n,$$
$$rS_n = \quad ar + ar^2 + \cdots + ar^n + ar^{n+1}.$$

Subtracting the second equation from the first and solving for S_n, we find

$$S_n = \frac{a(1 - r^{n+1})}{1 - r} \qquad (\text{if } r \neq 1). \tag{1}$$

The sum of the entire series is the limit

$$\sum_{i=0}^{\infty} ar^i = \lim_{n\to\infty} S_n$$
$$= \lim_{n\to\infty} \frac{a(1 - r^{n+1})}{1 - r} = \frac{a}{1 - r} \lim_{n\to\infty} (1 - r^{n+1})$$
$$= \frac{a}{1 - r}\left(1 - \lim_{n\to\infty} r^{n+1}\right).$$

(We used limit properties 1, 2, and 6.) If $|r| < 1$, then $\lim_{n\to\infty} r^{n+1} = 0$ (property 9), so in this case, $\sum_{i=1}^{\infty} ar^i$ is convergent, and its sum is $a/(1 - r)$. If $|r| > 1$ or $r = -1$, $\lim_{n\to\infty} r^{n+1}$ does not exist (property 9), so if $a \neq 0$, the series diverges. Finally, if $r = 1$, then $S_n = a + ar + \cdots + ar^n = a(n + 1)$, so if $a \neq 0$, the series diverges.

Geometric Series

If $|r| < 1$ and a is any number, then $a + ar + ar^2 + \cdots = \sum_{i=0}^{\infty} ar^i$ converges and the sum is $a/(1 - r)$.

If $|r| \geqslant 1$ and $a \neq 0$, then $\sum_{i=0}^{\infty} ar^i$ diverges.

Example 3 Sum the series: (a) $1 + \frac{1}{3} + \frac{1}{9} + \frac{1}{27} + \frac{1}{81} + \cdots$, (b) $\sum_{n=0}^{\infty} \frac{1}{6^{n/2}}$, and (c) $\sum_{i=1}^{\infty} \frac{1}{5^i}$.

Solution (a) This is a geometric series with $r = \frac{1}{3}$ and $a = 1$. (Note that a is the first term and r is the ratio of any term to the preceding one.) Thus

$$1 + \frac{1}{3} + \frac{1}{9} + \cdots = \sum_{i=0}^{\infty}\left(\frac{1}{3}\right)^i = \frac{1}{1 - 1/3} = \frac{3}{2}.$$

(b) $\sum_{n=0}^{\infty}[1/(6^{n/2})] = 1 + (1/\sqrt{6}\,) + (1/\sqrt{6}\,)^2 + \cdots = a/(1-r)$, where $a = 1$ and $r = 1/\sqrt{6}$, so the sum is $1/(1 - 1/\sqrt{6}\,) = (6 + \sqrt{6}\,)/5$. (Note that the index here is n instead of i.)

(c) $\sum_{i=1}^{\infty} 1/5^i = 1/5 + 1/5^2 + \cdots = (1/5)/(1 - 1/5) = 1/4$. (We may also think of this as the series $\sum_{i=0}^{\infty} 1/5^i$ with the first term removed. The sum is thus $1/[1 - (1/5)] - 1 = 1/4$.) ▲

The following example shows how a geometric series may arise in a physical problem.

Example 4 A bouncing ball loses half of its energy on each bounce. The height reached on each bounce is proportional to the square root of the energy. Suppose that the ball is dropped vertically from a height of one meter. How far does it travel? (Fig. 12.1.2.)

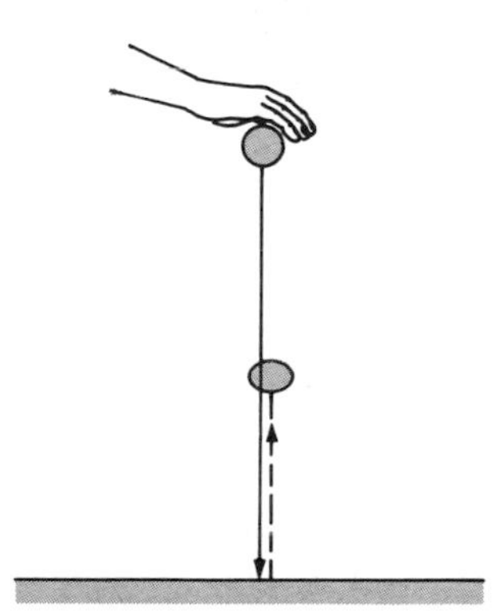

Figure 12.1.2. Find the total distance travelled by the bouncing ball.

Solution Each bounce is $1/\sqrt{2}$ as high as the previous one. After the ball falls from a height of 1 meter, it rises to $1/\sqrt{2}$ meters on the first bounce, $(1/\sqrt{2}\,)^2 = 1/2$ meter on the second, and so forth. The total distance travelled, in meters, is $1 + 2(1/\sqrt{2}\,) + 2(1/\sqrt{2}\,)^2 + 2(1/\sqrt{2}\,)^3 + \cdots$, which is

$$1 + \sum_{i=0}^{\infty} 2\left(\frac{1}{\sqrt{2}}\right)\cdot\left(\frac{1}{\sqrt{2}}\right)^i = 1 + \frac{\sqrt{2}}{1 - 1/\sqrt{2}} = 3 + 2\sqrt{2} = 5.828 \text{ meters. } ▲$$

Two useful general rules for summing series are presented in the box on the following page. To prove the validity of these rules, one simply notes that the identities

$$\sum_{i=1}^{n}(a_i + b_i) = \sum_{i=1}^{n} a_i + \sum_{i=1}^{n} b_i \quad \text{and} \quad \sum_{i=1}^{n} ca_i = c\sum_{i=1}^{n} a_i$$

are satisfied by the partial sums. Taking limits as $n \to \infty$ and applying the sum and constant multiple rules for limits of sequences results in the rules in the box.

Algebraic Rules For Series

Sum rule

If $\sum_{i=1}^{\infty} a_i$ and $\sum_{i=1}^{\infty} b_i$ converge, then $\sum_{i=1}^{\infty}(a_i + b_i)$ converges and

$$\sum_{i=1}^{\infty}(a_i + b_i) = \sum_{i=1}^{\infty} a_i + \sum_{i=1}^{\infty} b_i .$$

Constant multiple rule

If $\sum_{i=1}^{\infty} a_i$ converges and c is any real number, then $\sum_{i=1}^{\infty} ca_i$ converges and

$$\sum_{i=1}^{\infty} ca_i = c\sum_{i=1}^{\infty} a_i .$$

Example 5 Sum the series $\displaystyle\sum_{i=0}^{\infty} \frac{3^i - 2^i}{6^i}$.

Solution We may write the ith term as

$$\frac{3^i}{6^i} - \frac{2^i}{6^i} = \left(\frac{1}{2}\right)^i - \left(\frac{1}{3}\right)^i = \left(\frac{1}{2}\right)^i + (-1)\left(\frac{1}{3}\right)^i.$$

Since the series $\sum_{i=0}^{\infty}(1/2)^i$ and $\sum_{i=0}^{\infty}(1/3)^i$ are convergent, with sums 2 and $\frac{3}{2}$ respectively, the algebraic rules imply that

$$\sum_{i=0}^{\infty} \frac{3^i - 2^i}{6^i} = \sum_{i=0}^{\infty}\left[\left(\frac{1}{2}\right)^i + (-1)\left(\frac{1}{3}\right)^i\right]$$

$$= \sum_{i=0}^{\infty}\left(\frac{1}{2}\right)^i + (-1)\sum_{i=0}^{\infty}\left(\frac{1}{3}\right)^i = 2 - \frac{3}{2} = \frac{1}{2}. \ \blacktriangle$$

Example 6 Show that the series $1\frac{1}{2} + 3\frac{3}{4} + 7\frac{7}{8} + 15\frac{15}{16} + \cdots$ diverges. [*Hint*: Write it as the difference of a divergent and a convergent series.]

Solution The series is $\sum_{i=0}^{\infty}[2^i - (\frac{1}{2})^i]$. If it were convergent, we could add to it the convergent series $\sum_{i=0}^{\infty}(\frac{1}{2})^i$, and the result would have to converge by the sum rule; but the resulting series is $\sum_{i=0}^{\infty}[2^i - (\frac{1}{2})^i + (\frac{1}{2})^i] = \sum_{i=0}^{\infty}2^i$, which diverges because $2 > 1$, so the original series must itself be divergent. $\blacktriangle$

The sum rule implies that *we may change* (*or remove—that is, change to zero*) *finitely many terms of a series without affecting its convergence.* In fact, changing finitely many terms of the series $\sum_{i=1}^{\infty} a_i$ is equivalent to adding to it a series whose terms are all zero beyond a certain point. Such a finite series is always convergent, so adding it to the convergent series produces a convergent result. Of course, the sum of the new series is *not* the same as that of the old one, but rather is the sum of the finite number of added terms plus the sum of the original series.

Example 7 Show that

$$1 + 2 + 3 + 4 + \frac{1}{4} + \frac{1}{4^2} + \frac{1}{4^3} + \frac{1}{4^4} + \cdots$$

is convergent and find its sum.

Solution The series $1/4 + 1/4^2 + 1/4^3 + 1/4^4 + \cdots$ is a geometric series with sum $(1/4)/(1 - 1/4) = 1/3$; thus the given series is convergent with sum $1 + 2 +$

$3 + 4 + 1/3 = 10\frac{1}{3}$. To use the sum rule as stated, one can write

$$1 + 2 + 3 + 4 + \frac{1}{4} + \frac{1}{4^2} + \frac{1}{4^3} + \cdots$$
$$= (1 + 2 + 3 + 4 + 0 + 0 + 0 + \cdots)$$
$$+ \left(0 + 0 + 0 + 0 + \frac{1}{4} + \frac{1}{4^2} + \frac{1}{4^3} + \cdots\right). \blacktriangle$$

We can obtain a simple necessary condition for convergence by recalling that $a_i = S_i - S_{i-1}$. If $\lim_{i\to\infty} S_i$ exists, then $\lim_{i\to\infty} S_{i-1}$ has the same value. Hence, using properties 1 and 2 of limits of sequences, we find $\lim_{i\to\infty} a_i = \lim_{i\to\infty} S_i - \lim_{i\to\infty} S_{i-1} = 0$. In other words, if the series $\sum_{i=1}^{\infty} a_i$ converges, then, the "move" from one partial sum to the next must approach zero (see Fig. 12.1.1).

The *i*th Term Test

If $\sum_{i=1}^{\infty} a_i$ converges, then $\lim_{i\to\infty} a_i = 0$.
If $\lim_{i\to\infty} a_i \neq 0$, then $\sum_{i=1}^{\infty} a_i$ diverges.
If $\lim_{i\to\infty} a_i = 0$, the test is inconclusive: the series could converge or diverge, and further analysis is necessary.

The ith-term test can be used to show that a series diverges, such as the one in Example 6, but it cannot be used to establish convergence.

Example 8 Test for convergence: (a) $\sum_{i=1}^{\infty} \frac{i}{1+i}$; (b) $\sum_{i=1}^{\infty} (-1)^i \frac{i}{\sqrt{1+i}}$; (c) $\sum_{i=1}^{\infty} \left(\frac{1}{i}\right)$.

Solution (a) Here $a_i = \frac{i}{1+i} = \frac{1}{1/i+1} \to 1$ as $i \to \infty$. Since a_i does not tend to zero, the series must diverge.

(b) Here $|a_i| = i/\sqrt{1+i} = \sqrt{i}/\sqrt{1/i+1} \to \infty$ as $i \to \infty$. Thus a_i does not tend to zero, so the series diverges.

(c) Here $a_i = 1/i$, which tends to zero as $i \to \infty$, so our test is inconclusive. ▲

As an example of the "further analysis" necessary when $\lim_{i\to\infty} a_i = 0$, we consider the series

$$1 + \frac{1}{2} + \frac{1}{3} + \frac{1}{4} + \cdots = \sum_{i=1}^{\infty} \frac{1}{i}$$

from part (c) of Example 8, called the *harmonic series*. We show that the series diverges by noticing a pattern:

$$\begin{array}{ll}
1 & \\
\frac{1}{2} & \\
\frac{1}{3} + \frac{1}{4} & > \frac{1}{4} + \frac{1}{4} = \frac{1}{2} \\
\frac{1}{5} + \frac{1}{6} + \frac{1}{7} + \frac{1}{8} & > \frac{1}{8} + \frac{1}{8} + \frac{1}{8} + \frac{1}{8} = \frac{1}{2} \\
\frac{1}{9} + \cdots + \frac{1}{16} & > \frac{1}{16} + \cdots + \frac{1}{16} = \frac{1}{2} \\
\frac{1}{17} + \cdots + \frac{1}{32} & > \frac{1}{32} + \cdots + \frac{1}{32} = \frac{1}{2} \\
\vdots & \quad \vdots \qquad \vdots \qquad \vdots
\end{array}$$

and so on. Thus the partial sum S_4 is greater than $1 + \frac{1}{2} + \frac{1}{2} = 1 + \frac{2}{2}$, $S_8 > 1 + \frac{1}{2} + \frac{1}{2} + \frac{1}{2} = 1 + \frac{3}{2}$ and, in general $S_{2^n} > 1 + n/2$, which becomes arbitrarily large as n becomes large. Therefore, *the harmonic series diverges*.

Example 9 Show that the series (a) $\frac{1}{2}+\frac{1}{4}+\frac{1}{6}+\frac{1}{8}+\cdots$ and (b) $\sum_{i=1}^{\infty} 1/(1+i)$ diverge.

Solution (a) This series is $\sum_{i=1}^{\infty}(1/2i)$. If it converged, so would twice the series $\sum_{i=1}^{\infty} 2\cdot(1/2i)$, by the constant multiple rule; but $\sum_{i=1}^{\infty}(2\cdot 1/2i)=\sum_{i=1}^{\infty} 1/i$, which we have shown to diverge.

(b) This series is $\frac{1}{2}+\frac{1}{3}+\frac{1}{4}+\cdots$, which is the harmonic series with the first term missing; therefore this series diverges too. ▲

Supplement to Section 12.1: Zeno's Paradox

Zeno's paradox concerns a race between Achilles and a tortoise. The tortoise begins with a head start of 10 meters, and Achilles ought to overtake it. After a certain elapsed time from the start, Achilles reaches the point A where the tortoise started, but the tortoise has moved ahead to point B (Fig. 12.1.3).

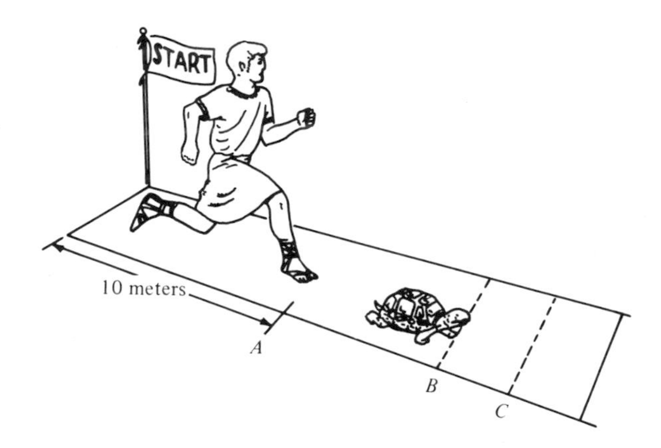

Figure 12.1.3. Will the runner overtake the tortoise?

After a certain further interval of time, Achilles reaches point B, but the tortoise has moved ahead to a point C, and so on forever. Zeno concludes from this argument that Achilles can never pass the tortoise. Where is the fallacy?

The resolution of the paradox is that although the number of time intervals being considered is infinite, the sum of their lengths is finite, so Achilles can overtake the tortoise in a finite time. The word *forever* in the sense of infinitely many terms is confused with "forever" in the sense of the time in the problem, resulting in the apparent paradox.

Exercises for Section 12.1

Write down the first four partial sums for the series in Exercises 1–4.

1. $\frac{1}{2}+\frac{1}{3}+\frac{1}{4}+\frac{1}{5}+\cdots$
2. $1-\frac{1}{2}+\frac{1}{4}-\frac{1}{8}+\frac{1}{16}-\cdots$
3. $\displaystyle\sum_{i=1}^{\infty}\left(\frac{2}{3}\right)^{i}$
4. $\displaystyle\sum_{n=1}^{\infty}\frac{n}{3^{n}}$

Sum the series in Exercises 5–8.

5. $1+\dfrac{1}{7}+\dfrac{1}{7^2}+\dfrac{1}{7^3}+\cdots$
6. $2+\dfrac{2}{9}+\dfrac{2}{9^2}+\dfrac{2}{9^3}+\cdots$
7. $\displaystyle\sum_{i=1}^{\infty}\left(\frac{7}{8}\right)^{i}$
8. $\displaystyle\sum_{n=1}^{\infty}\left(\frac{13}{15}\right)^{n}$

9. You wish to draw \$10,000 out of a Swiss bank account at age 65, and thereafter you want to draw $\frac{3}{4}$ as much each year as the preceding one. Assuming that the account earns no interest, how much money must you start with to be prepared for an arbitrarily large life span?

10. A decaying radioactive source emits $\frac{9}{10}$ as much radiation each year as the previous one. Assuming that 2000 roentgens are given off in the first year, what is the total emission over all time?

Sum the series (if they converge) in Exercises 11–20.

11. $\displaystyle\sum_{j=1}^{\infty} \frac{1}{13^j}$

12. $\displaystyle\sum_{k=1}^{\infty} \left(\frac{4}{5}\right)^k$

13. $\displaystyle\sum_{i=0}^{\infty} \frac{2^{3i+4}}{3^{2i+5}}$

14. $\displaystyle\sum_{l=0}^{\infty} \frac{4^{4l+2}}{5^{3l+80}}$

15. $\displaystyle\sum_{j=-3}^{\infty} \left(\frac{1}{3}\right)^j$

16. $\displaystyle\sum_{i=4}^{\infty} 5\left(\frac{1}{3}\right)^{i+1/2}$

17. $\displaystyle\sum_{n=1}^{\infty} \frac{2^n+3^n}{6^n}$

18. $\displaystyle\sum_{k=1}^{\infty} \frac{3^{2k}+1}{27^k}$

19. $\displaystyle\sum_{n=5}^{\infty} \frac{2^{n+1}}{3^{n-2}}$

20. $\displaystyle\sum_{i=1}^{\infty} \left[\left(\frac{1}{2}\right)^i + \left(\frac{1}{3}\right)^{2i} + \left(\frac{1}{4}\right)^{3i+1}\right]$

21. Show that $\sum_{i=1}^{\infty}(1 + 1/2^i)$ diverges.

22. Show that $\sum_{i=0}^{\infty}(3^i + 1/3^i)$ diverges.

23. Sum $2 + 4 + \frac{1}{2} + \frac{1}{4} + \frac{1}{8} + \cdots$.

24. Sum $1 + 1/2 + 1/3 + 1/3^2 + 1/3^3 + \cdots$.

Test the series in Exercises 25–30 for convergence.

25. $\displaystyle\sum_{i=1}^{\infty} \frac{i}{\sqrt{i+1}}$

26. $\displaystyle\sum_{i=1}^{\infty} \frac{\sqrt{i}+1}{\sqrt{i}+8}$

27. $\displaystyle\sum_{i=1}^{\infty} \frac{3}{5+5i}$

28. $\displaystyle\sum_{i=1}^{\infty} \frac{6}{7+7i}$

29. $1 + \underbrace{\tfrac{1}{2} + \tfrac{1}{4}}_{2} + \underbrace{\tfrac{1}{4} + \tfrac{1}{8} + \tfrac{1}{8} + \tfrac{1}{8}}_{4} + \underbrace{\tfrac{1}{8} + \tfrac{1}{16} + \cdots}_{8}$

30. $1 + \underbrace{\tfrac{1}{4} + \tfrac{1}{4}}_{2} + \underbrace{\tfrac{1}{16} + \tfrac{1}{16} + \tfrac{1}{16} + \tfrac{1}{16}}_{4} + \underbrace{\tfrac{1}{64} + \tfrac{1}{64} + \cdots}_{8}$

31. Show that the series $\sum_{j=1}^{\infty}(1 - 2^{-j})/j$ diverges.

32. Show that the series $\frac{1}{3} + \frac{1}{5} + \frac{1}{7} + \frac{1}{9} + \cdots$ diverges.

33. Give an example to show that $\sum_{i=1}^{\infty}(a_i + b_i)$ may converge while both $\sum_{i=1}^{\infty} a_i$ and $\sum_{i=1}^{\infty} b_i$ diverge.

34. Comment on the formula $1 + 2 + 4 + 8 + \cdots = 1/(1-2) = -1$.

35. A *telescoping series*, like a geometric series, can be summed. A series $\sum_{n=1}^{\infty} a_n$ is telescoping if its nth term a_n can be expressed as $a_n = b_{n+1} - b_n$ for some sequence b_n.
   (a) Verify that $a_1 + a_2 + a_3 + \cdots + a_n = b_{n+1} - b_1$; therefore the series converges exactly when $\lim_{n\to\infty} b_{n+1}$ exists, and $\sum_{n=1}^{\infty} a_n = \lim_{n\to\infty} b_{n+1} - b_1$.
   (b) Use partial fraction methods to write $a_n = 1/[n(n+1)]$ as $b_{n+1} - b_n$ for some sequence b_n. Then evaluate the sum of the series $\sum_{n=1}^{\infty} 1/[n(n+1)]$.

36. An experiment is performed, during which time successive excursions of a deflected plate are recorded. Initially, the plate has amplitude b_0. The plate then deflects downward to form a "dish" of depth b_1, then a "dome" of height b_2, and so on. (See Fig. 12.1.4.) The a's and b's are related by $a_1 = b_0 - b_1, a_2 = b_1 - b_2, a_3 = b_2 - b_3 \ldots$. The value a_n measures the amplitude "lost" at the nth oscillation (due to friction, say).

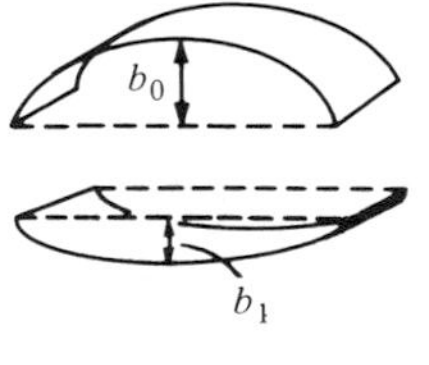

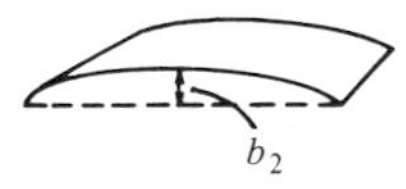

Figure 12.1.4. The deflecting plate in Exercise 36.

   (a) Find $\sum_{n=1}^{\infty} a_n$. Explain why $b_0 - \sum_{n=1}^{\infty} a_n$ is the "average height" of the oscillating plate after a large number of oscillations.
   (b) Suppose the "dishes" and "domes" decay to zero, that is, $\lim_{n\to\infty} b_{n+1} = 0$. Show that $\sum_{n=1}^{\infty} a_n = b_0$, and explain why this is physically obvious.

37. The joining of the transcontinental railroads occurred as follows. The East and West crews were setting track 12 miles apart, the East crew working at 5 miles per hour, the West crew working at 7 miles per hour. The official with the Golden Spike travelled feverishly by carriage back and forth between the crews until the rails joined. His speed was 20 miles per hour, and he started from the East.
   (a) Assume the carriage reversed direction with no waiting time at each encounter with an East or West crew. Let t_k be the carriage transit time for trip k. Verify that $t_{2n+2} = r^{n+1} \cdot (12/13)$, and $t_{2n+1} = r^n \cdot (12/27)$, where $r = (13/27) \cdot (15/25)$, $n = 0, 1, 2, 3, \ldots$.
   (b) Since the crews met in one hour, the total time for the carriage travel was one hour, i.e., $\lim_{n\to\infty}(t_1 + t_2 + t_3 + t_4 + \cdots + t_n) = 1$. Verify this formula using a geometric series.

12.2 The Comparison Test and Alternating Series

A series with positive terms converges if its terms approach zero quickly enough.

Most series, unlike the geometric series, cannot be summed explicitly. If we can prove that a given series converges, we can approximate its sum to any desired accuracy by adding up enough terms.

One way to tell whether a series converges or diverges is to compare it with a series which we already know to converge or diverge. As a fringe benefit of such a "comparison test," we sometimes get an estimate of the difference between the nth partial sum and the exact sum. Thus if we want to find the sum with a given accuracy, we know how many terms to take.

The comparison test for series is similar to that for integrals (Section 11.3). The test is simplest to understand for series with non-negative terms. Suppose that we are given series $\sum_{i=1}^{\infty} a_i$ and $\sum_{i=1}^{\infty} b_i$ such that $0 \leqslant a_i \leqslant b_i$ for all i:

$$\textit{if } \sum_{i=1}^{\infty} b_i \textit{ converges, then so does } \sum_{i=1}^{\infty} a_i .$$

The reason for this is easy to see on an intuitive level. The partial sums $S_n = \sum_{i=1}^{n} a_n$ are moving to the right on the real number line since $a_i \geqslant 0$. They must either march off to ∞ or approach a limit. (The proof of this sentence requires a careful study of the real numbers, but we will take it for granted here. Consult the Supplement to this section and the theoretical references listed in the Preface.) However, $\sum_{i=1}^{n} a_i \leqslant \sum_{i=1}^{n} b_i \leqslant \sum_{i=1}^{\infty} b_i$, since $a_i \leqslant b_i$ and the partial sums $\sum_{i=1}^{n} b_i$ are marching to the right toward their limit. Hence all the S_n's are bounded by the fixed number $\sum_{i=1}^{\infty} b_i$, and so they cannot go to ∞.

Example 1 Show that $\displaystyle\sum_{i=1}^{\infty} \frac{3}{2^i + 4}$ converges.

Solution We know that $\sum_{i=1}^{\infty} (3/2^i)$ is convergent since it is a geometric series with $a = 3$ and $r = \frac{1}{2} < 1$; but

$$0 < \frac{3}{2^i + 4} < \frac{3}{2^i},$$

so the given series converges by the comparison test. ▲

For series $\sum_{i=1}^{\infty} a_i$ with terms that can be either positive or negative, we replace the condition $0 \leqslant a_i \leqslant b_i$ by $|a_i| \leqslant b_i$. Then if $\sum_{i=1}^{\infty} b_i$ converges, so must $\sum_{i=1}^{\infty} |a_i|$, by the test above. The following fact is true for any series:

$$\textit{if } \sum_{i=1}^{\infty} |a_i| \textit{ converges, so does } \sum_{i=1}^{\infty} a_i, \textit{ and } \left| \sum_{i=1}^{\infty} a_i \right| \leqslant \sum_{i=1}^{\infty} |a_i|.$$

A careful proof of this fact is given at the end of this section; for now we simply observe that the convergence of $\sum_{i=1}^{\infty} |a_i|$ implies that the absolute values $|a_i|$ approach zero quickly, and the possibility of varying signs in the a_i's can only help in convergence. Therefore, if $0 \leqslant |a_i| \leqslant b_i$ and $\sum b_i$ converges, then $\sum |a_i|$ converges, and therefore so does $\sum a_i$. (We sometimes drop the "$i = 1$" and "∞" from $\sum$ if there is no danger of confusion.) This leads to the following test.

Comparison Test

Let $\sum_{i=1}^{\infty} a_i$ and $\sum_{i=1}^{\infty} b_i$ be series such that $|a_i| \leqslant b_i$. If $\sum_{i=1}^{\infty} b_i$ is convergent, then so is $\sum_{i=1}^{\infty} a_i$.

Example 2 Show that $\displaystyle\sum_{i=1}^{\infty} \frac{(-1)^i}{i3^{i+1}}$ converges.

Solution We can compare the series with $\sum_{i=1}^{\infty} 1/3^i$. Let $a_i = (-1)^i/(i3^{i+1})$ and $b_i = 1/3^i$. Since $i3^{i+1} = (3i) \cdot 3^i > 3^i$, we have

$$|a_i| = \frac{1}{i3^{i+1}} < \frac{1}{3^i} = b_i .$$

Therefore, since $\sum_{i=1}^{\infty} b_i$ converges (it is a geometric series), so does $\sum_{i=1}^{\infty} a_i$. ▲

If the terms of two series $\sum a_i$ and $\sum b_i$ "resemble" one another, we may expect that one of the series converges if the other does. This is the case when the ratio a_i/b_i approaches a limit, as can be deduced from the comparison test. For instance, suppose that $\lim_{i\to\infty}(|a_i|/b_i) = M < \infty$, with all $b_i > 0$. Then for large enough i, we have $|a_i|/b_i < M + 1$, or $|a_i| < (M+1)b_i$. Now if $\sum b_i$ converges, so does $\sum (M+1)b_i$, by the constant multiple rule for series, and hence $\sum a_i$ converges by the comparison test.[1]

Example 3 Test for convergence: $\displaystyle\sum_{i=1}^{\infty} \frac{1}{2^i - i}$.

Solution We cannot compare directly with $\sum_{i=1}^{\infty} 1/2^i$, since $1/(2^i - i)$ is *greater* than $1/2^i$. Instead, we look at the ratios a_i/b_i with $a_i = 1/(2^i - i)$ and $b_i = 1/2^i$. We have

$$\lim_{i\to\infty} \frac{a_i}{b_i} = \lim_{i\to\infty} \frac{1}{1 - i/2^i} = \frac{1}{1-0} = 1$$

($\lim_{i\to\infty}(i/2^i) = 0$ by l'Hôpital's rule). Since $\sum_{i=1}^{\infty}(1/2^i)$ converges, so does $\sum_{i=1}^{\infty}[1/(2^i - i)]$. ▲

The following tests can both be similarly justified using the original comparison test.

Ratio Comparison Tests

Let $\sum_{i=1}^{\infty} a_i$ and $\sum_{i=1}^{\infty} b_i$ be series, with $b_i > 0$ for all i.

If (1) $|a_i| \leqslant b_i$ for all i, or if $\lim_{i\to\infty}(|a_i|/b_i) < \infty$ and
(2) $\sum_{i=1}^{\infty} b_i$ is convergent, then $\sum_{i=1}^{\infty} a_i$ is convergent.

If (1) $a_i \geqslant b_i$ for all i, or if $\lim_{i\to\infty}(a_i/b_i) > 0$ and
(2) $\sum_{i=1}^{\infty} b_i$ is divergent, then $\sum_{i=1}^{\infty} a_i$ is divergent.

To choose b_i in applying the ratio comparison test, you should look for the "dominant terms" in the expression for a_i.

[1] Strictly speaking, to apply the comparison test we should have $|a_i| < (M+1)b_i$ for all i, not just sufficiently large i; but, as we saw earlier, the convergence or divergence of a series $\sum a_i$ is not affected by the values of its "early" terms, but only the behavior of a_i for large i. Of course, the *sum* of the series depends on all the terms.

Example 4 Show that $\sum_{i=1}^{\infty} \frac{2}{4+i}$ diverges.

Solution As $i \to \infty$, the dominant term in the denominator $4+i$ is i, that is, if i is very large (like 10^6), 4 is very small by comparison. Hence we are led to let $a_i = 2/(4+i)$, $b_i = 1/i$. Then

$$\lim_{i\to\infty} \frac{a_i}{b_i} = \lim_{i\to\infty} \frac{2/(4+i)}{1/i} = \lim_{i\to\infty} \frac{2i}{4+i} = \lim_{i\to\infty} \frac{2}{(4/i)+1} = \frac{2}{0+1} = 2.$$

Since $2>0$, and $\sum_{i=1}^{\infty} 1/i$ is divergent, it follows that $\sum_{i=1}^{\infty}[2/(4+i)]$ is divergent as well. ▲

The next example illustrates how one may estimate the difference between a partial sum and the full series. We sometimes refer to this difference as a *tail* of the series; it is equal to the sum of all the terms not included in the partial sum.

Example 5 Find the partial sum $\sum_{i=1}^{3} \frac{(-1)^i}{i3^{i+1}}$ (see Example 2) and estimate the difference between this partial sum and the sum of the entire series.

Solution The sum of the first three terms is

$$-\frac{1}{3^2\cdot 1} + \frac{1}{3^3\cdot 2} - \frac{1}{3^4\cdot 3} = -\frac{1}{9} + \frac{1}{54} - \frac{1}{243} = -\frac{47}{486} \approx -0.0967.$$

The difference between the full sum of a series and the nth partial sum is given by $\sum_{i=1}^{\infty} a_i - \sum_{i=1}^{n} a_i = \sum_{i=n+1}^{\infty} a_i$. To estimate this tail in our example, we write

$$\begin{aligned}
\left|\sum_{i=1}^{\infty} \frac{(-1)^i}{3^{i+1}i} - (-0.0967)\right| &= \left|\sum_{i=4}^{\infty} \frac{(-1)^i}{3^{i+1}i}\right| \\
&\leqslant \sum_{i=4}^{\infty} \frac{1}{3^{i+1}i} \qquad \left(\text{since } \left|\sum a_i\right| \leqslant \sum |a_i|\right) \\
&\leqslant \sum_{i=4}^{\infty} \frac{1}{3^{i+1}} \qquad (\text{since } i>1) \\
&= \frac{1}{3^5}\left(1 + \frac{1}{3} + \frac{1}{3^2} + \cdots\right) \\
&= \frac{1}{3^5}\left(\frac{1}{1-1/3}\right) = \frac{1}{3^5}\cdot\frac{3}{2} = \frac{1}{162} \cong 0.0062.
\end{aligned}$$

Thus the error is no more than 0.0062. We may therefore conclude that $\sum_{i=1}^{\infty}[(-1)^i/(3^{i+1}i)]$ lies in the interval $[-0.0967-0.0062, -0.0967+0.0062] = [-0.103, -0.090]$. ▲

The second kind of series which we will treat in this section is called an *alternating series*. To illustrate, recall that we saw in Section 12.1 that the harmonic series

$$1 + \tfrac{1}{2} + \tfrac{1}{3} + \tfrac{1}{4} + \cdots$$

is divergent even though $\lim_{i\to\infty}(1/i) = 0$. If we put a minus sign in front of every other term to obtain the series

$$1 - \tfrac{1}{2} + \tfrac{1}{3} - \tfrac{1}{4} + \cdots,$$

we might hope that the alternating positive and negative terms "neutralize" one another and cause the series to converge. The alternating series test will indeed guarantee convergence. First we need the following definition.

Alternating Series

A series $\sum_{i=1}^{\infty} a_i$ is called *alternating* if the terms a_i are alternately positive and negative and if the absolute values $|a_i|$ are decreasing to zero; that is, if:

1. $a_1 > 0$, $a_2 < 0$, $a_3 > 0$, $a_4 < 0$, and so on (or $a_1 < 0$, $a_2 > 0, \ldots$);
2. $|a_1| \geqslant |a_2| \geqslant |a_3| \geqslant \cdots$;
3. $\lim_{i\to\infty}|a_i| = 0$.

Conditions 1, 2, and 3 are often easy to verify.

Example 6 Is the series $1 - \frac{1}{2} + \frac{1}{3} - \frac{1}{4} \cdots$ alternating?

Solution The terms alternate in sign, $+ - + - \cdots$, so condition 1 holds. Since the ith term $a_i = (-1)^{i+1}(1/i)$ has absolute value $1/i$, and $1/i > 1/(i+1)$, the terms are decreasing in absolute value, so condition 2 holds. Finally, since $\lim_{i\to\infty}|a_i| = \lim_{i\to\infty}(1/i) = 0$, condition 3 holds. Thus the series is alternating. ▲

Later in this section, we will prove that *every alternating series converges*. The proof is based on the idea that the partial sums $S_n = \sum_{i=1}^{n} a_i$ oscillate back and forth and get closer and closer together, so that they must close in on a limiting value S. This argument also shows that the sum S lies between any two successive partial sums, so that the tail corresponding to the partial sum S_n is less than $|a_{n+1}|$, the size of the first omitted term. (See Fig. 12.2.1.)

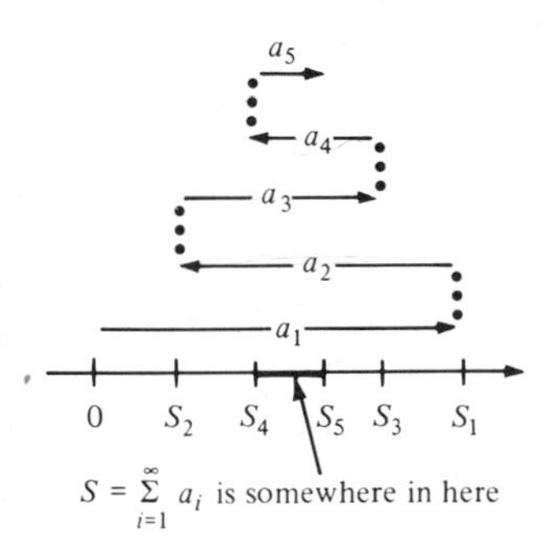

Figure 12.2.1. An alternating series converges, no matter how slowly the terms approach zero. The sum lies between each two successive partial sums.

Alternating Series Test

1. If $\sum_{i=1}^{\infty} a_i$ is a series such that the a_i alternate in sign, are decreasing in absolute value, and tend to zero, then it converges.
2. The error made in approximating the sum by $S_n = \sum_{i=1}^{n} a_i$ is not greater than $|a_{n+1}|$.

Example 7 Show that the series $1 - \frac{1}{2} + \frac{1}{3} - \frac{1}{4} + \frac{1}{5} - \cdots$ converges, and find its sum with an error of no more than 0.04.

Solution By Example 6, the series is alternating; therefore, by the alternating series test, it converges. To make the error at most $0.04 = \frac{1}{25}$, we must add up all the terms through $\frac{1}{24}$. Using a calculator, we find

$$S_{24} = 1 - \tfrac{1}{2} + \tfrac{1}{3} - \cdots - \tfrac{1}{24} \approx 0.6727.$$

(Since the sum lies between S_{24} and $S_{25} \approx 0.7127$, an even better estimate is the midpoint $\frac{1}{2}(S_{24} + S_{25}) = 0.6927$, which can differ from the sum by at most 0.02.) ▲

Example 8 Test for convergence:

(a) $\sum_{i=1}^{\infty} \frac{(-1)^i}{(1+i)^2}$;

(b) $\frac{2}{2} - \frac{1}{2} + \frac{2}{3} - \frac{1}{3} + \frac{2}{4} - \frac{1}{4} + \frac{2}{5} - \frac{1}{5} + \frac{2}{6} - \frac{1}{6} + \cdots$.

Solution (a) The terms alternate in sign since $(-1)^i = 1$ if i is even and $(-1)^i = -1$ if i is odd. The absolute values, $1/(1+i)^2$, are decreasing and converge to zero. Thus the series is alternating, so it converges.

(b) The terms alternate in sign and tend to zero, but the absolute values are not monotonically decreasing. Thus the series is not an alternating one and the alternating series test does not apply. If we group the terms by twos, we find that the series becomes

$$\left(\tfrac{2}{2} - \tfrac{1}{2}\right) + \left(\tfrac{2}{3} - \tfrac{1}{3}\right) + \left(\tfrac{2}{4} - \tfrac{1}{4}\right) + \left(\tfrac{2}{5} - \tfrac{1}{5}\right) + \cdots = \tfrac{1}{2} + \tfrac{1}{3} + \tfrac{1}{4} + \tfrac{1}{5} + \cdots$$

which diverges. (Notice that the nth partial sum of the "grouped" series is the $2n$th partial sum of the original series.) ▲

We noted early in this section that a series $\sum_{i=1}^{\infty} a_i$ always converges if its terms go to zero quickly enough so that the series $\sum_{i=1}^{\infty} |a_i|$ of absolute values is convergent. Such a series $\sum_{i=1}^{\infty} a_i$ is said to be *absolutely convergent*. On the other hand, a series like $1 - \frac{1}{2} + \frac{1}{3} - \frac{1}{4} + \cdots$, is convergent only due to the alternating signs of its terms; the series of absolute values, $1 + \frac{1}{2} + \frac{1}{3} + \cdots$, is divergent (it is the harmonic series). When $\sum_{i=1}^{\infty} a_i$ converges but $\sum_{i=1}^{\infty} |a_i|$ diverges, the series $\sum_{i=1}^{\infty} a_i$ is said to be *conditionally convergent*.

Example 9 Discuss the convergence of the series $\sum_{i=1}^{\infty} \frac{(-1)^i \sqrt{i}}{i+4}$.

Solution Let $a_i = (-1)^i \sqrt{i}/(i+4)$. We notice that for i large, $|a_i|$ appears to behave like $b_i = 1/\sqrt{i}$. The series $\sum_{i=1}^{\infty} b_i$ diverges by comparison with the harmonic series. To make the comparison between $|a_i|$ and b_i precise, look at the ratios: $\lim_{i\to\infty}(|a_i|/b_i) = \lim_{i\to\infty}[i/(i+4)] = 1$, so $\sum_{i=1}^{\infty} |a_i|$ diverges as well; hence our series is not *absolutely* convergent.

The series does look like it could be alternating: the terms alternate in sign and $\lim_{i\to\infty} a_i = 0$. To see whether the absolute values $|a_i|$ form a decreasing sequence, it is convenient to look at the function $f(x) = \sqrt{x}/(x+4)$. The derivative is

$$f'(x) = \frac{(1/2\sqrt{x})(x+4) - \sqrt{x}\cdot 1}{(x+4)^2} = \frac{2/\sqrt{x} - \sqrt{x}/2}{(x+4)^2} = \frac{4-x}{2\sqrt{x}\,(x+4)^2},$$

which is negative for $x > 4$, so $f(x)$ is decreasing for $x > 4$. Since $|a_i| = f(i)$, we have $|a_4| > |a_5| > |a_6| > \cdots$ which implies that our series $\sum a_i$, with its first three terms omitted, is alternating. It follows that the series is convergent; since it is not absolutely convergent, it is conditionally convergent. ▲

Absolute and Conditional Convergence

A series $\sum_{i=1}^{\infty} a_i$ is called *absolutely convergent* if $\sum_{i=1}^{\infty} |a_i|$ is convergent. Every absolutely convergent series converges.

A series may converge without being absolutely convergent; such a series is called *conditionally convergent*.

Supplement to Section 12.2: A Discussion of the Proofs of the Comparison and Alternating Series Tests

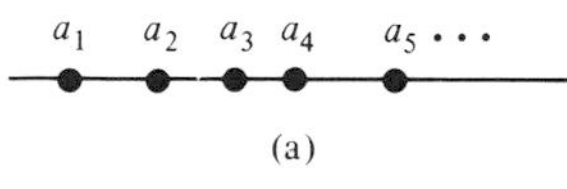

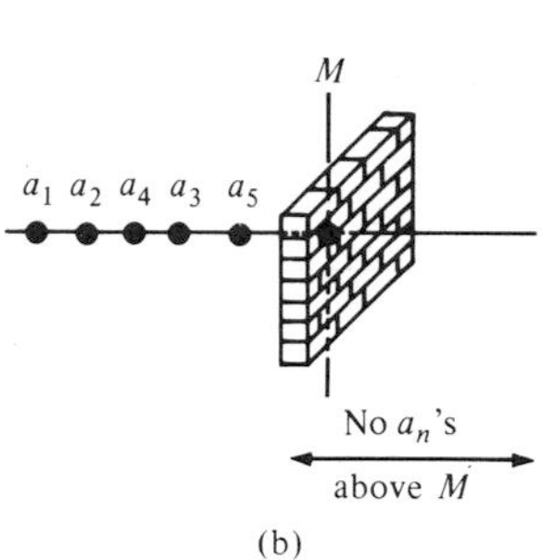

Figure 12.2.2. (a) An increasing sequence; (b) a sequence bounded above by M.

The key convergence property we need involves increasing sequences. It is similar to the existence of $\lim_{x\to\infty} f(x)$ if f is increasing and bounded above, which we used in Section 11.3 to establish the comparison test for integrals.

A sequence $a_1, a_2, \dots$ of real numbers is called *increasing* in case $a_1 \leqslant a_2 \leqslant \cdots$. The sequence is said to be *bounded above* if there is a number M such that $a_n \leqslant M$ for all n. (See Fig. 12.2.2.)

For example, let $a_n = n/(n+1)$. Let us show that a_n is increasing and is bounded above by M if M is any number $\geqslant 1$. To prove that it is increasing, we must show that $a_n \leqslant a_{n+1}$—that is, that

$$\frac{n}{n+1} \leqslant \frac{n+1}{(n+1)+1} \quad \text{or} \quad n(n+2) \leqslant (n+1)$$

or

$$n^2 + 2n \leqslant n^2 + 2n + 1 \quad \text{or} \quad 0 \leqslant 1.$$

Reversing the steps gives a proof that $a_n \leqslant a_{n+1}$; i.e., the sequence is increasing. Since $n < n+1$, we have $a_n = n/(n+1) < 1$, so $a_n < M$ if $M \geqslant 1$.

We will accept without proof the following property of the real numbers (see the references listed in the Preface).

Increasing Sequence Property

If a_n is an increasing sequence which is bounded above, then a_n converges to some number a as $n \to \infty$. (Similarly, a decreasing sequence bounded below converges.)

The increasing sequence property expresses a simple idea: if the sequence is increasing, the numbers a_n increase, but they can never exceed M. What else could they do but converge? Of course, the limit a satisfies $a_n \leqslant a$ for all n.

For example, consider

$$a_1 = 0.3, \qquad a_2 = 0.33, \qquad a_3 = 0.333$$

and so forth. These a_n's are increasing (in fact, strictly increasing) and are bounded above by 0.4, so we know that they must converge. In fact, the increasing sequence property shows that any infinite decimal expansion converges and so represents a real number.

To prove the comparison test for series with positive terms, we apply the increasing sequence property to the sequence of partial sums. If $\sum_{i=1}^{\infty} a_i$ is a series with $a_i \geqslant 0$ for each i, then since the partial sums S_n satisfy $S_n - S_{n-1} = a_n \geqslant 0$, they must be an increasing sequence (see Fig. 12.2.3). If the partial sums are bounded above, the sequence must have a limit, and so the series must converge.

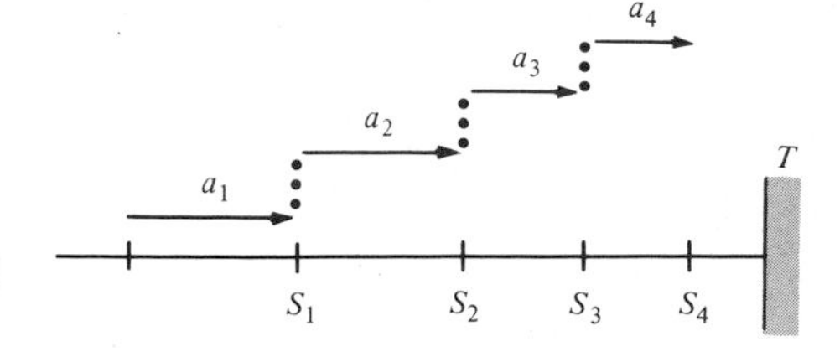

Figure 12.2.3. The partial sums of the series $\sum_{i=1}^{\infty} a_i$ are increasing and bounded above by T.

Now we may simply repeat the argument presented earlier in this section. If $0 \leqslant a_i \leqslant b_i$ for all i, and $T_n = \sum_{i=1}^{n} b_n$, then $S_n \leqslant T_n$. If the partial sums T_n approach a limit T, then they are bounded above by T, and so $S_n \leqslant T$ for all n. Thus $\lim_{n\to\infty} S_n$ exists and is less than or equal to T, i.e., $\sum_{i=1}^{\infty} a_n \leqslant \sum_{i=1}^{\infty} b_n$.

To complete the proof of the general comparison test, we must show that whenever $\sum_{i=1}^{\infty}|a_i|$ converges, so does $\sum_{i=1}^{\infty} a_i$; in other words, *every absolutely convergent series converges*. Suppose, then, that $\sum |a_i|$ converges.

We define two new series, $\sum_{i=1}^{\infty} b_i$ and $\sum_{i=1}^{\infty} c_i$, by the formulas

$$b_i = \begin{Bmatrix} |a_i| & \text{if } a_i \geqslant 0 \\ 0 & \text{if } a_i < 0 \end{Bmatrix} = \begin{Bmatrix} a_i & \text{if } a_i \geqslant 0 \\ 0 & \text{if } a_i < 0 \end{Bmatrix},$$

$$c_i = \begin{Bmatrix} |a_i| & \text{if } a_i \leqslant 0 \\ 0 & \text{if } a_i > 0 \end{Bmatrix} = \begin{Bmatrix} -a_i & \text{if } a_i \leqslant 0 \\ 0 & \text{if } a_i > 0 \end{Bmatrix},$$

These are the "positive and negative parts" of the series $\sum_{i=1}^{\infty} a_i$. It is easy to check that $a_i = b_i - c_i$. The series $\sum_{i=1}^{\infty} b_i$ and $\sum_{i=1}^{\infty} c_i$ are both convergent; in fact, since $b_i \leqslant |a_i|$, we have $\sum_{i=1}^{n} b_i \leqslant \sum_{i=1}^{n}|a_i| \leqslant \sum_{i=1}^{\infty}|a_i|$, which is finite since we assumed the series $\sum_{i=1}^{\infty} a_i$ to be absolutely convergent. Since $b_i \geqslant 0$ for all i, $\sum_{i=1}^{\infty} b_i$ is convergent. The same argument proves that $\sum_{i=1}^{\infty} c_i$ is convergent. The sum and constant multiple rules now apply to give the convergence of $\sum_{i=1}^{\infty} a_i = \sum_{i=1}^{\infty} b_i - \sum_{i=1}^{\infty} c_i$.

Finally, we note that, by the triangle inequality,

$$\left|\sum_{i=1}^{n} a_i\right| \leqslant \sum_{i=1}^{n} |a_i| \leqslant \sum_{i=1}^{\infty} |a_i|.$$

Since this is true for all n, and

$$\left|\sum_{i=1}^{\infty} a_i\right| = \left|\lim_{n\to\infty} \sum_{i=1}^{n} a_i\right| = \lim_{n\to\infty} \left|\sum_{i=1}^{n} a_i\right|$$

(the absolute value function is continuous), it follows that $|\sum_{i=1}^{\infty} a_i| \leqslant \sum_{i=1}^{\infty}|a_i|$. (Here we again used the fact that if $b_n \leqslant M$ for all n and b_n converges to b, then $b \leqslant M$).

We conclude this section with a proof that every alternating series converges.

Let $\sum_{i=1}^{\infty} a_i$ be an alternating series. If we let $b_i = (-1)^{i+1} a_i$, then all the b_i are positive, and our series is $b_1 - b_2 + b_3 - b_4 + b_5 \cdots$. In addition, we have $b_1 > b_2 > b_3 > \cdots$, and $\lim_{i\to\infty} b_i = 0$. Each even partial sum S_{2n} can be grouped as $(b_1 - b_2) + (b_3 - b_4) + \cdots + (b_{n-1} - b_n)$, which is a series of positive terms, so we have $S_2 \leqslant S_4 \leqslant S_6 \leqslant \cdots$. On the other hand, the odd partial sums S_{2n+1} can be grouped as $b_1 - (b_2 - b_3) - (b_4 - b_5) - \cdots - (b_{2n} - b_{2n+1})$, which is a sum of negative terms (except for the first), so we have $S_1 \geqslant S_3 \geqslant S_5 \geqslant \cdots$. Next, we note that $S_{2n+1} = S_{2n} + b_{2n+1} \geqslant S_{2n}$. Thus the even partial sums S_{2n} form an increasing sequence which is bounded above by any member of the decreasing sequence of odd partial sums. (See Fig. 12.2.1.) By the increasing sequence property, the sequence S_{2n} approaches a limit, S_{even}. Similarly, the decreasing sequence S_{2n+1} approaches a limit, S_{odd}.

Thus we have $S_2 \leqslant S_4 \leqslant S_6 \leqslant \cdots S_{2n} \leqslant \cdots \leqslant S_{\text{even}} \leqslant S_{\text{odd}} \cdots \leqslant S_{2n+1} \leqslant \cdots \leqslant S_3 \leqslant S_1$. Now $S_{2n+1} - S_{2n}$ is a_{2n+1}, which approaches zero as $n \to \infty$; the difference $S_{\text{odd}} - S_{\text{even}}$ is less than $S_{2n+1} - S_{2n}$, so it must be zero; i.e., $S_{\text{odd}} = S_{\text{even}}$. Call this common value S. Thus $|S_{2n} - S| \leqslant |S_{2n} - S_{2n+1}|$

$= b_{2n+1} = |a_{2n+1}|$ and $|S_{2n+1} - S| \leqslant |S_{2n+1} - S_{2n+2}| = b_{2n+2} = |a_{2n+2}|$, so each difference $|S_n - S|$ is less than $|a_{n+1}|$. Since $a_{n+1} \to 0$, we must have $S_n \to S$ as $n \to \infty$.

This argument also shows that each tail of an alternating series is no greater than the first term omitted from the partial sum.

Exercises for Section 12.2

Show that the series in Exercises 1–8 converge, using the comparison test for series with positive terms.

1. $\sum_{i=1}^{\infty} \frac{8}{3^i + 2}$
2. $\sum_{i=1}^{\infty} \frac{9}{4^i + 6}$
3. $\sum_{i=1}^{\infty} \frac{1}{3^i - 1}$
4. $\sum_{i=1}^{\infty} \frac{2}{4^i - 3}$
5. $\sum_{i=1}^{\infty} \frac{(-1)^i}{3^i + 2}$
6. $\sum_{i=1}^{\infty} \frac{(-1)^i}{4^i + 1}$
7. $\sum_{i=1}^{\infty} \frac{\sin i}{2^i - 1}$
8. $\sum_{i=1}^{\infty} \frac{\cos(\pi i)}{3^i - 1}$

Show that the series in Exercises 9–12 diverge, by using the comparison test.

9. $\sum_{i=1}^{\infty} \frac{3}{2 + i}$
10. $\sum_{i=1}^{\infty} \frac{-8}{5 + 7i}$
11. $\sum_{i=1}^{\infty} \frac{8}{6i - 1}$
12. $\sum_{i=1}^{\infty} \frac{3}{2i - 1}$

Test the series in Exercises 13–34 for convergence.

13. $\sum_{n=1}^{\infty} \frac{3}{4^n + 2}$
14. $\sum_{n=1}^{\infty} \left(\frac{-4}{2^n + 3} \right)^n$
15. $\sum_{i=1}^{\infty} \frac{1}{2^i + 3^i}$
16. $\sum_{i=1}^{\infty} \frac{(1/2)^i}{i + 6}$
17. $\sum_{i=1}^{\infty} \frac{1}{3i + 1/i}$
18. $\sum_{i=1}^{\infty} \frac{2}{2i + 1}$
19. $\sum_{n=1}^{\infty} \frac{4^n + 5^n}{2^n 3^n}$
20. $\sum_{n=1}^{\infty} \frac{\sqrt{3 + n}}{4^n}$
21. $\sum_{i=1}^{\infty} \frac{1 + (-1)^i}{8i + 2^{i+1}}$
22. $\sum_{i=1}^{\infty} \frac{(-2)^i}{3^i + 1}$
23. $\sum_{i=1}^{\infty} \frac{1}{\sqrt{i} + 2}$
24. $\sum_{i=1}^{\infty} \frac{1}{\sqrt{i} + 1}$
25. $\sum_{i=1}^{\infty} \frac{3i}{2^i}$
26. $\sum_{j=1}^{\infty} \frac{2^j}{j}$
27. $\sum_{i=1}^{\infty} \left(\frac{1}{i} + \frac{2}{i^2} + \frac{3}{i^3} \right)$
28. $\sum_{i=1}^{\infty} \frac{3}{1 + 3^i}$
29. $\sum_{j=1}^{\infty} \frac{\sin j}{2^j}$
30. $\sum_{n=1}^{\infty} e^{-n}$
31. $\sum_{i=2}^{\infty} \frac{1}{\ln i}$
32. $\sum_{i=1}^{\infty} \left(\frac{i}{i + 2} \right)^i$
33. $\frac{1}{3} + \frac{1}{5} + \frac{1}{9} + \frac{1}{17} + \frac{1}{33} + \frac{1}{65} + \cdots + \frac{1}{2^n + 1} + \cdots$
34. $1 + \frac{1}{3} + \frac{1}{7} + \frac{1}{15} + \cdots + \frac{1}{2^n - 1} + \cdots$

Find the sum of the series in Exercises 35–38 with an error of no more than 0.01.

35. $\sum_{j=1}^{\infty} \frac{1}{j4^j}$
36. $\sum_{k=0}^{\infty} \frac{k}{2^k}$ [*Hint:* Compare with $\sum_{k=0}^{\infty} \left(\frac{2}{3} \right)^k$.]
37. $\sum_{n=1}^{\infty} \frac{2^n - 1}{5^n + 1}$
38. $\sum_{p=1}^{\infty} \frac{(-1)^p}{2^p + p}$

Test the series in Exercises 39–50 for convergence and absolute convergence.

39. $\sum_{n=1}^{\infty} \frac{1}{\sqrt{n}}$
40. $\sum_{n=1}^{\infty} \frac{(-1)^n}{\sqrt{n}}$
41. $\sum_{k=1}^{\infty} \frac{k}{k + 1}$
42. $\sum_{k=1}^{\infty} \frac{(-1)^k k}{k + 1}$
43. $\sum_{i=1}^{\infty} \frac{\cos \pi i}{2^i}$
44. $\sum_{n=1}^{\infty} \frac{(-1)^n}{8n + 2}$
45. $1 - \frac{1}{2} + \frac{2}{3} - \frac{3}{4} + \frac{4}{5} - \cdots$
46. $1 - \frac{1}{2} + \frac{1}{4} - \frac{1}{8} + \frac{1}{16} - \cdots$
47. $\sum_{i=1}^{\infty} (-1)^i \frac{i}{i^2 + 1}$
48. $\sum_{i=1}^{\infty} a_i$, where $a_i = 1/(2^i)$ if i is even and $a_i = 1/i$ if i is odd.
49. $\sum_{n=1}^{\infty} (-1)^n \ln[(n + 1)/n]$. [*Hint*: First prove that $\ln(1 + a) \geqslant a/2$ for small $a > 0$.]
50. $\sum_{n=1}^{\infty} (-1)^{n+1} \ln[(n + 3)/n]$. (See the hint in 49.)

Estimate the sum of the series in Exercises 51–54 with an error of no more than that specified.

51. $\sum_{i=1}^{\infty} \frac{(-1)^i}{3^i + 1}$; 0.01
52. $\sum_{n=1}^{\infty} \frac{(-1)^n n}{4n^3 + 1}$; 0.005
53. $\sum_{n=1}^{\infty} \left(\frac{(-1)^n}{2n} + \frac{1}{5^n} \right)$; 0.02
54. $1 - \frac{1}{3} + \frac{1}{5} - \frac{1}{7} + \cdots$; 0.02

55. Test for convergence: $\frac{1}{2} + \frac{1}{2} - \frac{1}{4} - \frac{1}{4} + \frac{1}{6} + \frac{1}{6} - \frac{1}{8} - \frac{1}{8} + \cdots$.
56. Does the series $\frac{1}{2} + \frac{1}{3} - \frac{1}{4} - \frac{1}{5} + \frac{1}{6} + \frac{1}{7} - \cdots$ converge?

Exercises 57 and 58 deal with an application of the increasing sequence test to inductively defined sequences. For example, let a_n be defined as follows:

$$a_0 = 0, \qquad a_1 = \sqrt{3},$$

$$a_2 = \sqrt{3 + a_1} = \sqrt{3 + \sqrt{3}},$$

$$a_3 = \sqrt{3 + a_2} = \sqrt{3 + \sqrt{3 + \sqrt{3}}} \quad \ldots,$$

and, in general, $a_n = \sqrt{3 + a_{n-1}}$. If we attempt to write out a_n "explicitly," we quickly find ourselves in a notational nightmare. However, numerical computation suggests that the sequence may be convergent:

$a_1 = 1.73205$	$a_2 = 2.17533$	$a_3 = 2.27493$
$a_4 = 2.29672$	$a_5 = 2.30146$	$a_6 = 2.30249$
$a_7 = 2.30271$	$a_8 = 2.30276$	$a_9 = 2.30277$
$a_{10} = 2.30278$	$a_{11} = 2.30278$	$a_{12} = 2.30278 \ldots$

The sequence appears to be converging to a number $l \approx 2.30278\ldots$, but the numerical evidence only *suggests* that the sequence converges. The increasing sequence test enables us to prove this.

57. Let the sequence a_n be defined inductively by the rules $a_0 = 0$, $a_n = \sqrt{4 + a_{n-1}}$.
    (a) Write out a_1, a_2, and a_3 in terms of square roots.
    (b) Calculate a_1 through a_{12} and guess the value of $\lim_{n\to\infty} a_n$ to four significant figures.

★58. (a) Prove by induction on n that for the sequence in Exercise 57, we have $a_n > a_{n-1}$ and $a_n < 5$.
    (b) Conclude that the limit $l = \lim_{n\to\infty} a_n$ exists.
    (c) Show that l must satisfy the equation $l = \sqrt{4 + l}$.
    (d) Solve the equation in (c) for l and evaluate l to four significant figures. Compare the result with Exercise 57(b).

Show that the sequences in Exercises 59–62 are increasing (or decreasing) and bounded above (or below).

★59. $a_n = \dfrac{2n}{n+3}$

★60. $a_n = \dfrac{n}{n^2+1}$

★61. $a_n = \dfrac{1}{2n} - \dfrac{1}{n+1}$

★62. $b_n = n \sin\left(\dfrac{1}{n}\right)$

★63. Let $B > 0$ and $a_0 = 1$; $a_{n+1} = \frac{1}{2}(a_n + B/a_n)$. Show that $a_n \to \sqrt{B}$.

★64. Let $a_{n+1} = 3 - (1/a_n)$; $a_0 = 1$. Prove that the sequence is increasing and bounded above. What is $\lim_{n\to\infty} a_n$?

★65. Let $a_{n+1} = \frac{1}{2}a_n + \sqrt{a_n}$; $a_0 = 1$. Prove that a_n is increasing and bounded above. What is $\lim_{n\to\infty} a_n$?

★66. Let $a_{n+1} = \frac{1}{2}(1 + a_n)$, and $a_0 = 1$. Show that $\lim_{n\to\infty} a_n = 1$.

★67. Give an alternative proof that $\lim_{n\to\infty} r^n = 0$ if $0 < r < 1$ as follows. Show that r^n decreases and is bounded below by zero. If the limit is l, show that $rl = l$ and conclude that $l = 0$. Why does the limit exist?

★68. Suppose that $a_0 = 1$, $a_{n+1} = 1 + 1/(1 + a_n)$. Show that a_n converges and find the limit.

★69. The celebrated example due to Karl Weierstrass of a *nowhere differentiable continuous function* $f(x)$ in $-\infty < x < \infty$ is given by

$$f(x) = \sum_{n=0}^{\infty} \left(\frac{3}{4}\right)^n \phi(4^n x),$$

where $\phi(x+2) = \phi(x)$, and $\phi(x)$ on $0 \leqslant x \leqslant 2$ is the "triangle" through $(0,0), (1,1), (2,0)$. By construction, $0 \leqslant \phi(4^n x) \leqslant 1$. Verify by means of the comparison test that the series converges for any value of x. [See *Counterexamples in Analysis* by B. R. Gelbaum and J. M. H. Olmsted, Holden-Day, San Francisco (1964), p. 38 for the proof that f is nowhere differentiable.]

★70. Prove that $a_n = (1 + 1/n)^n$ is increasing and bounded above as follows:
    (a) If $0 \leqslant a \leqslant b$, prove that

$$\frac{b^{n+1} - a^{n+1}}{b - a} < (n+1)b^n.$$

    That is, prove $b^n[(n+1)a - nb] < a^{n+1}$.
    (b) Let $a = 1 + [1/(n+1)]$ and $b = 1 + (1/n)$ and deduce that a_n is increasing.
    (c) Let $a = 1$ and $b = 1 + (1/2n)$ and deduce that $(1 + 1/2n)^{2n} < 4$.
    (d) Use parts (b) and (c) to show that $a_n < 4$.

Conclude that a_n converges to some number (the number is e—see Section 6.3).

12.3 The Integral and Ratio Tests

The integral test establishes a connection between infinite series and improper integrals.

The sum of any infinite series may be thought of as an improper integral. Namely, given a series $\sum_{i=1}^{\infty} a_i$, we define a step function $g(x)$ on $[1, \infty)$ by the formulas:

$$g(x) = a_1 \qquad (1 \leqslant x < 2)$$
$$g(x) = a_2 \qquad (2 \leqslant x < 3)$$
$$\vdots$$
$$g(x) = a_i \qquad (i \leqslant x < i+1)$$
$$\vdots$$

Since $\int_i^{i+1} g(x) = a_i$, the partial sum $\sum_{i=1}^{n} a_i$ is equal to $\int_1^{n+1} g(x)\,dx$, and the sum $\sum_{i=1}^{\infty} a_i = \lim_{n \to \infty} \sum_{i=1}^{n} a_i$ exists if and only if the integral $\int_1^{\infty} g(x)\,dx = \lim_{b \to \infty} \int_1^b g(x)\,dx$ does.

By itself, this relation between series and integrals is not very useful. However, suppose now, as is often the case, that the formula which defines the term a_i as a function of i makes sense when i is a real number, not just an integer. In other words, suppose that there is a function $f(x)$, defined for all x satisfying $1 \leqslant x < \infty$, such that $f(i) = a_i$ when $i = 1, 2, 3, \ldots$. Suppose further that f satisfies these conditions:

1. $f(x) > 0$ for all x in $[1, \infty)$;
2. $f(x)$ is decreasing on $[1, \infty)$.

For example, if $a_i = 1/i$, the harmonic series, we may take $f(x) = 1/x$.

We may now compare $f(x)$ with the step function $g(x)$. When x satisfies $i \leqslant x < i+1$, we have

$$0 \leqslant f(x) \leqslant f(i) = a_i = g(x).$$

Hence $0 \leqslant f(x) \leqslant g(x)$. (See Fig. 12.3.1.)

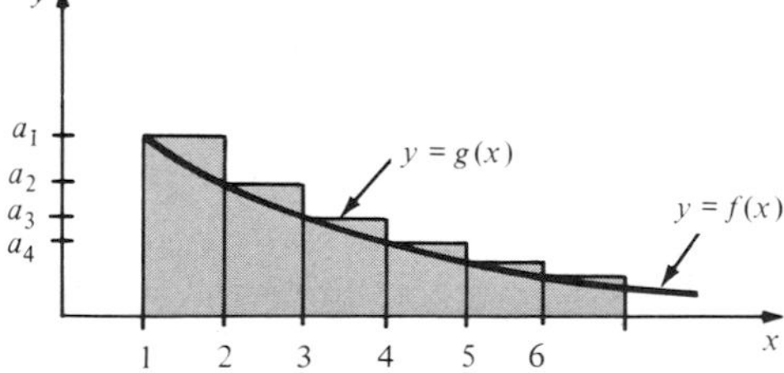

Figure 12.3.1. The area under the graph of f is less than the shaded area, so $0 \leqslant \int_1^{n+1} f(x)\,dx \leqslant \sum_{i=1}^{n} a_i$.

It follows that, for any n,

$$0 \leqslant \int_1^{n+1} f(x)\,dx \leqslant \int_1^{n+1} g(x)\,dx = \sum_{i=1}^{n} a_i. \tag{1}$$

We conclude that if the series $\sum_{i=1}^{\infty} a_i$ converges, then the integrals $\int_1^{n+1} f(x)\,dx$ are bounded above by the sum $\sum_{i=1}^{\infty} a_i$, so that the indefinite integral $\int_1^{\infty} f(x)\,dx$ converges (see Section 11.3).

In other words, if the integral $\int_1^{\infty} f(x)\,dx$ *diverges*, then so does the series $\sum_{i=1}^{\infty} a_i$.

Example 1 Show that

$$1+\frac{1}{2}+\cdots+\frac{1}{n} \geqslant \ln(n+1)$$

and so obtain a new proof that the harmonic series diverges.

Solution We take our function $f(x)$ to be $1/x$. Then, from formula (1) above, we get

$$1+\frac{1}{2}+\cdots+\frac{1}{n}=\sum_{i=1}^{n}\frac{1}{i} \geqslant \int_1^{n+1}\frac{1}{x}\,dx=\ln(n+1).$$

Since $\lim_{n\to\infty}\ln(n+1)=\infty$, the integral $\int_1^{\infty}(1/x)\,dx$ diverges; hence the series $\sum_{i=1}^{\infty}(1/i)$ diverges, too. ▲

We would like to turn around the preceding argument to show that if $\int_1^{\infty}f(x)\,dx$ converges, then $\sum_{i=1}^{\infty}a_i$ converges as well. To do so, we draw the rectangles with height a_i to the *left* of $x=i$ rather than the right; see Fig. 12.3.2. This procedure defines a step function $h(x)$ on $[1,\infty)$ defined by

$$h(x)=a_{i+1} \qquad (i \leqslant x < i+1).$$

Now we have $\int_i^{i+1}h(x)\,dx=a_{i+1}$, so $\sum_{i=2}^{n}a_i=\int_1^n h(x)\,dx$. If x satisfies $i \leqslant x < i+1$, we have

$$f(x) \geqslant f(i+1)=a_{i+1}=h(x) \geqslant 0.$$

Hence $f(x) \geqslant h(x) \geqslant 0$. (See Fig. 12.3.2.) Thus

$$\int_1^n f(x)\,dx \geqslant \int_1^n h(x)\,dx=\sum_{i=2}^{n}a_i \geqslant 0. \tag{2}$$

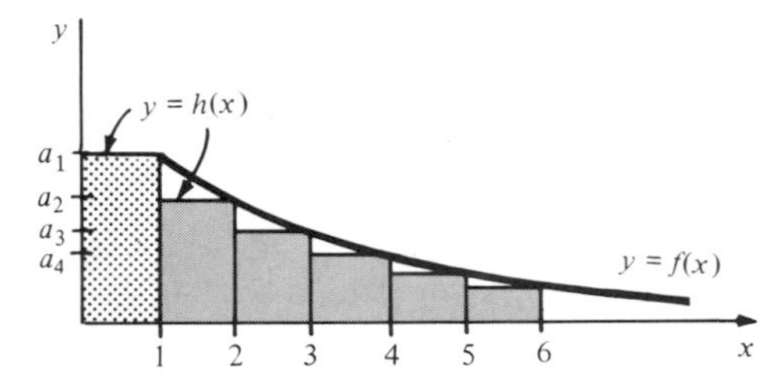

Figure 12.3.2. The area under the graph of f is greater than the shaded area, so $0 \leqslant \sum_{i=2}^{n}a_i \leqslant \int_1^n f(x)\,dx$.

If the integral $\int_1^{\infty}f(x)\,dx$ converges, then the partial sums $\sum_{i=1}^{n}a_i=a_1+\sum_{i=2}^{n}a_i$ are bounded above by $a_1+\int_1^{\infty}f(x)\,dx$, and therefore the series $\sum_{i=1}^{\infty}a_i$ is convergent (see the Supplement to Section 12.2).

Integral Test

To test the convergence of a series $\sum_{n=1}^{\infty}a_i$ of positive decreasing terms, find a positive, decreasing function $f(x)$ on $[1,\infty)$ such that $f(i)=a_i$.

If $\int_1^{\infty}f(x)\,dx$ converges, so does $\sum_{i=1}^{\infty}a_i$.

If $\int_1^{\infty}f(x)\,dx$ diverges, so does $\sum_{i=1}^{\infty}a_i$.

Example 2 Show that $1+\frac{1}{4}+\frac{1}{9}+\frac{1}{16}+\cdots$ converges.

Solution This series is $\sum_{i=1}^{\infty}(1/i^2)$. We let $f(x)=1/x^2$; then

$$\int_1^{\infty}\frac{1}{x^2}\,dx=\lim_{b\to\infty}\int_1^{b}\frac{1}{x^2}\,dx=\lim_{b\to\infty}\left(1-\frac{1}{b}\right)=1.$$

The indefinite integral converges, so the series does, too. ▲

Example 3 Show that $\sum_{m=2}^{\infty} \frac{1}{m\sqrt{\ln m}}$ diverges, but $\sum_{m=2}^{\infty} \frac{1}{m(\ln m)^2}$ converges.

Solution Note that the series start at $m = 2$ rather than $m = 1$. We consider the integral

$$\int_2^{\infty} \frac{1}{x(\ln x)^p}\,dx = \lim_{b\to\infty} \int_2^b (\ln x)^{-p} \frac{1}{x}\,dx$$

$$= \lim_{b\to\infty} \left. \frac{(\ln x)^{-p+1}}{-p+1} \right|_2^b$$

$$= \frac{1}{-p+1} \lim_{b\to\infty} \left[(\ln b)^{-p+1} - (\ln 2)^{-p+1} \right].$$

The limit is finite if $p = 2$ and infinite if $p = \frac{1}{2}$, so the integral converges if $p = 2$ and diverges if $p = \frac{1}{2}$. It follows that $\sum_{m=2}^{\infty}[1/(m\sqrt{\ln m})]$ diverges and $\sum_{m=2}^{\infty}[1/m(\ln m)^2]$ converges. ▲

Examples 1 and 2 are special cases of a result called the p-series test, which arises from the integral test with $f(x) = 1/x^p$. We recall that $\int_1^{\infty} x^n\,dx$ converges if $n < -1$ and diverges if $n \geqslant -1$ (see Example 2, Section 11.3). Thus we arrive at the test in the following box.

p-Series

If $p \leqslant 1$, then $\sum_{i=1}^{\infty} \frac{1}{i^p}$ diverges.

If $p > 1$, then $\sum_{i=1}^{\infty} \frac{1}{i^p}$ converges.

The p-series are often useful in conjunction with the comparison test.

Example 4 Test for convergence:

$$\text{(a)}\ \sum_{i=1}^{\infty} \frac{1}{1+i^2}; \qquad \text{(b)}\ \sum_{j=1}^{\infty} \frac{j^2+2j}{j^4-3j^2+10}; \qquad \text{(c)}\ \sum_{n=1}^{\infty} \frac{3n+\sqrt{n}}{2n^{3/2}+2}.$$

Solution (a) We compare the given series with the convergent p series $\sum_{i=1}^{\infty} 1/i^2$. Let $a_i = 1/(1+i^2)$ and $b_i = 1/i^2$. Then $0 < a_i < b_i$ and $\sum_{i=1}^{\infty} b_i$ converges, so $\sum_{i=1}^{\infty} a_i$ does, too.

(b) Let $a_j = (j^2+2j)/(j^4-3j^2+10)$ and $b_j = j^2/j^4 = 1/j^2$. Then

$$\lim_{j\to\infty} \left| \frac{a_j}{b_j} \right| = \lim_{j\to\infty} \left| \frac{1+2/j}{1-3/j^2+10/j^4} \right| = 1.$$

Since $\sum_{j=1}^{\infty} b_j$ converges, so does $\sum_{j=1}^{\infty} a_j$, by the ratio comparison test.

(c) Take $a_n = (3n+\sqrt{n})/(2n^{3/2}+2)$ and $b_n = n/(n^{3/2}) = 1/\sqrt{n}$. Then

$$\lim_{n\to\infty} \frac{3+(1/\sqrt{n})}{2+(2/n^{3/2})} = \frac{3}{2}.$$

Since $\sum_{n=1}^{\infty} b_n$ diverges, so does $\sum_{n=1}^{\infty} a_n$. ▲

What is the error in approximating a p-series by a partial sum? Let us show that $\sum_{n=1}^{N}(1/n^p)$ approximates $\sum_{n=1}^{\infty}(1/n^p)$ with error which does not exceed $1/[(p-1)N^{p-1}]$.

Indeed, just as in the proof of formula (2), we have

$$\sum_{n=N+1}^{\infty} \frac{1}{n^p} \leqslant \int_N^{\infty} \frac{1}{x^p}\, dx.$$

The left-hand side is the error:

$$\sum_{n=N+1}^{\infty} \frac{1}{n^p} = \sum_{n=1}^{\infty} \frac{1}{n^p} - \sum_{n=1}^{N} \frac{1}{n^p} \leqslant \int_N^{\infty} \frac{1}{x^p}\, dx = \frac{1}{(p-1)N^{p-1}}.$$

$$\text{Thus, error} \leqslant \frac{1}{(p-1)N^{p-1}}. \tag{3}$$

Example 5 It is known that $\sum_{n=1}^{\infty}(1/n^2) = \pi^2/6$. Use this equation[2] to calculate $\pi^2/6$ with error less than 0.05.

Solution By equation (3), the error in stopping at N terms is at most $1/N$. To have error $< 0.05 = \frac{1}{20}$, we must take 20 terms (note that 100 terms are needed to get two decimal places!). We find:

$$1 = 1,$$
$$1 + \tfrac{1}{4} = 1.25,$$
$$1 + \tfrac{1}{4} + \tfrac{1}{9} = 1.36,$$
$$1 + \tfrac{1}{4} + \tfrac{1}{9} + \tfrac{1}{16} = 1.42,$$
$$1 + \tfrac{1}{4} + \tfrac{1}{9} + \tfrac{1}{16} + \tfrac{1}{25} = 1.46,$$

and so forth, obtaining 1.49, 1.51, 1.53, 1.54, 1.55, 1.56, ... Finally, $1 + \frac{1}{4} + \cdots + \frac{1}{400} = 1.596\ldots$. (Notice the "slowness" of the convergence.) We may compare this with the exact value $\pi^2/6 = 1.6449\ldots$. ▲

The idea used in the preceding example can be used to estimate the tail of a series whenever convergence is proven by the integral test. (See Exercise 11.)

Another important test for convergence is called the *ratio test*. This test provides a general way to compare a series with a geometric series, but it formulates the hypotheses in a way which is particularly convenient, since no explicit comparison is needed. Here is the test.

Ratio Test

Let $\sum_{i=1}^{\infty} a_i$ *be a series. Suppose that* $\lim_{i\to\infty}\left|\frac{a_i}{a_{i-1}}\right|$ *exists.*

1. *If* $\lim_{i\to\infty}\left|\frac{a_i}{a_{i-1}}\right| < 1$, *then the series converges (absolutely).*
2. *If* $\lim_{i\to\infty}\left|\frac{a_i}{a_{i-1}}\right| > 1$, *then the series diverges.*
3. *If* $\lim_{i\to\infty}\left|\frac{a_i}{a_{i-1}}\right| = 1$, *the test is inconclusive.*

[2] For a proof using only elementary calculus, see Y. Matsuoka, "An Elementary Proof of the Formula $\sum_{k=1}^{\infty} 1/k^2 = \pi^2/6$," *American Mathematical Monthly* **68**(1961): 485–487 (reprinted in T. M. Apostol (ed.), *Selected Papers on Calculus*, Math. Assn. of America (1969), p. 372). The formula may also be proved using Fourier series; see for instance J. Marsden, Elementary Classical Analysis, Freeman (1974), Ch. 10.

Do not confuse this test, in which ratios of successive terms in the *same* series are considered, with the ratio comparison test in Section 12.2, where we took the ratios of terms in two *different* series.

Proof of the ratio test By definition of the limit, $|a_i/a_{i-1}|$ will be close to its limit l for i large. To prove part 1, let $l<1$ and let $r=(l+1)/2$ be the midpoint between l and 1, so that $l<r<1$. Thus there is an N such that

$$\left|\frac{a_i}{a_{i-1}}\right|<r \qquad \text{if} \quad i>N.$$

We will show this implies that the given series converges.

We have $|a_{N+1}/a_N|<r$ so $|a_{N+1}|<|a_N|r$, $|a_{N+2}/a_{N+1}|<r$; hence $|a_{N+2}|<|a_{N+1}|r<|a_N|r^2$ and, in general, $|a_{N+j}|<|a_N|r^j$; but $\sum_{j=1}^{\infty}|a_N|r^j=|a_N|\sum_{j=1}^{\infty}r^j$ is a convergent geometric series since $r<1$. Hence, by the comparison test, $\sum_{j=1}^{\infty}|a_{N+j}|$ converges. Since we have omitted only $|a_1|$, $|a_2|, \ldots, |a_N|$, the series $\sum_{j=1}^{\infty}|a_j|$ converges as well and part 1 is proved.

For part 2 we find, as in part 1, that $|a_{N+j}|>|a_N|r^j$, where $r=(l+1)/2$ is now greater than 1. As $j\to\infty$, $r^j\to\infty$, so $|a_{N+j}|\to\infty$. Thus the series cannot converge, since its terms do not converge to zero.

To prove part 3, we consider the p-series with $a_i=i^p$. The ratio is $|a_i/a_{i-1}|=[i/(i-1)]^p$, and $\lim_{i\to\infty}[i/(i-1)]^p=[\lim_{i\to\infty}(i/(i-1))]^p=1^p=1$ for all $p>0$; but the p-series is convergent if $p>1$ and divergent if $p\leqslant 1$, so the ratio test does not give any useful information for these series. ■

Example 6 Test for convergence: $2+\frac{2^2}{2^8}+\frac{2^3}{3^8}+\frac{2^4}{4^8}+\cdots=2+\frac{1}{64}+\frac{8}{6561}+\frac{1}{4096}+\cdots$.

Solution We have $a_i=2^i/i^8$. The ratio a_i/a_{i-1} is

$$\frac{2^i}{i^8}\cdot\frac{(i-1)^8}{2^{i-1}}=2\cdot\left(\frac{i-1}{i}\right)^8,$$

so

$$\lim_{i\to\infty}\frac{a_i}{a_{i-1}}=2\left[\lim_{i\to\infty}\left(\frac{i-1}{i}\right)\right]^8=2\cdot 1^8=2$$

which is greater than 1, and so the series diverges. ▲

Example 7 Test for convergence:

(a) $\sum_{n=1}^{\infty}\frac{1}{n!}$, where $n!=n(n-1)\cdots 3\cdot 2\cdot 1$

(b) $\sum_{j=1}^{\infty}\frac{b^j}{j!}$, b any constant

Solution (a) Here $a_n=1/n!$, so

$$\frac{a_n}{a_{n-1}}=\frac{1/n(n-1)\cdots 3\cdot 2\cdot 1}{1/(n-1)(n-2)\cdots 3\cdot 2\cdot 1}=\frac{1}{n}.$$

Thus $|a_n/a_{n-1}|=1/n\to 0<1$, so we have convergence.

(b) Here $a_j=b^j/j!$, so

$$\frac{a_j}{a_{j-1}}=\frac{b^j/j!}{b^{j-1}/(j-1)!}=\frac{b}{j}.$$

Thus $|a_j/a_{j-1}|=b/j\to 0$, so we have convergence. In this example, note that the numerator b^j and the denominator $j!$ tend to infinity, but the denominator does so much faster. In fact, since the series converges, $b^j/j!\to 0$ as $j\to\infty$. ▲

Let us show that if $|a_n/a_{n-1}| < r < 1$ for $n > N$, then the error made in approximating $\sum_{n=1}^{\infty} a_n$ by $\sum_{n=1}^{N} a_n$ is no greater than $|a_N|r/(1-r)$. In short,

$$\text{error} \leqslant \frac{|a_N|r}{1-r}. \tag{4}$$

Indeed, $\sum_{n=1}^{\infty} a_n - \sum_{n=1}^{N} a_n = \sum_{n=N+1}^{\infty} a_n$. As in the proof of the ratio test, $|a_{N+1}| < |a_N|r$, and, in general, $|a_{N+j}| < |a_N|r^j$, so $\sum_{j=1}^{\infty}|a_{N+j}| \leqslant |a_N|r/(1-r)$ by the formula for the sum of a geometric series and the comparison test. Hence the error is no greater than $|a_N|r/(1-r)$.

Example 8 What is the error made in approximating $\displaystyle\sum_{n=1}^{\infty} \frac{1}{n!}$ by $\displaystyle\sum_{n=1}^{4} \frac{1}{n!}$?

Solution Here $|a_n/a_{n-1}| = 1/n$, which is $< \frac{1}{5}$ if $n > 4 = N$. By inequality (4), the error is no more than $a_4/5(1-1/5) = 1/4 \cdot 4! = 1/96 < 0.0105$. The error becomes small very quickly if N is increased. ▲

Our final test is similar in spirit to the ratio test, in that it is also proved by comparison with a geometric series.

Root Test

Let $\displaystyle\sum_{i=1}^{\infty} a_i$ be a given series, and suppose that $\displaystyle\lim_{i\to\infty} |a_i|^{1/i}$ exists.

1. If $\displaystyle\lim_{i\to\infty} |a_i|^{1/i} < 1$, then $\displaystyle\sum_{i=1}^{\infty} a_i$ converges absolutely.
2. If $\displaystyle\lim_{i\to\infty} |a_i|^{1/i} > 1$, then $\displaystyle\sum_{i=1}^{\infty} a_i$ diverges.
3. If $\displaystyle\lim_{i\to\infty} |a_i|^{1/i} = 1$, the test is inconclusive.

To prove 1, let $l = \lim_{n\to\infty}(|a_n|^{1/n})$ and let $r = (1+l)/2$ be the midpoint of 1 and l, so $l < r < 1$. From the definition of the limit, there is an N such that $|a_n|^{1/n} < r < 1$ if $n > N$. Hence $|a_n| < r^n$ if $n > N$. Thus, by direct comparison of $\sum_{n=N+1}^{\infty}|a_n|$ with the geometric series $\sum_{n=N+1}^{\infty} r^n$, which converges since $r < 1$, $\sum_{n=N+1}^{\infty}|a_n|$ converges. Since we have neglected only finitely many terms, the given series converges.

Cases 2 and 3 are left as exercises (see Exercises 37 and 38).

Example 9 Test for convergence: (a) $\displaystyle\sum_{n=1}^{\infty} \frac{1}{n^n}$ and (b) $\displaystyle\sum_{n=1}^{\infty} \frac{3^n}{n^2}$.

Solution (a) Here $a_n = 1/n^n$, so $|a_n|^{1/n} = 1/n$. Thus $\lim_{n\to\infty}|a_n|^{1/n} = 0 < 1$. Thus, by the root test (with i replaced by n), the series converges (absolutely). [This example can also be done by the comparison test: $\sum(1/n^n) < 1/n^2$ for $n \geqslant 2$.]
(b) Here $a_n = 3^n/n^2$, so $|a_n|^{1/n} = 3/n^{2/n}$; but $\lim_{n\to\infty} n^{2/n} = 1$, since $\ln(n^{2/n}) = 2(\ln n)/n \to 0$ as $n \to \infty$ (by l'Hôpital's rule). Thus $\lim_{n\to\infty}|a_n|^{1/n} = 3 > 1$, so the series diverges. ▲

The tests we have covered enable us to deal with a wide variety of series. Of course, if the series is geometric, it may be summed. Otherwise, either the ratio test, the root test, comparison with a p-series, the integral test, or the alternating series test will usually work.

Example 10 Test for convergence: (a) $\sum_{n=1}^{\infty} \frac{n^n}{n!}$ and (b) $\sum_{n=1}^{\infty} \frac{1}{n^2 - \ln n}$.

Solution (a) We use the ratio test. Here, $a_n = n^n/n!$, so

$$\left|\frac{a_n}{a_{n-1}}\right| = \frac{n^n}{(n-1)^{n-1}} \cdot \frac{(n-1)!}{n!} = \frac{n \cdot n^{n-1}}{(n-1)^n} \cdot \frac{1}{n} = \left(\frac{n}{n-1}\right)^{n-1}$$

$$= \frac{1}{(1-1/n)^{n-1}} = \frac{1-1/n}{(1-1/n)^n}.$$

The numerator approaches 1 while the denominator approaches e^{-1} (see Section 6.4), so $\lim_{n\to\infty}|a_n/a_{n-1}| = e > 1$, and the series diverges.

(b) We expect the series to behave like $\sum_{n=1}^{\infty}(1/n^2)$, so we use the ratio comparison test, with $a_i = 1/(i^2 - \ln i)$ and $b_i = 1/i^2$. The ratio between the terms in the two series is

$$\frac{a_i}{b_i} = \frac{i^2}{i^2 - \ln i} = \frac{1}{1 - \ln i/i^2}.$$

Since $\lim_{i\to\infty}[(\ln i)/i^2] = 0$ (by l'Hôpital's rule), $\lim_{i\to\infty} a_i/b_i = 1$. The p-series $\sum_{i=1}^{\infty} b_i = \sum_{i=1}^{\infty}(1/i^2)$ converges, so the series

$$\sum_{i=1}^{\infty} a_i = \sum_{i=1}^{\infty} \frac{1}{i^2 - \ln i} \text{ converges, too. } \blacktriangle$$

Exercises for Section 12.3

Use the integral test to determine the convergence or divergence of the series in Exercises 1–4.

1. $\sum_{i=1}^{\infty} \frac{i}{i^2+1}$
2. $\sum_{i=1}^{\infty} \frac{1}{i^2+4}$
3. $\sum_{i=2}^{\infty} \frac{1}{i(\ln i)^{3/2}}$
4. $\sum_{i=2}^{\infty} \frac{1}{i(\ln i)^{2/3}}$

Use the p-series test and a comparison test to test the series in Exercises 5–8 for convergence or divergence.

5. $\sum_{n=1}^{\infty} \frac{\cos n}{n^2}$
6. $\sum_{n=1}^{\infty} \frac{\sin n}{n^{3/2}}$
7. $\sum_{n=1}^{\infty} \frac{n}{n^3+4}$
8. $\sum_{n=1}^{\infty} \frac{n}{n^2+4}$

Estimate the sums in Exercises 9 and 10 to within 0.05.

9. $\sum_{n=1}^{\infty} \frac{\cos n}{n^3}$
10. $\sum_{n=1}^{\infty} \frac{\sin n}{n^4}$

11. Let $f(x)$ be a positive decreasing function on $[1,\infty)$ such that $\int_1^\infty f(x)\,dx$ converges. Show that

$$\left|\sum_{n=1}^{\infty} f(n) - \sum_{n=1}^{N} f(n)\right| \leqslant \int_N^\infty f(x)\,dx.$$

12. Estimate $\sum_{n=1}^{\infty}[(1+n^2)/(1+n^8)]$ to within 0.02. (Use the comparison test *and* the integral test.)

Use the ratio test to determine the convergence or divergence of the series in Exercises 13–16.

13. $\sum_{n=1}^{\infty} \frac{2\sqrt{n}}{3^n}$
14. $\sum_{n=1}^{\infty} \frac{3^n}{2\sqrt{n}}$
15. $\sum_{i=1}^{\infty} \frac{i^3 \cdot 3^i}{i!}$
16. $\sum_{n=1}^{\infty} \frac{2n^2+n!}{n^5+(3n)!}$

Estimate the sums in Exercises 17 and 18 to within 0.05.

17. $\sum_{n=0}^{\infty} \frac{\pi^{2n+1}}{(2n+1)!}$
18. $\sum_{n=0}^{\infty} \frac{(\pi/2)^{2n+1}}{(2n+1)!}$

19. Estimate $\sum_{n=1}^{\infty}(1/n!)$: (a) To within 0.05. (b) To within 0.005. (c) How many terms would you need to calculate to get an accuracy of five decimal places?

20. (a) Show that $\sum_{n=1}^{\infty} \frac{\sin(\pi n/2)}{n!}$ converges.
(b) Estimate the sum to within 0.01.

Use the root test to determine the convergence or divergence of the series in Exercises 21–24.

21. $\sum_{n=1}^{\infty} \frac{3^n}{n^n}$
22. $\sum_{n=1}^{\infty} \frac{n^n}{2^n}$
23. $\sum_{n=1}^{\infty} \frac{2^n}{n^3}$
24. $\sum_{n=1}^{\infty} \frac{n^2}{2^n}$

Test for convergence in Exercises 25–36.

25. $\sum_{i=1}^{\infty} \frac{1}{i^4}$
26. $\sum_{j=3}^{\infty} \frac{j^2+\cos j}{j^4+\sin j}$
27. $\sum_{n=1}^{\infty} \frac{(-1)^n(n+1)}{2n+1}$
28. $\sum_{m=1}^{\infty} \frac{(-1)^m(m+1)}{m^2+1}$
29. $\sum_{k=2}^{\infty} \frac{\cos k\pi}{\ln k}$
30. $\sum_{j=1}^{\infty} (-1)^j \sin\left(\frac{\pi}{4j}\right)$
31. $\sum_{j=1}^{\infty} \frac{(j+1)^{100}}{j!}$
32. $\sum_{n=1}^{\infty} \frac{\sqrt{n}+n+n^{3/2}}{\sqrt{n}+n+n^{5/2}+n^3}$

33. $\sum_{r=0}^{\infty} \frac{2^r}{2^r + 3^r}$

34. $\sum_{s=1}^{\infty} \frac{s - \ln s}{s^2 + \ln s}$

35. $\sum_{t=2}^{\infty} \frac{(-1)^t}{(\ln t)^{1/2}}$

36. $\sum_{t=1}^{\infty} \frac{(-1)^t}{t^{1/4}}$

In Exercises 37 and 38, complete the proof of the root test by showing the following.

★37. If $\lim_{n\to\infty}|a_n|^{1/n} > 1$, then $\sum_{n=1}^{\infty} a_n$ diverges.

★38. If $\lim_{n\to\infty}|a_n|^{1/n} = 1$, the test is inconclusive. (You may use the fact that $\lim_{n\to\infty} n^{1/n} = 1$.)

★39. For which values of p does $\sum_{i=1}^{\infty}[\sin(1/i)]^p$ converge?

★40. For which values of p does $\sum_{n=2}^{\infty}[1/n(\ln n)^p]$ converge?

★41. For which p does $\sum_{n=2}^{\infty}(1/n^p \ln n)$ converge?

★42. For which values of p and q is the series $\sum_{n=2}^{\infty} 1/[n^p(\ln n)^q]$ convergent?

★43. (a) Let $f(x)$ be positive and decreasing on $[1, \infty)$, and suppose that $f(i) = a_i$ for $i = 1, 2, 3, \ldots$. Show that

$$S - \frac{1}{2} f(n) \leqslant \sum_{i=1}^{\infty} a_i \leqslant S + \frac{1}{2} f(n),$$

where

$$S = \sum_{i=1}^{n} f(i) + \frac{1}{2}\int_n^{n+1} f(x)\,dx + \int_{n+1}^{\infty} f(x)\,dx.$$

[*Hint:* Look at the proof of the integral test; show that $\int_{n+1}^{\infty} f(x)\,dx \leqslant \sum_{i=n+1}^{\infty} a_i \leqslant \int_n^{\infty} f(x)\,dx$.]

(b) Estimate $\sum_{n=1}^{\infty} 1/n^4$ to within 0.0001. How many terms did you use? How much work do you save by using the method of part (a) instead of the formula: error $< 1/(p-1)N^{p-1}$?

★44. Using Fourier analysis, it is possible to show that

$$\frac{\pi^4}{96} = 1 + \frac{1}{3^4} + \frac{1}{5^4} + \frac{1}{7^4} + \cdots .$$

(a) Show directly that the series on the right is convergent, by means of the integral test.

(b) Determine how many terms are needed to compute $\pi^4/96$ accurate to 20 digits.

★45. A bar of length L is loaded by a weight W at its midpoint. At $t = 0$ the load is removed. The deflection $y(t)$ at the midpoint, measured from the straight profile $y = 0$, is given by

$$y(t) = \frac{2WL^2}{\pi^4 EI}\left[\cos r + \frac{\cos(9r)}{3^4} + \frac{\cos(25r)}{5^4} + \cdots\right],$$

where $r = \left(\frac{\pi^2}{L^2}\sqrt{\frac{EIg}{\gamma\Omega}}\right)t$. The numbers $E, I, g, \gamma, \Omega, L$ are positive constants.

(a) Show by substitution that the bracketed terms are the first three terms of the infinite series

$$\sum_{n=0}^{\infty} \frac{\cos\left[(2n+1)^2 r\right]}{(2n+1)^4}.$$

(b) Make accurate graphs of the first three partial sums

$$S_1(r) = \cos(r),$$

$$S_2(r) = \cos(r) + \frac{1}{3^4}\cos(9r),$$

$$S_3(r) = \cos(r) + \frac{1}{3^4}\cos(9r) + \frac{1}{5^4}\cos(25r).$$

Up to a magnification factor, these graphs approximate the motion of the midpoint of the bar.

(c) Using the integral test and the comparison test, show that the series converges.

12.4 Power Series

Many functions can be expressed as "polynomials with infinitely many terms."

A series of the form $\sum_{i=0}^{\infty} a_i(x - x_0)^i$, where the a_i's and x_0 are constants and x is a variable, is called a *power series* (since we are summing the powers of $(x - x_0)$). In this section, we show how a power series may be considered as a function of x, defined on a certain interval. In the next section, we begin with an arbitrary function and show how to find the power series which represents it (if there is such a series).

We first consider power series in which $x_0 = 0$; that is, those of the form

$$f(x) = a_0 + a_1 x + a_2 x^2 + a_3 x^3 + \cdots = \sum_{i=0}^{\infty} a_i x^i,$$

where the a_i are given constants. The domain of f can be taken to consist of those x for which the series converges.

If there is an integer N such that $a_i = 0$ for all $i > N$, then the power series is equal to a finite sum, $\sum_{i=0}^{N} a_i x^i$, which is just a polynomial of degree N. In general, we may think of a power series as a polynomial of "infinite degree"; we will see that as long as they converge, power series may be manipulated (added, subtracted, multiplied, divided, differentiated) just like ordinary polynomials.

The simplest power series, after a polynomial, is the geometric series

$$f(x) = 1 + x + x^2 + \cdots,$$

which converges when $|x| < 1$; the sum is the function $1/(1-x)$. Thus we have written $1/(1-x)$ as a power series:

$$\frac{1}{1-x} = 1 + x + x^2 + x^3 + \cdots \qquad \text{if} \quad |x| < 1.$$

Convergence of general power series may often be determined by a test similar to the ratio test.

Ratio Test for Power Series

Let $\sum_{i=0}^{\infty} a_i x^i$ be a power series. Assume that

$$\lim_{i\to\infty} \left| \frac{a_i}{a_{i-1}} \right| = l$$

exists. Let $R = 1/l$; if $l = 0$, let $R = \infty$, and if $l = \infty$, let $R = 0$. Then:

1. If $|x| < R$, the power series converges absolutely.
2. If $|x| > R$, the power series diverges.
3. If $x = \pm R$, the power series could converge or diverge.

To prove part 1, we use the ratio test for series of numbers; the ratio of successive terms for $\sum_{i=0}^{\infty} a_i x^i$ is

$$\left| \frac{a_i x^i}{a_{i-1} x^{i-1}} \right| = \left| \frac{a_i}{a_{i-1}} \right| |x|.$$

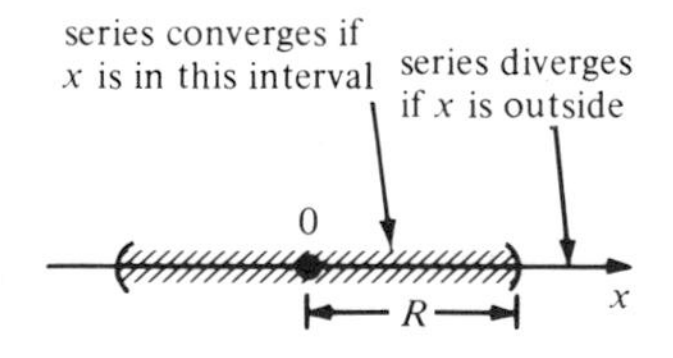

Figure 12.4.1. R is the radius of convergence of $\sum_{i=0}^{\infty} a_i x^i$.

By hypothesis, this converges to $l \cdot |x| < l \cdot R = 1$. Hence, by the ratio test, the series converges absolutely when $|x| < R$. The proof of part 2 is similar, and the examples below will show that at $x = \pm R$, either convergence or divergence can occur.

The number R in this test is called the *radius of convergence* of the series (see Fig. 12.4.1). One can show that a number R (possibly infinity) with the three properties in the preceding box exists for any power series, even if $\lim_{i\to\infty} |a_i / a_{i-1}|$ does not exist.

Example 1 For which x does $\sum_{i=0}^{\infty} \frac{i}{i+1} x^i$ converge?

Solution Here $a_i = i/(i+1)$. Then

$$\frac{a_i}{a_{i-1}} = \frac{i/(i+1)}{(i-1)/i} = \frac{i^2}{(i+1)(i-1)}$$
$$= \frac{1}{(1+1/i)(1-1/i)} \to 1 \quad \text{as} \quad i \to \infty.$$

Hence $l = 1$. Thus the series converges if $|x| < 1$ and diverges if $|x| > 1$. If $x = 1$, then $\lim_{i\to\infty}[i/(i+1)]x^i = 1$, so the series diverges at $x = 1$ since the terms do not go to zero. If $x = -1$, $\lim_{i\to\infty}|[i/(i+1)]x^i| = \lim_{i\to\infty}[i/(i+1)] = 1$, so again the series diverges. ▲

Example 2 Determine the radius of convergence of $\sum_{k=0}^{\infty} \frac{k^5}{(k+1)!} x^k$.

Solution To use the ratio test, we look at

$$l = \lim_{k\to\infty} \left| \frac{a_k}{a_{k-1}} \right|.$$

Here $a_k = k^5/(k+1)!$, so

$$l = \lim_{k\to\infty} \left| \frac{k^5}{(k+1)!} \cdot \frac{k!}{(k-1)^5} \right|$$
$$= \lim_{k\to\infty} \left(\frac{k}{k-1} \right)^5 \cdot \lim_{k\to\infty} \frac{1}{k+1} = 1 \cdot 0 = 0.$$

Thus $l = 0$, so $R = \infty$ and the radius of convergence is infinite (that is, the series converges for all x). ▲

Example 3 For which x do the following series converge? (a) $\sum_{i=1}^{\infty} \frac{x^i}{i}$ (b) $\sum_{i=1}^{\infty} \frac{x^i}{i^2}$ (c) $\sum_{i=0}^{\infty} \frac{x^i}{i!}$ (By convention, we define $0! = 1$.)

Solution (a) We have $a_i = 1/i$, so

$$l = \lim_{i\to\infty} \left| \frac{a_i}{a_{i-1}} \right| = \lim_{i\to\infty} \left(\frac{i-1}{i} \right) = 1;$$

the series therefore converges for $|x| < 1$ and diverges for $|x| > 1$. When $x = 1$, $\sum_{i=1}^{\infty} x^i/i$ is the divergent harmonic series; for $x = -1$, the series is alternating, so it converges.
(b) We have $a_i = 1/i^2$, so

$$l = \lim_{i\to\infty} \frac{(i-1)^2}{i^2} = 1$$

and the radius of convergence is again 1. This time, when $x = 1$, we get the p-series $\sum_{i=1}^{\infty}(1/i^2)$, which converges since $p = 2 > 1$. The series for $x = -1$, $\sum_{i=1}^{\infty}[(-1)^i/i^2]$, converges absolutely, so is also convergent.
(c) Here $a_i = 1/i!$, so $|a_i/a_{i-1}| = (i-1)!/i! = 1/i \to 0$ as $i \to \infty$. Thus $l = 0$, so the series converges for all x. ▲

Series of the form $\sum_{i=0}^{\infty} a_i(x - x_0)^i$ are also called power series; their theory is essentially the same as for the case $x_0 = 0$ already studied, because $\sum_{i=0}^{\infty} a_i(x - x_0)^i$ may be written as $\sum_{i=0}^{\infty} a_i w^i$, where $w = x - x_0$.

Example 4 For which x does the series $\displaystyle\sum_{n=0}^{\infty} \frac{4^n}{\sqrt{2n+5}}(x+5)^n$ converge?

Solution This series is of the form $\sum_{i=0}^{\infty} a_i(x - x_0)^i$, with $a_i = 4^i/\sqrt{2i+5}$ and $x_0 = -5$. We have

$$l = \lim_{i\to\infty} \frac{a_i}{a_{i-1}} = \lim_{i\to\infty} \frac{4^i}{\sqrt{2i+5}} \cdot \frac{\sqrt{2(i-1)+5}}{4^{i-1}} = \lim_{i\to\infty} 4\sqrt{\frac{2i+3}{2i+5}} = 4,$$

so the radius of convergence is $\frac{1}{4}$. Thus the series converges for $|x+5| < \frac{1}{4}$ and diverges for $|x+5| > \frac{1}{4}$. When $x = -5\frac{1}{4}$, the series becomes $\sum_{i=0}^{\infty}[(-1)^i/\sqrt{2i+5}\,]$, which converges because it is alternating. When $x = -4\frac{3}{4}$, the series is $\sum_{i=0}^{\infty}[1/\sqrt{2i+5}\,]$, which diverges by the ratio comparison test with $\sum_{i=1}^{\infty}(1/\sqrt{2i}\,)$ (or by the integral test). Thus our power series converges when $-5\frac{1}{4} \leqslant x < -4\frac{3}{4}$. ▲

In place of the ratio test, one can sometimes use the root test in the same way.

Root Test for Power Series

Let $\displaystyle\sum_{i=0}^{\infty} a_i x^i$ be a given power series. Assume that $\displaystyle\lim_{i\to\infty} |a_i|^{1/i} = \rho$ exists. Then the radius of convergence is $R = 1/\rho$.

Indeed, if $|x| < R$, $\lim_{i\to\infty}|a_i x^i|^{1/i} = \lim_{i\to\infty}|a_i|^{1/i}|x| = \rho|x| < \rho R = 1$, so the power series converges by the root test.

Example 5 Find the radius of convergence of the series $\displaystyle\sum_{i=1}^{\infty} \frac{x^i}{(2+1/i)^i}$.

Solution $\rho = \lim_{i\to\infty}|a_i|^{1/i} = \lim_{i\to\infty}(1/(2+1/i)^i)^{1/i} = \lim_{i\to\infty}\{1/[2+(1/i)]\} = \frac{1}{2}$, so the radius of convergence is $R = 2$. ▲

Let $f(x) = \sum_{i=0}^{\infty} a_i x^i$, defined where the series converges. By analogy with ordinary polynomials, we might guess that

$$f'(x) = \sum_{i=1}^{\infty} i a_i x^{i-1}$$

and that

$$\int f(x)\,dx = \sum_{i=0}^{\infty} \frac{a_i x^{i+1}}{i+1} + C.$$

In fact, this is true. The proof is contained in (the moderately difficult) Exercises 41–45 at the end of the section.

Example 6 If $f(x) = \displaystyle\sum_{n=0}^{\infty} \frac{x^n}{n!}$, show that $f'(x) = f(x)$. Conclude that $f(x) = e^x$.

Solution By Example 3(c), the series for $f(x)$ converges for all x. Then $f'(x) = \sum_{i=1}^{\infty}(ix^{i-1}/i)! = \sum_{i=1}^{\infty}[x^{i-1}/(i-1)!] = \sum_{i=0}^{\infty}(x^i/i!) = f(x)$. By the uniqueness of the solution of the differential equation $f'(x) = f(x)$ (see Section 8.2), $f(x)$ must be ce^x for some c. Since $f(0) = 1$, c must be 1, and so $f(x) = e^x$. ▲

Differentiation and Integration of Power Series

To differentiate or integrate a power series within its radius of convergence R, differentiate or integrate it term by term: if $|x - x_0| < R$,

$$\frac{d}{dx}\sum_{i=0}^{\infty} a_i(x - x_0)^i = \sum_{i=1}^{\infty} ia_i(x - x_0)^{i-1},$$

$$\text{and} \quad \int\left[\sum_{i=0}^{\infty} a_i(x - x_0)^i\right]dx = \sum_{i=0}^{\infty}\frac{a_i}{i+1}(x - x_0)^{i+1} + C.$$

(The resulting series converge if $|x - x_0| < R$.)

Example 7 Let $f(x) = \sum_{i=0}^{\infty}\frac{i}{i+1}x^i$. Find a series expression for $f'(x)$. Where is it valid?

Solution By Example 1, $f(x)$ converges for $|x| < 1$. Thus $f'(x)$ also converges if $|x| < 1$, and we may differentiate term by term:

$$f'(x) = \sum_{i=0}^{\infty}\frac{i^2}{i+1}x^{i-1}, \qquad |x| < 1 \text{ (this series diverges at } x = \pm 1\text{)}.$$

(Notice that $f'(x)$ is again a power series, so it too can be differentiated. Since this can be repeated, we conclude that f can be differentiated as many times as we please. We say that f is *infinitely differentiable*.) ▲

Example 8 Write down power series for $x/(1 + x^2)$ and $\ln(1 + x^2)$. Where do they converge?

Solution First, we expand $1/(1 + x^2)$ as a geometric series using the general formula $1/(1 - r) = 1 + r + r^2 + \cdots$, with r replaced by $-x^2$, obtaining $1 - x^2 + x^4 - \cdots$. Multiplying by x gives $x/(1 + x^2) = x - x^3 + x^5 - \cdots$, which converges for $|x| < 1$. (It diverges for $x = \pm 1$.)

Now we observe that $(d/dx)\ln(1 + x^2) = 2x/(1 + x^2)$, so

$$\begin{aligned}\ln(1 + x^2) &= 2\int\frac{x}{1 + x^2}\,dx = 2\int(x - x^3 + x^5 - \cdots)\,dx \\ &= 2\left(\frac{x^2}{2} - \frac{x^4}{4} + \frac{x^6}{6} - \cdots\right) = x^2 - \frac{x^4}{2} + \frac{x^6}{3} - \frac{x^8}{4} + \cdots.\end{aligned}$$

(The integration constant was dropped because $\ln(1 + 0^2) = 0$.) This series converges for $|x| < 1$, and also for $x = \pm 1$, because there it is alternating. ▲

The operations of addition and multiplication by a constant may be performed term by term on power series, just as on polynomials. This may be proved using the limit theorems. The operations of multiplication and division proceed by the same methods one uses for polynomials, but are more subtle to justify. We state the results in the following box.

Algebraic Operations on Power Series

Let $f(x) = \sum_{i=0}^{\infty} a_i x^i$, with radius of convergence R.

Let $g(x) = \sum_{i=0}^{\infty} b_i x^i$, with radius of convergence S.

If T is the smaller of R and S, then

$$f(x) + g(x) = \sum_{i=0}^{\infty} (a_i + b_i)x^i \quad \text{for} \quad |x| < T;$$

$$cf(x) = \sum_{i=0}^{\infty} (ca_i)x^i \quad \text{for} \quad |x| < R;$$

$$f(x)g(x) = \sum_{i=0}^{\infty} \left(\sum_{j=0}^{i} a_j b_{i-j} \right) x^i \quad \text{for} \quad |x| < T.$$

If $b_0 \neq 0$, then $f(x)/g(x) = \sum_{i=0}^{\infty} c_i x^i$ for x near zero, where the c_i's may be determined by long division. The determination of the radius of convergence of f/g requires further analysis.

Example 9 Write down power series of the form $\sum_{i=0}^{\infty} a_i x^i$ for $2/(3-x)$, $5/(4-x)$, and $(23-7x)/[(3-x)(4-x)]$. What are their radii of convergence?

Solution We may write

$$\frac{2}{3-x} = \frac{2}{3}\left(\frac{1}{1-x/3}\right) = \frac{2}{3}\sum_{i=0}^{\infty}\left(\frac{x}{3}\right)^i = \sum_{i=0}^{\infty} \frac{2}{3^{i+1}} x^i.$$

The ratio of successive coefficients is $(1/3^{i+1})/(1/3^i) = 1/3$, so the radius of convergence is 3.

Similarly,

$$\frac{5}{4-x} = \sum_{i=0}^{\infty} \frac{5}{4^{i+1}} x^i$$

with radius of convergence 4. Finally, we may use partial fractions (Section 10.2) to write $(23-7x)/[(3-x)(4-x)] = 2/(3-x) + 5/(4-x)$, so we have

$$\frac{23-7x}{(3-x)(4-x)} = \sum_{i=0}^{\infty}\left(\frac{2}{3^{i+1}} + \frac{5}{4^{i+1}}\right)x^i.$$

By the preceding box, the radius of convergence of this series is at least 3. In fact, a limit computation shows that the ratio of successive coefficients approaches $\frac{1}{3}$, so the radius of convergence is exactly 3. ▲

In practice, we do not use the formula for $f(x)g(x)$ in the box above, but merely multiply the series for f and g term by term; in the product, we collect the terms involving each power of x.

Example 10 Write down the terms through x^4 in the series for $e^x/(1-x)$.

Solution We have $e^x = 1 + x + x^2/2 + x^3/6 + x^4/24 + \cdots$ (from Example 6) and $1/(1-x) = 1 + x + x^2 + x^3 + x^4 + \cdots$. We multiply terms in the first series by terms in the second series, in all possible ways.

	1	x	$\frac{x^2}{2}$	$\frac{x^3}{6}$	$\frac{x^4}{24}$	$\cdots$
1	1	x	$\frac{x^2}{2}$	$\frac{x^3}{6}$	$\frac{x^4}{24}$	$\cdots$
x	x	x^2	$\frac{x^3}{6}$	$\frac{x^4}{6}$	$\cdots$	
x^2	x^2	x^3	$\frac{x^4}{2}$	$\cdots$		
x^3	x^3	x^4	$\cdots$			
x^4	x^4	$\cdots$				

(Since we want the product series only through x^4, we may neglect the terms in higher powers of x.) Reading along diagonals from lower left to upper right, we collect the powers of x to get

$$\frac{e^x}{1-x} = 1 + (x+x) + \left(x^2 + x^2 + \frac{x^2}{2}\right) + \left(x^3 + x^3 + \frac{x^3}{2} + \frac{x^3}{6}\right)$$
$$+ \left(x^4 + x^4 + \frac{x^4}{2} + \frac{x^4}{6} + \frac{x^4}{24}\right) + \cdots$$
$$= 1 + 2x + \frac{5}{2}x^2 + \frac{8}{3}x^3 + \frac{65}{24}x^4 + \cdots . \blacktriangle$$

Exercises for Section 12.4

For which x do the series in Exercises 1–10 converge?

1. $\sum_{i=0}^{\infty} \frac{2}{i+1} x^i$
2. $\sum_{i=0}^{\infty} (2i+1)x^i$
3. $\sum_{n=1}^{\infty} \frac{3}{n^2} x^n$
4. $\sum_{n=1}^{\infty} \frac{(-1)^n 2^n}{n(n+1)} x^n$
5. $\sum_{i=1}^{\infty} \frac{5i+1}{i}(x-1)^i$
6. $\sum_{r=0}^{\infty} \frac{r!}{3^{2r}}(x+2)^r$
7. $\sum_{n=2}^{\infty} \frac{1}{n!\sin(\pi/n)} x^n$
8. $\sum_{i=14}^{\infty} \frac{i(i+3)}{i^3-4i+7} x^i$
9. $\sum_{n=1}^{\infty} \frac{x^n}{2^n+4^n}$
10. $\sum_{s=1}^{\infty} \left(\frac{2^s+1}{8s^7}\right)^{3/2} x^s$

Find the radius of convergence of the series in Exercises 11–14.

11. $x + \frac{x^2}{2!} + \frac{x^3}{3!} + \cdots$
12. $1 + \frac{x}{2} + \frac{2!}{4!}x^2 + \frac{3!}{6!}x^3 + \cdots$
13. $\frac{5x}{2} + \frac{10x^2}{4} + \frac{15x^3}{8} + \frac{20x^4}{16} + \cdots$
14. $1 + \frac{x}{2} + \frac{x^2}{3} + \frac{x^3}{4} + \cdots$

Find the radius of convergence R of the series $\sum_{n=0}^{\infty} a_n x^n$ in Exercises 15–18 for the given choices of a_n. Discuss convergence at $\pm R$.

15. $a_n = 1/(n+1)^n$
16. $a_n = (-1)^n/(n+1)$
17. $a_n = (n^2+n^3)/(1+n)^5$
18. $a_n = n$

Use the root test to determine the radius of convergence of the series in Exercises 19–22.

19. $\sum_{n=1}^{\infty} \frac{x^n}{(3+1/n)^n}$
20. $\sum_{n=1}^{\infty} \frac{2^n x^n}{n^n}$
21. $\sum_{n=1}^{\infty} (-1)^n n^n x^n$
22. $\sum_{n=1}^{\infty} \frac{2x^n}{1+5^n}$
23. Let $f(x) = x - x^3/3! + x^5/5! - \cdots$. Show that f is defined and is differentiable for all x. Show that $f''(x) + f(x) = 0$. Use the uniqueness of solutions of this equation (Section 8.1) to show that $f(x) = \sin x$.
24. By differentiating the result of Exercise 23, find a series representation for $\cos x$.
25. Let $f(x) = \sum_{i=1}^{\infty}(i+1)x^i$.
    (a) Find the radius of convergence of this series.
    (b) Find the series for $\int_0^x f(t)\,dt$.
    (c) Use the result of part (b) to sum the series $f(x)$.
    (d) Sum the series $\frac{2}{2} + \frac{3}{4} + \frac{4}{8} + \frac{5}{16} + \cdots$.
26. (a) Write a power series representing the integral of $1/(1-x)$ for $|x| < 1$. (b) Write a power series for $\ln x = \int (dx/x)$ in powers of $1-x$. Where is it valid?

Write power series representations for the functions in Exercises 27–30.

27. e^{-x^2}. (Use Example 6.)
28. $(d/dx)e^{-x^2}$.

29. $\tan^{-1}x$ and its derivative [*Hint:* Do the derivative first.]
30. The second derivative of $1/(1-x)$.
31. Find the series for $1/[(1-x)(2-x)]$ by writing
$$\frac{1}{(1-x)(2-x)} = \frac{A}{1-x} + \frac{B}{2-x}$$
and adding the resulting geometric series.
32. Find the series for $x/(x^2-4x+3)$. (See Exercise 31).
33. Using the result of Exercise 23, write the terms through x^6 in a power series expansion of $\sin^2 x$.
34. Find the terms through x^6 in the series for $\sin^3 x/x$.
35. Find series $f(x)$ and $g(x)$ such that the series $f(x)+g(x)$ is not identically zero but has a larger radius of convergence than either $f(x)$ or $g(x)$.
36. Find series $f(x)$ and $g(x)$, each of them having radius of convergence 2, such that $f(x)+g(x)$ has radius of convergence 3.
37. (a) By dividing the series for $\sin x$ by that for $\cos x$, find the terms through x^5 in the series for $\tan x$.
    (b) Find the terms through x^4 in the series for $\sec^2 x = (d/dx)\tan x$.
    (c) Using the result of part (b), find the terms through x^4 in the series for $1/\sec^2 x$.
38. Find the terms through x^5 in the series for
$$\tanh x = \frac{e^x - e^{-x}}{e^x + e^{-x}}$$
39. Find a power series which converges just when $-1 < x \leqslant 1$.
40. Why can't $x^{1/3}$ be represented in the form of a series $\sum_{i=0}^{\infty} a_i x^i$, convergent near $x=0$?

Exercises 41–45 contain the proof of the results on the differentiation and integration of power series. For simplicity, we consider only the case $x_0 = 0$. Refer to the following theorem.

Theorem *Suppose that $\sum_{i=0}^{\infty} a_i x^i$ converges for some particular value of x, say $x = x_0$. Then:*

1. *There is an integer N such that $\sqrt[i]{|a_i|} < 1/|x_0|$ for all $i \geqslant N$.*
2. *If $|y| < |x_0|$, then $\sum_{i=0}^{\infty} a_i y^i$ converges absolutely.*

Proof For part 1, suppose that $\sqrt[i]{|a_i|} \geqslant 1/|x_0|$ for arbitrarily large values of i. Then for these values of i we have $|a_i| \geqslant 1/|x_0|^i$, and $|a_i x_0^i| \geqslant 1$; but then we could not have $a_i x_0^i \to 0$, as is required for convergence.

For part 2, let $r = |y|/|x_0|$, so that $|r| < 1$. By part 1, $|a_i y^i| = |a_i|\,|x_0|^i r^i < r^i$ for all $i \geqslant N$. By the comparison test, the series $\sum_{i=N}^{\infty} a_i y^i$ converges absolutely; it follows that the entire series converges absolutely as well. ■

★41. Prove that the series $f(x) = \sum_{i=0}^{\infty} a_i x^i$, $g(x) = \sum_{i=1}^{\infty} i a_i x^{i-1}$, and $h(x) = \sum_{i=0}^{\infty}[a_i/(i+1)]x^{i+1}$ all have the same radius of convergence. [*Hint:* Use the theorem above, the comparison test, and the fact that $\sum_{i=0}^{\infty} i r^i$ converges absolutely if $|r| < 1$.]

★42. Prove that if $0 < R_1 < R$, where R is the radius of convergence of $f(x) = \sum_{i=0}^{\infty} a_i x^i$, then given any $\varepsilon > 0$, there is a positive number M such that, for every number N greater than M, the difference $|f(x) - \sum_{i=0}^{N} a_i x^i|$ is less than ε for all x in the interval $[-R_1, R_1]$. [*Hint:* Compare $\sum_{i=N+1}^{\infty} a_i x^i$ with a geometric series, using the theorem above.]

★43. Prove that if $|x_0| < R$, where R is the radius of convergence of $f(x) = \sum_{i=0}^{\infty} a_i x^i$, then f is continuous at x_0. [*Hint:* Use Exercise 42, together with the fact that the polynomial $\sum_{i=0}^{N} a_i x^i$ is continuous. Given $\varepsilon > 0$, write $f(x) - f(x_0)$ as a sum of terms, each of which is less than $\varepsilon/3$, by choosing N large enough and $|x - x_0|$ less than some δ.]

★44. Prove that if $|x| < R$, where R is the radius of convergence of $f(x) = \sum_{i=0}^{\infty} a_i x^i$, then the integral $\int_0^x f(t)\,dt$ (which exists by Exercise 43) is equal to $\sum_{i=0}^{\infty}[a_i/(i+1)]x^{i+1}$. [*Hint:* Use the result of Exercise 42 to show that the difference $|\int_0^x f(t)\,dt - \sum_{i=0}^{\infty}[a_i x^i/(i+1)]|$ is less than any positive number ε.]

★45. Prove that if $f(x) = \sum_{i=0}^{\infty} a_i x^i$ and $g(x) = \sum_{i=1}^{\infty} i a_i x^{i-1}$ have radius of convergence R, then $f'(x) = g(x)$ on $(-R, R)$. [*Hint:* Apply the result of Exercise 44 to $\int_0^x g(t)\,dt$; then use the alternative version of the fundamental theorem of calculus.]

12.5 Taylor's Formula

The power series which represents a function is determined by the derivatives of the function at a single point.

Up until now, we have used various makeshift methods to find power series expansions for specific functions. In this section, we shall see how to do this systematically. The idea is to assume the existence of a power series and to identify the coefficients one by one.

If $f(x) = \sum_{i=0}^{\infty} a_i(x - x_0)^i$ is convergent for $x - x_0$ small enough, we can find the coefficient a_0 simply by setting $x = x_0$: $f(x_0) = \sum_{i=0}^{\infty} a_i(x_0 - x_0)^i = a_0$. Differentiating and *then* substituting $x = x_0$, we can find a_1. Writing out the series explicitly will clarify the procedure:

$$f(x) = a_0 + a_1(x - x_0) + a_2(x - x_0)^2 + a_3(x - x_0)^3 + \cdots, \text{ so } f(x_0) = a_0\,;$$

$$f'(x) = a_1 + 2a_2(x - x_0) + 3a_3(x - x_0)^2 + 4a_4(x - x_0)^3 + \cdots,$$

$$\text{so } f'(x_0) = a_1\,.$$

Similarly, by taking more and more derivatives before we substitute, we find

$$f''(x) = 2a_2 + 3 \cdot 2a_3(x - x_0) + 4 \cdot 3a_4(x - x_0)^2 + \cdots \qquad \text{so } f''(x_0) = 2a_2\,;$$

$$f'''(x) = 3 \cdot 2a_3 + 4 \cdot 3 \cdot 2a_4(x - x_0) + \cdots \qquad \text{so } f'''(x_0) = 3 \cdot 2a_3\,;$$

$$f''''(x) = 4 \cdot 3 \cdot 2a_4 + \cdots \qquad \text{so } f''''(x_0) = 4 \cdot 3 \cdot 2a_4\,;$$

etc.

Solving for the a_i's, we have $a_0 = f(x_0)$, $a_1 = f'(x_0)$, $a_2 = f''(x_0)/2$, $a_3 = f'''(x_0)/2 \cdot 3$, and, in general, $a_i = f^{(i)}(x_0)/i!$. Here $f^{(i)}$ denotes the ith derivative of f, and we recall that $i! = i \cdot (i-1) \cdots 3 \cdot 2 \cdot 1$, read "$i$ factorial." (We use the conventions that $f^{(0)} = f$ and $0! = 1$.)

This argument shows that if a function $f(x)$ can be written as a power series in $(x - x_0)$, then this series *must* be

$$\sum_{i=0}^{\infty} \frac{f^{(i)}(x_0)}{i!}(x - x_0)^i.$$

For any f, this series is called the *Taylor series* of f about the point $x = x_0$. (This formula is responsible for the factorials which appear in so many important power series.)

The point x_0 is often chosen to be zero, in which case the series becomes

$$\sum_{i=0}^{\infty} \frac{f^{(i)}(0)}{i!} x^i$$

and is called the *Maclaurin*[3] *series* of f.

[3] Brook Taylor (1685–1731) and Colin Maclaurin (1698–1746) participated in the development of calculus following Newton and Leibniz. According to the *Guinness Book of World Records*, Maclaurin has the distinction of being the youngest full professor of all time at age 19 in 1717. He was recommended by Newton. Another mathematician-physicist, Lord Kelvin, holds the record for the youngest and fastest graduation from college—between October 1834 and November 1834, at age 10.

Taylor and Maclaurin Series

If f is infinitely differentiable on some interval containing x_0, the series

$$\sum_{i=0}^{\infty} \frac{f^{(i)}(x_0)}{i!}(x-x_0)^i$$

is called the *Taylor series* of f at x_0.

When $x_0 = 0$, the series has the simpler form

$$\sum_{i=0}^{\infty} \frac{f^{(i)}(0)}{i!}x^i$$

and is called the *Maclaurin series* of f.

Example 1 Write down the Maclaurin series for $\sin x$.

Solution We have

$$\begin{aligned}
f(x) &= \sin x, & f(0) &= 0;\\
f'(x) &= \cos x, & f'(0) &= 1;\\
f''(x) &= -\sin x, & f''(0) &= 0;\\
f^{(3)}(x) &= -\cos x, & f^{(3)}(0) &= -1;\\
f^{(4)}(x) &= \sin x, & f^{(4)}(0) &= 0;
\end{aligned}$$

and the pattern repeats from here on. Hence the Maclaurin series is

$$\frac{f(0)}{0!}x^0 + \frac{f'(0)}{1!}x + \frac{f^{(2)}(0)}{2!}x^2 + \cdots = x - \frac{x^3}{3!} + \frac{x^5}{5!} - \frac{x^7}{7!} + \cdots . \blacktriangle$$

Example 2 Find the terms through x^3 in the Taylor series for $1/(1+x^2)$ at $x_0 = 1$.

Solution *Method* 1. We differentiate $f(x)$ three times:

$$\begin{aligned}
f(x) &= \frac{1}{1+x^2}, & f(1) &= \frac{1}{2}, & a_0 &= f(1) = \frac{1}{2};\\
f'(x) &= \frac{-2x}{(1+x^2)^2}, & f'(1) &= -\frac{1}{2}, & a_1 &= f'(1) = -\frac{1}{2};\\
f''(x) &= \frac{6x^2-2}{(1+x^2)^3}, & f''(1) &= \frac{1}{2}, & a_2 &= \frac{f''(1)}{2!} = \frac{1}{4};\\
f'''(x) &= \frac{-24x^3+24x}{(1+x^2)^4}, & f'''(1) &= 0, & a_3 &= \frac{f'''(1)}{3!} = 0;
\end{aligned}$$

so the Taylor series begins

$$\frac{1}{1+x^2} = \frac{1}{2} - \frac{1}{2}(x-1) + \frac{1}{4}(x-1)^2 + 0\cdot(x-1)^3 + \cdots .$$

Method 2. Write

$$\frac{1}{1+x^2} = \frac{1}{1+[(x-1)+1]^2} = \frac{1}{2+2(x-1)+(x-1)^2}$$

$$= \frac{1}{2}\left[\frac{1}{1+(x-1)+\frac{1}{2}(x-1)^2}\right]$$

$$= \frac{1}{2}\left[1-(x-1)-\frac{1}{2}(x-1)^2 + \left((x-1)+\frac{(x-1)^2}{2}\right)^2 - \cdots\right] \quad \text{(geometric series)}$$

$$= \frac{1}{2}\left[1-(x-1)-\frac{1}{2}(x-1)^2+(x-1)^2+\cdots\right]$$

$$= \frac{1}{2}-\frac{1}{2}(x-1)+\frac{1}{4}(x-1)^2+\cdots. \blacktriangle$$

Notice that we can write the Taylor series for any function which can be differentiated infinitely often, but we do not yet know whether the series converges to the given function. To understand when this convergence takes place, we proceed as follows. Using the fundamental theorem of calculus, write

$$f(x) = f(x_0) + \int_{x_0}^{x} f'(t)\,dt. \tag{1}$$

We now use integration by parts with $u = f'(t)$ and $v = x - t$. The result is

$$\int_{x_0}^{x} f'(t)\,dt = -\int_{x_0}^{x} u\,dv = -\left(uv\Big|_{x_0}^{x} - \int_{x_0}^{x} v\,du\right)$$

$$= f'(x_0)(x-x_0) + \int_{x_0}^{x}(x-t)f''(t)\,dt.$$

Thus we have proved the identity

$$f(x) = f(x_0) + f'(x_0)(x-x_0) + \int_{x_0}^{x}(x-t)f''(t)\,dt. \tag{2}$$

Note that the first two terms on the right-hand side of formula (2) equal the first two terms in the Taylor series of f. If we integrate by parts again with

$$u = f''(t) \quad \text{and} \quad v = \frac{(x-t)^2}{2},$$

we get

$$\int_{x_0}^{x}(x-t)f''(t)\,dt = -\int_{x_0}^{x} u\,dv = -uv\Big|_{x_0}^{x} + \int_{x_0}^{x} v\,du$$

$$= \frac{f''(x_0)}{2}(x-x_0)^2 + \int_{x_0}^{x}\frac{(x-t)^2}{2}f'''(t)\,dt;$$

so, substituting into formula (2),

$$f(x) = f(x_0) + f'(x_0)(x-x_0) + \frac{f''(x_0)}{2}(x-x_0)^2 + \int_{x_0}^{x}\frac{(x-t)^2}{2}f'''(t)\,dt. \tag{3}$$

Repeating the procedure n times, we obtain the formula

$$f(x) = f(x_0) + f'(x_0)(x-x_0) + \frac{f''(x_0)}{2}(x-x_0)^2 + \cdots + \frac{f^{(n)}(x_0)}{n!}(x-x_0)^n + \int_{x_0}^{x}\frac{(x-t)^n}{n!}f^{(n+1)}(t)\,dt \tag{4}$$

which is called *Taylor's formula with remainder in integral form*. The expression

$$R_n(x) = \int_{x_0}^{x} \frac{(x-t)^n}{n!} f^{(n+1)}(t)\,dt \tag{5}$$

is called the *remainder*, and formula (4) may be written in the form

$$f(x) = \sum_{i=0}^{n} \frac{f^{(i)}(x_0)}{i!}(x-x_0)^i + R_n(x). \tag{6}$$

By the second mean value theorem of integral calculus (Review Exercise 40, Chapter 9), we can write

$$R_n(x) = f^{(n+1)}(c)\left[\int_{x_0}^{x} \frac{(x-t)^n}{n!}\,dt\right] = f^{(n+1)}(c)\frac{(x-x_0)^{n+1}}{(n+1)!} \tag{7}$$

for some point c between x_0 and x. Substituting formula (7) into formula (6), we have

$$f(x) = \sum_{i=0}^{n} \frac{f^{(i)}(x_0)}{i!}(x-x_0)^i + \frac{f^{(n+1)}(c)}{(n+1)!}(x-x_0)^{n+1}. \tag{8}$$

Formula (8), which is called *Taylor's formula with remainder in derivative form*, reduces to the usual mean value theorem when we take $n=0$; that is,

$$f(x) = f(x_0) + f'(c)(x-x_0)$$

for some c between x_0 and x.

If $R_n(x) \to 0$ as $n \to \infty$, then formula (6) tells us that the Taylor series of f will converge to f.

The following box summarizes our discussion of Taylor series.

Convergence of Taylor Series

1. If $f(x) = \sum_{i=0}^{\infty} a_i(x-x_0)^i$ is a convergent power series on an open interval I centered at x_0, then f is infinitely differentiable and $a_i = f^{(i)}(x_0)/i!$, so
$$f(x) = \sum_{i=0}^{\infty} \frac{f^{(i)}(x_0)}{i!}(x-x_0)^i.$$
2. If f is infinitely differentiable on an open interval I centered at x_0, and if $R_n(x) \to 0$ as $n \to \infty$ for x in I, where $R_n(x)$ is defined by formula (5), then the Taylor series of f converges on I and equals f:
$$f(x) = \sum_{i=0}^{\infty} \frac{f^{(i)}(x_0)}{i!}(x-x_0)^i.$$

Example 3 (a) Expand the function $f(x) = 1/(1+x^2)$ in a Maclaurin series.
(b) Use part (a) to find $f'''''(0)$ and $f''''''(0)$ without calculating derivatives of f directly.
(c) Integrate the series in part (a) to prove that

$$\tan^{-1}x = x - \frac{x^3}{3} + \frac{x^5}{5} - \frac{x^7}{7} + \cdots \qquad \text{for} \quad |x| < 1.$$

★(d) Justify the formula of Euler:

$$\frac{\pi}{4} = 1 - \frac{1}{3} + \frac{1}{5} - \frac{1}{7} + \cdots.$$

Solution (a) We expand $1/(1 + x^2)$ as a geometric series:

$$\frac{1}{1+x^2} = \frac{1}{1-(-x^2)} = 1 + (-x^2) + (-x^2)^2 + (-x^2)^3 + \cdots$$
$$= 1 - x^2 + x^4 - x^6 + \cdots$$

which is valid if $|-x^2| < 1$; that is, if $|x| < 1$. By the box above this is the Maclaurin series of $f(x) = 1/(1 + x^2)$.

(b) We find that $f'''''(0)/5!$ is the coefficient of x^5. Hence, as this coefficient is zero, $f'''''(0) = 0$. Likewise, $f''''''(0)/6!$ is the coefficient of x^6; thus $f''''''(0) = -6!$. This is *much* easier than calculating the sixth derivative of $f(x)$.

(c) Integrating from zero to x (justified in Section 12.4) gives

$$\int_0^x \frac{dt}{1+t^2} = x - \frac{x^3}{3} + \frac{x^5}{5} - \frac{x^7}{7} + \cdots;$$

but we know that the integral of $1/(1 + t^2)$ is $\tan^{-1}t$, so

$$\tan^{-1}x = x - \frac{x^3}{3} + \frac{x^5}{5} - \frac{x^7}{7} + \cdots \qquad \text{for } |x| < 1.$$

(d) If we set $x = 1$ and use $\tan^{-1}1 = \pi/4$, we get Euler's formula:

$$\frac{\pi}{4} = 1 - \frac{1}{3} + \frac{1}{5} - \frac{1}{7} + \cdots;$$

but this is not quite justified, since the series for $\tan^{-1}x$ is valid only for $|x| < 1$. (It is plausible, though, since $1 - \frac{1}{3} + \frac{1}{5} - \frac{1}{7} + \cdots$, being an alternating series, converges.) To justify Euler's formula, we may use the finite form of the geometric series expansion:

$$\frac{1}{1+t^2} = 1 - t^2 + t^4 + \cdots + (-1)^n t^{2n} + (-1)^{n+1}\frac{t^{2n+2}}{1+t^2}.$$

Integrating from 0 to 1, we have

$$\frac{\pi}{4} = \tan^{-1}1 = 1 - \frac{1}{3} + \frac{1}{5} - \cdots + \frac{(-1)^n}{2n+1} + (-1)^{n+1}\int_0^1 \frac{t^{2n+2}}{1+t^2}\,dt.$$

We will be finished if we can show that the last term goes to zero as $n \to \infty$. We have

$$0 \leqslant \int_0^1 \frac{t^{2n+2}}{1+t^2}\,dt \leqslant \int_0^1 t^{2n+2}\,dt = \frac{1}{2n+3}.$$

Since $\lim_{n\to\infty}[1/(2n+3)] = 0$, the limit of

$$(-1)^{n+1}\int_0^1 \frac{t^{n+2}}{1+t^2}\,dt$$

is zero as well (by the comparison test on p. 543). ▲

There is a simple test which guarantees that the remainder of a Taylor series tends to zero.

Taylor Series Test

To prove that a function $f(x)$ equals its Taylor series

$$\sum_{i=0}^{\infty} \frac{f^{(i)}(x_0)}{i!}(x - x_0)^i \qquad \text{on} \quad I,$$

it is sufficient to show:

1. f is infinitely differentiable on I;
2. the derivatives of f grow no faster than a constant C times the powers of a constant M; that is, for x in I,

$$|f^{(n)}(x)| \leqslant CM^n, \qquad n = 0, 1, 2, 3, \dots .$$

To justify this, we must show that $R_n(x) \to 0$. By formula (7),

$$|R_n(x)| = \left| f^{(n+1)}(c) \frac{(x - x_0)^{n+1}}{(n+1)!} \right| \leqslant \frac{CM^{n+1}|x - x_0|^{n+1}}{(n+1)!}.$$

For any number b, however, $b^n/n! \to 0$, since $\sum_{i=0}^{\infty}(b^i/i!)$ converges by Example 7, Section 12.3. Choosing $b = M|x - x_0|$, we can conclude that $R_n(x) \to 0$, so the Taylor series converges to f.

Example 4 Prove that:

(a) $e^x = 1 + x + \dfrac{x^2}{2} + \dfrac{x^3}{3!} + \cdots$ for all x.

(b) $\sin x = x - \dfrac{x^3}{3!} + \dfrac{x^5}{5!} - \dfrac{x^7}{7!} + \cdots$ for x in $(-\infty, \infty)$.

(c) $1 = \dfrac{\pi}{2} - \dfrac{\pi^3}{2^3 \cdot 3!} + \dfrac{\pi^5}{2^5 \cdot 5!} - \dfrac{\pi^7}{2^7 \cdot 7!} + \cdots .$

Solution (a) Let $f(x) = e^x$. Since $f^{(n)}(x) = e^x$, f is infinitely differentiable. Since all the derivatives at $x_0 = 0$ are 1, the Maclaurin series of e^x is $\sum_{n=0}^{\infty}(x^n/n!)$. To establish equality, it suffices to show $|f^{(n)}(x)| \leqslant CM^n$ on any finite interval I; but $f^{(n)}(x) = e^x$, independent of n, so in fact we can choose $M = 1$ and C the maximum of e^x on I.

(b) Since $f'(x) = \cos x$, $f''(x) = -\sin x, \dots,$ we see that f is infinitely differentiable. Notice that $f^{(n)}(x)$ is $\pm\cos x$ or $\pm\sin x$, so $|f^{(n)}(x)| \leqslant 1$. Thus we can choose $C = 1$, $M = 1$. Hence $\sin x$ equals its Maclaurin series, which was shown in Example 1 to be $x - x^3/3! + x^5/5! - \cdots$.

(c) Let $x = \pi/2$ in part (b). ▲

Some discussion of the limitations of Taylor series is in order. Consider, for example, the function $f(x) = 1/(1 + x^2)$, whose Maclaurin series is $1 - x^2 + x^4 - x^6 + \cdots$. Even though the function f is infinitely differentiable on the whole real line, its Maclaurin series converges only for $|x| < 1$. If we wish to represent $f(x)$ for x near 1 by a series, we may use a Taylor series with $x_0 = 1$ (see Example 2).

Another instructive example is the function $g(x) = e^{-1/x^2}$, where $g(0) = 0$. This function is infinitely differentiable, but all of its derivatives at $x = 0$ are equal to zero (see Review Exercise 123). Thus the Maclaurin series of g is $\sum_{i=0}^{\infty} 0 \cdot x^i$, which converges (it is zero) for all x, but not to the function g. There also exist infinitely differentiable functions with Taylor series having radius of convergence zero.[4] In each of these examples, the hypothesis that $R_n(x) \to 0$ as $n \to \infty$ fails, so the assertion in the box above is not contradicted. It simply does not apply. (Functions which satisfy $R_n(x) \to 0$, and so equal their Taylor series for x close to x_0, are important objects of study; these functions are called *analytic*).

The following box contains the most basic series expansions. They are worth memorizing.

Some Important Taylor and Maclaurin Series

Geometric:

$$\frac{1}{1-x} = 1 + x + x^2 + \cdots = \sum_{i=0}^{\infty} x^i, \qquad R = 1.$$

Binomial:

$$(1+x)^{\alpha} = 1 + \alpha x + \frac{\alpha(\alpha-1)}{2!}x^2 + \cdots$$

$$= \sum_{i=0}^{\infty} \frac{\alpha(\alpha-1)\cdots(\alpha-i+1)}{i!}x^i, \qquad R = 1.$$

Sine:

$$\sin x = x - \frac{x^3}{3!} + \frac{x^5}{5!} - \cdots = \sum_{i=0}^{\infty} \frac{(-1)^i x^{2i+1}}{(2i+1)!}, \qquad R = \infty.$$

Cosine:

$$\cos x = 1 - \frac{x^2}{2!} + \frac{x^4}{4!} - \cdots = \sum_{i=0}^{\infty} (-1)^i \frac{x^{2i}}{(2i)!}, \qquad R = \infty.$$

Exponential:

$$e^x = 1 + x + \frac{x^2}{2!} + \frac{x^3}{3!} + \cdots = \sum_{i=0}^{\infty} \frac{x^i}{i!}, \qquad R = \infty.$$

Logarithm:

$$\ln x = (x-1) - \frac{(x-1)^2}{2} + \frac{(x-1)^3}{3} - \cdots$$

$$= \sum_{i=1}^{\infty} (-1)^{i+1} \frac{(x-1)^i}{i}, \qquad R = 1.$$

$$\ln(1+x) = x - \frac{x^2}{2} + \frac{x^3}{3} - \cdots = \sum_{i=1}^{\infty} (-1)^{i+1} \frac{x^i}{i}, \qquad R = 1.$$

The only formula in the box which has not yet been justified is the binomial series. It may be proved by evaluating the derivatives of $f(x) = (1+x)^{\alpha}$ at $x = 0$ and verifying convergence by the *method* of the test in the box entitled Taylor series test. (See Review Exercise 124.) If $\alpha = n$ is a positive integer, the series terminates and we get the binomial formula

$$(1+x)^n = 1 + \binom{n}{1}x + \binom{n}{2}x^2 + \cdots + \binom{n}{n}x^n,$$

where

$$\binom{n}{k} = \frac{n(n-1)\cdots(n-k+1)}{k!}$$

is the number of ways of choosing k objects from a collection of n objects.

[4] See B. R. Gelbaum and J. M. H. Olmsted, *Counterexamples in Analysis*, Holden-Day, San Francisco (1964), p. 68.

Example 5 Expand $\sqrt{1+x^2}$ about $x_0 = 0$.

Solution The binomial series, with $\alpha = \frac{1}{2}$ and x^2 in place of x, gives

$$(1+x^2)^{1/2} = 1 + \tfrac{1}{2}x^2 + \frac{(\frac{1}{2})(\frac{1}{2}-1)}{2!}x^4 + \frac{(\frac{1}{2})(\frac{1}{2}-1)(\frac{1}{2}-2)}{3!}x^6 + \cdots$$

$$= 1 + \frac{1}{2}x^2 - \frac{1}{8}x^4 + \frac{1}{16}x^6 - \cdots, \text{ valid for } |x| <. \blacktriangle$$

Taylor's formula with remainder,

$$f(x) = \sum_{i=0}^{n} \frac{f^{(i)}(x_0)}{i!}(x-x_0)^i + R_n(x)$$

can be used to obtain approximations to $f(x)$; we can estimate the accuracy of these approximations using the formula

$$R_n(x) = \frac{f^{(n+1)}(c)}{(n+1)!}(x-x_0)^{n+1}$$

(for some c between x and x_0) and estimating $f^{(n+1)}$ on the interval between x and x_0. The partial sum of the Taylor series,

$$\sum_{i=0}^{n} \frac{f^{(i)}(x_0)}{i!}(x-x_0)^i$$

is a polynomial of degree n in x called the nth *Taylor* (*or Maclaurin* if $x_0 = 0$) *polynomial for* f *at* x_0, or the nth-order approximation to f at x_0. The first Taylor polynomial,

$$f(x_0) + f'(x_0)(x-x_0)$$

is just the *linear approximation* to $f(x)$ at x_0; the formula for the remainder $R_1(x) = [f''(c)/2](x-x_0)^2$ shows that we can estimate the error in the first-order approximation in terms of the size of the second derivative f'' on the interval between x and x_0.

A useful consequence of Taylor's theorem is that for many functions we can improve upon the linear approximation by using Taylor polynomials of higher order.

Example 6 Sketch the graph of $\sin x$ along with the graphs of its Maclaurin polynomials of degree 1, 2, and 3. Evaluate the polynomials at $x = 0.02$, 0.2, and 2, and compare with the exact value of $\sin x$.

Solution The Maclaurin polynomials of order 1, 2, and 3 are x, $x + 0x^2$, and $x - x^3/6$. They are sketched in Fig. 12.5.1. Evaluating at $x = 0.02$, 0.2, 2, and 20 gives the results shown in the table below.

x	$x - x^3/6$	$\sin x$
0.02	0.0199986667	0.0199986667
0.2	0.1986666	0.1986693
2	0.666666	0.909
20	−1313	0.912

▲

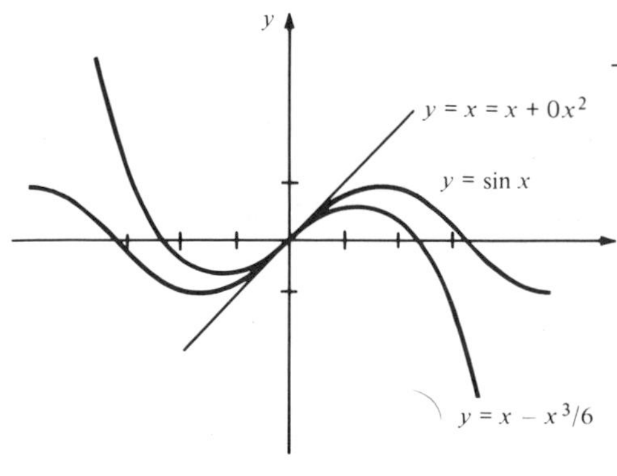

Figure 12.5.1. The first- and third-order approximations to $\sin x$.

The Maclaurin polynomials through degree 71 for $\sin x$ are shown in Fig. 12.5.2.[5] Notice that as n increases, the interval on which the nth Taylor polynomial is a good approximation to $\sin x$ becomes larger and larger; if we go beyond this interval, however, the polynomials of higher degree "blow up" more quickly than the lower ones.

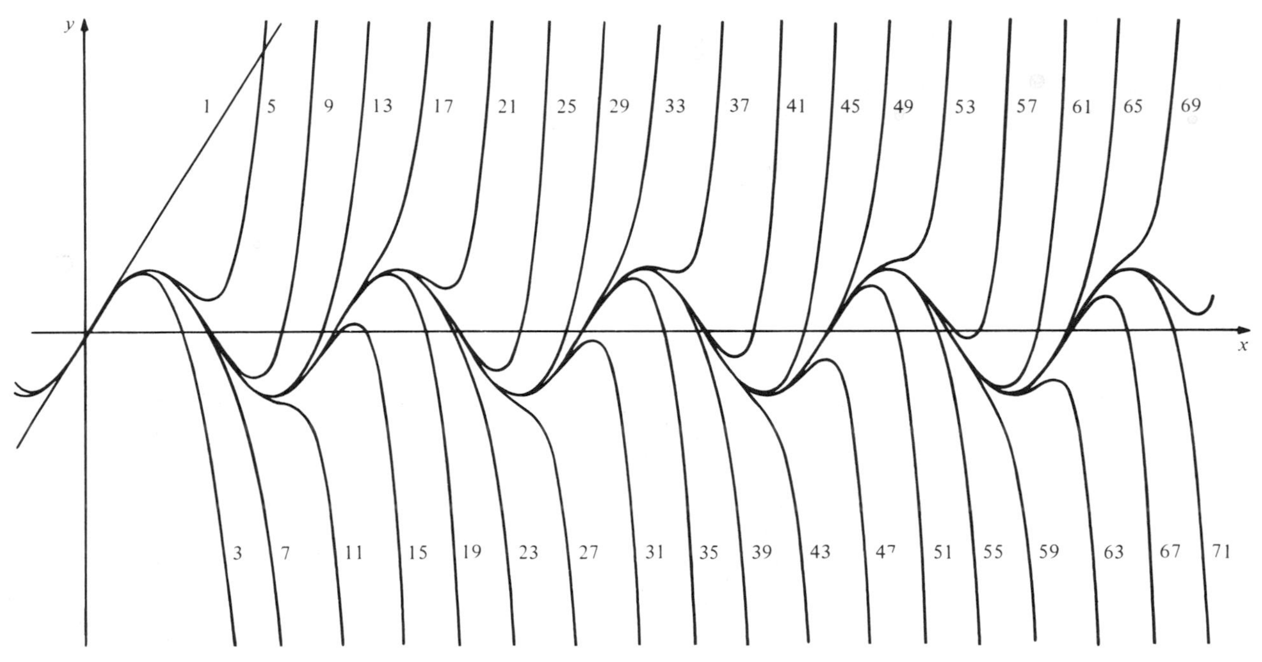

Figure 12.5.2. The Maclaurin polynomials for $\sin x$ through order 71. (The graphs to the left of the y axis are obtained by rotating the figure through 180°.)

The following example shows how errors may be estimated.

Example 7 Write down the Taylor polynomials of degrees 1 and 2 for $\sqrt[3]{x}$ at $x_0 = 27$. Use these polynomials to approximate $\sqrt[3]{28}$, and estimate the error in the second-order approximation by using the formula for $R_2(x)$.

Solution Let $f(x) = x^{1/3}$, $x_0 = 27$, $x = 28$. Then $f'(x) = \frac{1}{3}x^{-2/3}$, $f''(x) = -\frac{2}{9}x^{-5/3}$, and $f'''(x) = \frac{10}{27}x^{-8/3}$. Thus $f(27) = 3$, $f'(27) = \frac{1}{27}$, and $f''(27) = -(2/3^7)$, so the Taylor polynomials of degree 1 and 2 are, respectively,

$$3 + \frac{1}{27}(x - 27) \quad \text{and} \quad 3 + \frac{1}{27}(x - 27) - \frac{1}{3^7}(x - 27)^2.$$

Evaluating these at $x = 28$ gives 3.0370 . . . and 3.0365798 . . . for the first- and second-order approximations. The error in the second-order approximation is at most $1/3!$ times the largest value of $(10/27)x^{-8/3}$ on [27, 28], which is

$$\frac{1}{6}\,\frac{10}{27}\,\frac{1}{3^8} = \frac{5}{3^{12}} \leqslant 0.00001. \qquad (\text{Actually, } \sqrt[3]{28} = 3.0365889\ldots.)\ \blacktriangle$$

[5] We thank H. Ferguson for providing us with this computer-generated figure.

Example 8 By integrating a series for e^{-x^2}, calculate $\int_0^1 e^{-x^2}\,dx$ to within 0.001.

Solution Substituting $-x^2$ for x in the series for e^x gives

$$e^{-x^2} = 1 - x^2 + \frac{x^4}{2!} - \frac{x^6}{3!} + \frac{x^8}{4!} - \cdots.$$

Integrating term by term gives

$$\int_0^x e^{-t^2}\,dt = x - \frac{x^3}{3} + \frac{x^5}{10} - \frac{x^7}{42} + \cdots,$$

and so

$$\int_0^1 e^{-x^2}\,dx = 1 - \tfrac{1}{3} + \tfrac{1}{10} - \tfrac{1}{42} + \tfrac{1}{216} - \tfrac{1}{1320} + \cdots.$$

This is an alternating series, so the error is no greater than the first omitted term. To have accuracy 0.001, we should include $\frac{1}{216}$. Thus, within 0.001,

$$\int_0^1 e^{-x^2}\,dx \approx 1 - \frac{1}{3} + \frac{1}{10} - \frac{1}{42} + \frac{1}{216} \approx 0.747.$$

This method has an advantage over the methods in Section 11.5: to increase accuracy, we need only add on another term. Rules like Simpson's, on the other hand, require us to start over. (See Review Exercise 84 for Chapter 11.) Of course, if we have numerical data, or a function with an unknown or complicated series, using Simpson's rule may be necessary. ▲

Example 9 Calculate $\sin(\pi/4 + 0.06)$ to within 0.0001 by using the Taylor series about $x_0 = \pi/4$. How many terms would have been necessary if you had used the Maclaurin series?

Solution With $f(x) = \sin x$, and $x_0 = \pi/4$, we have

$$f(x) = \sin x, \qquad f(x_0) = \frac{1}{\sqrt{2}};$$

$$f'(x) = \cos x, \qquad f'(x_0) = \frac{1}{\sqrt{2}};$$

$$f''(x) = -\sin x, \qquad f''(x_0) = -\frac{1}{\sqrt{2}};$$

$$f'''(x) = -\cos x, \qquad f'''(x_0) = -\frac{1}{\sqrt{2}};$$

$$f''''(x) = \sin x, \qquad f''''(x_0) = \frac{1}{\sqrt{2}};$$

and so on. We have

$$R_n(x) = \frac{f^{(n+1)}(c)(x - x_0)^{n+1}}{(n+1)!}$$

for c between $\pi/4$ and $\pi/4 + 0.06$. Since $f^{(n+1)}(c)$ has absolute value less than 1, we have $|R_n(x)| \leqslant (0.06)^{n+1}/(n+1)!$. To make $|R_n(x)|$ less than 0.0001, it suffices to choose $n = 2$. The second-order approximation to $\sin x$ is

$$\frac{1}{\sqrt{2}} + \frac{1}{\sqrt{2}}\left(x - \frac{\pi}{4}\right) - \frac{1}{2\sqrt{2}}\left(x - \frac{\pi}{4}\right)^2.$$

Evaluating at $x = \pi/4 + 0.06$ gives 0.7483.

If we had used the Maclaurin polynomial of degree n, the error estimate would have been $|R_n(x)| \leqslant (\pi/4 + 0.06)^{n+1}/(n+1)!$. To make $|R_n(x)|$ less than 0.0001 would have required $n = 6$. ▲

Finally, we show how Taylor series can be used to evaluate limits in indeterminate form. The method illustrated below is sometimes more efficient than l'Hôpital's rule when that rule must be applied several times.

Example 10 Evaluate $\lim_{x\to 0} \dfrac{\sin x - x}{x^3}$ using a Maclaurin series.

Solution Since $\sin x = x - x^3/3! + x^5/5! - \cdots$, $\sin x - x = -x^3/3! + x^5/5! - \cdots$, and so $(\sin x - x)/x^3 = -1/6 + x^2/5! - \cdots$. Since this power series converges, it is continuous at $x = 0$, and so

$$\lim_{x\to 0} \frac{\sin x - x}{x^3} = -\frac{1}{6}. \blacktriangle$$

Example 11 Use Taylor series to evaluate

(a) $\displaystyle \lim_{x\to 0} \frac{\sin x - x}{\tan x - x}$

(compare Example 4, Section 11.2) and

(b) $\displaystyle \lim_{x\to 1} \frac{\ln x}{e^x - e}$.

Solution (a)
$$\frac{\sin x - x}{\tan x - x} = \frac{(\sin x)(\cos x) - x\cos x}{\sin x - x\cos x}$$
$$= \frac{(x - x^3/6 + \cdots)(1 - x^2/2 + \cdots) - x(1 - x^2/2 + \cdots)}{(x - x^3/6 + \cdots) - x(1 - x^2/2 + \cdots)}$$
$$= \frac{x - x^3/2 - x^3/6 + \cdots - x + x^3/2 + \cdots}{x - x^3/6 + \cdots - x + x^3/2 + \cdots}$$
$$= \frac{-x^3/6 + \cdots}{(1/3)x^3 + \cdots} = -\frac{-1/6 + \cdots}{1/3 + \cdots} \qquad \text{(dividing by } x^3\text{)}.$$

Since the terms denoted "$+ \cdots$" tend to zero as $x \to 0$, we get

$$\lim_{x\to 0} \frac{\sin x - x}{\tan x - x} = -\frac{1/6}{1/3} = -\frac{1}{2}.$$

(b)
$$\lim_{x\to 1} \frac{\ln x}{e^x - e} = \lim_{x\to 1} \frac{\ln x}{e(e^{x-1} - 1)}$$
$$= \frac{1}{e} \lim_{x\to 1} \frac{(x-1) - (1/2)(x-1)^2 + \cdots}{1 + (x-1) + (1/2)(x-1)^2 + \cdots - 1}$$
$$= \frac{1}{e} \lim_{x\to 1} \frac{1 - (1/2)(x-1) + \cdots}{1 + (1/2)(x-1) + \cdots}$$
$$= \frac{1}{e} \cdot \frac{1}{1} = \frac{1}{e}. \blacktriangle$$

For the last example, l'Hôpital's rule would have been a little easier to use.

Exercises for Section 12.5

Write down the Maclaurin series for the functions in Exercises 1–4.

1. $\sin 3x$
2. $\cos 4x$
3. $\cos x + e^{-2x}$
4. $\sin 2x - e^{-4x}$

Find the terms through x^3 in the Taylor series at $x_0 = 1$ for the functions in Exercises 5–8.

5. $1/(1 + x^2 + x^4)$
6. $1/\sqrt{2 - x^2}$
7. e^x
8. $\tan(\pi x/4)$
9. (a) Expand $f(x) = 1/(1 + x^2 + x^4)$ in a Maclaurin series through the terms in x^6, using a geometric series. (b) Use (a) to calculate $f''''''(0)$.
10. Expand $g(x) = e^{x^2}$ in a Maclaurin series as far as necessary to calculate $g^{(8)}(0)$ and $g^{(9)}(0)$.

Establish the equalities in Exercises 11–14 for a suitable domain in x.

11. $\ln(1 + x) = x - x^2/2 + x^3/3 - \cdots$.
12. $e^{1+x} = e + ex + \dfrac{ex^2}{2!} + \dfrac{ex^3}{3!} + \cdots$.
13. $\sqrt{x} = 1 + \frac{1}{2}(x - 1) - \frac{1}{8}(x - 1)^2 + \frac{1}{16}(x - 1)^3 - \cdots$.
14. $$\sin x = \frac{1}{\sqrt{2}}\left[1 + \left(x - \frac{\pi}{4}\right) - \frac{(x - \pi/4)^2}{2!} - \frac{(x - \pi/4)^3}{3!} + \cdots\right].$$
15. (a) Write out the Maclaurin series for the function $1/\sqrt{1 + x^2}$.(Use the binomial series.) (b) What is $(d^{20}/dx^{20})(1/\sqrt{1 + x^2})|_{x=0}$?
16. (a) Using the binomial series, write out the Maclaurin series for $g(x) = \sqrt{1 + x} + \sqrt{1 - x}$. (b) Find $g^{(20)}(0)$ and $g^{(2001)}(0)$.
17. Sketch the graphs of the Maclaurin polynomials through degree 4 for $\cos x$.
18. Sketch the graphs of the Maclaurin polynomials through degree 4 for $\tan x$.
19. Calculate $\ln(1.1)$ to within 0.001 by using a power series.
20. Calculate $e^{\ln 2 + 0.02}$ to within 0.0001 using a Taylor series about $x_0 = \ln 2$. How many terms would have been necessary if you had used the Maclaurin series?
21. Use the power series for $\ln(1 + x)$ to calculate $\ln 2\frac{1}{2}$, correct to within 0.1. [*Hint:* $2\frac{1}{2} = \frac{3}{2} \cdot \frac{5}{3}$.]
22. Continue the work of Example 7 by finding the third-, fourth-, fifth- (and so on) order approximations to $\sqrt[3]{28}$. Stop when the round-off errors on your calculator become greater than the remainder of the series.
23. Using the Maclaurin expansion for $1/(1 + x)$, approximate $\int_0^{1/2}[dx/(1 + x)]$ to within 0.01.
24. Use a binomial expansion to approximate $\int_0^{1/4}\sqrt{1 + x^3}\, dx$ to within 0.01.
25. (a) Use the second-order approximation at x_0 to derive the approximation
$$\int_{x_0 - R}^{x_0 + R} f(x)\, dx \approx 2Rf(x_0) + \frac{2f''(x_0)}{3!}R^3.$$
Find an estimate for the error.
(b) Using the formula given in part (a), find an approximate value for $\int_{-1/2}^{1/2}(dx/\sqrt{1 + x^2})$. Compare the answer with that obtained from Simpson's rule with $n = 4$.
26. (a) Can we use the binomial expansion of $\sqrt{1 + x}$ to obtain a convergent series for $\sqrt{2}$. Why or why not?
(b) Writing $2 = \frac{9}{4} \cdot \frac{8}{9}$, we have $\sqrt{2} = \frac{3}{2}\sqrt{8/9}$. Use this equation, together with the binomial expansion, to obtain an approximation to $\sqrt{2}$ correct to two decimal places.
(c) Use the method of part (b) to obtain an approximation to $\sqrt{3}$ correct to two decimal places.

Evaluate the limits in Exercises 27–30 using Maclaurin series.

27. $\lim_{x\to 0}\dfrac{\sin 2x - 2x}{x^3}$
28. $\lim_{x\to 0}\dfrac{\sqrt{1 + x} - \sqrt{1 - x}}{x}$
29. $\lim_{x\to 0}\left(\dfrac{1}{x \sin x} - \dfrac{1}{x^2}\right)$ (use a common denominator).
30. $\lim_{x\to \pi}\dfrac{1 + \cos x}{(x - \pi)^2}$

Expand each of the functions in Exercises 31–36 as a Maclaurin series and determine for what x it is valid.

31. $\dfrac{1}{1 - x}$
32. $\dfrac{1}{1 + x}$
33. $\dfrac{1}{1 - x} - \dfrac{1}{1 + x}$
34. $\dfrac{1}{2}\left(\dfrac{1}{1 - x} + \dfrac{1}{1 + x}\right)$
35. $\dfrac{1}{1 - x^2}$
36. $\dfrac{1}{1 - x^2} - \dfrac{1}{1 + x}$
37. Find the Maclaurin series for $f(x) = (1 + x^2)^2$ in two ways:
(a) by multiplying out the polynomial;
(b) by taking successive derivatives and evaluating them at $x = 0$ (without multiplying out).
38. Write down the Taylor series for $\ln x$ at $x_0 = 2$.
39. Find a power series expansion for $\int_1^x \ln t\, dt$. Compare this with the expansion for $x \ln x$. What is your conclusion?
40. Using the Taylor series for $\sin x$ and $\cos x$, find the terms through x^6 in the series for $(\sin x)^2 + (\cos x)^2$.

Let $f(x) = a_0 + a_1x + a_2x^2 + \cdots$. Find a_0, a_1, a_2, and a_3 for each of the functions in Exercises 41–44.

41. $\sec x$

42. $\sqrt{1 - x^2}$

43. $(d/dx)\sqrt{1 - x^2}$

44. e^{1+x}

Find Maclaurin expansions through the term in x^5 for each of the functions in Exercises 45–48.

45. $(1 - \cos x)/x^2$

46. $\dfrac{x - \sin 3x}{x^3}$

47. $\dfrac{1 - x}{1 + x}$

48. $\dfrac{d^2}{dx^2}\dfrac{1}{\sqrt{1 + x^2}}$

49. Find the Taylor polynomial of degree 4 for $\ln x$ at: (a) $x_0 = 1$; (b) $x_0 = e$; (c) $x_0 = 2$.

50. Find a power series expansion for a function $f(x)$ such that $f(0) = 0$ and $f'(x) - f(x) = x$. (Write $f(x) = a_0 + a_1x + a_2x^2 + \cdots$ and solve for the a_i's one after another.) (b) Find a formula for the function whose series you found in part (a).

Find the first four nonvanishing terms in the power series expansion for the functions in Exercises 51–54.

51. $\ln(1 + e^x)$

52. e^{x^2+x}

53. $\sin(e^x)$

54. $e^x \cos x$

55. An engineer is about to compute sin(36°), when the batteries in her hand calculator give out. She quickly grabs a backup unit, only to find it is made for statistics and does not have a "sin" key. Unperturbed, she enters 3.1415926, divides by 5, and enters the result into the memory, called "x" hereafter. Then she computes $x(1 - x^2/6)$ and uses it for the value of sin(36°).
   (a) What was her answer?
   (b) How good was it?
   (c) Explain what she did in the language of Taylor series expansions.
   (d) Describe a similar method for computing tan(10°).

56. An automobile travels on a straight highway. At noon it is 20 miles from the next town, travelling at 50 miles per hour, with its acceleration kept between 20 miles per hour per hour and -10 miles per hour per hour. Use the formula $x(t) = x(0) + x'(0)t + \int_0^t (t - s)x''(s)\,ds$ to estimate the auto's distance from the town 15 minutes later.

★57. (a) Let
$$f(x) = \begin{cases} (\sin x)/x, & x \neq 0, \\ 1, & x = 0. \end{cases}$$
Find $f'(0)$, $f''(0)$, and $f'''(0)$.
(b) Find the Maclaurin expansion for $(\sin x)/x$.

★58. Using Taylor's formula, prove the following inequalities:
   (a) $e^x - 1 \geqslant x$ for $x \geqslant 0$.
   (b) $6x - x^3 + x^5/20 \geqslant 6\sin x \geqslant 6x - x^3$ for $x \geqslant 0$.
   (c) $x^2 - x^4/12 \leqslant 2 - 2\cos x \leqslant x^2$ for $x \geqslant 0$.

★59. Prove that $\ln 2 = 1 - \frac{1}{2} + \frac{1}{3} - \frac{1}{4} + \cdots$.

★60. (a) Write the Maclaurin series for the functions $1/\sqrt{1 - x^2}$ and $\sin^{-1}x$. Where do they converge?
   (b) Find the terms through x^3 in the series for $\sin^{-1}(\sin x)$ by substituting the series for $\sin x$ in the series for $\sin^{-1}x$; that is, if $\sin^{-1}x = a_0 + a_1x + a_2x^2 + \cdots$, then
$$\sin^{-1}(\sin x) = a_0 + a_1\left(x - \frac{x^3}{3!} + \frac{x^5}{5!} - \cdots\right) + a_2\left(x - \frac{x^3}{3!} + \frac{x^5}{5!} - \cdots\right)^2 + \cdots.$$
   (c) Use the substitution method of part (b) to *obtain* the first five terms of the series for $\sin^{-1}x$ by using the relation $\sin^{-1}(\sin x) = x$ and solving for a_0 through a_5.
   (d) Find the terms through x^5 of the Maclaurin series for the inverse function $g(s)$ of $f(x) = x^3 + x$. (Use the relation $g(f(x)) = x$ and solve for the coefficients in the series for g.)

12.6 Complex Numbers

Complex numbers provide a square root for -1.

This section is a brief introduction to the algebra and geometry of complex numbers; i.e., numbers of the form $a + b\sqrt{-1}$. We show the utility of complex numbers by comparing the series expansions for $\sin x$, $\cos x$, and e^x derived in the preceding section. This leads directly to Euler's formula relating the numbers 0, 1, e, π, and $\sqrt{-1}$: $e^{\pi\sqrt{-1}} + 1 = 0$. Applications of complex numbers to second-order differential equations are given in the next section. Section 12.8, on series solutions, can, however, be read before this one.

If we compare the three power series

$$\sin x = x - \frac{x^3}{3!} + \frac{x^5}{5!} - \cdots, \tag{1}$$

$$\cos x = 1 - \frac{x^2}{2!} + \frac{x^4}{4!} - \cdots, \tag{2}$$

$$e^x = 1 + x + \frac{x^2}{2!} + \frac{x^3}{3!} + \cdots, \tag{3}$$

it looks as if $\sin x$ and $\cos x$ are almost the "odd and even parts" of e^x. If we write the series

$$e^{-x} = 1 - x + \frac{x^2}{2!} - \frac{x^3}{3!} + \cdots \tag{4}$$

subtract equation (4) from equation (3) and divide by 2, we get

$$\frac{e^x - e^{-x}}{2} = x + \frac{x^3}{3!} + \frac{x^5}{5!} + \cdots. \tag{5}$$

Similarly, adding equations (3) and (4) and dividing by 2, gives

$$\frac{e^x + e^{-x}}{2} = 1 + \frac{x^2}{2!} + \frac{x^4}{4!} + \cdots. \tag{6}$$

These are the Maclaurin series of the hyperbolic functions $\sinh x$ and $\cosh x$; they are just missing the alternating signs in the series for $\sin x$ and $\cos x$.

Can we get the right signs by an appropriate substitution other than changing x to $-x$? Let us try changing x to ax, where a is some constant. We have, for example,

$$\cosh ax = \frac{e^{ax} + e^{-ax}}{2} = 1 + a^2\frac{x^2}{2!} + a^4\frac{x^4}{4!} + a^6\frac{x^6}{6!} + \cdots.$$

This would become the series for $\cos x$ if we had $a^2 = a^6 = a^{10} = \cdots = -1$ and $a^4 = a^8 = a^{12} = \cdots = 1$. In fact, all these equations would follow from the one relation $a^2 = -1$.

We know that the square of any real number is positive, so that the equation $a^2 = -1$ has no real solutions. Nevertheless, let us pretend that there is a solution, which we will denote by the letter i, for "imaginary." Then we would have $\cosh ix = \cos x$.

Example 1 What is the relation between $\sinh ix$ and $\sin x$?

Solution Since $i^2 = -1$, we have $i^3 = -i$, $i^4 = (-i)\cdot i = 1$, $i^5 = i$, $i^6 = -1$, etc., so substituting ix for x in (5) gives

$$\sinh ix = \frac{e^{ix} - e^{-ix}}{2} = ix - i\frac{x^3}{3!} + i\frac{x^5}{5!} - i\frac{x^7}{7!} + \cdots.$$

Comparing this with equation (1), we find that $\sinh ix = i \sin x$. ▲

The sum of the two series (5) and (6) is the series (3), i.e., $e^x = \cosh x + \sinh x$. Substituting ix for x, we find

$$e^{ix} = \cosh ix + \sinh ix$$

or

$$e^{ix} = \cos x + i \sin x. \qquad (7)$$

Formula (7) is called *Euler's formula*. Substituting π for x, we find that

$$e^{i\pi} = -1,$$

and adding 1 to both sides gives

$$e^{i\pi} + 1 = 0, \qquad (8)$$

a formula composed of seven of the most important symbols in mathematics: 0, 1, +, =, e, i, and π.

Example 2 Using formula (7), express the sine and cosine functions in terms of exponentials.

Solution Substituting $-x$ for x in equation (7) and using the symmetry properties of cosine and sine, we obtain

$$e^{-ix} = \cos x - i \sin x.$$

Adding this equation to (7) and dividing by 2 gives

$$\cos x = \frac{e^{ix} + e^{-ix}}{2},$$

while subtracting the equations and dividing by $2i$ gives

$$\sin x = \frac{e^{ix} - e^{-ix}}{2i}. \ ▲$$

Example 3 Find $e^{i(\pi/2)}$ and $e^{2\pi i}$.

Solution Using formula (7), we have

$$e^{i(\pi/2)} = \cos\frac{\pi}{2} + i \sin\frac{\pi}{2} = i$$

and

$$e^{2\pi i} = \cos 2\pi + i \sin 2\pi = 1. \ ▲$$

Since there is no real number having the property $i^2 = -1$, all of the calculations above belong so far to mathematical "science fiction." In the following paragraphs, we will see how to construct a number system in which -1 does have a square root; in this new system, all the calculations which we have done above will be completely justified.

When they were first introduced, square roots of negative numbers were deemed merely to be symbols on paper with no real existence (whatever that means) and therefore "imaginary." These imaginary numbers were not taken seriously until the cubic and quartic equations were solved in the sixteenth century (in the formula in the Supplement to Section 3.4 for the roots of a cubic equation, the symbol $\sqrt{-3}$ appears and must be contended with, even if all the roots of the equation are real.) A proper way to define square roots of negative numbers was finally obtained through the work of Girolamo Cardano around 1545 and Bombelli in 1572, but it was only with the work of L. Euler, around 1747, that their importance was realized. A way to understand imaginaries in terms of real numbers was discovered by Wallis, Wessel, Argand, Gauss, Hamilton, and others in the early nineteenth century.

To define a number system which contains $i = \sqrt{-1}$, we note that such a system ought to contain all expressions of the form $a + b\sqrt{-1} = a + bi$, where a and b are ordinary real numbers. Such expressions should obey the laws

$$(a + bi) + (c + di) = (a + c) + (b + d)i$$

and

$$(a + bi)(c + di) = ac + adi + bci + bdi^2 = (ac - bd) + (ad + bc)i.$$

Thus the sum and product of two of these expressions are expressions of the same type.

All the data in the "number" $a + bi$ is carried by the *pair* (a, b) of real numbers, which may be considered a point in the xy plane. Thus we define our new number system, the complex numbers, by imposing the desired operations on pairs of real numbers.

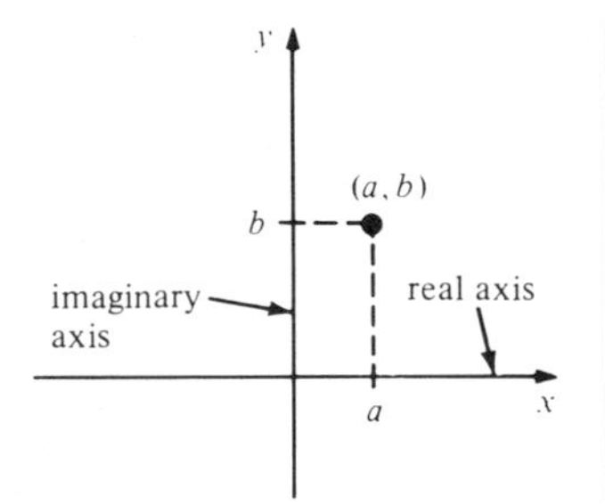

Figure 12.6.1. A complex number is just a point (a, b) in the plane.

Complex Numbers

A *complex number* is a point (a, b) in the xy plane. Complex numbers are added and multiplied as follows:

$$(a, b) + (c, d) = (a + c, b + d),$$

$$(a, b)(c, d) = (ac - bd, ad + bc).$$

The point $(0, 1)$ is denoted by the symbol i, so that $i^2 = (-1, 0)$ (using $a = 0$, $c = 0$, $b = 1$, $d = 1$ in the definition of muliplication). The x axis is called the *real axis* and the y axis is the *imaginary axis*. (See Fig. 12.6.1.)

It is convenient to denote the point $(a, 0)$ just by a since we are thinking of points on the real axis as ordinary real numbers. Thus, in this notation, $i^2 = -1$. Also,

$$(a, b) = (a, 0) + (0, b) = (a, 0) + (b, 0)(0, 1)$$

as is seen from the definition of multiplication. Replacing $(a, 0)$ and $(b, 0)$ by a and b, and $(0, 1)$ by i, we see that

$$(a, b) = a + bi.$$

Since two points in the plane are equal if and only if their coordinates are equal, we see that

$$a + ib = c + id \quad \text{if and only if} \quad a = c \quad \text{and} \quad b = d.$$

Thus, if $a + ib = 0$, both a and b must be zero.

We now see that sense can indeed be made of the symbol $a + ib$, where $i^2 = -1$. The notation $a + ib$ is much easier to work with than ordered pairs, so we now revert to the old notation $a + ib$ and dispense with ordered pairs in our calculations. However, the geometric picture of plotting $a + ib$ as the point (a, b) in the plane is very useful and will be retained.

It can be verified, although we shall not do it, that the usual laws of algebra hold for complex numbers. For example, if we denote complex numbers by single letters such as $z = a + ib$, $w = c + id$, and $u = e + if$, we have

$$z(w+u) = zw + zu,$$

$$z(wu) = (zw)u,$$

etc.

Example 4 (a) Plot the complex number $8 - 6i$. (b) Simplify $(3+4i)(8+2i)$. (c) Factor $x^2 + x + 3$. (d) Find $\sqrt{i}$.

Solution (a) $8 - 6i$ corresponds to the point $(8, -6)$, plotted in Fig. 12.6.2.

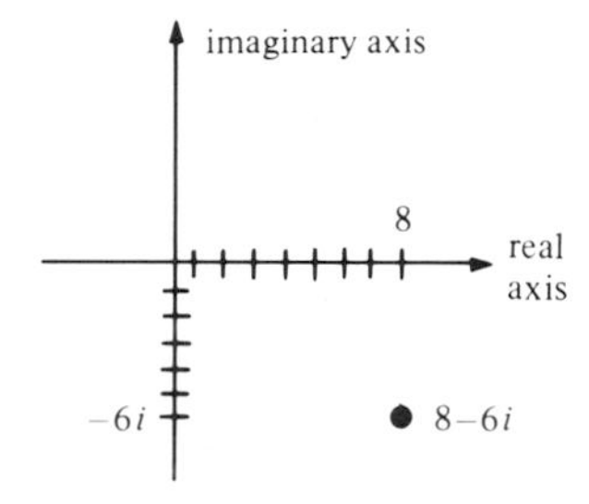

Figure 12.6.2. The point 8–6*i* plotted in the *xy* plane.

(b)
$$\begin{aligned}(3+4i)(8+2i) &= 3\cdot 8 + 3\cdot 2i + 4\cdot 8i + 2\cdot 4i^2 \\ &= 24 + 6i + 32i - 8 \\ &= 16 + 38i.\end{aligned}$$

(c) By the quadratic formula, the roots of $x^2 + x + 3 = 0$ are given by $(-1 \pm \sqrt{1-12}\,)/2 = (-1/2) \pm (\sqrt{11}\,/2)i$. We may factor using these two roots: $x^2 + x + 3 = [x + (1/2) - (\sqrt{11}\,/2)i][x + (1/2) + (\sqrt{11}\,/2)i]$. (You may check by multiplying out.)

(d) We seek a number $z = a + ib$ such that $z^2 = i$; now $z^2 = a^2 - b^2 + 2abi$, so we must solve $a^2 - b^2 = 0$ and $2ab = 1$. Hence $a = \pm b$, so $b = \pm(1/\sqrt{2}\,)$. Thus there are two numbers whose square is i, namely, $\pm[(1/\sqrt{2}\,) + (i/\sqrt{2}\,)]$, i.e., $\sqrt{i} = \pm(1/\sqrt{2}\,)(1+i) = \pm(\sqrt{2}\,/2)(1+i)$. Although for positive real numbers, there is a "preferred" square root (the positive one), this is not the case for a general complex number. ▲

Example 5 (a) Show that if $z = a + ib \neq 0$, then

$$\frac{1}{z} = \frac{a - ib}{a^2 + b^2} = \frac{a}{a^2+b^2} - \frac{b}{a^2+b^2}i$$

is a complex number whose product with z equals 1; thus, $1/z$ is the inverse of z, and we can divide by nonzero complex numbers.

(b) Write $1/(3+4i)$ in the form $a + bi$.

Solution (a)
$$\begin{aligned}\left(\frac{a-ib}{a^2+b^2}\right)(a+ib) &= \left(\frac{1}{a^2+b^2}\right)(a-ib)(a+ib) \\ &= \left(\frac{1}{a^2+b^2}\right)(a^2 + aib - iba - b^2i^2) \\ &= \left(\frac{1}{a^2+b^2}\right)(a^2+b^2) = 1.\end{aligned}$$

Hence $z\left(\dfrac{a-ib}{a^2+b^2}\right) = 1$, so $(a-ib)/(a^2+b^2)$ can be denoted $1/z$. Note that $z \neq 0$ means that not both a and b are zero, so $a^2 + b^2 \neq 0$ and division by the real number $a^2 + b^2$ is legitimate.

(b) $1/(3+4i) = (3-4i)/(3^2+4^2) = (3/25) - (4/25)i$ by the formula in (a). ▲

Terminology for Complex Numbers

If $z = a + ib$ is a complex number, then:

(i) a is called the *real part* of z;
(ii) b is called the *imaginary part* of z (note that the imaginary part is itself a real number);
(iii) $a - ib$ is called the *complex conjugate* of z and is denoted $\bar{z}$;
(iv) $r = \sqrt{a^2 + b^2}$ is called the *length* or *absolute value* of z and is denoted $|z|$;
(v) θ defined by $a = r\cos\theta$ and $b = r\sin\theta$ is called the *argument* of z.

The notions in the box above are illustrated in Fig. 12.6.3. Note that the real and imaginary parts are simply the x and y coordinates, the complex conjugate is the reflection in the x axis, and the absolute value is (by Pythagoras' theorem) the length of the line joining the origin and z. The argument of z is the angle this line makes with the x axis. Thus, (r, θ) are simply the polar coordinates of the point (a, b).

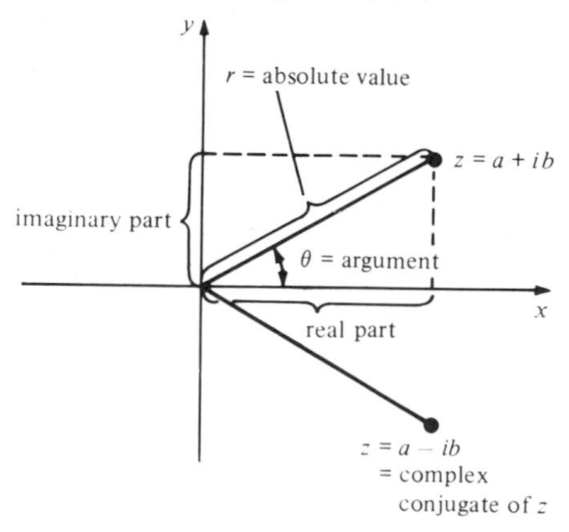

Figure 12.6.3. Illustrating various quantities attached to a complex number.

The terminology and notation above simplify manipulations with complex numbers. For example, notice that

$$z \cdot \bar{z} = (a + ib)(a - ib) = a^2 + b^2 = |z|^2,$$

so that $1/z = \bar{z}/|z|^2$ which reproduces the result of Example 5(a). Notice that we can remember this by:

$$\frac{1}{z} = \frac{1}{z}\frac{\bar{z}}{\bar{z}} = \frac{\bar{z}}{z\bar{z}} = \frac{\bar{z}}{|z|^2}.$$

Example 6 (a) Find the absolute value and argument of $1 + i$.
(b) Find the real parts of $1/i$, $1/(1 + i)$, and $(8 + 2i)/(1 - i)$.

Solution (a) The real part is 1, and the imaginary part is 1. Thus the absolute value is $\sqrt{1^2 + 1^2} = \sqrt{2}$, and the argument is $\tan^{-1}(1/1) = \pi/4$.
(b) $1/i = (1/i)(-i/-i) = -i/1 = -i$, so the real part of $1/i = -i$ is zero. $1/(1 + i) = (1 - i)/(1 + i)(1 - i) = (1 - i)/2$, so the real part of $1/(1 + i)$ is $1/2$. Finally,

$$\frac{8 + 2i}{1 - i} = \frac{(8 + 2i)}{(1 - i)}\frac{1 + i}{1 + i} = \frac{8 + 10i - 2}{2} = \frac{6 + 10i}{2} = 3 + 5i,$$

so the real part of $(8 + 2i)/(1 - i)$ is 3. ▲

Properties of Complex Numbers

(i) $\overline{z_1 z_2} = \overline{z_1} \cdot \overline{z_2}, \quad \overline{z_1/z_2} = \overline{z_1}/\overline{z_2}$;
(ii) z is real if and only if $z = \bar{z}$;
(iii) $|z_1 z_2| = |z_1| \cdot |z_2|$, $|z_1/z_2| = |z_1|/|z_2|$; and
(iv) $|z_1 + z_2| \leqslant |z_1| + |z_2|$ (triangle inequality).

The proofs of these properties are left to the examples and exercises.

Example 7 (a) Prove property (i) of complex numbers.
(b) Express $\overline{(1+i)^{100}}$ without a bar.

Solution (a) Let $z_1 = a + ib$ and $z_2 = c + id$, so $\bar{z}_1 = a - ib$, $\bar{z}_2 = c - id$. From $z_1 z_2 = (ac - bd) + (ad + bc)i$, we get $\overline{z_1 z_2} = (ac - bd) - (ad + bc)i$; we also have $\overline{z_1} \cdot \overline{z_2} = (a - ib)(c - id) = (ac - bd) - ibc - aid = \overline{z_1 z_2}$. For the quotient, write $z_2 \cdot z_1/z_2 = z_1$ so by the rule just proved, $\overline{z_2} \cdot \overline{(z_1/z_2)} = \bar{z}_1$. Dividing by $\bar{z}_2$ gives the result.
(b) Since the complex conjugate of a product is the product of the complex conjugates (proved in (a)), we similarly have $\overline{z_1 z_2 z_3} = \overline{z_1 z_2}\bar{z}_3 = \bar{z}_1\bar{z}_2\bar{z}_3$ and so on for any number of factors. Thus $\overline{z^n} = \bar{z}^n$, and hence $\overline{(1+i)^{100}} = (\overline{1+i})^{100} = (1-i)^{100}$. ▲

Example 8 Given $z = a + ib$, construct iz geometrically and discuss.

Solution If $z = a + ib$, $iz = ai - b = -b + ia$. Thus in the plane, $z = (a, b)$ and $iz = (-b, a)$. This point $(-b, a)$ is on the line perpendicular to the line Oz since the slopes are negative reciprocals. See Fig. 12.6.4. Since iz has the same length as z, we can say that iz is obtained from z by a rotation through 90°. ▲

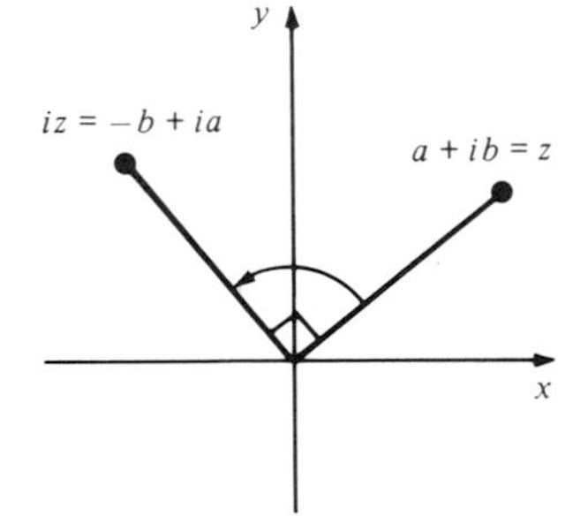

Figure 12.6.4. The number iz is obtained from z by a 90° rotation about the origin.

Using the algebra of complex numbers, we can define $f(z)$ when f is a rational function and z is a complex number.

Example 9 If $f(z) = (1 + z)/(1 - z)$ and $z = 1 + i$, express $f(z)$ in the form $a + bi$.

Solution Substituting $1 + i$ for z, we have

$$f(1+i) = \frac{1 + 1 + i}{1 - (1 + i)} = \frac{2 + i}{-i} = -1 - \frac{2}{i} = -1 + 2i. \ ▲$$

How can we define more general functions of complex numbers, like e^z? One way is to use power series, writing

$$e^z = 1 + z + \frac{z^2}{2!} + \frac{z^3}{3!} + \cdots = \sum_{n=0}^{\infty} \frac{z^n}{n!}.$$

To make sense of this, we would have to define the limit of a sequence of complex numbers so that the sum of the infinite series could be taken as the limit of its sequence of partial sums. Fortunately, this is possible, and in fact the whole theory of infinite series carries over to the complex numbers. This approach would take us too far afield,[6] though, and we prefer to take the approach of *defining* the particular function e^{ix}, for x real, by Euler's formula

$$e^{ix} = \cos x + i \sin x. \tag{9}$$

Since $e^{x+y} = e^x e^y$, we expect a similar law to hold for e^{ix}.

Example 10 (a) Show that

$$e^{i(x+y)} = e^{ix} e^{iy} \tag{10}$$

(b) Give a definition of e^z for $z = x + iy$.

Solution (a) The right-hand side of equation (10) is

$$\begin{aligned}(\cos x + i \sin x)&(\cos y + i \sin y)\\ &= \cos x \cos y - \sin x \sin y + i(\sin x \cos y + \sin y \cos x)\\ &= \cos(x + y) + i \sin(x + y) = e^{i(x+y)}\end{aligned}$$

by equation (9) and the addition formulae for sin and cos.
(b) We would like to have $e^{z_1+z_2} = e^{z_1}e^{z_2}$ for any complex numbers, so we should define $e^{x+iy} = e^x \cdot e^{iy}$, i.e., $e^{x+iy} = e^x(\cos y + i \sin y)$. [With this definition, the law $e^{z_1+z_2} = e^{z_1}e^{z_2}$ can then be proved for all z_1 and z_2.] ▲

Equation (10) contains all the information in the trigonometric addition formulas. This is why the use of e^{ix} is so convenient: the laws of exponents are easier to manipulate than the trigonometric identities.

Example 11 (a) Calculate $\overline{e^{i\theta}}$ and $|e^{i\theta}|$. (b) Calculate $e^{i\pi/2}$ and $e^{i\pi}$. (c) Prove that

$$1 + \cos\theta + \cos 2\theta + \cdots + \cos n\theta = \frac{1}{2} + \frac{1}{2}\left(\frac{\cos n\theta - \cos(n+1)\theta}{1 - \cos\theta}\right)$$

by considering $1 + e^{i\theta} + e^{2i\theta} + \cdots + e^{ni\theta}$.

Solution (a) $e^{i\theta} = \cos\theta + i \sin\theta$, so by definition of the complex conjugate we should change the sign of the imaginary part:

$$\overline{e^{i\theta}} = \cos\theta - i \sin\theta = \cos(-\theta) + i \sin(-\theta) = e^{-i\theta},$$

since $\cos(-\theta) = \cos\theta$ and $\sin(-\theta) = -\sin\theta$. Thus $|e^{i\theta}| = \sqrt{\cos^2\theta + \sin^2\theta} = 1$ using the general definition $|z| = \sqrt{a^2 + b^2}$, where $z = a + ib$.
(b) $e^{i\pi/2} = \cos(\pi/2) + i \sin(\pi/2) = i$ and $e^{i\pi} = \cos\pi + i \sin\pi = -1$.
(c) Since $\cos n\theta$ is the real part of $e^{in\theta}$, we are led to consider $1 + e^{i\theta} + e^{i2\theta} + \cdots + e^{in\theta}$. Recalling that $1 + r + \cdots + r^n = (1 - r^{n+1})/(1 - r)$, we get

$$\begin{aligned}1 + e^{i\theta} + e^{i2\theta} + \cdots + e^{in\theta} &= \frac{1 - e^{i(n+1)\theta}}{1 - e^{i\theta}}\\ &= \frac{1 - e^{i(n+1)\theta}}{1 - e^{i\theta}} \cdot \frac{1 - e^{-i\theta}}{1 - e^{-i\theta}}\end{aligned}$$

[6] See a text on complex variables such as J. Marsden, *Basic Complex Analysis*, Freeman, New York (1972) for a thorough treatment of complex series.

$$= \frac{1 - e^{-i\theta} - e^{i(n+1)\theta} + e^{in\theta}}{2 - (e^{i\theta} + e^{-i\theta})}$$

$$= \frac{1 - e^{-i\theta} - e^{i(n+1)\theta} + e^{in\theta}}{2(1 - \cos\theta)}.$$

Taking the real part of both sides gives the result. ▲

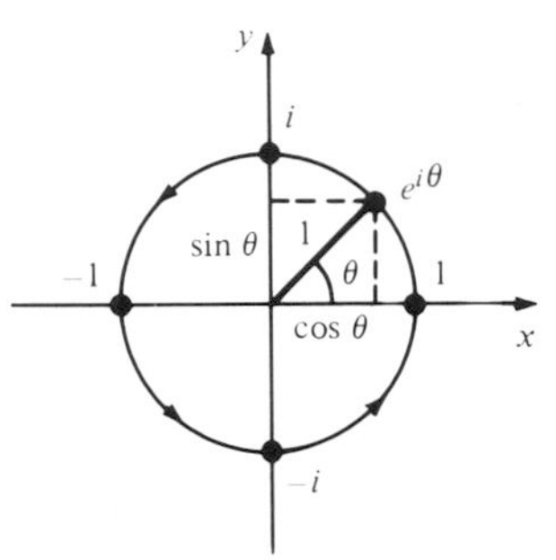

Figure 12.6.5. As θ goes from 0 to 2π, the point $e^{i\theta}$ goes once around the unit circle in the complex plane.

Let us push our analysis of e^{ix} a little further. Notice that $e^{i\theta} = \cos\theta + i\sin\theta$ represents a point on the unit circle with argument θ. As θ ranges from 0 to 2π, this point moves once around the circle (Fig. 12.6.5). (This is the same basic geometric picture we used to introduce the trigonometric functions in Section 5.1).

Recall that if $z = a + ib$, and r, θ are the polar coordinates of (a,b), then $a = r\cos\theta$ and $b = r\sin\theta$. Thus

$$z = r\cos\theta + ir\sin\theta = r(\cos\theta + i\sin\theta) = re^{i\theta}.$$

Hence we arrive at the following.

Polar Representation of Complex Numbers

If $z = a + ib$ and if (r,θ) are the polar coordinates of (a,b), i.e., the absolute value and argument of z, then

$$z = re^{i\theta}.$$

This representation is very convenient for algebraic manipulations. For example,

$$\text{if} \quad z_1 = r_1e^{i\theta_1}, \quad z_2 = r_2e^{i\theta_2}, \qquad \text{then} \quad z_1z_2 = r_1r_2e^{i(\theta_1+\theta_2)},$$

which shows how the absolute value and arguments behave when we take products; i.e., it shows that $|z_1z_2| = |z_1|\,|z_2|$ and that the argument of z_1z_2 is the sum of the arguments of z_1 and z_2.

Let us also note that if $z = re^{i\theta}$, then $z^n = r^ne^{in\theta}$. Thus if we wish to solve $z^n = w$ where $w = \rho e^{i\phi}$, we must have $r^n = \rho$, i.e., $r = \sqrt[n]{\rho}$ (remember that r, ρ are non-negative) and $e^{in\theta} = e^{i\phi}$, i.e., $e^{i(n\theta-\phi)} = 1$, i.e., $n\theta = \phi + 2\pi k$ for an integer k (this is because $e^{it} = 1$ exactly when t is a multiple of 2π—see Fig. 12.6.5). Thus $\theta = \phi/n + 2\pi k/n$. When $k = n$, $\theta = \phi/n + 2\pi$, so $e^{i\theta} = e^{i\phi/n}$. Thus we get the same value for $e^{i\theta}$ when $k = 0$ and $k = n$, and we need take only $k = 0, 1, 2, \ldots, n-1$. Hence we get the following formula for the nth roots of a complex number.

De Moivre's Formula[7]

The numbers z such that $z^n = w = \rho e^{i\phi}$, i.e., the nth roots of w, are given by

$$\sqrt[n]{\rho}\, e^{i(\phi/n + 2\pi k/n)}, \qquad k = 0, 1, 2, \ldots, n-1.$$

[7] Abraham DeMoivre (1667–1754), of French descent, worked in England around the time of Newton.

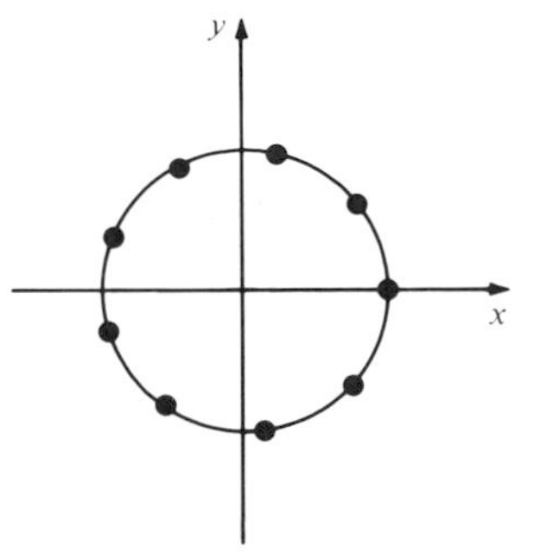

Figure 12.6.6. The ninth roots of 1.

For example, the ninth roots of 1 are the complex numbers $e^{i2\pi k/9}$, for $k = 0, 1, \ldots, 8$, which are 9 points equally spaced around the unit circle. See Fig. 12.6.6.

It is shown in more advanced books that any nth degree polynomial $a_0 + a_1z + \cdots + a_nz^n$ has at least one complex root[8] z_1 and, as a consequence, that the polynomial can be completely factored:

$$a_0 + a_1z + \cdots + a_nz^n = (z - z_1) \cdots (z - z_n).$$

For example,

$$z^2 + z + 1 = \left(z + \frac{1 + \sqrt{3}\,i}{2}\right)\left(z + \frac{1 - \sqrt{3}\,i}{2}\right),$$

although $z^2 + z + 1$ cannot be factored using only real numbers.

Example 12 (a) Redo Example 8 using the polar representation. (b) Give a geometric interpretation of multiplication by $(1 + i)$.

Solution (a) Since $i = e^{i\pi/2}$, $iz = re^{i(\theta + \pi/2)}$ if $z = re^{i\theta}$. Thus iz has the same magnitude as z but its argument is increased by $\pi/2$. Hence iz is z rotated by 90°, in agreement with the solution to Example 8.

(b) Since $(1 + i) = \sqrt{2}\,e^{i\pi/4}$, multiplication of a complex number z by $(1 + i)$ rotates z through an angle $\pi/4 = 45°$ and multiplies its length by $\sqrt{2}$. ▲

Example 13 Find the 4th roots of $1 + i$.

Solution $1 + i = \sqrt{2}\,e^{i\pi/4}$, since $1 + i$ has $r = \sqrt{2}$ and $\theta = \pi/4$. Hence the fourth roots are, according to DeMoivre's formula,

$$\sqrt[8]{2}\,e^{i((\pi/16) + (\pi k/2))}, \qquad k = 0, 1, 2, 3,$$

i.e.,

$$\sqrt[8]{2}\,e^{i\pi/16}, \quad \sqrt[8]{2}\,e^{i\pi 9/16}, \quad \sqrt[8]{2}\,e^{i\pi \cdot 17/16}, \quad \text{and} \quad \sqrt[8]{2}\,e^{i\pi \cdot 25/16}. \ ▲$$

Exercises for Section 12.6

Express the quantities in Exercises 1–4 in the form $a + bi$.

1. $e^{-\pi i/2}$
2. $e^{\pi i/4}$
3. $e^{(3\pi/2)i}$
4. $e^{-i\pi}$

Plot the complex numbers in Exercises 5–12 as points in the xy plane.

5. $4 + 2i$
6. $-1 + i$
7. $3i$
8. $-(2 + i)$
9. $-\frac{2}{3}i$
10. $3 + 7i$
11. $0.1 + 0.2i$
12. $0 + 1.5i$

Simplify the expressions in Exercises 13–20.

13. $(1 + 2i) - 3(5 - 2i)$
14. $(4 - 3i)(8 + i) + (5 - i)$
15. $(2 + i)^2$
16. $\dfrac{1}{(3 + i)}$
17. $\dfrac{1}{5 - 3i}$
18. $\dfrac{2i}{1 - i}$
19. $\dfrac{(1 + i)(3 - 2i)}{8 + i}$
20. $\dfrac{(2 + 2i) + 6i}{(1 + 2i)(-4i)}$

Write the solutions of the equations in Exercises 21–26 in the form $a + bi$, where a and b are real numbers and $i = \sqrt{-1}$.

21. $z^2 + 3 = 0$
22. $z^2 - 2z + 5 = 0$
23. $z^2 + \frac{1}{3}z + \frac{1}{2} = 0$
24. $z^3 + 2z^2 + 2z + 1 = 0$ [*Hint:* factor]
25. $z^2 - 7z - 1 = 0$
26. $z^3 - 3z^2 + 3z - 1 = 0$ [*Hint:* factor]

Using the method of Example 4(d), find the quantities in Exercises 27–30.

27. $\sqrt{8i}$
28. $\sqrt{9i}$
29. $\sqrt{-16i}$
30. $\sqrt{\sqrt{i}}$

[8] See any text in complex variables, such as J. Marsden, *op. cit.* The theorem referred to is called the "fundamental theorem of algebra." It was first proved by Gauss in his doctoral thesis in 1799.

Find the imaginary part of the complex numbers in Exercises 31–36.

31. $\dfrac{1+i}{i}$
32. $\dfrac{2-3i}{1+3i}$
33. $\dfrac{10+5i}{(1+2i)^2}$
34. $(1-8i)\left(2+\frac{1}{4}i\right)^{-1}$
35. $\dfrac{1/2+(3/5)i}{7/8-i}$
36. $\dfrac{(3/4)i}{9/4+(1/5)i}$

Find the complex conjugate of the complex numbers in Exercises 37–46.

37. $5+2i$
38. $1-bi$
39. $\sqrt{3}+\frac{1}{2}i$
40. $1/i$
41. $\dfrac{2-i}{3i}$
42. $i(1+i)$
43. $\dfrac{3-5i}{4+8i}$
44. $\dfrac{1}{2i}\left(\dfrac{1+i}{1-i}\right)$
45. 3
46. $\dfrac{10+i}{7+4i}$

Find the absolute value and argument of the complex numbers in Exercises 47–58. Plot.

47. $-1-i$
48. $7+2i$
49. 2
50. $4i$
51. $\frac{1}{2}-\frac{2}{3}i$
52. $3-2i$
53. $-5+7i$
54. $-10+\frac{1}{2}i$
55. $-8-2i$
56. $5+5i$
57. $1.2+0.7i$
58. $50+10i$

59. Prove property (iii) of complex numbers.
60. Prove property (iv) of complex numbers.
61. Express $\overline{(8-3i)^4}$ without a bar.
62. Express $\overline{(2+3i)^2(8-i)^3}$ without a bar.

In Exercises 63–66, draw an illustration of the addition of the pairs of complex numbers, i.e., plot both along with their sums.

63. $1+\frac{1}{2}i, 3-i$
64. $-8-2i, 5-i$
65. $-3+4i, 6i$
66. $7, 4i$

67. Find $|(1+i)(2-i)(\sqrt{2}\,i)|$.
68. If $z=x+iy$, express x and y in terms of z and $\bar{z}$.
69. If $z=x+iy$ with x and y real, what is $|e^z|$ and the argument of e^z?
70. Find the real and imaginary parts of $(x+iy)^3$ as polynomials in x and y.

Write the numbers in Exercises 71–76 in the form $a+bi$.

71. $e^{i\pi/3}$
72. $e^{1-\pi i/2}$
73. $e^{1-\pi i/2}$
74. e^{1+2i}
75. $e^{1+\pi i/2}$
76. $e^{(1-\pi/6)i}$

77. If $f(z)=1/z^2$, express $f(2+i)$ in the form $a+bi$.
78. Express $f(i)$ in the form $a+bi$, if $f(z)=z^2+2z+1$.
79. (a) Using a trigonometric identity, show that $e^{ix}e^{-ix}=1$. (b) Show that $e^{-z}=1/e^z$ for all complex numbers z.
80. Show that $e^{3z}=(e^z)^3$ for all complex z.
81. Prove that $e^{i(\theta+3\pi/2)}=-ie^{i\theta}$.
82. Prove that

$$\sin\theta+\sin 2\theta+\cdots+\sin n\theta = \left(\cot\frac{\theta}{2}\right)\left(\frac{1}{2}+\frac{1}{2}\left(\frac{\sin n\theta-\sin(n+1)\theta}{\sin\theta}\right)\right).$$

83. Prove that $(\cos\theta+i\sin\theta)^n=\cos n\theta+i\sin n\theta$, if n is an integer.
84. Use Exercise 83 to find the real part of $\left(\dfrac{1}{\sqrt{2}}+\dfrac{i}{\sqrt{2}}\right)^3$ and the imaginary part of $\left(\dfrac{1}{2}+i\dfrac{\sqrt{3}}{2}\right)^9$.

Find the polar representation (i.e., $z=re^{i\theta}$) of the complex numbers in Exercises 85–94.

85. $1+i$
86. $\dfrac{1}{i}$
87. $(2+i)^{-1}$
88. $\sqrt{3}$
89. $7-3i$
90. $4+i^3$
91. $-\frac{1}{2}-3i$
92. $\dfrac{(2+5i)}{(1-i)}$
93. $(3+4i)^2$
94. $-1+\frac{1}{3}i$

95. Find the fifth roots of $\frac{1}{2}-\frac{1}{2}\sqrt{3}\,i$ and $1+2i$. Sketch.
96. Find the fourth roots of i and $\sqrt{i}$. Sketch.
97. Find the sixth roots of $\sqrt{5}+3i$ and $3+\sqrt{5}\,i$. Sketch.
98. Find the third roots of $1/7$ and $i/7$. Sketch.
99. Give a geometric interpretation of division by $1-i$.
100. (a) Give a geometric interpretation of multiplication by an arbitrary complex number $z=re^{i\theta}$.
(b) What happens if we divide?
101. Prove that if $z^6=1$ and $z^{10}=1$, then $z=\pm 1$.
102. Suppose we know that $z^7=1$ and $z^{41}=1$. What can we say about z?
103. Let $z=re^{i\theta}$. Prove that $\bar{z}=re^{-i\theta}$.
104. (a) Let $f(z)=az^3+bz^2+cz+d$, where a, b, c, and d are real numbers. Prove that $f(\bar{z})=\overline{f(z)}$.
(b) Does equality still hold if a, b, c, and d are allowed to be arbitrary complex numbers?

Factor the polynomials in Exercises 105–108, where z is complex. [*Hint:* Find the roots.]

105. z^2+2z+i
106. $z^2+2iz-4$
107. $z^2+2iz+4-4i$
108. $3z^2+z-e^{i\pi/3}$

109. (a) Write $\tan i\theta$ in the form $a+bi$ where a and b are real functions of θ.
(b) Write $\tan i\theta$ in the form $re^{i\phi}$.
110. Let $z=f(t)$ be a *complex valued* function of the *real* variable t. If $z=x+iy=g(t)+ih(t)$, where g and h are real valued, we *define* $dz/dt=f'(t)$ to be $(dx/dt)+i(dy/dt)=g'(t)+ih'(t)$.
(a) Show that $(d/dt)(Ce^{i\omega t})=i\omega Ce^{i\omega t}$, if C is

any complex number and ω is any real number.

(b) Show that $z = Ce^{i\omega t}$ satisfies the *spring equation* (see Section 8.1): $z'' + \omega^2 z = 0$.

(c) Show that $z = De^{-i\omega t}$ also satisfies the spring equation.

(d) Find C and D such that $Ce^{i\omega t} + De^{-i\omega t} = f(t)$ satisfies $f(0) = A$, $f'(0) = B$. Express the resulting function $f(t)$in terms of sines and cosines.

(e) Compare the result of (d) with the results in Section 8.1.

111. Let z_1 and z_2 be nonzero complex numbers. Find an algebraic relation between z_1 and z_2 which is equivalent to the fact that the lines from the origin through z_1 and z_2 are perpendicular.

112. Let $w = f(z) = (1 + (z/2))/(1 - (z/2))$.
(a) Show that if the real part of z is 0, then $|w| = 1$.
(b) Are all points on the circle $|w| = 1$ in the range of f? [*Hint:* Solve for z in terms of w.]

113. (a) Show that, if $z^n = 1$, n a positive integer, then either $z^{n-1} + z^{n-2} + \cdots + z + 1 = 0$ or $z = 1$.
(b) Show that, if $z^{n-1} + z^{n-2} + \cdots + z + 1 = 0$, then $z^n = 1$.
(c) Find all the roots of the equation $z^3 + z^2 + z + 1 = 0$.

114. Describe the motion in the complex plane, as the real number t goes from $-\infty$ to ∞, of the point $z = e^{i\omega t}$, when
(a) $\omega = i$,
(b) $\omega = 1 + i$,
(c) $\omega = -i$,
(d) $\omega = -1 - i$,
(e) $\omega = 0$,
(f) $\omega = 1$,
(g) $\omega = -1$.

115. Describe the motion in the complex plane, as the real number t varies, of the point given by $z = 93{,}000{,}000\ e^{2\pi(t/365)} + 1{,}000{,}000 e^{2\pi i(t/29)}$. What astronomical phenomenon does this represent?

116. What is the relation between e^z and $e^{\bar{z}}$?

★117. (a) Find *all* complex numbers z for which $e^z = 1$. (b) How might you define $\ln(-1)$? What is the difficulty here?

★118. (a) Find λ such that the function $x = e^{\lambda t}$ satisfies the equation $x'' - 2x' + 2x = 0$; $x' = dx/dt$.
(b) Express the function $e^{\lambda t} + e^{-\lambda t}$ in terms of sines, cosines, and *real* exponents.
(c) Show that the function in (b) satisfies the differential equation in (a).

12.7 Second-Order Linear Differential Equations

The nature of the solutions of $ay'' + by' + cy = 0$ depends on whether the roots of $ar^2 + br + c = 0$ are real or complex.

We shall now use complex numbers to study second-order differential equations more general than the spring equation discussed in Section 8.1.

We begin by studying the equation

$$ay'' + by' + cy = 0, \tag{1}$$

where y is an unknown function of x, $y' = dy/dx$, $y'' = d^2y/dx^2$, and a, b, c are constants. We assume that $a \neq 0$; otherwise equation (1) would be a first-order equation, which we have already studied in Sections 8.2 and 8.6.

We look for solutions of equation (1) in the form

$$y = e^{rx}, \qquad r \text{ a constant.} \tag{2}$$

Substituting equation (2) into equation (1) gives

$$ar^2e^{rx} + bre^{rx} + ce^{rx} = 0,$$

which is equivalent to

$$ar^2 + br + c = 0, \tag{3}$$

since $e^{rx} \neq 0$. Equation (3) is called the *characteristic equation* of equation (1). By the quadratic formula, it has roots

$$r = \frac{-b \pm \sqrt{b^2 - 4ac}}{2a},$$

which we shall denote by r_1 and r_2. Thus, $y = e^{r_1 x}$ and $y = e^{r_2 x}$ are solutions of equation (1).

By analogy with the spring equation, we expect the general solution of equation (1) to involve two arbitrary constants. In fact, $y = c_1 e^{r_1 x} + c_2 e^{r_2 x}$ is a solution of equation (1) for constants c_1 and c_2; indeed, note that if y_1 and y_2 solve equation (1), so does $c_1 y_1 + c_2 y_2$ since

$$a(c_1 y_1 + c_2 y_2)'' + b(c_1 y_1 + c_2 y_2)' + c(c_1 y_1 + c_2 y_2)$$
$$= c_1(ay_1'' + by_1' + cy_1) + c_2(ay_2'' + by_2' + cy_2) = 0.$$

If r_1 and r_2 are distinct, then one can show that $y = c_1 e^{r_1 x} + c_2 e^{r_2 x}$ is the *general solution*; i.e., any solution has this form for particular values of c_1 and c_2. (See the Supplement to this section for the proof.)

Second-Order Equations: Distinct Roots

If $ar^2 + br + c = 0$ has distinct roots r_1 and r_2, then the general solution of

$$ay'' + by' + cy = 0$$

is

$$y = c_1 e^{r_1 x} + c_2 e^{r_2 x}, \qquad c_1, c_2 \text{ constants.}$$

Example 1 Consider the equation $2y'' - 3y' + y = 0$. (a) Find the general solution, and (b) Find the particular solution satisfying $y(0) = 1$, $y'(0) = 0$.

Solution (a) The characteristic equation is $2r^2 - 3r + 1 = 0$, which factors: $(2r - 1)(r - 1) = 0$. Thus $r_1 = 1$ and $r_2 = \frac{1}{2}$ are the roots, and so

$$y = c_1 e^x + c_2 e^{x/2}$$

is the general solution.

(b) Substituting $y(0) = 1$ and $y'(0) = 0$ in the preceding formula for y gives

$$c_1 + c_2 = 1,$$
$$c_1 + \tfrac{1}{2}c_2 = 0.$$

Subtracting gives $\frac{1}{2}c_2 = 1$, so $c_2 = 2$ and hence $c_1 = -1$. Thus

$$y = 2e^{x/2} - e^x$$

is the particular solution sought. ▲

If the roots of the characteristic equation are distinct but complex, we can convert the solution to sines and cosines using the relation $e^{ix} = \cos x + i \sin x$, which was established in Section 12.6. Differentiating a complex valued function is carried out by differentiating the real and imaginary parts separately. One finds that $(d/dt)Ce^{rt} = Cre^{rt}$ for any complex numbers C and r (see Exercise 110 in Section 12.6). Thus, the results in the above box still work if r_1, r_2, C_1 and C_2 are complex.

Example 2 Find the general solution of $y'' + 2y' + 2y = 0$.

Solution The characteristic equation is $r^2 + 2r + 2 = 0$, whose roots are

$$r = \frac{-2 \pm \sqrt{4 - 8}}{2} = -1 \pm i.$$

Thus

$$
\begin{aligned}
y &= c_1 e^{(-1+i)x} + c_2 e^{(-1-i)x} \\
&= c_1 e^{-x} e^{ix} + c_2 e^{-x} e^{-ix} \\
&= e^{-x}\left[c_1(\cos x + i\sin x) + c_2(\cos x - i\sin x)\right] \\
&= e^{-x}(C_1\cos x + C_2\sin x),
\end{aligned}
$$

where $C_1 = c_1 + c_2$ and $C_2 = i(c_1 - c_2)$. If we desire a real (as opposed to complex) solution, C_1 and C_2 should be real. (Although we used complex numbers as a helpful tool in our computations, the final answer involves only real numbers and can be verified directly.) ▲

For the spring equation $y'' + \omega^2 y = 0$, the characteristic equation is $r^2 + \omega^2 = 0$, which has roots $r = \pm i\omega$, so the general solution is

$$
\begin{aligned}
y &= c_1 e^{i\omega x} + c_2 e^{-i\omega x} \\
&= C_1 \cos\omega x + C_2 \sin\omega x,
\end{aligned}
$$

where C_1 and C_2 are as in Example 2. Thus we recover the same general solution that we found in Section 8.1.

If the roots of the characteristic equation are equal ($r_1 = r_2$), then we have so far only the solution $y = c_1 e^{r_1 x}$, where c_1 is an arbitrary constant. We still expect another solution, since the general solution of a second-order equation should involve two arbitrary constants. To find the second solution, we may use either of two methods.

Method 1. Reduction of Order. We seek another solution of the form

$$y = ve^{r_1 x}. \tag{4}$$

where v is now a function rather than a constant. To see what equation is satisfied by v, we substitute equation (4) into equation (1). Noting that

$$y' = v'e^{r_1 x} + r_1 v e^{r_1 x},$$

and

$$y'' = v''e^{r_1 x} + 2r_1 v' e^{r_1 x} + r_1^2 v e^{r_1 x},$$

substitution into (1) gives

$$a\left(v'' + 2r_1 v' + r_1^2 v\right)e^{r_1 x} + b(v' + r_1 v)e^{r_1 x} + cve^{r_1 x} = 0;$$

but $e^{r_1 x} \neq 0$, $ar_1^2 + br_1 + c = 0$, and $2ar_1 + b = 0$ (since r_1 is a repeated root), so this reduces to $av'' = 0$. Hence $v = c_1 + c_2 x$, so equation (4) becomes

$$y = (c_1 + c_2 x)e^{r_1 x}. \tag{5}$$

This argument actually proves that equation (5) is the *general solution* to equation (1) in the case of a repeated root. (The reason for the name "reduction of order" is that for more general equations $y'' + b(x)y' + c(x)y = 0$, if one solution $y_1(x)$ is known, one can find another one of the form $v(x)y_1(x)$, where $v'(x)$ satisfies a *first* order equation—see Exercise 48.)

Method 2. Root Splitting. If $ay'' + by' + cy = 0$ has a repeated root r_1, the characteristic equation is $(r - r_1)(r - r_1) = 0$. Now consider the new equation $(r - r_1)(r - (r_1 + \varepsilon)) = 0$ which has distinct roots r_1 and $r_2 = r_1 + \varepsilon$ if $\varepsilon \neq 0$. The corresponding differential equation has solutions $e^{r_1 x}$ and $e^{(r_1+\varepsilon)x}$. Hence $(1/\varepsilon)(e^{(r_1+\varepsilon)x} - e^{r_1 x})$ is also a solution. Letting $\varepsilon \to 0$, we get the solution $(d/dr)e^{rx}|_{r=r_1} = xe^{r_1 x}$ for the given equation. (If you are suspicious of this reasoning, you may verify directly that $xe^{r_1 x}$ satisfies the given equation).

Second-Order Equations: Repeated Roots

If $ar^2 + br + c = 0$ has a repeated root $r_1 = r_2$, then the general solution of

$$ay'' + by' + cy = 0$$

is

$$y = (c_1 + c_2x)e^{r_1x}, \tag{5}$$

where c_1 and c_2 are constants.

Example 3 Find the solution of $y'' - 4y' + 4 = 0$ satisfying $y'(0) = -1$ and $y(0) = 3$.

Solution The characteristic equation is $r^2 - 4r + 4 = 0$, or $(r-2)^2 = 0$, so $r_1 = 2$ is a repeated root. Thus the general solution is given by equation (5):

$$y = (c_1 + c_2x)e^{2x}.$$

Thus $y'(x) = 2c_1e^{2x} + c_2e^{2x} + 2c_2xe^{2x}$. The data $y(0) = 3$, $y'(0) = -1$ give

$$c_1 = 3 \quad \text{and} \quad 2c_1 + c_2 = -1,$$

so $c_1 = 3$ and $c_2 = -7$. Thus $y = (3 - 7x)e^{2x}$. ▲

Now we shall apply the preceding methods to study damped harmonic motion. In Figure 12.7.1 we show a weight hanging from a spring; recall from

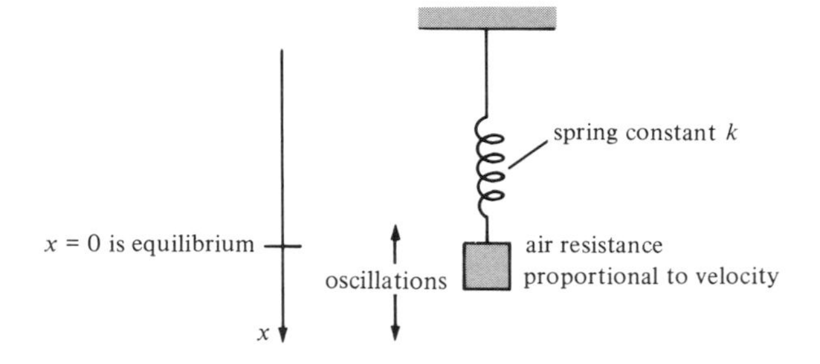

Figure 12.7.1. The physical set up for damped harmonic motion.

Section 8.1 that the equation of motion of the spring is $m(d^2x/dt^2) = F$, where F is the total force acting on the weight. The force due to the spring is $-kx$, just as in Section 8.1. (The force of gravity determines the equilibrium position, which we have called $x = 0$; see Exercise 51.) We also suppose that the force of air resistance is proportional to the velocity. Thus $F = -kx - \gamma(dx/dt)$, so the equation of motion becomes

$$m\frac{d^2x}{dt^2} = -kx - \gamma\frac{dx}{dt}, \tag{6}$$

where $\gamma > 0$ is a constant. (Can you see why there is a minus sign before γ?). If we rewrite equation (6) as

$$\frac{d^2x}{dt^2} + \beta\frac{dx}{dt} + \omega^2x = 0, \tag{7}$$

where $\beta = \gamma/m$ and $\omega^2 = k/m$, it has the form of equation (1) with $a = 1$, $b = \beta$, and $c = \omega^2$. To solve it, we look at the characteristic equation

$$r^2 + \beta r + \omega^2 = 0 \qquad \text{which has roots} \qquad r = \frac{-\beta \pm \sqrt{\beta^2 - 4\omega^2}}{2}.$$

If $\beta^2 > 4\omega^2$ (i.e., $\beta > 2\omega$), then there are two real roots and so the solution is $x = c_1e^{r_1t} + c_2e^{r_2t}$, where r_1 and r_2 are the two roots $\frac{1}{2}(-\beta \pm \sqrt{\beta^2 - 4\omega^2})$. Note that r_1 and r_2 are both negative, so the solution tends to zero as $t \to \infty$, although it will cross the t axis once if c_1 and c_2 have opposite signs; this case is called the *overdamped case*. A possible solution is sketched in Fig. 12.7.2.

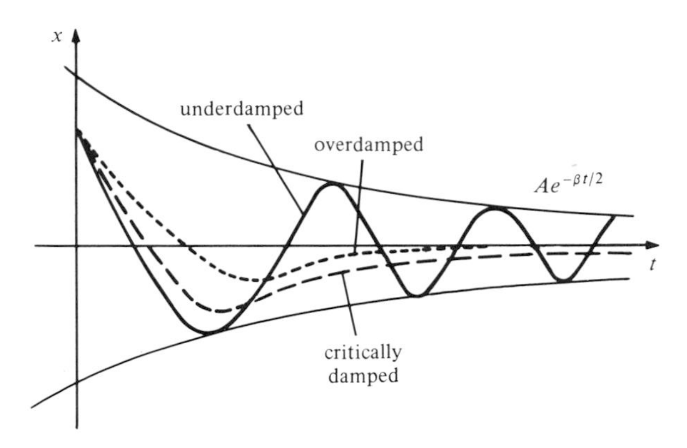

Figure 12.7.2. Damped harmonic motion.

If $\beta^2 = 4\omega^2$, there is a repeated root $r_1 = -\beta/2$, so the solution is $x = (c_1 + c_2t)e^{-\beta t/2}$. This case is called *critically damped*. Here the solution also tends to zero as $t \to \infty$, although it may cross the t axis once if c_1 and c_2 have opposite signs (this depends on the initial conditions). A possible trajectory is given in Figure 12.7.2.

Finally, if $\beta^2 < 4\omega^2$, then the roots are complex. If we let $\bar{\omega} = \frac{1}{2}\sqrt{4\omega^2 - \beta^2} = \omega\sqrt{1 - \beta^2/4\omega^2}$, then the solution is

$$x = e^{-\beta t/2}(c_1\cos\bar{\omega}t + c_2\sin\bar{\omega}t)$$

which represents *underdamped oscillations* with frequency $\bar{\omega}$. (Air resistance slows down the motion so the frequency $\bar{\omega}$ is lower than ω.) These solutions may be graphed by utilizing the techniques of Section 8.1; write $x = Ae^{-\beta t/2}\cos(\bar{\omega}t - \theta)$, where (A,θ) are the polar coordinates of c_1 and c_2. A typical graph is shown in Fig. 12.7.2.At $t = 0$, $0 = x = c_1$.

Example 4 Consider a spring with$\beta = \pi/4$ and $\omega = \pi/6$.

(a) Is it over, under, or critically damped?

(b) Find and sketch the solution with $x(0) = 0$ and $x'(0) = 1$, for $t \geqslant 0$.

(c) Find and sketch the solution with the same initial conditions but with $\beta = \pi/2$.

Solution (a) Here $\beta^2 - 4\omega^2 = \pi^2/16 - 4\pi^2/36 = -7\pi^2/36 < 0$, so the spring is underdamped.

(b) The effective frequency is $\bar{\omega} = \omega\sqrt{1 - \beta^2/4\omega^2} = (\pi/6)\sqrt{1 - 9/16} = \pi\sqrt{7}/24$, so the general solution is

$$x = e^{-\pi t/8}\left(c_1\cos\left(\frac{\pi\sqrt{7}\,t}{24}\right) + c_2\sin\left(\frac{\pi\sqrt{7}\,t}{24}\right)\right).$$

At $t = 0$, $0 = x = c_1$. Thus,

$$x = c_2e^{-\pi t/8}\sin\left(\frac{\pi\sqrt{7}}{24}t\right).$$

Hence

$$x' = c_2\left[\left(-\frac{\pi}{8}\right)e^{-\pi t/8}\sin\left(\frac{\pi\sqrt{7}}{24}t\right) + \frac{\pi\sqrt{7}}{24}e^{-\pi t/8}\cos\left(\frac{\pi\sqrt{7}}{24}t\right)\right].$$

At $t = 0$, $x' = 1$, so $1 = c_2[\pi\sqrt{7}/24]$, and hence $c_2 = 24/\pi\sqrt{7}$. Thus the solution is

$$x = \frac{24}{\pi\sqrt{7}}e^{-\pi t/8}\sin\left(\frac{\pi\sqrt{7}}{24}t\right).$$

This is a sine wave multiplied by the decaying factor $e^{-\pi t/8}$; it is sketched in Fig. 12.7.3. The first maximum occurs when $x' = 0$; i.e., when $\tan((\pi\sqrt{7}/24)t) = 8\sqrt{7}/24$, or $t \approx 2.09$, at which point $x \approx 0.84$.

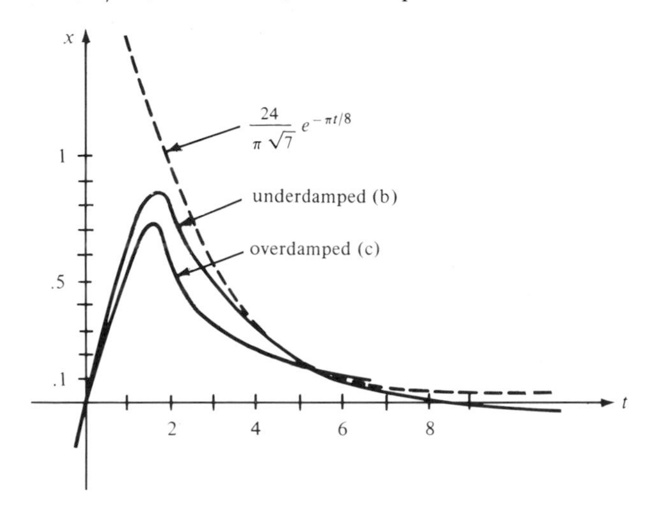

Figure 12.7.3. Graph of the solution to Example 4.

(c) For $\beta = \pi/2$, we have $\beta^2 - 4\omega^2 = \pi^2/4 - 4\pi^2/36 = 5\pi^2/36 > 0$, so the spring is overdamped. The roots r_1 and r_2 are

$$\frac{1}{2}\left(-\beta \pm \sqrt{\beta^2 - 4\omega^2}\right) = \frac{1}{2}\left(\frac{-\pi}{2} \pm \frac{\sqrt{5}}{6}\right) = \frac{\pi}{4}\left(-1 \pm \frac{\sqrt{5}}{3}\right),$$

so the solution is of the form $x = c_1e^{(\pi/4)(-1+\sqrt{5}/3)t} + c_2e^{(\pi/4)(-1-\sqrt{5}/3)t}$. At $t = 0$, $x = 0$, so $c_1 + c_2 = 0$ or $c_1 = -c_2$. Also, at $t = 0$,

$$1 = x' = c_1\frac{\pi}{4}\left(-1 + \frac{\sqrt{5}}{3}\right) + c_2\frac{\pi}{4}\left(-1 - \frac{\sqrt{5}}{3}\right) = c_1\frac{\pi\sqrt{5}}{6},$$

so $c_1 = 6/\pi\sqrt{5}$ and $c_2 = -6/\pi\sqrt{5}$. Thus our solution is

$$x = c_1(e^{r_1t} - e^{r_2t}) = \frac{6}{\pi\sqrt{5}}e^{(\pi/4)(-1+\sqrt{5}/3)t} - \frac{6}{\pi\sqrt{5}}e^{(\pi/4)(-1-\sqrt{5}/3)t}$$

$$= \frac{6}{\sqrt{5}}e^{\pi t/4}\left(e^{\sqrt{5}t/3} - e^{-\sqrt{5}t/3}\right).$$

The derivative is $x' = c_1(r_1e^{r_1t} - r_2e^{r_2t}) = c_1(r_1 - r_2e^{(r_2-r_1)t})e^{r_1t}$ which vanishes when

$$\frac{r_1}{r_2} = e^{(r_2-r_1)t} \quad \text{or} \quad \frac{-1+\sqrt{5}/3}{-1-\sqrt{5}/3} = e^{-(\pi\sqrt{5}/6)t}, \quad \text{or} \quad t \approx 1.64;$$

at this point, $x \approx 0.731$. See Fig. 12.7.3. ▲

In the preceding discussion we have seen how to solve the equation (1): $ay'' + by' + cy = 0$. Let us now study the problem of solving

$$ay'' + by' + cy = F(x), \tag{8}$$

where $F(x)$ is a given function of x. We call equation (1) the *homogeneous equation*, while equation (8) is called the *nonhomogeneous equation*. Using what we know about equation (1), we can find the general solution to equation (8) provided we can find just one particular solution.

Nonhomogeneous Equations: Particular and General Solutions

If $y_h = c_1y_1 + c_2y_2$ is the general solution to the homogeneous equation (1) and if y_p is a particular solution to the nonhomogeneous equation (8), then

$$y = y_p + c_1y_1 + c_2y_2 = y_p + y_h \tag{9}$$

is the general solution to the nonhomogeneous equation.

To see that equation (9) is a solution of equation (8), note that

$$\begin{aligned} a(y_p + y_h)'' &+ b(y_p + y_h)' + c(y_p + y_h) \\ &= (ay_p'' + by_p' + cy_p) + (ay_h'' + by_h' + cy_h) \\ &= F(x) + 0 = F(x). \end{aligned}$$

To see that equation (9) is the *general* solution of equation (8), note that if $\tilde{y}$ is any solution to equation (9), then $\tilde{y} - y_p$ solves equation (1) by a calculation similar to the one just given. Hence $\tilde{y} - y_p$ must equal y_h for suitable c_1 and c_2 since y_h is the general solution to equation (1). Thus $\tilde{y}$ has the form of equation (9), so equation (9) is the general solution.

Sometimes equation (8) can be solved by inspection; for example, if $F(x) = F_0$ is a constant and if $c \neq 0$, then $y = F_0/c$ is a particular solution.

Example 5 (a) Solve $2y'' - 3y' + y = 10$. (b) Solve $2x'' - 3x' + x = 8\cos(t/2)$ (where x is a function of t). (c) Solve $2y'' - 3y' + y = 2e^{-x}$.

Solution (a) Here a particular solution is $y = 10$. From Example 1,

$$y_h = c_1e^x + c_2e^{x/2}.$$

Thus the general solution, given by equation (9), is

$$y = c_1e^x + c_2e^{x/2} + 10.$$

(b) When the right-hand side is a trigonometric function, we can try to find a particular solution which is a combination of sines and cosines of the same frequency, since they reproduce linear combinations of each other when differentiated. In this case, $8\cos(t/2)$ appears, so we try

$$x_p = A\cos\left(\frac{t}{2}\right) + B\sin\left(\frac{t}{2}\right),$$

where A and B are constants, called *undetermined coefficients*. Then

$$\begin{aligned} 2x_p'' - 3x_p' + x_p &= 2\left[-\frac{A}{4}\cos\left(\frac{t}{2}\right) - \frac{B}{4}\sin\left(\frac{t}{2}\right)\right] \\ &\quad - 3\left[-\frac{A}{2}\sin\left(\frac{t}{2}\right) + \frac{B}{2}\cos\left(\frac{t}{2}\right)\right] \\ &\quad + \left[A\cos\left(\frac{t}{2}\right) + B\sin\left(\frac{t}{2}\right)\right] \\ &= \left(\frac{1}{2}A - \frac{3}{2}B\right)\cos\left(\frac{t}{2}\right) + \left(\frac{3}{2}A + \frac{1}{2}B\right)\sin\left(\frac{t}{2}\right). \end{aligned}$$

For this to equal $8\cos(t/2)$, we choose A and B such that

$$\tfrac{1}{2}A - \tfrac{3}{2}B = 8, \quad \text{and} \quad \tfrac{3}{2}A + \tfrac{1}{2}B = 0.$$

The second equation gives $B = -3A$ which, upon substitution in the first, gives $\frac{1}{2}A + \frac{9}{2}A = 8$. Thus $A = \frac{8}{5}$ and $B = -\frac{24}{5}$, so

$$x_p = \frac{8}{5}\left[\cos\left(\frac{t}{2}\right) - 3\sin\left(\frac{t}{2}\right)\right],$$

and the general solution is

$$x = c_1e^t + c_2e^{t/2} + \frac{8}{5}\left[\cos\left(\frac{t}{2}\right) - 3\sin\left(\frac{t}{2}\right)\right].$$

A good way to check your arithmetic is to substitute this solution into the original differential equation.

(c) Here we try $y_p = Ae^{-x}$ since e^{-x} reproduces itself, up to a factor, when differentiated. Then

$$2y_p'' - 3y_p' + y_p = 2Ae^{-x} + 3Ae^{-x} + Ae^{-x}.$$

For this to match $2e^x$, we require $6A = 2$ or $A = \frac{1}{3}$. Thus $y_p = \frac{1}{3}e^{-x}$ is a particular solution, and so the general solution is

$$y = c_1e^x + c_2e^{x/2} + \tfrac{1}{3}e^{-x}. \blacktriangle$$

The technique used in parts (b) and (c) of this example is called the *method of undetermined coefficients*. This method works whenever the right-hand side of equation (8) is a polynomial, an exponential, sums of sines and cosines (of the same frequency), or products of these functions.

There is another method called *variation of parameters* or *variation of constants* which always enables us to find a particular solution of equation (8) in terms of integrals. This method proceeds as follows. We seek a solution of the form

$$y = v_1y_1 + v_2y_2 \tag{10}$$

where y_1 and y_2 are solutions of the homogeneous equation (1) and v_1 and v_2 are functions of x to be found. Note that equation (10) is obtained by replacing the *constants* (or *parameters*) c_1 and c_2 in the general solution to the homogeneous equation by *functions*. This is the reason for the name "variation of parameters." (Note that a similar procedure was used in the method of reduction of order—see equation (4).) Differentiating v_1y_1 using the product rule gives

$$(v_1y_1)' = v_1'y_1 + v_1y_1',$$

and

$$(v_1y_1)'' = v_1''y_1 + 2v_1'y_1' + v_1y_1'',$$

and similarly for v_2y_2. Substituting these expressions into equation (8) gives

$$a\left[(v_1''y_1 + 2v_1'y_1' + v_1y_1'') + (v_2''y_2 + 2v_2'y_2' + v_2y_2'')\right] \\ + b\left[(v_1'y_1 + v_1y_1') + (v_2'y_2 + v_2y_2')\right] + c(v_1y_1 + v_2y_2) = F.$$

Simplifying, using (1) for y_1 and y_2, we get

$$a\left[v_1''y_1 + 2v_1'y_1' + v_2''y_2 + 2v_2'y_2')\right] + b\left[v_1'y_1 + v_2'y_2\right] = F. \tag{11}$$

This is only one condition on the two functions v_1 and v_2, so we are free to impose a second condition; we shall do so to make things as simple as possible. As our second condition, we require that the coefficient of b vanish (identically, as a function of x):

$$v_1'y_1 + v_2'y_2 = 0. \tag{12}$$

This implies, on differentiation, that $v_1''y_1 + v_1'y_1' + v_2''y_2 + v_2'y_2' = 0$, so equation (11) simplifies to

$$v_1'y_1' + v_2'y_2' = \frac{F}{a}. \tag{13}$$

Equations (12) and (13) can now be solved algebraically for v_1' and v_2' and the resulting expressions integrated to give v_1 and v_2. (Even if the resulting integrals cannot be evaluated, we have succeeded in expressing our solution in terms of integrals; the problem is then generally regarded as "solved").

Variation of Parameters

A particular solution of the nonhomogeneous equation (8) is given by

$$y_p = v_1y_1 + v_2y_2,$$

where y_1 and y_2 are solutions of the homogeneous equation and where v_1 and v_2 are found by solving equations (12) and (13) algebraically for v_1' and v_2' and then integrating.

Combining the two preceding boxes, one has a recipe for finding the general solution to the nonhomogeneous equation.

Example 6 (a) Find the general solution of $2y'' - 3y' + y = e^{2x} + e^{-2x}$ using variation of parameters. (b) Find the general solution of $2y'' - 3y' + y = 1/(1 + x^2)$ (express your answer in terms of integrals).

Solution (a) Here $y_1 = e^x$ and $y_2 = e^{x/2}$ from Example 1. Thus, equations (12) and (13) become

$$v_1'e^x + v_2'e^{x/2} = 0,$$

$$v_1'e^x + \frac{1}{2}v_2'e^{x/2} = \frac{e^{2x} + e^{-2x}}{2},$$

respectively. Subtracting,

$$\frac{1}{2}v_2'e^{x/2} = -\left(\frac{e^{2x} + e^{-2x}}{2}\right),$$

and so

$$v_2' = -e^{-x/2}(e^{2x} + e^{-2x}) = -e^{3x/2} - e^{-5x/2}.$$

Similarly,

$$v_1' = -e^{-x/2}v_2' = e^x + e^{-3x},$$

and so integrating, dropping the constants of integration,

$$v_2 = -\tfrac{2}{3}e^{3x/2} + \tfrac{2}{5}e^{-5x/2},$$

$$v_1 = e^x - \tfrac{1}{3}e^{-3x}.$$

Hence a particular solution is

$$y_p = v_1y_1 + v_2y_2 = e^{2x} - \tfrac{1}{3}e^{-2x} - \tfrac{2}{3}e^{2x} + \tfrac{2}{5}e^{-2x} = \tfrac{1}{3}e^{2x} + \tfrac{1}{15}e^{-2x},$$

and so the general solution is

$$y = c_1e^x + c_2e^{x/2} + \tfrac{1}{3}e^{2x} + \tfrac{1}{15}e^{-2x}.$$

The reader can check that the method of undetermined coefficients gives the same answer.

(b) Here equations (12) and (13) become

$$v_1'e^x + v_2'e^{x/2} = 0,$$

$$v_1'e^x + \frac{1}{2}v_2'e^{x/2} = \frac{1}{1+x^2},$$

respectively. Solving,

$$v_2' = -\frac{2e^{-x/2}}{1+x^2}, \qquad \text{so} \quad v_2 = -2\int \frac{e^{-x/2}}{1+x^2}\,dx$$

and

$$v_1' = \frac{2e^{-x}}{1+x^2}, \qquad \text{so} \quad v_1 = 2\int \frac{e^{-x}}{1+x^2}\,dx.$$

Thus the general solution is

$$y = c_1e^x + c_2e^{-x} + 2e^x\int \frac{e^{-x}}{1+x^2}\,dx - 2e^{x/2}\int \frac{e^{-x/2}}{1+x^2}\,dx. \ \blacktriangle$$

Let us now apply the above method to the problem of *forced oscillations*. Imagine that our weight on a spring is subject to a periodic external force $F_0\cos\Omega t$; the spring equation (6) then becomes

$$m\frac{d^2x}{dt^2} = -kx - \gamma\frac{dx}{dt} + F_0\cos\Omega t. \tag{14}$$

A periodic force can be directly applied to our bobbing weight by, for example, an oscillating magnetic field. In many engineering situations, equation (14) is used to model the phenomenon of *resonance*; the response of a ship to a periodic swell in the ocean and the response of a bridge to the periodic steps of a marching army are examples of this phenomenon. When the forcing frequency is close to the natural frequency, large oscillations can set in—this is resonance.[9] We shall see this emerge in the subsequent development.

Let us first study the case in which there is no damping: $\gamma = 0$, so that $m(d^2x/dt^2) + kx = F_0\cos\Omega t$. This is called a *forced oscillator equation*. A particular solution can be found by trying $x_p = C\cos\Omega t$ and solving for C. We find $x_p = [F_0/m(\omega^2 - \Omega^2)]\cos\Omega t$, where $\omega = \sqrt{k/m}$ is the frequency of the unforced oscillator. Thus the general solution is

[9] For further information on resonance and how it was involved in the Tacoma bridge disaster of 1940, see M. Braun, *Differential Equations and their Applications*, Second Edition, 1981, Springer-Verlag, New York, Section 2.6.1.

$$x = A\cos\omega t + B\sin\omega t + \frac{F_0}{m(\omega^2 - \Omega^2)}\cos\Omega t, \tag{15}$$

where A and B are determined by the initial conditions.

Example 7 Find the solution of $d^2x/dt^2 + 9x = 5\cos 2t$ with $x(0) = 0$, $x'(0) = 0$, and sketch its graph.

Solution We try a particular solution of the form $x = C\cos 2t$; substituting into the equation gives $-4C\cos 2t + 9C\cos 2t = 5\cos 2t$, so C must be 1. On the other hand, the solution of the homogeneous equation $d^2x/dt^2 + 9x = 0$ is $A\cos 3t + B\sin 3t$, and therefore the general solution of the given equation is $x(t) = A\cos 3t + B\sin 3t + \cos 2t$. For this solution, $x(0) = A + 1$ and $x'(0) = 3B$, so if $x(0) = 0$ and $x'(0) = 0$, we must have $A = -1$ and $B = 0$. Thus, our solution is $x(t) = -\cos 3t + \cos 2t$.

To graph this function, we will use the product formula

$$\sin Rt \sin St = \tfrac{1}{2}\left[\cos(R-S)t - \cos(R+S)t\right].$$

To recover $-\cos 3t + \cos 2t$, we must have $R + S = 3$ and $R - S = 2$, so $R = \frac{5}{2}$ and $S = \frac{1}{2}$. Thus

$$x(t) = 2\sin\left(\tfrac{5}{2}t\right)\sin\left(\tfrac{1}{2}t\right).$$

We may think of this as a rapid oscillation, $\sin\frac{5}{2}t$, with variable amplitude $2\sin\frac{1}{2}t$, as illustrated in Fig. 12.7.4. The function is periodic with period 2π, with a big peak coming at π, 3π, 5π, etc., when $-\cos 3t$ and $\cos 2t$ are simultaneously equal to 1. ▲

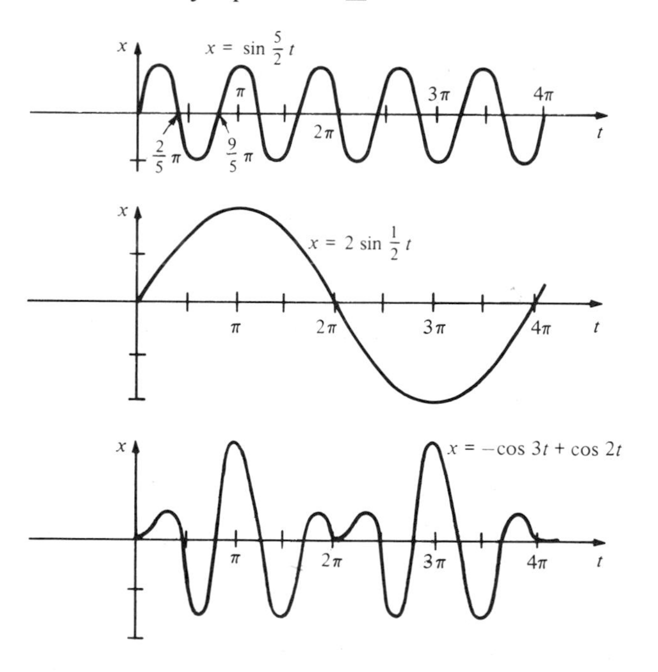

Figure 12.7.4. $x(t) = -\cos 3t + \cos 2t = 2\sin\frac{1}{2}t\sin\frac{5}{2}t$.

If in equation (15), $x(0) = 0$ and $x'(0) = 0$, then we find, as in Example 7, that

$$\begin{aligned} x(t) &= \frac{F_0}{m(\omega^2 - \Omega^2)}(\cos\Omega t - \cos\omega t) \\ &= \frac{2F_0}{m(\omega^2 - \Omega^2)}\left[\sin\left(\frac{(\omega - \Omega)t}{2}\right)\right]\left[\sin\left(\frac{(\omega + \Omega)t}{2}\right)\right]. \end{aligned}$$

If $\omega - \Omega$ is small, then this is the product of a relatively rapidly oscillating function $[\sin((\omega + \Omega)t/2)]$ with a slowly oscillating one $[\sin((\omega - \Omega)t/2)]$. The slowly oscillating function "modulates" the rapidly oscillating one as shown in Fig. 12.7.5. The slow rise and fall in the amplitude of the fast oscillation is the phenomenon of *beats*. It occurs, for example, when two musical instruments are played slightly out of tune with one another.

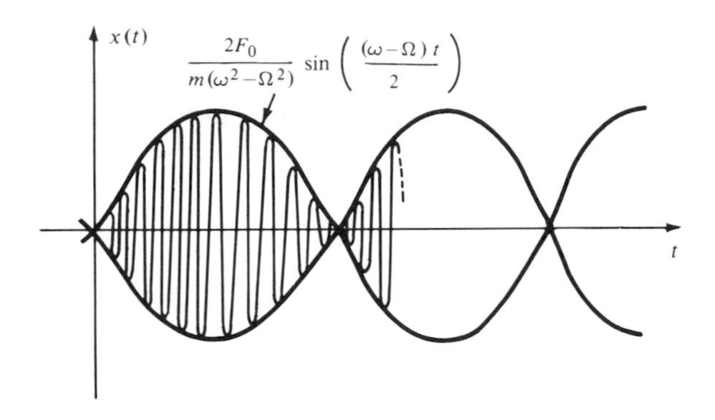

Figure 12.7.5. Beats.

The function (15) is the solution to equation (14) in the case where $\gamma = 0$ (no damping). The general case ($\gamma \neq 0$) is solved similarly. The method of undetermined coefficients yields a particular solution of the form $x_p = \alpha\cos\Omega t + \beta\sin\Omega t$, which is then added to the general solution of the homogeneous equation found by the method of Example 4. We state the result of such a calculation in the following box and ask the reader to verify it in Review Exercise 110.

Damped Forced Oscillations

The solution of

$$m\frac{d^2x}{dt^2} = -kx - \gamma\frac{dx}{dt} + F_0\cos\Omega t$$

is

$$x(t) = c_1e^{r_1t} + c_2e^{r_2t} + \frac{F_0}{\sqrt{m^2(\omega^2-\Omega^2)^2 + \gamma^2\omega^2}}\cos(\Omega t - \delta), \quad (16)$$

where c_1 and c_2 are constants determined by the initial conditions, $\omega = \sqrt{k/m}$, r_1 and r_2 are roots of the characteristic equation $mr^2 + \gamma r + k = 0$ [if r_1 is a repeated root, replace $c_1e^{r_1t} + c_2e^{r_2t}$ by $(c_1 + c_2t)e^{r_1t}$], and

$$\delta = \tan^{-1}\left(\frac{\gamma\Omega}{m(\omega^2-\Omega^2)}\right),$$

$$\cos\delta = \frac{m(\omega^2-\Omega^2)}{\sqrt{m^2(\omega^2-\Omega^2)^2+\gamma^2\Omega^2}}, \qquad \sin\delta = \frac{\gamma\Omega}{\sqrt{m^2(\omega^2-\Omega^2)^2+\gamma^2\Omega^2}}$$

In equation (16), as $t\to\infty$, the solution $c_1e^{r_1t} + c_2e^{r_2t}$ tends to zero (if $\gamma > 0$) as we have seen. This is called the *transient part*; the solution thus approaches the *oscillatory part*,

$$\frac{F_0}{\sqrt{m^2(\omega^2-\Omega^2)^2+\gamma^2\Omega^2}}\cos(\Omega t-\delta),$$

which oscillates with a modified amplitude at the forcing frequency Ω and with the *phase shift* δ. If we vary ω, the amplitude is largest when $\omega=\Omega$; this is the *resonance* phenomenon.

Example 8 Consider the equation

$$\frac{d^2x}{dt^2}+8\frac{dx}{dt}+25x=2\cos t.$$

(a) Write down the solution with $x(0)=0$, $x'(0)=0$.
(b) Discuss the behavior of the solution for large t.

Solution (a) The characteristic equation is

$$r^2+8r+25=0$$

which has roots $r=(-8\pm\sqrt{64-100})/2=-4\pm 3i$. Also, $m=1$, $\omega=5$, $\Omega=1$, $F_0=2$, and $\gamma=8$; so

$$\frac{F_0}{\sqrt{m^2(\omega^2-\Omega^2)^2+\gamma^2\Omega^2}}=\frac{2}{\sqrt{576+64}}=\frac{2}{\sqrt{640}}=\frac{1}{4\sqrt{10}}\approx 0.079$$

and $\delta=\tan^{-1}(\frac{8}{24})=\tan^{-1}(\frac{1}{3})\approx 0.322$. The general solution is given by equation (16); writing sines and cosines in place of the complex exponentials, we get

$$x(t)=e^{-4t}(A\cos 3t+B\sin 3t)+\frac{1}{4\sqrt{10}}\cos(t-\delta).$$

At $t=0$ we get

$$\begin{aligned}0=x(0)&=A+\frac{1}{4\sqrt{10}}\cos\delta\\&=A+\frac{1}{4\sqrt{10}}\cdot\frac{24}{\sqrt{640}}\\&=A+\frac{3}{40};\end{aligned}$$

so $A=-\frac{3}{40}$. Computing $x'(t)$ and substituting $t=0$ gives

$$\begin{aligned}0=x'(0)&=-4A+3B+\frac{1}{4\sqrt{10}}\sin\delta\\&=-4A+3B+\frac{1}{4\sqrt{10}}\cdot\frac{8}{\sqrt{640}}\\&=-4A+3B+\frac{1}{40}=\frac{12}{40}+3B+\frac{1}{40}.\end{aligned}$$

Thus $B=-\frac{13}{120}$, and our solution becomes

$$x(t)=-\frac{e^{-4t}}{120}(9\cos 3t+13\sin 3t)+0.079\cos(t-0.322).$$

(b) As $t\to\infty$ the transient part disappears and we get the oscillatory part $0.079\cos(t-0.322)$. ▲

Supplement to Section 12.7: Wronskians

In this section we have shown how to find solutions to equation (1): $ay'' + by' + cy = 0$; whether the roots of the characteristic equation $ar^2 + br + c = 0$ are real, complex, or coincident, we found two solutions y_1 and y_2. We then asserted that the *linear combination* $c_1 y_1 + c_2 y_2$ represents the *general* solution. In this supplement we shall prove this.

Suppose that y_1 and y_2 are solutions of equation (1); our goal is to show that every solution y of equation (1) can be written as $y = c_1 y_1 + c_2 y_2$. To do so, we try to find c_1 and c_2 by matching initial conditions at x_0:

$$y(x_0) = c_1 y_1(x_0) + c_2 y_2(x_0),$$

$$y'(x_0) = c_1 y_1'(x_0) + c_2 y_2'(x_0).$$

We can solve these equations for c_1 by multiplying the first equation by $y_2'(x_0)$, the second by $y_2(x_0)$, and subtracting:

$$c_1 = \frac{y(x_0)y_2'(x_0) - y_2(x_0)y'(x_0)}{y_1(x_0)y_2'(x_0) - y_2(x_0)y_1'(x_0)}. \tag{17a}$$

Similarly,

$$c_2 = \frac{y(x_0)y_1(x_0) - y_1'(x_0)y'(x_0)}{y_1(x_0)y_2'(x_0) - y_2(x_0)y_1'(x_0)}. \tag{17b}$$

These are valid as long as $y_1(x_0)y_2'(x_0) - y_2(x_0)y_1'(x_0) \neq 0$. The expression

$$W(x) = y_1(x)y_2'(x) - y_2(x)y_1'(x). \tag{18}$$

is called the *Wronskian* of y_1 and y_2 [named after the Polish mathematician Count Hoëné Wronski (1778–1853)]. (The expression (18) is a determinant—see Exercise 43, Section 13.6).

Two solutions y_1 and y_2 are said to be a *fundamental set* if their Wronskian does not vanish. It is an important fact that $W(x)$ *is either everywhere zero or nowhere zero*. To see this, we compute the derivative of W:

$$\begin{aligned} W'(x) &= \left[y_1'(x)y_2'(x) + y_1(x)y_2''(x) \right] - \left[y_2'(x)y_1'(x) + y_2(x)y_1''(x) \right] \\ &= y_1(x)y_2''(x) - y_2(x)y_1''(x). \end{aligned}$$

If y_1 and y_2 are solutions, we can substitute $-(b/a)y_1' - (c/a)y_1$ for y_1'' and similarly for y_2'' to get

$$\begin{aligned} W'(x) &= y_1(x)\left[-\frac{b}{a}\, y_2'(x) - \frac{c}{2}\, y_2(x) \right] - y_2(x)\left[-\frac{b}{a}\, y_1'(x) - \frac{c}{a}\, y_1(x) \right] \\ &= -\frac{b}{a}\left[y_1(x)y_2'(x) - y_2(x)y_1'(x) \right]. \end{aligned}$$

Thus

$$W'(x) = -\frac{b}{a}\, W(x).$$

Therefore, from Section 8.2, $W(x) = Ke^{-(b/a)x}$ for some constant K. We note that W is nowhere zero unless $K = 0$, in which case it is identically zero.

If y_1 and y_2 are a fundamental set, then equation (17) makes sense, and so we can find c_1 and c_2 such that $c_1 y_1 + c_2 y_2$ attains any given initial conditions. Such a specification of initial conditions gives a unique solution and determines y uniquely; therefore $y = c_1 y_1 + c_2 y_2$. In fact, the proof of uniqueness of a solution given its initial conditions also follows fairly easily from what we have done; see Exercise 46 for a special case and Exercise 47 for the general case. Thus, in summary, we have proved:

If y_1 and y_2 are a fundamental set of solutions for $ay'' + by' + cy = 0$, then $y = c_1y_1 + c_2y_2$ is the general solution.

To complete the justification of the claims about general solutions made earlier in this section, we need only check that in each specific case, the solutions form a fundamental set. For example, suppose that r_1 and r_2 are distinct roots of $ar^2 + br + c = 0$. We know that $y_1 = e^{r_1x}$ and $y_2 = e^{r_2x}$ are solutions. To check that they form a fundamental set, we compute

$$\begin{aligned} W(x) &= y_1(x)y_2'(x) - y_2(x)y_1'(x) \\ &= e^{r_1x} \cdot r_2e^{r_2x} - e^{r_2x} \cdot r_1e^{r_1x} \\ &= (r_2 - r_1)e^{(r_1+r_2)x}. \end{aligned}$$

This is nonzero since $r_2 \neq r_1$, so we have a fundamental set. One can similarly check the case of a repeated root (Exercise 45).

Exercises for Section 12.7

Find the general solution of the differential equations in Exercises 1–4.

1. $y'' - 4y' + 3y = 0$.
2. $2y'' - y = 0$.
3. $3y'' - 4y' + y = 0$.
4. $y'' - y' - 2y = 0$.

Find the particular solutions of the stated equations in Exercises 5–8 satisfying the given conditions.

5. $y'' - 4y' + 3y = 0$, $y(0) = 0$, $y'(0) = 1$.
6. $2y'' - y = 0$, $y(1) = 0$, $y'(1) = 1$.
7. $3y'' - 4y' + y = 0$, $y(0) = 1$, $y'(0) = 1$.
8. $y'' - y' - 2y = 0$, $y(1) = 0$, $y'(1) = 2$.

Find the general solution of the differential equations in Exercises 9–12.

9. $y'' - 4y' + 5y = 0$.
10. $y'' + 2y' + 5y = 0$.
11. $y'' - 6y' + 13y = 0$.
12. $y'' + 2y' + 26y = 0$.

Find the solution of the equations in Exercises 13–16 satisfying the given conditions.

13. $y'' - 6y' + 9y = 0$, $y(0) = 0$, $y'(0) = 1$.
14. $y'' - 8y' + 16y = 0$, $y(0) = -3$, $y'(0) = 0$.
15. $y'' - 2\sqrt{2}\,y' + 2y = 0$, $y(1) = 0$, $y'(1) = 1$.
16. $y'' - 2\sqrt{3}\,y' + 3y = 0$, $y(0) = 0$, $y'(0) = -1$.

In Exercises 17–20 consider a spring with β, ω, $x(0)$, and $x'(0)$ as given. (a) Determine if the spring is over, under, or critically damped. (b) Find and sketch the solution.

17. $\beta = \pi/16$, $\omega = \pi/2$, $x(0) = 0$, $x'(0) = 1$.
18. $\beta = 1$, $\omega = \pi/8$, $x(0) = 1$, $x'(0) = 0$.
19. $\beta = \pi/3$, $\omega = \pi/6$, $x(0) = 0$, $x'(0) = 1$.
20. $\beta = 0.03$, $\omega = \pi/2$, $x(0) = 1$, $x'(0) = 1$.

In Exercises 21–28, find the general solution to the given equation (y is a function of x or x is a function of t as appropriate).

21. $y'' - 4y' + 3y = 6x + 10$.
22. $y'' - 4y' + 3y = 2e^x$.
23. $3x'' - 4x' + x = 2\sin t$.
24. $3x'' - 4x' + x = e^t + e^{-t}$.
25. $y'' - 4y' + 5y = x + x^2$.
26. $y'' - 4y' + 5y = 10 + e^{-x}$.
27. $y'' - 2\sqrt{2}\,y' + 2y = \cos x + \sin x$.
28. $y'' - 2\sqrt{2}\,y' + 2y = \cos x - e^{-x}$.

In Exercises 29–32 find the general solution to the given equation using the method of variation of parameters.

29. $y'' - 4y' + 3y = 6x + 10$.
30. $y'' - 4y' + 3y = 2e^x$.
31. $3x'' - 4x' + x = 2\sin t$.
32. $3x'' - 4x' + x = e^t + e^{-t}$.

In Exercises 33–36 find the general solution to the given equation. Express your answer in terms of integrals if necessary.

33. $y'' - 4y' + 3y = \tan x$.
34. $y'' - 4y' + 3y = \dfrac{1}{x^2 + 2}$.
35. $y'' - 4y' + 5y = \dfrac{1}{1 + \cos^2 x}$.
36. $y'' - 4y' + 5y = \dfrac{e^x}{1 + x^2}$.

In Exercises 37–40, find the solution of the given forced oscillator equation satisfying the given initial conditions.

37. $x'' + 4x = 3\cos t$, $x(0) = 0$, $x'(0) = 0$.
38. $x'' + 9x = 4\sin 4t$, $x(0) = 0$, $x'(0) = 0$.
39. $x'' + 25x = \cos t$, $x(0) = 0$, $x'(0) = 1$.
40. $x'' + 25x = \cos 6t$, $x(0) = 1$, $x'(0) = 0$.

In Exercises 41–44, (a) write down the solution of the given equation with the stated initial conditions and (b) discuss the behavior of the solution for large t.

41. $\dfrac{d^2x}{dt^2} + 4\dfrac{dx}{dt} + 25x = 2\cos 2t$, $x(0) = 0$, $x'(0) = 0$.
42. $\dfrac{dx^2}{dt^2} + 2\dfrac{dx}{dt} + 36x = 4\cos 3t$, $x(0) = 0$, $x'(0) = 0$.
43. $\dfrac{d^2x}{dt^2} + \dfrac{dx}{dt} + 4x = \cos t$, $x(0) = 1$, $x'(0) = 0$.
44. $\dfrac{d^2x}{dt^2} + 2\dfrac{dx}{dt} + 9x = \cos 4t$, $x(0) = 1$, $x'(0) = 0$.
45. If r_1 is a repeated root of the characteristic equation, use the Supplement to this section to show that $y_1 = e^{r_1x}$ and $y_2 = xe^{r_1x}$ form a fundamental set and hence conclude that $y = c_1y_1 + c_2y_2$ is the general solution to equation (1).

★46. Suppose that in (1), $a > 0$ and $b^2 - 4ac < 0$. If y satisfies equation (1), prove $w(x) = e^{(b/2a)x}y(x)$ satisfies $w'' + [(4ac - b^2)/4a]w = 0$, which is a spring equation. Use this observation to do the following.
(a) Derive the general form of the solution to equation (1) if the roots are complex.
(b) Use existence and uniqueness results for the spring equation proved in Section 8.1 to prove corresponding results for equation (1) if the roots are complex.

★47. If we know that equation (1) admits a fundamental set y_1, y_2, show uniqueness of solutions to equation (1) with given values of $y(x_0)$ and $y'(x_0)$ as follows.
(a) Demonstrate that it is enough to show that if $y(x_0) = 0$ and $y'(x_0) = 0$, then $y(x) \equiv 0$.
(b) Use facts above the Wronskian proved in the Supplement in order to show that $y(x)y_1'(x) - y'(x)y_1(x) = 0$ and that $y(x)y_2'(x) - y'(x)y_2(x) = 0$.
(c) Solve the equations in (b) to show that $y(x) = 0$.

★48. (a) Generalize the method of reduction of order so it applies to the differential equation $a(x)y'' + b(x)y' + c(x)y = 0$, $a(x) \neq 0$. Thus, given one solution, develop a method for finding a second one.
(b) Show that x^r is a solution of *Euler's* equation $x^2y'' + \alpha xy' + \beta y = 0$ if $r^2 + (\alpha - 1)r + \beta = 0$.
(c) Use (a) to show that if $(\alpha - 1)^2 = 4\beta$, then $(\ln x)x^{(1-\alpha)/2}$ is a second solution.

★49. (a) Show that the basic facts about Wronskians, fundamental sets, and general solutions proved in the Supplement also apply to the equation in Exercise 48(a).
(b) Show that the solutions of Euler's equations found in Exercises 48(b) and 48(c) form a fundamental set. Write down the general solution in each case.

★50. (a) Generalize the method of variation of parameters to the equation $a(x)y'' + b(x)y' + c(x)y = F(x)$.
(b) Find the general solution to the equation $x^2y'' + 5xy' + 3y = xe^x$ (see Exercise 48; express your answer in terms of integrals if necessary).

★51. In Fig. 12.7.1, consider the motion relative to an arbitrarily placed x axis pointing downward.
(a) Taking all forces, including the constant force g of gravity into account, show that the equation of motion is

$$m\frac{d^2y}{dt^2} = -k(y - y_e) + g - \gamma\frac{dy}{dt},$$

where y_e is the equilibrium position of the spring in the absence of the mass.
(b) Make a change of variables $x = y + c$ to reduce this equation to equation (6).

★52. Show that solutions of equation (15) exhibit beats, without assuming that $x(0) = 0$ and $x'(0) = 0$.

★53. Find the general solution of $y'''' + y = 0$.

★54. Find the general solution of $y'''' - y = 0$.

★55. Find the general solution of $y'''' + y = e^x$.

★56. Find the general solution of $y'''' - y = \cos x$.

12.8 Series Solutions of Differential Equations

Power series solutions of differential equations can often be found by the method of undetermined coefficients.

Many differential equations cannot be solved by means of explicit formulas. One way of attacking such equations is by the numerical methods discussed in Section 8.5. In this section, we show how to use infinite series in a systematic way for solving differential equations.

Many equations arising in engineering and mathematical physics can be treated by this method. We shall concentrate on equations of the form $a(x)y'' + b(x)y' + c(x)y = f(x)$, which are similar to equation (1) in Section 12.7, except that a, b, and c are now functions of x rather than constants. The basic idea in the power series method is to consider the a_i's in a sum $y = \sum_{i=0}^{\infty} a_ix^i$ as undetermined coefficients and to solve for them in successive order.

Example 1 Find a power series solution of $y'' + xy' + y = 0$.

Solution If a solution has a convergent series of the form $y = a_0 + a_1x + a_2x^2 + \cdots = \sum_{i=0}^{\infty} a_i x^i$, we may use the results of Section 12.4 to write

$$y' = a_1 + 2a_2x + 3a_3x^2 + \cdots = \sum_{i=1}^{\infty} ia_i x^{i-1} \quad \text{and}$$

$$y'' = 2a_2 + 6a_2x + 12a_4x^2 + \cdots = \sum_{i=2}^{\infty} i(i-1)a_i x^{i-2}.$$

Therefore

$$y'' + xy' + y = \sum_{i=2}^{\infty} i(i-1)a_i x^{i-2} + \sum_{i=1}^{\infty} ia_i x^i + \sum_{i=0}^{\infty} a_i x^i = 0.$$

In performing manipulations with series, it is important to keep careful track of the summation index; writing out the first few terms explicitly usually helps. Thus,

$$\begin{aligned} y'' &+ xy' + y \\ &= (2a_2 + 6a_3x + 12a_4x^2 + \cdots) + (a_1x + 2a_2x^2 + \cdots) \\ &\quad + (a_0 + a_1x + a_2x^2 + \cdots). \end{aligned}$$

To write this in summation notation, we shift the summation index so all x's appear with the same exponent:

$$y'' + xy' + y = \sum_{i=0}^{\infty} (i+2)(i+1)a_{i+2}x^i + \sum_{i=1}^{\infty} ia_i x^i + \sum_{i=0}^{\infty} a_i x^i = 0.$$

(Check the first few terms from the explicit expression.) Now we set the coefficient of each x^i equal to zero in an effort to determine the a_i. The first two conditions are

$$2a_2 + a_0 = 0 \qquad \text{(constant term)},$$

$$6a_3 + 2a_1 = 0 \qquad \text{(coefficient of } x\text{)}.$$

Note that this determines a_2 and a_3 in terms of a_0 and a_1: $a_2 = -\frac{1}{2}a_0$, $a_3 = -\frac{1}{3}a_1$. For $i \geqslant 1$, equating the coefficient of x^i to zero gives

$$(i+2)(i+1)a_{i+2} + (i+1)a_i = 0$$

or

$$a_{i+2} = -\frac{1}{(i+2)}a_i.$$

Thus,

$$a_2 = -\frac{1}{2}a_0 \qquad a_3 = -\frac{1}{3}a_1$$

$$a_4 = -\frac{1}{4}a_2 = \frac{1}{4\cdot 2}a_0 \qquad a_5 = -\frac{1}{5}a_3 = \frac{1}{5\cdot 3}a_1$$

$$a_6 = -\frac{1}{6}a_4 = -\frac{1}{6\cdot 4\cdot 2}a_0 \qquad a_7 = -\frac{1}{7}a_5 = \frac{-1}{7\cdot 5\cdot 3}a_1$$

$$\vdots$$

Hence

$$a_{2n} = \frac{(-1)^n}{2n\cdot(2n-2)\cdot(2n-4)\ldots 4\cdot 2}a_0$$

$$= \frac{(-1)^n}{2\ldots 2\cdot n(n-1)\ldots 2\cdot 1}a_0 = \frac{(-1)^n}{2^n n!}a_0$$

and

$$a_{2n+1} = \frac{(-1)^n}{(2n+1)(2n-1)(2n-3)\ldots 5\cdot 3}a_1 = \frac{(-1)^n 2^n n!}{(2n+1)!}a_1.$$

Thus, we get (using $0! = 1$),

$$y = a_0\left(\sum_{i=0}^{\infty}(-1)^i\frac{x^{2i}}{2^i i!}\right) + a_1\left(\sum_{i=0}^{\infty}(-1)^i\frac{2^i i!\,x^{2i+1}}{(2i+1)!}\right). \tag{1}$$

What we have shown so far is that any convergent series solution must be of the form of equation (1). To show that equation (1) *really is* a solution, we must show that the given series converges; but this convergence follows from the ratio test. ▲

The constants a_0 and a_1 found in Example 1 are arbitrary and play the same role as the two arbitrary constants we found for the solutions of constant coefficient equations in the preceding section.

Example 2 Find the first four nonzero terms in the power series solution of $y'' + x^2y = 0$ satisfying $y(0) = 0$, $y'(0) = 1$.

Solution Let $y = a_0 + a_1x + a_2x^2 + \cdots$. The initial conditions $y(0) = 0$ and $y'(0) = 1$ can be put in immediately if we set $a_0 = 0$ and $a_1 = 1$, so that $y = x + a_2x^2 + \cdots$. Then

$$y'' = 2a_2 + 3\cdot 2a_3x + 4\cdot 3a_4x^2 + 5\cdot 4a_5x^3 + \cdots + (i+1)ia_{i+1}x^{i-1} + \cdots$$

and so

$$x^2y = x^3 + a_2x^4 + a_3x^5 + \cdots + a_{i-3}x^{i-1} + \cdots.$$

Setting $y'' + x^2y = 0$ gives

$$a_2 = 0 \qquad \text{(constant term)},$$

$$a_3 = 0 \qquad \text{(coefficient of } x\text{)},$$

$$a_4 = 0 \qquad \text{(coefficient of } x^2\text{)},$$

$$a_5 = -\frac{1}{5\cdot 4} \qquad \text{(coefficient of } x^3\text{)},$$

$$a_6 = 0 = a_7 = a_8 \qquad \text{(coefficients of } x^4, x^5, x^6\text{)},$$

$$a_9 = -\frac{1}{9\cdot 8}a_5 = \frac{1}{9\cdot 8\cdot 5\cdot 4} \qquad \text{(coefficient of } x^7\text{), etc.}$$

Thus, the first four nonzero terms are

$$y = x - \frac{1}{5\cdot 4}x^5 + \frac{1}{9\cdot 8\cdot 5\cdot 4}x^9 - \frac{1}{13\cdot 12\cdot 9\cdot 8\cdot 5\cdot 4}x^{13} + \cdots.$$

[The recursion relation is

$$a_{i+1} = -\frac{1}{i(i+1)}a_{i-3}$$

and the general term is

$$(-1)^j\frac{1}{(4j+1)(4j)\ldots 9\cdot 8\cdot 5\cdot 4}x^{4j+1}.$$

The ratio test shows that this series converges.] ▲

Example 3 **(Legendre's equation)**[10] Find the recursion relation and the first few terms for the solution of $(1-x^2)y''-2xy'+\lambda y=0$ as a power series.

Solution We write $y=\sum_{i=0}^{\infty}a_i x^i$ and get

$$y = a_0 + a_1x + a_2x^2 + \cdots + a_ix^i + \cdots,$$

$$y' = a_1 + 2a_2x + 3a_3x^2 + \cdots + ia_ix^{i-1} + \cdots,$$

$$-2xy' = -2a_1x - 2\cdot 2a_2x^2 - 2\cdot 3a_3x^3 - \cdots - 2ia_ix^i - \cdots,$$

$$y'' = 2a_2 + 3\cdot 2a_3x + 4\cdot 3a_4x^2 + \cdots + i(i-1)a_ix^{i-2} + \cdots,$$

$$-x^2y'' = -2a_2x^2 - 3\cdot 2a_3x^3 - \cdots - i(i-1)a_ix^i - \cdots.$$

Thus, setting $(1-x^2)y''-2xy'+\lambda y=0$ gives

$$2a_2 + \lambda a_0 = 0 \qquad \text{(constant term)},$$

$$3\cdot 2a_3 - 2a_1 + \lambda a_1 = 0 \qquad \text{(coefficient of } x),$$

$$4\cdot 3a_4 - 2a_2 - 4a_2 + \lambda a_2 = 0 \qquad \text{(coefficient of } x^2),$$

$$5\cdot 4a_5 - 3\cdot 2a_3 - 2\cdot 3a_3 + \lambda a_3 = 0 \qquad \text{(coefficient of } x^3),$$

$$\vdots$$

Solving,

$$a_2 = -\frac{\lambda}{2}a_0, \qquad a_3 = \frac{2-\lambda}{3\cdot 2}a_1,$$

$$a_4 = \frac{6-\lambda}{4\cdot 3}a_2 = -\frac{6-\lambda}{4\cdot 3}\frac{\lambda}{2}a_0,$$

$$a_5 = \frac{12-\lambda}{5\cdot 4}a_3 = \frac{12-\lambda}{5\cdot 4}\cdot\frac{2-\lambda}{3\cdot 2}a_1, \text{ etc.}$$

Thus, the solution is

$$y = a_0\left(1 - \frac{\lambda}{2}x^2 - \frac{(6-\lambda)}{4\cdot 3\cdot 2}x^4 + \cdots\right)$$

$$+ a_1\left(x + \frac{2-\lambda}{3\cdot 2}x^3 + \frac{(12-\lambda)(2-\lambda)}{5\cdot 4\cdot 3\cdot 2}x^5 + \cdots\right).$$

The recursion relation comes from setting the coefficient of x^i equal to zero:

$$(i+2)(i+1)a_{i+2} - i(i-1)a_i - 2ia_i + \lambda a_i = 0,$$

so

$$a_{i+2} = \frac{i(i+1)-\lambda}{(i+2)(i+1)}a_i.$$

From the ratio test one sees that the series solution has a radius of convergence of at least 1. It is exactly 1 unless there is a nonnegative integer n such that $\lambda = n(n+1)$, in which case the series can terminate: if n is even, set $a_1=0$; if n is odd, set $a_0=0$. Then the solution is a polynomial of degree n called *Legendre's polynomial*; it is denoted $P_n(x)$. The constant is fixed by demanding $P_n(1)=1$. ▲

[10] This equation occurs in the study of wave phenomena and quantum mechanics using spherical coordinates (see Section 14.5).

Example 4 **(Hermite's equation)**[11] Find the recursion relation and the first few terms for the solution of $y'' - 2xy' + \lambda y = 0$ as a power series.

Solution Again write $y = a_0 + a_1 x + a_2 x^2 + \cdots + a_i x^i + \cdots$, so

$$\lambda y = \lambda a_0 + \lambda a_1 x + \lambda a_2 x^2 + \cdots + \lambda a_i x^i + \cdots,$$

$$-2xy' = -2a_1 x - 4a_2 x^2 - 2\cdot 3a_3 x^3 - \cdots - 2ia_i x^i - \cdots,$$

and

$$y'' = 2a_2 + 3\cdot 2a_3 x + 4\cdot 3a_4 x^2 + \cdots + i(i-1)a_i x^{i-2} + \cdots.$$

Setting the coefficients of powers of x to zero in $y'' - 2xy' + y = 0$, we get

$$2a_2 + \lambda a_0 = 0 \qquad \text{(constant term)},$$

$$3\cdot 2a_3 - 2a_1 + \lambda a_1 = 0 \qquad \text{(coefficient of } x\text{)},$$

$$4\cdot 3a_4 - 4a_2 + \lambda a_2 = 0 \qquad \text{(coefficient of } x^2\text{)},$$

and in general

$$(i+2)(i+1)a_{i+2} - 2ia_i + \lambda a_i = 0.$$

Thus

$$a_2 = -\frac{\lambda}{2}a_0, \qquad a_3 = \frac{2-\lambda}{3\cdot 2}a_1,$$

$$a_4 = \frac{4-\lambda}{4\cdot 3}a_2 = -\frac{\lambda(4-\lambda)}{4\cdot 3\cdot 2}a_0,$$

and in general,

$$a_{i+2} = \frac{2i-\lambda}{(i+2)(i+1)}a_i.$$

Thus

$$a_5 = \frac{6-\lambda}{5\cdot 4}a_3 = \frac{(6-\lambda)(2-\lambda)}{5!}a_1,$$

etc., and so

$$y = a_0\left(1 - \frac{\lambda}{2}x^2 - \frac{(4-\lambda)\lambda}{4!}x^4 - \frac{(8-\lambda)(4-\lambda)\lambda}{6!}x^6 - \cdots\right)$$

$$+ a_1\left(x + \frac{(2-\lambda)}{3!}x^3 + \frac{(6-\lambda)(2-\lambda)}{5!}x^5 + \frac{(10-\lambda)(6-\lambda)(2-\lambda)}{7!} + \cdots\right)$$

This series converges for all x. If λ is an even integer, one of the series, depending on whether or not λ is a multiple of 4, terminates, and so we get a polynomial solution (called a Hermite polynomial). ▲

Sometimes the power series method runs into trouble—it may lead to only one solution, or the solution may not converge (see below and Exercise 23 for examples). To motivate the method that follows, which is due to Georg Frobenius (1849–1917), we consider Euler's equation:

$$x^2 y'' + \alpha xy' + \beta y = 0.$$

Here we could try $y = a_0 + a_1 x + a_2 x^2 + \cdots$ as before, but as we will now show, this leads nowhere. To be specific, we choose $\alpha = \beta = 1$. Write

[11] This equation arises in the quantum mechanics of a harmonic oscillator.

$$y = a_0 + a_1x + a_2x^2 + a_3x^3 + \cdots,$$

$$xy' = a_1x + 2a_2x^2 + 3a_3x^3 + \cdots,$$

and

$$x^2y'' = 2a_2x^2 + 3 \cdot 2a_3x^3 + \cdots.$$

Setting $x^2y'' + xy' + y = 0$, we get

$$a_0 = 0 \qquad \text{(constant term)},$$

$$2a_1 = 0 \qquad \text{(coefficient of } x\text{)},$$

$$5a_2 = 0 \qquad \text{(coefficient of } x^2\text{)},$$

$$10a_3 = 0 \qquad \text{(coefficient of } x^3\text{), etc.,}$$

and so all of the a_i's are zero and we get only a trivial solution.

The difficulty can be traced to the fact that the coefficient of y'' vanishes at the point $x = 0$ about which we are expanding our solution. One can, however, try to find a solution of the form x^r. Letting $y = x^r$, where r need not be an integer, we get

$$y' = rx^{r-1} \qquad \text{so} \quad \alpha xy' = \alpha rx^r$$

and

$$y'' = r(r-1)x^{r-2} \qquad \text{so} \quad x^2y'' = r(r-1)x^r.$$

Thus, Euler's equation is satisfied if

$$r(r-1) + \alpha r + \beta = 0$$

which is a quadratic equation for r with, in general, two solutions. (See Exercise 48 of Section 12.7 for the case when the roots are coincident.)

Frobenius' idea is that, by analogy with the Euler equation, we should look for solutions of the form $y = x^r\sum_{i=0}^{\infty}a_ix^i$ whenever the coefficient of y'' in a second-order equation vanishes at $x = 0$. Of course, r is generally not an integer; otherwise we would be dealing with ordinary power series.

Example 5 Find the first few terms in the general solution of $4xy'' - 2y'' + y = 0$ using the Frobenius method.

Solution We write

$$y = a_0x^r + a_1x^{r+1} + a_2x^{r+2} + \cdots,$$

so $$-2y' = -2ra_0x^{r-1} - 2(r+1)a_1x^r - 2(r+2)a_2x^{r+1} - \cdots \quad \text{and}$$

$$4xy'' = 4r(r-1)a_0x^{r-1} + 4(r+1)ra_1x^r + 4(r+2)(r+1)a_2x^{r+1} + \cdots.$$

Thus to make $4xy'' - 2y' + y = 0$, we set

$$a_0\left[4r(r-1) - 2r\right] = 0 \qquad \text{(coefficient of } x^{r-1}\text{)}.$$

If a_0 is to be allowed to be nonzero (which we desire, to avoid the difficulty encountered in our discussion of Euler's equation), we set $4r(r-1) - 2r = 0$. Thus $r(4r-6) = 0$, so $r = 0$ or $r = \frac{3}{2}$. First, we take the case $r = 0$:

$$y = a_0 + a_1x + a_2x^2 + a_3x^3 + \cdots,$$

$$-2y' = -2a_1 - 2 \cdot 2a_2x - 3 \cdot 2a_3x^2 - \cdots,$$

$$4xy'' = 4 \cdot 2a_2x + 4 \cdot 6a_3x^2 + 4 \cdot 12a_4x^3 + \cdots.$$

Then $4xy'' - 2y' + y = 0$ gives

$$a_0 - 2a_1 = 0 \qquad \text{(constant term)},$$

$$4 \cdot 2a_2 - 2 \cdot 2a_2 + a_1 = 0 \qquad \text{(coefficient of } x),$$

$$4 \cdot 6a_3 - 3 \cdot 2a_3 + a_2 = 0 \qquad \text{(coefficient of } x^2);$$

so $a_1 = \frac{1}{2}a_0$, $a_2 = -\frac{1}{4}a_1 = -\frac{1}{8}a_0$, and $a_3 = -\frac{1}{18}a_2 = \frac{1}{144}a_0$.

Thus

$$y = a_0\left(1 + \tfrac{1}{2}x - \tfrac{1}{8}x^2 + \tfrac{1}{144}x^3 - \cdots\right).$$

For the case $r = \frac{3}{2}$, we have

$$y = a_0x^{3/2} + a_1x^{5/2} + a_2x^{7/2} + a_3^{9/2} + \cdots,$$

$$-2y' = -3a_0x^{1/2} - 5a_1x^{3/2} - 7a_2x^{5/2} - 9a_3x^{7/2} - \cdots,$$

and

$$4xy'' = 3a_0x^{1/2} + 5 \cdot 3a_1x^{3/2} + 7 \cdot 5a_2x^{5/2} + 9 \cdot 7a_3x^{7/2} + \cdots.$$

Equating coefficients of $4x'' - 2y' + y = 0$ to zero gives

$$3a_0 - 3a_0 = 0 \qquad \text{(coefficient of } x^{1/2}),$$

$$5 \cdot 3a_1 - 5a_1 + a_0 = 0 \qquad \text{(coefficient of } x^{3/2}),$$

$$7 \cdot 5a_2 - 7a_2 + a_1 = 0 \qquad \text{(coefficient of } x^{5/2}),$$

$$9 \cdot 7a_3 - 9a_3 + a_2 = 0 \qquad \text{(coefficient of } x^{7/2});$$

so

$$a_1 = -\frac{1}{10}a_0, \quad a_2 = -\frac{a_1}{28} = \frac{1}{280}a_0,$$

and

$$a_3 = -\frac{a_2}{54} = -\frac{1}{280 \cdot 54}a_0.$$

Thus

$$y = a_0\left(x^{3/2} - \frac{1}{10}x^{5/2} + \frac{1}{280}x^{7/2} - \frac{1}{280 \cdot 54}x^{9/2} + \cdots\right).$$

The general solution is a linear combination of the two we have found:

$$y = c_1\left(1 + \tfrac{1}{2}x - \tfrac{1}{8}x^2 + \frac{1}{144}x^3 - \cdots\right)$$

$$+ c_2\left(x^{3/2} - \frac{1}{10}x^{5/2} + \frac{1}{280}x^{7/2} - \cdots\right). \blacktriangle$$

The equation that determines r, obtained by setting the coefficient of the lowest power of x in the equation to zero, is called the *indicial equation*.

The Frobenius method requires modification in two cases. First of all, if the indicial equation has a repeated root r_1, then there is one solution of the form $y_1(x) = a_0x^{r_1} + a_1x^{r_1+1} + \cdots$ and there is a second of the form $y_2(x) = y_1(x)\ln x + b_0x^{r_1} + b_1x^{r_1+1} + \cdots$. This second solution can also be found by the method of reduction of order. (See Exercise 48, Section 12.7.) Second, if the roots of the indicial equation differ by an integer, the method may again lead to problems: one may or may not be able to find a genuinely new solution. If $r_2 = r_1 + N$, then the second series $b_0x^{r_2} + b_1x^{r_2+1} + \cdots$ is of the same form, $a_0x^{r_1} + a_1x^{r_1+1} + \cdots$, with the first N coefficients set equal to zero. Thus it would require very special circumstances to obtain a second solution this way. (If the method fails, one can use reduction of order, but this may lead to a complicated computation).

We conclude with an example where the roots of the indicial equation differ by an integer.

Example 6 Find the general solution of Bessel's equation[12] $x^2y'' + xy' + (x^2 - k^2)y = 0$ with $k = \frac{1}{2}$.

Solution We try $y = x^r\sum_{i=0}^{\infty}a_ix^i$. Then

$$y = \sum_{i=0}^{\infty} a_i x^{i+r},$$

$$x^2y = \sum_{i=0}^{\infty} a_i x^{i+r+2} = \sum_{i=2}^{\infty} a_{i-2}x^{i+r},$$

$$y' = \sum_{i=0}^{\infty} (i+r)a_i x^{i+r-1} = \sum_{i=-1}^{\infty} (i+r+1)a_{i+1}x^{i+r},$$

$$xy' = \sum_{i=0}^{\infty} (i+r)a_i x^{i+r},$$

$$y'' = \sum_{i=-2}^{\infty} (i+r+1)(i+r+2)a_{i+2}x^{i+r},$$

and

$$x^2y'' = \sum_{i=0}^{\infty} (i+r-1)(i+r)a_i x^{i+r}.$$

Setting the coefficient of x^r in $x^2y'' + xy' + (x^2 - k^2)y = 0$ equal to zero, we get

$$0 = (r-1)ra_0 + ra_0 - \tfrac{1}{4}a_0,$$

so the indicial equation is $0 = (r-1)r + r - \frac{1}{4} = r^2 - \frac{1}{4}$, the roots of which are $r_1 = -\frac{1}{2}$ and $r_2 = \frac{1}{2}$, which differ by the integer 1.

Setting the coefficient of x^{r+1} equal to zero gives

$$0 = r(r+1)a_1 + (r+1)a_1 - \tfrac{1}{4}a_1 = \left[(r+1)^2 - \tfrac{1}{4}\right]a_1,$$

and the general recursion relation arising from the coefficient of x^{i+r}, $i \geqslant 2$, is

$$0 = r(r+i)a_i + (r+i)a_i - \tfrac{1}{4}a_i + a_{i-2} = \left((r+i)^2 - \tfrac{1}{4}\right)a_i + a_{i-2}.$$

Let us work first with the root $r_1 = -\frac{1}{2}$. Since $-\frac{1}{2}$ and $+\frac{1}{2}$ are both roots of the indicial equation, the coefficients a_0 and a_1 are arbitrary. The recursion relation is

$$a_i = -\frac{a_{i-2}}{(-1/2+i)^2 - 1/4} = -\frac{a_{i-2}}{i^2 - i} = -\frac{a_{i-2}}{i(i-1)}$$

for $k \geqslant 2$. Thus

$$a_2 = -\frac{a_0}{2\cdot 1}, \qquad a_4 = -\frac{a_2}{4\cdot 3} = \frac{a_0}{4\cdot 3\cdot 2\cdot 1},$$

$$a_6 = -\frac{a_4}{6\cdot 5} = -\frac{a_0}{6!}, \ldots, \qquad a_{2k} = \frac{a_0(-1)^k}{(2k)!}.$$

[12] This equation was extensively studied by F. W. Bessel (1784–1846), who inaugurated modern practical astronomy at Königsberg Observatory.

Similarly

$$a_3 = -\frac{a_1}{3 \cdot 2}, \qquad a_5 = -\frac{a_3}{5 \cdot 4} = \frac{a_1}{5!}, \ \ldots, \qquad a_{2k+1} = \frac{a_1(-1)^k}{(2k+1)!}.$$

Our general solution is then

$$x^{-1/2}\left[a_0\left(1 - \frac{x^2}{2!} + \frac{x^4}{4!} - \frac{x^6}{6!} + \cdots\right) + a_1\left(x - \frac{x^3}{3!} + \frac{x^5}{5!} - \cdots\right)\right]$$

which we recognize to be $a_0(\cos x)/\sqrt{x} + a_1(\sin x)/\sqrt{x}$.

Notice that in this case we have found the general solution from just one root of the indicial equation. ▲

Exercises for Section 12.8

In Exercises 1–4, find solutions of the given equation in the form of power series: $y = \sum_{i=0}^{\infty} a_i x^i$.

1. $y'' - xy' - y = 0$.
2. $y'' - 2xy' - 2y = 0$.
3. $y'' + 2xy' = 0$.
4. $y'' + xy' = 0$.

In Exercises 5–8, find the first three nonzero terms in the power series solution satisfying the given equation and initial conditions.

5. $y'' + 2xy' = 0$, $y(0) = 0$, $y'(0) = 1$.
6. $y'' + 2x^2y = 0$, $y(0) = 1$, $y'(0) = 0$.
7. $y'' + 2xy' + y = 0$, $y(0) = 0$, $y'(0) = 2$.
8. $y'' - 2xy' + y = 0$, $y(0) = 0$, $y'(0) = 1$.
9. *Airy's equation* is $y'' = xy$. Find the first few terms and the recursion relation for a power series solution.
10. *Tchebycheff's equation* is $(1 - x^2)y'' - xy' - \alpha^2 y = 0$. Find the first few terms and the recursion relation for a power series solution. What happens if $\alpha = n$ is an integer?

In Exercises 11–14, use the Frobenius method to find the first few terms in the general solution of the given equation.

11. $3xy'' - y' + y = 0$.
12. $2xy'' + (2 - x)y' - y = 0$.
13. $3x^2y'' + 2xy' + y = 0$.
14. $2x^2y'' - 2xy' + y = 0$.
15. Consider Bessel's equation of order k, namely, $x^2y'' + xy' + (x^2 - k^2)y = 0$.
    (a) Find the first few terms of a solution of the form $J_k(x) = a_0x^k + a_1x^{k+1} + \cdots$.
    (b) Find a second solution if k is not an integer.
16. *Laguerre functions* are solutions of the equation $xy'' + (1 - x)y' + \lambda y = 0$.
    (a) Find a power series solution by the Frobenius method.
    (b) Show that there is a polynomial solution if λ is an integer.
17. Verify that the power series solutions of $y'' + \omega^2 y = 0$ are just $y = A \cos \omega x + B \sin \omega x$.

★18. Find the first few terms of the general solution for Bessel's equation of order $\frac{3}{2}$.

★19. (a) Verify that the solution of Legendre's equation does not converge for all x unless $\lambda = n(n+1)$ for some nonnegative integer n.
(b) Compute $P_1(x)$, $P_2(x)$, and $P_3(x)$.

★20. Use Wronskians and Exercise 49 of Section 12.7 to show that the solution found in Example 1 is the general solution.

★21. Use Wronskians and Exercise 49 of Section 12.7 to show that the solution found in Example 2 is the general solution.

★22. Prove by induction that the Legendre polynomials are given by *Rodrigues' formula*:

$$P_n(x) = \frac{1}{2^n n!}\frac{d^n}{dx^n}(x^2 - 1)^n.$$

★23. (a) Solve $x^2y' + (x - 1)y - 1 = 0$, $y(0) = 1$ as a power series to obtain $y = \sum n!\,x^n$, which converges only at $x = 0$. (b) Show that the solution is

$$y = \frac{e^{-1/x}}{x}\int \frac{e^{1/x}}{x}\,dx.$$

Review Exercises for Chapter 12

In Exercises 1–8, test the given series for convergence. If it can be summed using a geometric series, do so.

1. $\sum_{i=1}^{\infty} \frac{1}{(12)^i}$
2. $\sum_{i=1}^{\infty} \frac{1}{100(i+1)}$
3. $\sum_{i=1}^{\infty} \frac{3^{i+1}}{5^{i-1}}$
4. $\sum_{i=1}^{\infty} \frac{8}{9^i}$
5. $1 + 2 + \frac{1}{3} + \frac{1}{3^2} + \frac{1}{3^3} + \cdots$
6. $100 + \frac{1}{9} + \frac{1}{9^2} + \frac{1}{9^3} + \cdots$
7. $\sum_{i=1}^{\infty} \frac{9}{10+11i}$
8. $\sum_{i=1}^{\infty} \frac{6}{7+8i}$

In Exercises 9–24, test the given series for convergence.

9. $\sum_{n=1}^{\infty} 5^{-n}$
10. $\sum_{n=1}^{\infty} \frac{4^n}{(2n+1)!}$
11. $\sum_{k=1}^{\infty} \frac{k}{3^k}$
12. $\sum_{n=1}^{\infty} \frac{2n}{n+3}$
13. $\sum_{n=1}^{\infty} \frac{(-1)^n n}{3^n}$
14. $\sum_{n=1}^{\infty} \frac{2n}{n^2+3}$
15. $\sum_{n=1}^{\infty} \frac{(-1)^{2n}}{n}$
16. $\sum_{j=0}^{\infty} \frac{(-1)^j j}{j^2+8}$
17. $\sum_{n=1}^{\infty} \frac{2^{n^2}}{n!}$
18. $\sum_{i=1}^{\infty} \frac{i}{i^3+8}$
19. $\sum_{n=1}^{\infty} ne^{-n^2}$
20. $\sum_{n=1}^{\infty} \frac{\sqrt{n}}{n^2 - \sin^2 99n}$
21. $\sum_{n=2}^{\infty} \frac{1}{(\ln n)^{\ln n}}$
22. $\sum_{n=1}^{\infty} \left(\frac{1}{n} - \frac{1}{\sqrt{n}}\right)$
23. $\sum_{n=1}^{\infty} \frac{n}{(n+1)!}$
24. $\sum_{n=1}^{\infty} \frac{n^2}{(n+1)!}$

Sum the series in Exercises 25–32 to within 0.05.

25. $1 - \frac{1}{4} + \frac{1}{16} - \frac{1}{32} + \cdots$
26. $\frac{1}{2} - \frac{2}{4} + \frac{3}{8} - \frac{4}{16} + \frac{5}{32} - \cdots$
27. $\sum_{i=1}^{\infty} \frac{(-1)^i}{2^i+3}$
28. $\sum_{n=1}^{\infty} \frac{(-1)^n n}{6n^2-1}$
29. $\sum_{n=1}^{\infty} \frac{1-n}{3^n}$
30. $\sum_{n=1}^{\infty} \frac{n^2}{n!}$
31. $\sum_{n=1}^{\infty} \frac{\cos n}{n^4+1}$
32. $\sum_{n=1}^{\infty} \frac{\cos n}{4^n+1}$

Tell whether each of the statements in Exercise 33–46 is true or false. Justify your answer.

33. If $a_n \to 0$, then $\sum_{n=1}^{\infty} a_n$ converges.
34. Every geometric series $\sum_{i=1}^{\infty} r^i$ converges.
35. Convergence or divergence of any series may be determined by the ratio test.
36. $\sum_{i=1}^{\infty} 1/2^i = 1$.
37. $e^{2x} = 1 + 2x + x^2 + x^3/3 + \cdots$.
38. If a series converges, it must also converge absolutely.
39. The error made in approximating a convergent series by a partial sum is no greater than the first term omitted.
40. $\cos x = \sum_{k=0}^{\infty} (-1)^k x^{2k}/(2k)!$.
41. If $\sum_{j=1}^{\infty} a_j$ and $\sum_{k=0}^{\infty} b_k$ are both convergent, then $\sum_{j=1}^{\infty} a_j + \sum_{k=0}^{\infty} b_k = b_0 + \sum_{i=1}^{\infty} (a_i + b_i)$.
42. $\sum_{i=1}^{\infty} (-1)^i [3/(i+2)]$ converges conditionally.
43. The convergence of $\sum_{n=1}^{\infty} a_n$ implies the convergence of $\sum_{n=1}^{\infty} (a_n + a_{n+1})$.
44. The convergence of $\sum_{n=1}^{\infty} (a_n + a_{n+1})$ implies the convergence of $\sum_{n=1}^{\infty} a_n$.
45. The convergence of $\sum_{n=1}^{\infty} (|a_n| + |b_n|)$ implies the convergence of $\sum_{n=1}^{\infty} |a_n|$.
46. The convergence of $\sum_{n=1}^{\infty} a_n$ implies the convergence of $\sum_{n=1}^{\infty} a_n^2$.

47. If $0 \leqslant a_n \leqslant ar^n$, $r < 1$, show that the error in approximating $\sum_{i=1}^{\infty} a_i$ by $\sum_{i=1}^{n} a_i$ is less than or equal to $ar^{n+1}/(1-r)$.
48. Determine how many terms are needed to compute the sum of $1 + r + r^2 + \cdots$ with error less than 0.01 when (a) $r = 0.5$ and (b) $r = 0.09$.

Find the sums of the series in Exercises 49–52.

49. $\sum_{n=1}^{\infty} 1/9^n$
50. $\sum_{n=1}^{\infty} \frac{1}{n(n+1)}$ [*Hint:* Use partial fractions.]
51. $\sum_{n=1}^{\infty} \frac{n}{(n+1)!}$ [*Hint:* Write the numerator as $n + 1 - 1$.]
52. $\sum_{n=1}^{\infty} \frac{1+n}{2^n}$ [*Hint:* Differentiate a certain power series.]

Find the radius of convergence of the series in Exercises 53–58.

53. $1 - \frac{x^2}{2!} + \frac{x^4}{3!} - \frac{x^6}{4!} + \cdots$
54. $1 + 3x + 5x^2 + 7x^3 + \cdots$
55. $\sum_{n=0}^{\infty} \frac{x^n}{(3n)!}$
56. $\sum_{n=0}^{\infty} \frac{(x - 1/2)^n}{(n+1)!}$
57. $\sum_{n=0}^{\infty} \frac{(-1)^n}{2^n} x^n$
58. $\sum_{n=0}^{\infty} \frac{n^n}{n!} x^n$

Find the Maclaurin series for the functions in Exercises 59–66.

59. $f(x) = \cos 3x + e^{2x}$
60. $g(x) = \frac{1}{1-x^3}$
61. $f(x) = \ln(1 + x^4)$
62. $g(x) = \frac{1}{\sqrt{1-x^4}}$
63. $f(x) = \frac{d}{dx}(\sin x - x)$
64. $g(k) = \frac{d^2}{dk^2}(\cos k^2)$
65. $f(x) = \int_0^x \frac{(e^t - 1)}{t}\, dt$
66. $g(y) = \int_0^y \sin t^2\, dt$

Find the Taylor expansion of each function in Exercises 67–70 about the indicated point, and find the radius of convergence.

67. e^x about $x = 2$
68. $1/x$ about $x = 1$
69. $x^{3/2}$ about $x = 1$
70. $\cos(\pi x)$ about $x = 1$

Find the limits in Exercises 71–74 using series methods.

71. $\lim_{x\to 0} \dfrac{1-\cos\pi x}{x^2}$.
72. $\lim_{x\to 0} [6\dfrac{\sin\pi x}{x^5} - \dfrac{6\pi - \pi^3 x^2}{x^4}]$.
73. $\lim_{x\to 0} \dfrac{(1+x)^{3/2} - (1-x)^{3/2}}{x^2}$.
74. $\lim_{x\to\pi} [1 + \cos x - \dfrac{(x-\pi)^2}{2} + \dfrac{(x-\pi)^4}{24}]$.

In Exercises 75–78 find the real part, the imaginary part, the complex conjugate, and the absolute value of the given complex number.

75. $3 + 7i$
76. $2 - 10i$
77. $\sqrt{2} - i$
78. $(2+i)/(2-i)$

In Exercises 79–82, plot the given complex numbers, indicating r and θ on your diagram, and write them in polar form $z = re^{i\theta}$.

79. $1 - i$
80. $\dfrac{1+i}{1-i}$
81. $ie^{\pi i/2}$
82. $(1+i)e^{i\pi/4}$
83. Solve for z: $z^2 - 2z + \pi i = 0$.
84. Solve for z: $z^8 = \sqrt{5} + 3i$.

Find the general solution of the differential equations in Exercises 85–96.

85. $y'' + 4y = 0$
86. $y'' - 4y = 0$
87. $y'' + 6y' + 5y = 0$
88. $y'' - 6y' - 2y = 0$
89. $y'' + 3y' - 10y = e^x + \cos x$
90. $y'' - 2y' - 3y = x^2 + \sin x$
91. $y'' - 6y' + 9y = \cos\left(\dfrac{x}{2}\right)$
92. $y'' - 10y + 25 = \cos(2x)$
93. $y'' + 4y = \dfrac{x}{\sqrt{x^2+1}}$. (Express your answer in terms of integrals.)
94. $y'' - 3y' - 3y = \dfrac{\sin x}{\sqrt{\cos^2 x + 1}}$. (Express your answer in terms of integrals.)
95. $y''' + 2y'' + 2y' = 0$
96. $y''' - 3y'' + 3y' + y = e^x$

In Exercises 97–100, identify the equation as a spring equation and describe the limiting behavior as $t \to \infty$.

97. $x'' + 9x + x' = \cos 2t$.
98. $x'' + 9x + 0.001x' = \sin(50t)$.
99. $x'' + 25x + 6x' = \cos(\pi t)$.
100. $x'' + 25x + 0.001x' = \cos(60\pi t)$.

In Exercises 101–104, find the first few terms of the general solution as a power series in x.

101. $y'' + 2xy = 0$
102. $y'' - (4\sin x)y = 0$
103. $y'' - 2x^2y' + 2y = 0$
104. $y'' + y' + xy = 0$

Find the first few terms in an appropriate series for at least one solution of the equations in Exercises 105–108.

105. $5x^2y'' + y' + y = 0$.
106. $xy'' + y' - 4y = 0$.
107. Bessel's equation with $k = 1$.
108. Legendre's equation with $\lambda = 3$.

109. The current I in the electric circuit shown in Figure 12.R.1 satisfies

$$L\frac{d^2I}{dt^2} + R\frac{dI}{dt} + \frac{I}{C} = \frac{dE}{dt},$$

where E is the applied voltage and L, R, C are constants.

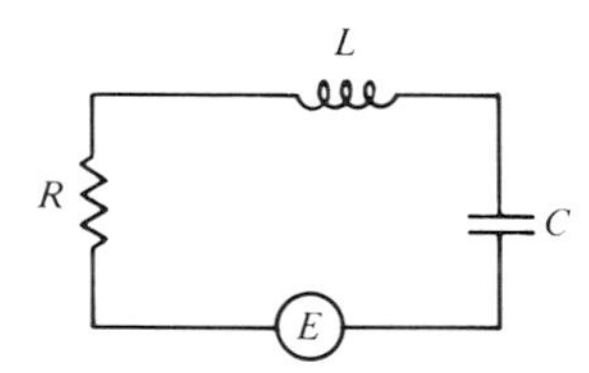

Figure 12.R.1. An electric circuit.

(a) Find the values of m, k, γ that make this equation a damped spring equation.
(b) Find $I(t)$ if $I(0) = 0$, $I'(0) = 0$ and $L = 5$, $C = 0.1$, $R = 100$, and $E = 2\cos(60\pi t)$.

110. Verify formula (16) in Section 12.7.
111. Verify that $\sum_{n=0}^{\infty} x^2(1+x^2)^{-n}$ is a convergent geometric series for $x \neq 0$ with sum $1 + x^2$. It also converges to 0 when $x = 0$. (This shows that the sum of an infinite series of continuous terms need not be continuous.)
112. A beam of length L feet supported at its ends carries a concentrated load of P lbs at its center. The maximum deflection D of the beam from equilibrium is

$$D = \frac{2L^3P}{EI\pi^4}\sum_{n=1}^{\infty}\frac{|\sin(n\pi/2)|}{n^4}.$$

(a) Use the formula $\sum_{n=1}^{\infty}(1/n^4) = \pi^4/90$ to show that

$$\sum_{k=1}^{\infty}\frac{1}{(2k)^4} = \left(\frac{1}{2^4}\right)\left(\frac{\pi^4}{90}\right).$$

[*Hint:* Factor out 2^{-4}.]

(b) Show that

$$\sum_{k=0}^{\infty}\frac{1}{(2k+1)^4} = \left(\frac{15}{16}\right)\left(\frac{\pi^4}{90}\right);$$

hence $D = (1/48)(L^3P/EI)$. [*Hint:* A series is the sum of its even and odd terms.]

(c) Use the first two nonzero terms in the series for D to obtain a simpler formula for D. Show that this result differs at most by 0.23% from the theoretical value.

113. The deflection $y(x,t)$ of a string from its straight profile at time t, measured vertically at location x along the string, $0 \leqslant x \leqslant L$, is

$$y(x,t) = \sum_{n=1}^{\infty} A_n \sin\left(\frac{n\pi x}{L}\right)\cos\left(\frac{n\pi ct}{L}\right),$$

where A_n, L and c are constants.

(a) Explain what this equation means in terms of limits of partial sums for x, t fixed.

(b) Initially (at $t = 0$), the deflection of the string is

$$\sum_{n=1}^{\infty} A_n \sin\left(\frac{n\pi x}{L}\right).$$

Find the deflection value as an infinite series at the midpoint $x = L/2$.

114. In the study of saturation of a two-phase motor servo, an engineer starts with a transfer function equation $V(s)/E(s) = K/(1 + s\tau)$, then goes to the first-order approximation $V(s)/E(s) = K(1 - s\tau)$, from which he obtains an approximate equation for the saturation dividing line.

(a) Show that $1/(1 + s\tau) = \sum_{n=0}^{\infty}(-s\tau)^n$, by appeal to the theory of geometric series. Which values of $s\tau$ are allowed?

(b) Discuss the replacement of $1/(1 + s\tau)$ by $1 - s\tau$; include an error estimate in terms of the value of $s\tau$.

115. Find the area bounded by the curves $xy = \sin x$, $x = 1$, $x = 2$, $y = 0$. Make use of the Taylor expansion of $\sin x$.

116. A wire of length L inches and weight w lbs/inch, clamped at its lower end at a small angle $\tan^{-1}P_0$ to the vertical, deflects $y(x)$ inches due to bending. The displacement $y(L)$ at the upper end is given by

$$y(L) = \frac{2P_0}{3L^{1/3}} \frac{\int_0^{L^{3/2}} u(az)\,dz}{u(aL^{3/2})},$$

where $a = \frac{2}{3}\sqrt{W/EI}$, and

$$u(az) = \frac{az^{-1/3}}{2} \sum_{k=0}^{\infty} (-1)^k \frac{a^{2k}z^{2k}}{2^{2k}k!\,\Gamma(k + 2/3)}.$$

The values of the *gamma function* Γ may be found in a mathematical table or on some calculators as $\Gamma(x) = (x - 1)!$ [$\Gamma(\frac{2}{3}) = 1.3541$, $\Gamma(\frac{5}{3}) = 0.9027$, $\Gamma(\frac{8}{3}) = 1.5046$, $\Gamma(\frac{11}{3}) = 4.0122$]. The function u is the *Bessel function* of order $-\frac{1}{3}$.

(a) Find the smallest positive root of $u(az) = 0$ by using the first four terms of the series.

(b) Evaluate $y(L)$ approximately by using the first four terms of the series.

117. (a) Use a power series for $\sqrt{1 + x}$ to calculate $\sqrt{5/4}$ correct to 0.01. (b) Use the result of part (a) to calculate $\sqrt{5}$. How accurate is your answer?

118. In each of the following, evaluate the indicated derivative:

(a) $f^{(12)}(0)$, where $f(x) = x/(1 + x^2)$;

(b) $f^{(10)}$, where $f(x) = x^6 e^{x+1}$.

119. Let

$$\frac{z}{e^z - 1} = 1 + B_1 z + \frac{B_2}{2!}z^2 + \frac{B_3}{3!}z^3 + \cdots.$$

Determine the numbers B_1, B_2, and B_3. (The B_i are known as the *Bernoulli numbers*.)

120. Show in the following two ways that $\sum_{n=1}^{\infty} na^n = a/(1 - a)^2$ for $|a| < 1$.

(a) Consider

$$S_n = a + 2a^2 + 3a^3 + \cdots + na^n,$$

$$aS_n = a^2 + 2a^3 + \cdots + (n-1)a^n + na^{n+1}$$

and subtract.

(b) Differentiate $\sum_{n=0}^{\infty} a^n = 1/(1 - a)$ with respect to a, and then subtract your answer from $\sum_{n=0}^{\infty} a^n = 1/(1 - a)$.

121. In highway engineering, a *transitional spiral* is defined to be a curve whose curvature varies directly as the arc length. Assume this curve starts at $(0,0)$ as the continuation of a road coincident with the negative x axis. Then the parametric equations of the spiral are

$$x = k\int_0^{\phi} \frac{\cos\theta}{\sqrt{\theta}}\,d\theta, \qquad y = k\int_0^{\phi} \frac{\sin\theta}{\sqrt{\theta}}\,d\theta.$$

(a) By means of infinite series methods, find the ratio x/y for $\phi = \pi/4$.

(b) Try to graph the transitional spiral for $k = 1$, using accurate graphs of $(\cos\theta)/\sqrt{\theta}$, $(\sin\theta)/\sqrt{\theta}$ and the area interpretation of the integral.

★122. The free vibrations of an elastic circular membrane can be described by infinite series, the terms of which involve trigonometric functions and *Bessel functions*. The series

$$\sum_{i=0}^{\infty} \frac{(-1)^i}{i!\,(n+i)!}(x/2)^{n+2i}$$

is called the *Bessel Function* $J_n(x)$; n is an integer $\geqslant 0$.

(a) Establish convergence by the ratio test.

(b) The frequencies of oscillation of the circular membrane are essentially solutions of the equation $J_n(x) = 0$, $x > 0$. Examine the equation $J_0(x) = 0$, and see if you can explain why $J_0(2.404) = 0$ is possible.

(c) Check that J_n satisfies Bessel's equation (Example 6, Section 12.8).

★123. Show that g defined by $g(x) = e^{-1/x^2}$ if $x \neq 0$ and $g(0) = 0$ is infinitely differentiable and $g^{(i)}(0) = 0$ for all i. [*Hint:* Use the definition of the derivative and the following lemma provable by l'Hôpital's rule: if $P(x)$ is any polynomial, then $\lim_{x \to 0} P(x)g(x) = 0$.]

★124. Let $f(x) = (1+x)^\alpha$, where α is a real number. Show by an induction argument that $f^{(i)}(x) = \alpha(\alpha-1)\ldots(\alpha-i+1)(1+x)^{\alpha-i}$, and hence show that $(1+x)^\alpha$ is analytic for $|x|<1$.

★125. True or false: The convergence of $\sum_{n=1}^{\infty} a_n^2$ and $\sum_{n=1}^{\infty} b_n^2$ implies absolute convergence of $\sum_{n=1}^{\infty} a_n b_n$.

★126. (a) Show that if the radius of convergence of $\sum_{n=1}^{\infty} a_n x^n$ is R, then the radius of convergence of $\sum_{n=1}^{\infty} a_n x^{2n}$ is $\sqrt{R}$.
(b) Find the radius of convergence of the series $\sum_{n=0}^{\infty} (\pi/4)^n x^{2n}$.

★127. Let $f(x) = \sum_{i=0}^{\infty} a_i x^i$ and $g(x) = f(x)/(1-x)$.
(a) By multiplying the power series for $f(x)$ and $1/(1-x)$, show that $g(x) = \sum_{i=0}^{\infty} b_i x^i$, where $b_i = a_0 + \cdots + a_i$ is the ith partial sum of the series $\sum_{i=0}^{\infty} a_i$.
(b) Suppose that the radius of convergence of $f(x)$ is greater than 1 and that $f(1) \neq 0$. Show that $\lim_{i\to\infty} b_i$ exists and is not equal to zero. What does this tell you about the radius of convergence of $g(x)$?
(c) Let $e^x/(1-x) = \sum_{i=0}^{\infty} b_i x^i$. What is $\lim_{i\to\infty} b_i$?

★128. (a) Find the second-order approximation at $T=0$ to the day-length function S (see the supplement to Chapter 5) for latitude 38° and your own latitude.
(b) How many minutes earlier (compared with $T=0$) does the sun set when $T = 1, 2, 10, 30$?
(c) Compare the results in part (b) with those obtained from the exact formula and with listings in your local newspaper.
(d) For how many days before and after June 21 is the second-order approximation correct to within 1 minute? Within 5 minutes?

★129. Prove that e is irrational, as follows: if $e = a/b$ for some integers a and b, let $k > b$ and let $\alpha = k!(e - 2 - \frac{1}{2!} - \frac{1}{3!} - \cdots - \frac{1}{k!})$. Show that α is an integer and that $\alpha < 1/k$ to derive a contradiction.

Chapter 7 Answers

7.1 Calculating Integrals

1. $x^3 + x^2 - 1/2x^2 + C$
3. $e^x + x^2 + C$
5. $-(\cos 2x)/2 + 3x^2/2 + C$
7. $-e^{-x} + 2\sin x + 5x^3/3 + C$
9. $1084/9$
11. $105/2$
13. $844/5$
15. $1/12$
17. 0
19. 6
21. $3\pi/4$
23. $\pi/12$
25. 1
27. $(e^6 - e^3)/3 + 3(2^{5/3} - 1)/5$
29. $\ln 5$
31. $4\ln 2 + 61/24$
33. 400
35. $116/15$
37. (b) $e^{(e^2)} - e + 3$
39. (a) 11
(b) -8
(c) Note that $\int_5^7 f(t)\,dt$ is negative
41. $-2t\sqrt{e^{t^2}} + \sin 5t^4$
43. 3
45. (a) 0
(b) $5/6$
(c) $\begin{cases} -\cos x & \text{if } 0 < x \leqslant 2 \\ \cos(2) - 2\cos x & \text{if } 2 < x \leqslant \pi \end{cases}$

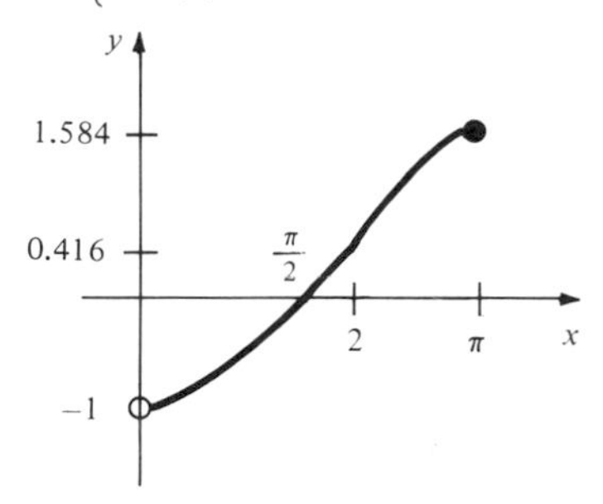

47. $2 + \tan^{-1}2 - \dfrac{1}{2}\ln 5$
49. 16.4
51. $(1/2)(e^2 - 1)$
53. $16/3 - \pi$
55.

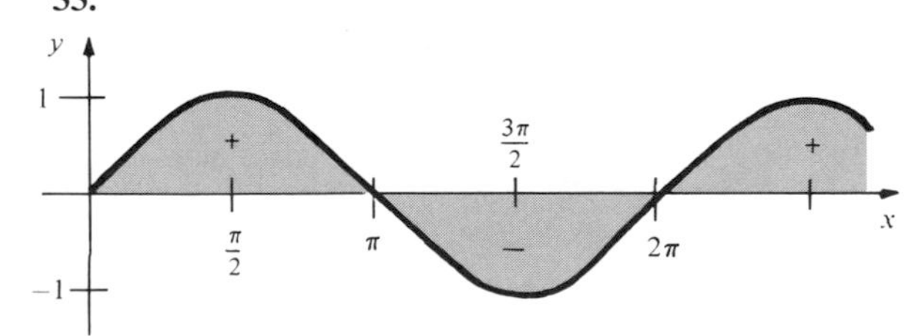

57. $\pi/4$
59. (a) Differentiate the right-hand side.
(b) Integrate both sides of the identity.
(c) $1/8$
61. Use the fact that $\tan^{-1}a$ and $\tan^{-1}b$ lie in the interval $(-\pi/2, \pi/2)$
63. 16,000,014 meters
65. (a) Evaluate the integral.
(b) $A = \$45{,}231.46$
67. (a) $R(t) = 2000e^{t/2} - 2000$, $C(t) = 1000t - t^2$
(b) \$57,279.90
69. $1 + \ln(2) - \ln(1 + e) \approx 0.380$

7.2 Integration by Substitution

1. $\frac{2}{5}(x^2 + 4)^{5/2} + C$
3. $-1/4(y^8 + 4y - 1) + C$
5. $-1/2\tan^2\theta + C$
7. $\sin(x^2 + 2x)/2 + C$
9. $(x^4 + 2)^{1/2}/2 + C$
11. $-3(t^{4/3} + 1)^{-1/2}/2 + C$
13. $-\cos^4(r^2)/4 + C$
15. $\tan^{-1}(x^4)/4 + C$
17. $-\cos(\theta + 4) + C$
19. $(x^5 + x)^{101}/101 + C$
21. $\sqrt{t^2 + 2t + 3} + C$
23. $(t^2 + 1)^{3/2}/3 + C$
25. $\sin\theta - \sin^3\theta/3 + C$
27. $\ln|\ln x| + C$
29. $2\sin^{-1}(x/2) + x\sqrt{4 - x^2}/2 + C$
31. $\ln(1 + \sin\theta) + C$
33. $-\cos(\ln t) + C$
35. $-3(3 + 1/x)^{4/3}/4 + C$
37. $(\sin^2 x)/2 + C$
39. m a non-negative integer and n an odd positive integer, or n a non-negative integer and m an odd positive integer.

7.3 Changing Variables in the Definite Integral

1. $2(3\sqrt{3} - 1)/3$
3. $(5\sqrt{5} - 1)/3$
5. $2[(25)^{9/4} - (9)^{9/4}]/9$
7. $1/7$
9. $(e - 1)/2$
11. $-1/3$
13. 0
15. 1
17. $\ln(\sqrt{2}\cos(\pi/8))$
19. $1/2$
21. $4 - \tan^{-1}(3) + \pi/4$
23. (a) $\pi/2$
(b) $\pi/4$
(c) $\pi/8$
25. The substitution is not helpful in evaluating the integral.
27. $(\sqrt{2}/2)[\tan^{-1}2\sqrt{2} - \tan^{-1}(\sqrt{2}/2)]$
29. $(1/\sqrt{3})\ln[(4 + 3\sqrt{2})/(1 + \sqrt{3})]$
31. Let $u = x - t$.
33. $(5\sqrt{2} - 2\sqrt{5})/10$
35. $(\pi/27)(145\sqrt{145} - 10\sqrt{10})$
37. (a) $1/3$
(b) Yes.

7.4 Integration By Parts

1. $(x+1)\sin x + \cos x + C$
3. $x\sin 5x/5 + \cos 5x/25 + C$
5. $(x^2-2)\sin x + 2x\cos x + C$
7. $(x+1)e^x + C$
9. $x\ln(10x) - x + C$
11. $(x^3/9)(3\ln x - 1) + C$
13. $e^{3s}(9s^2 - 6s + 2)/27 + C$
15. $(x^3-4)^{1/3}(x^3+12)/4 + C$
17. $t^2\sin t^2 + \cos t^2 + C$
19. $-(1/x)\sin(1/x) - \cos(1/x) + C$
21. $-[\ln(\cos x)]^2/2 + C$
23. $x\cos^{-1}(2x) - \sqrt{1-4x^2}\,/2 + C$
25. $y\sqrt{1/y-1} - \tan^{-1}\sqrt{1/y-1} + C$
27. $\sin^2 x/2 + C$
29. The integral becomes more complicated.
31. $(16+\pi)/5$
33. $3(3\ln 3 - 2)$
35. $\sqrt{2}\,[(\pi/4)^2 + 3\pi/4 - 2]/2 - 1.$
37. $\sqrt{3}\,/8 - \pi/24$
39. $e-2$
41. $-(e^{2\pi} - e^{-2\pi})/4$
43. $\frac{3}{5}(2^{2/3}(2^{2/3}+1)^{5/2} - 2^{5/2} + \frac{2}{7}[2^{7/2} - (2^{2/3}+1)^{7/2}]) \approx 4.025$
45. $(\pi - 4)/8\sqrt{2} - 1/2$
47. $\int_0^1 \sqrt{2-x^2}\,dx - \int_0^{\sqrt{2}} \sqrt{2-x^2}\,dx = -\int_1^{\sqrt{2}} \sqrt{2-x^2}\,dx$ is $-1/8$ the area of a circle of radius $\sqrt{2}$ corrected by the area of a triangle (draw a graph).
49. $(-2\pi\cos 2\pi a)/a + (\sin 2\pi a)/a^2$. (This tends to zero as a tends to ∞. Neighboring oscillations tend to cancel one another.)
51. (b) $(5e^{3\pi/10} - 3)/34$
53. (a) Use integration by parts, writing $\cos^n x = \cos^{n-1}x \times \cos x$.
55. $2\pi^2$
57. (a) $Q = \int EC(\alpha^2/\omega + \omega)e^{-\alpha t}\sin(\omega t)\,dt$
    (b) $Q(t) = EC\{1 - e^{\alpha t}[\cos(\omega t) + \alpha\sin(\omega t)/\omega]\}$
59.

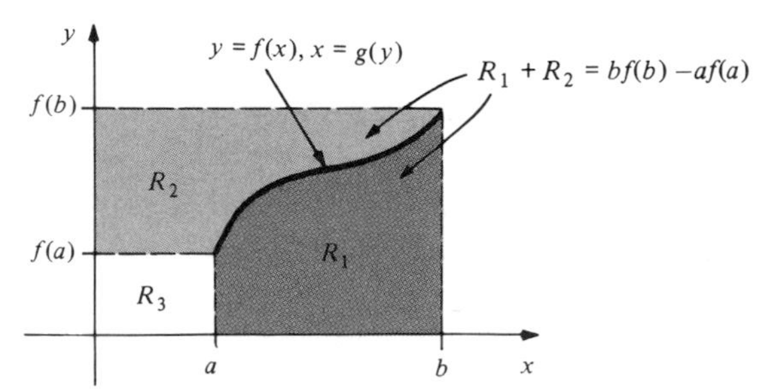

61. (a) $a_0 = 2$, all others are zero.
    (b) $a_0 = 2\pi$, $b_n = -2/n$ if $n \neq 0$, all others are zero.
    (c) $a_0 = 8\pi^2/3$, $a_n = 4/n^2$ if $n \neq 0$, $b_0 = 0$, $b_n = -4\pi/n$ if $n \neq 0$.
    (d) $a_4 = b_2 = b_3 = 1$, all others are zero.

Review Exercises for Chapter 7

1. $x^2/2 - \cos x + C$
3. $x^4/4 + \sin x + C$
5. $e^x - x^3/3 - \ln|x| + \sin x + C$
7. $e^\theta + \theta^3/3 + C$
9. $-\cos(x^3/3) + C$
11. $e^{(x^3)}/3 + C$
13. $(x+2)^6/6 + C$
15. $e^{4x^3}/12 + C$
17. $-\cos^3 2x + C$
19. $x^2\tan^{-1}x/2 - x/2 + \tan^{-1}x/2 + C$
21. $\sin^{-1}(t/2) + t^3/3 + C$
23. $xe^{4x}/4 - e^{4x}/16 + C$
25. $x^2\sin x + 2x\cos x - 2\sin x + C$
27. $(e^{-x}\sin x - e^{-x}\cos x)/2 + C$
29. $x^3\ln 3x/3 - x^3/9 + C$
31. $(2/5)(x-2)(x+3)^{3/2} + C$
33. $x\sin 3x/3 + \cos 3x/9 + C$
35. $3x\sin 2x/2 + 3\cos 2x/4 + C$
37. $x^2e^{x^2}/2 - e^{x^2}/2 + C$
39. $x^2(\ln x)^2/2 - x^2(\ln x)/2 + x^2/4 + C$
41. $2e^{\sqrt{x}}(\sqrt{x} - 1) + C$
43. $\sin x\ln|\sin x| - \sin x + C$
45. $x\tan^{-1}x - \ln(1+x^2)/2 + C$
47. -1
49. $\pi/25$
51. $\sin(1) - \sin(1/2)$
53. $(\pi^2/32 + 1/2)\tan^{-1}(\pi/4) - \pi/8$
55. $(4\sqrt{2} - 2)/3 + (2\sqrt{2} - 2)a$
57. $33\sqrt{3}\,/5$
59. $399/4$

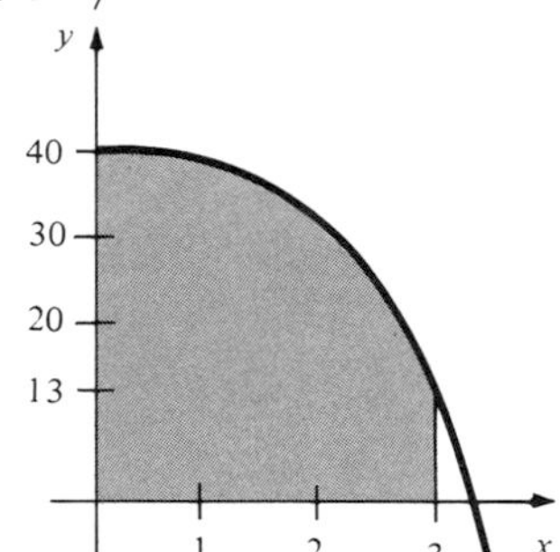

61. 6

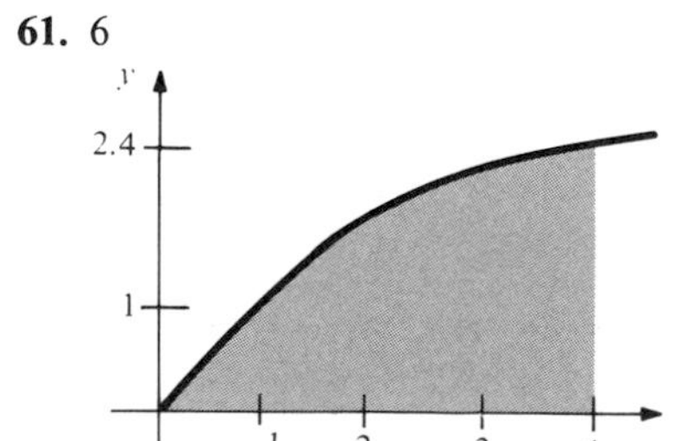

63. $(2-\sqrt{2})/2$

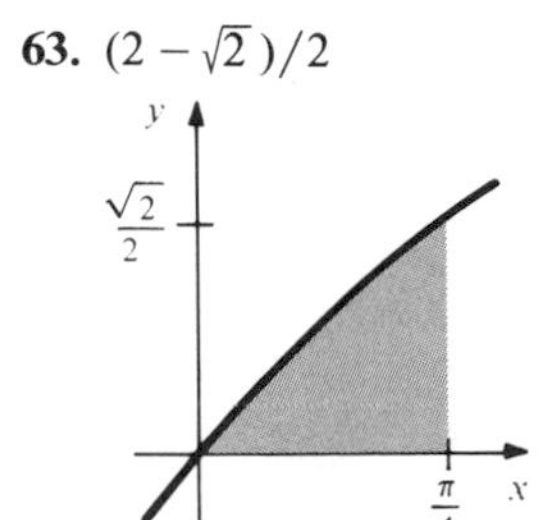

65. ln 2

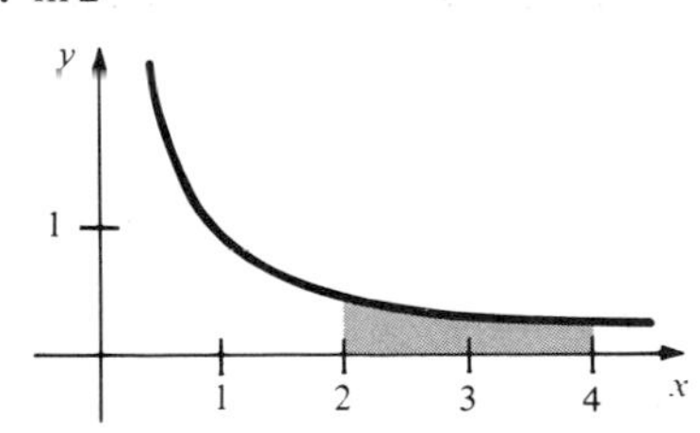

67. $2/(n+1)$

69. 18.225

71. (a) 90008.46 liters
(b) 3000.28 liters/minute

73. $\frac{4}{3}[\sin(\pi x/2)\sin(\pi x)/\pi + \cos(\pi x/2)\cos(\pi x)/2\pi] + C$

75. $\sin^{-1}x - \sqrt{1-x^2} + C$

77. (a) $(\ln x)^2/2 + C$
(b) $(2/9)(-\sqrt{3}/3+1)$

79. $(x^{n+1}\ln x^{n+1} - x^{n+1})/(n+1)^2 + C$

81. (a) $(100/26)(\sin 5t/5 + \cos 5t + e^{-25t})$
(b) Substitute $t = 1.01$ in part (a).

83. (a) $m^2 + n^2 + mn + 2m + 2n + 1 = 0$. (b) The discriminant is negative. (c) Yes; for example $x^{-1/2}$ and $x^{(-3\pm\sqrt{5})/4}$.

85. $xe^{ax}[b\sin(bx) + a\cos(bx)]/(a^2+b^2) + e^{ax}[(b^2-a^2)\cos(bx) - 2ab\sin(bx)]/(a^2+b^2)^2 + C$

Chapter 8 Answers

8.1 Oscillations

1. $\cos(3t) = \cos\left[3\left(t+\frac{2\pi}{3}\right)\right]$

3. $\cos(6t) + \sin(3t)$
$= \cos\left[6\left(t+\frac{2\pi}{3}\right)\right] + \sin\left[3\left(t+\frac{2\pi}{3}\right)\right]$

5. $\cos 3t - 2\sin 3t/3$

7. $-\frac{1}{6}\sqrt{3}\sin(2\sqrt{3}\,t)$

9. $2\pi/3$, 3, $1/3$

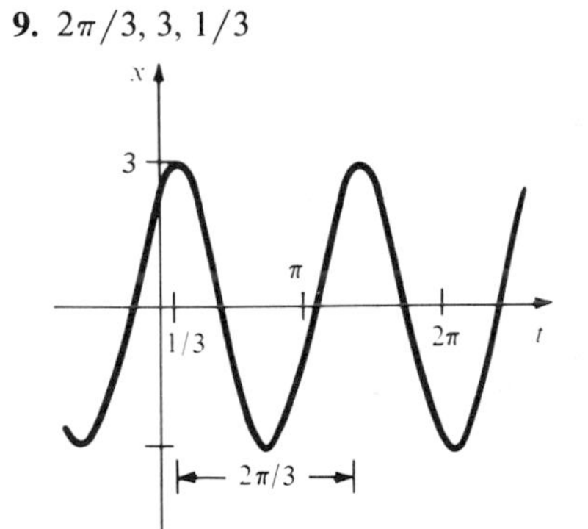

11. 2π, 4, -1

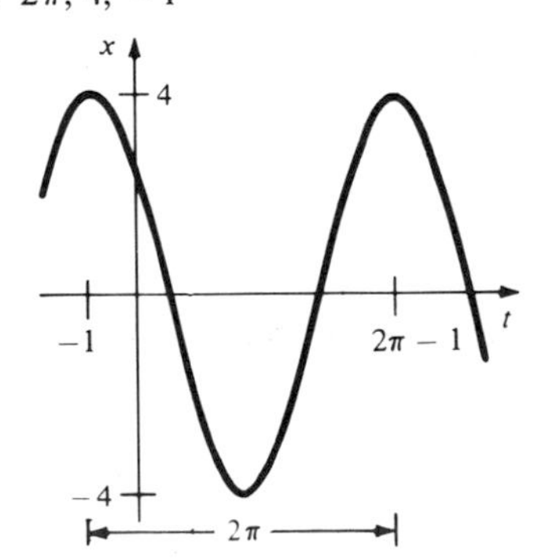

13. $-\cos 2t$

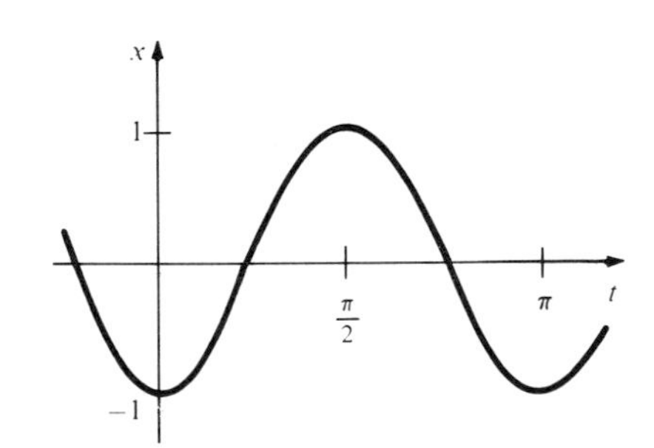

15. $\sqrt{26}\cos(5t - \tan^{-1}(1/5))$

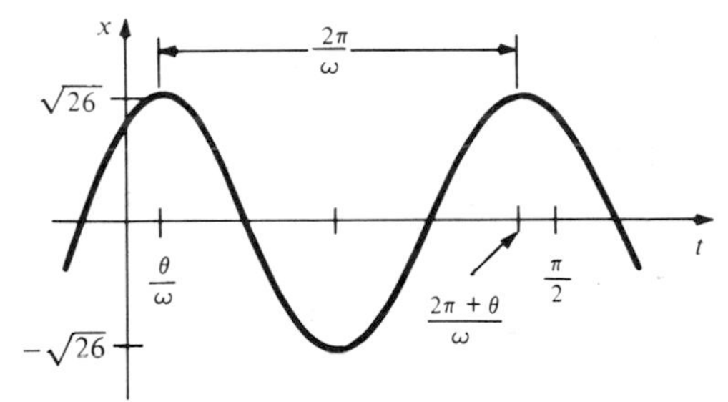

Phase shift $= \frac{\theta}{\omega} = \frac{\tan^{-1}(\frac{1}{5})}{5}$

Period $= \frac{2\pi}{\omega} = \frac{2\pi}{5}$

17. $\cos 2t + (3/2)\sin 2t$

19. $2\cos 4x$

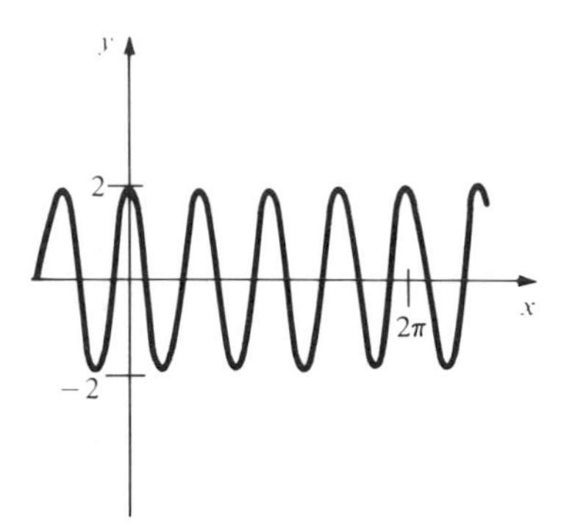

21. (a) $16\pi^2$
(b)

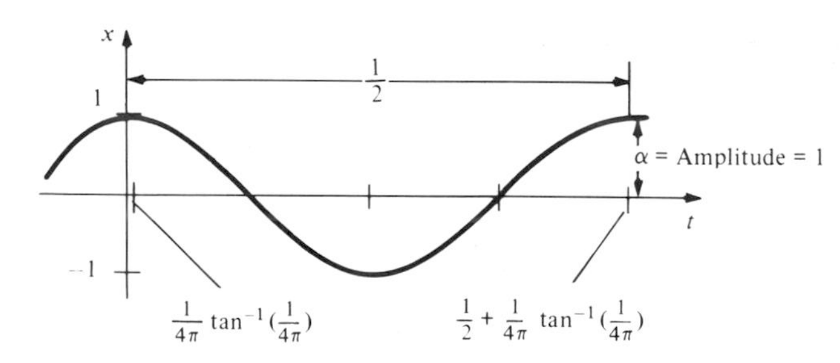

23. The frequency decreases by a factor of $\sqrt{3}$.

25. (a) $27(d^2x/dt^2) = -3x + 2x^3$
(b) $27(d^2x/dt^2) = -3x$
(c) 6π

27. (a) $x_0 = \dfrac{x_2 + x_1\sqrt[3]{k_2/k_1}}{1 + \sqrt[3]{k_2/k_1}}$
(b) $f'(x_0) > 0$

29. There is no restriction on b.

31. Multiply (9) by $\omega \sin \omega t$ and (10) by $\cos \omega t$ and add.

33. (a) $V''(x_0) > 0$, so the second derivative test applies.
(b) Compute dE/dt using the sum and chain rules.
(c) Since E is constant, if it is initially small, the sum of $\frac{1}{2}m\left(\frac{dx}{dt}\right)^2$ and $V(x)$ must remain small, so both dx/dt and $x - x_0$ remain small.

8.2 Growth and Decay

1. $dT/dt = -0.11(T-20)$

3. $dQ/dt = -(0.00028)Q$

5. $2e^{-3t}$

7. e^{3t}

9. $2e^{8t-8}$

11. $2e^{6-2t}$

13. 7.86 minutes

15. 2,476 years

17. e^{3t}

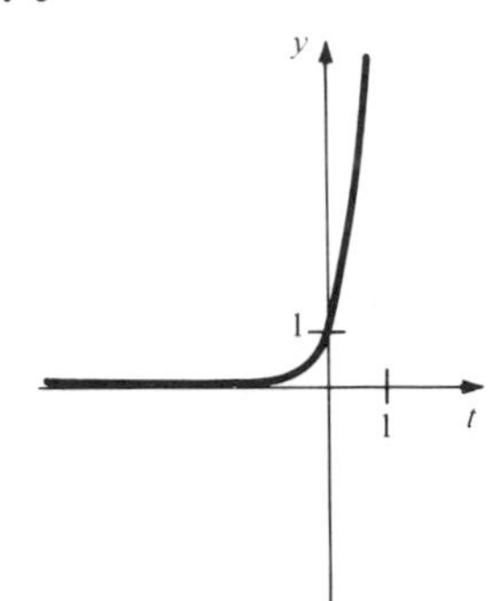

19. e^{8t+1}

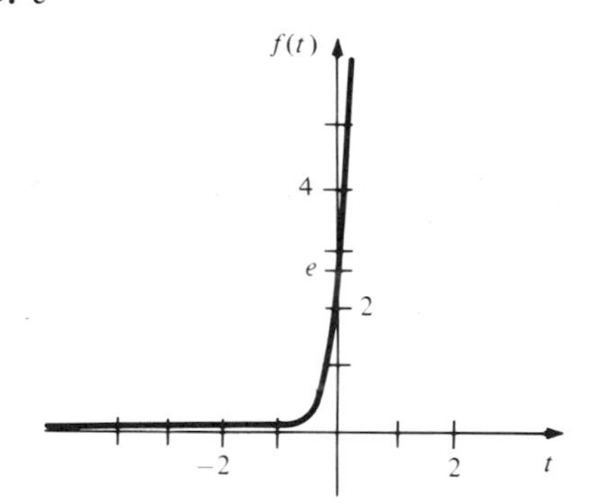

21. Increasing

23. Decreasing

25. 33,000 years

27. 173,000 years

29. 1.5×10^9 years

31. 2,880 years

33. 49 minutes

35. 4.3 minutes

37. 18.5 years

39. The annual percentage rate is $100(e^{r/100} - 1) \approx 18.53\%$.

41. (a) $300\, e^{-0.3t}$
(b) 2000; 2000 books will eventually be sold.
(c)

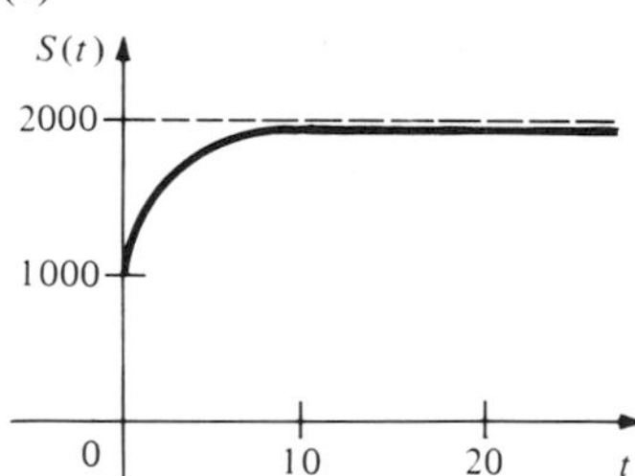

43. K is the distance the water must rise to fill the tank.

45. (a) Verify by differentiation.
(b) $a(t) = t(e^{-1/t} + 1 - e^{-1})$

47. $(2m/\delta)\ln 2$

8.3 The Hyperbolic Functions

1. Divide (3) by $\cosh^2 t$.

3. Proceed as in Example 2.

5. $\dfrac{d}{dx}(\cosh x) = \dfrac{1}{2}\dfrac{d}{dx}(e^x + e^{-x}) = \dfrac{1}{2}(e^x - e^{-x})$
$= \sinh x$.

7. Use the reciprocal rule and Exercise 5.

9. $(3x^2 + 2x)\cosh(x^3 + x^2 + 2)$
11. $\cosh x \sinh 5x + 5\cosh 5x \sinh x$
13. $8 \sin 8x \cosh(\cos 8x)$
15. $4 \sinh x \cosh x$
17. $-3\,\mathrm{csch}^2 3x$
19. $(2\,\mathrm{sech}^2 2x)\exp(\tanh 2x)$
21. $[\sinh x(1 + \tanh x) - \mathrm{sech}\, x]/(1 + \tanh x)^2$
23. $(\sinh x)(\int[dx/(1 + \tanh^2 x)]) +$ $\cosh x/(1 + \tanh^2 x)$
25. $(\sinh 3t)/3$
27. $2\cosh\sqrt{3}\, t$
29. $\cosh 3t + (\sinh 3t)/3$
31. $2\cosh 6t$
33.

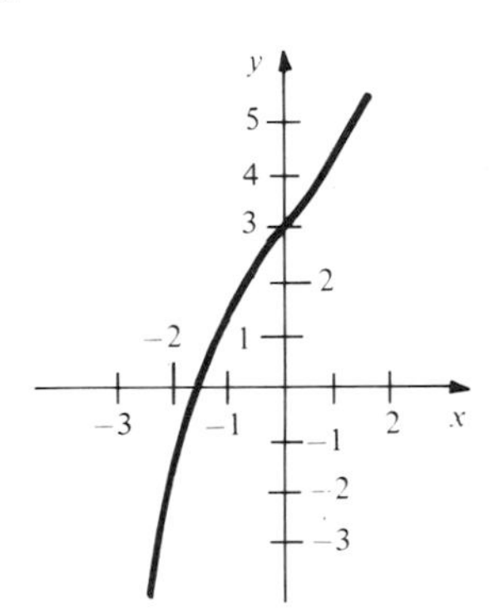

35.

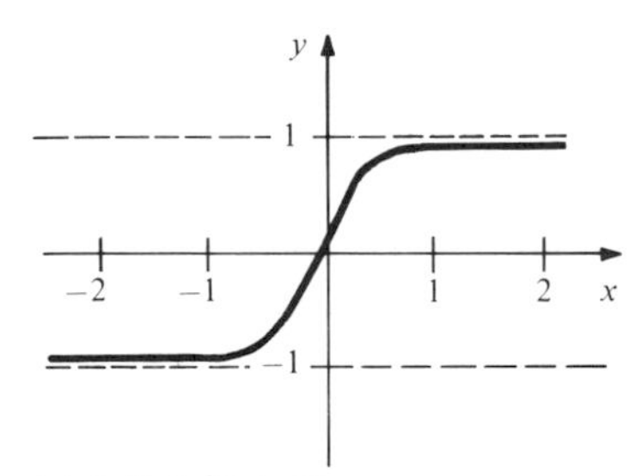

37. $(\sinh 3x)/3 + C$
39. $\ln|\sinh x| + C$
41. $(\sinh 2x)/4 - x/2 + C$
43. $e^{2x}/4 - x/2 + C$
45. $\cosh^3 x/3 + C$
47. $[y - \cosh(x + y)]/[\cosh(x + y) - x]$
49. $-3y\,\mathrm{sech}^2 3xy/(\cosh y + 3x\,\mathrm{sech}^2 3xy)$
51. (a) $x_0 = 1$
    (b) $d^2x/dt^2 = 2(x - 1)$
53. Use the definitions of $\sinh x$ and $\cosh x$. (Don't expand the nth power!)

8.4 The Inverse Hyperbolic Functions

1. $2x/\sqrt{x^4 + 4x^2 + 3}$
3. $(3 - \sin x)\Big/\sqrt{(3x + \cos x)^2 + 1}$
5. $\tan^{-1}(x^2 - 1) + 2/(2 - x^2)$
7. $[(1 + 1/\sqrt{x^2 - 1}\,)(\sinh^{-1}x + x) -$ $(x + \cosh^{-1}x)(1 + 1/\sqrt{x^2 + 1}\,)]/(\sinh^{-1}x + x)^2$
9. $[\exp(1 + \sinh^{-1}x)]/\sqrt{x^2 + 1}$
11. $-3\sin 3x/\sqrt{\cos^2 3x + 1}$
13. 0.55
15. 1.87
17. Let $y = \cosh^{-1}x$, so $x = \frac{1}{2}(e^y + e^{-y})$. Multiply by $2e^y$, solve the resulting quadratic equation for e^y and take logs.
19. Let $y = \mathrm{sech}^{-1}x$ so $x = 2/(e^y + e^{-y})$. Invert and proceed as in Exercise 17.
21. $$\frac{d}{dx}\tanh^{-1}x = \frac{1}{\frac{d}{dy}\tanh y} = \frac{1}{\mathrm{sech}^2 y} = \frac{1}{1 - \tanh^2 y} = \frac{1}{1 - x^2}.$$
23. $$\frac{d}{dx}\mathrm{sech}^{-1}x = \frac{1}{\frac{d}{dy}\mathrm{sech}\, y} = \frac{1}{-\mathrm{sech}\, y \tanh y} = \frac{-1}{x\sqrt{1 - \mathrm{sech}^2 y}} = \frac{-1}{x\sqrt{1 - x^2}}$$
25. Differentiate the right hand side.
27. Differentiate the right hand side.
29. $(1/4)\ln|(1 + 2x)/(1 - 2x)| + C$
31. $(1/2)\ln(2x + \sqrt{4x^2 + 1}\,) + C$
33. $\ln(\sin x + \sqrt{\sin^2 x + 1}\,) + C$
35. $(1/2)\ln|(1 + e^x)/(1 - e^x)| + C$
37. No

8.5 Separable Differential Equations

1. $y = \sin x + 1$
3. $y = \exp(x^2 - 2x + 1) - 1$
5. $y = -2x$
7. $e^y(y - 1) = (1/2)\ln(x^2 + 1)$
9. $y = 2x + 1$
11. $y = \exp(-\sin x) + 1$
13.

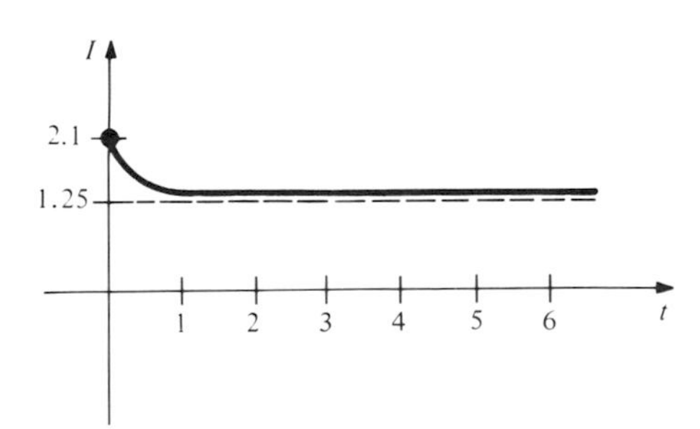

15. (a) $Q = EC(1 - \exp(-t/RC))$
    (b) $t = RC\ln(100)$
17. Verify that the equations hold with $dx/dt = 0$ and $dy/dt = 0$.
19. $P = P_0 A\exp(P_0 kt)/[1 + A\exp(P_0 kt)]$
21. As T_0 increases, $\cosh\left(\frac{mgx}{T_0}\right) \to 1$, so $y \to h$, which represents a straight cable.

23. (a) $y' = -y/x$
(b) $y' = x/y$; $y^2 = x^2 + C$.

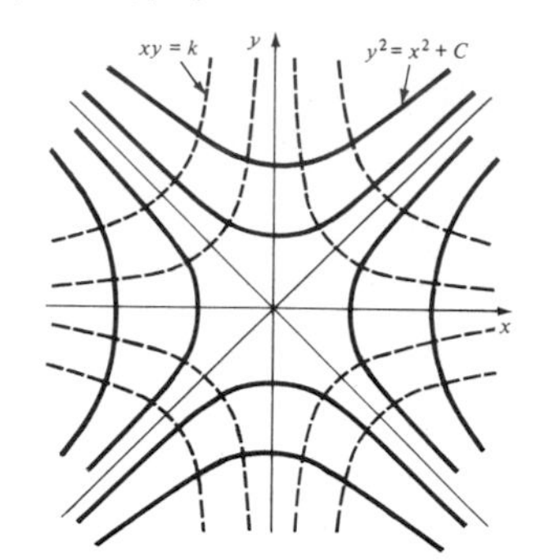

25. (a)

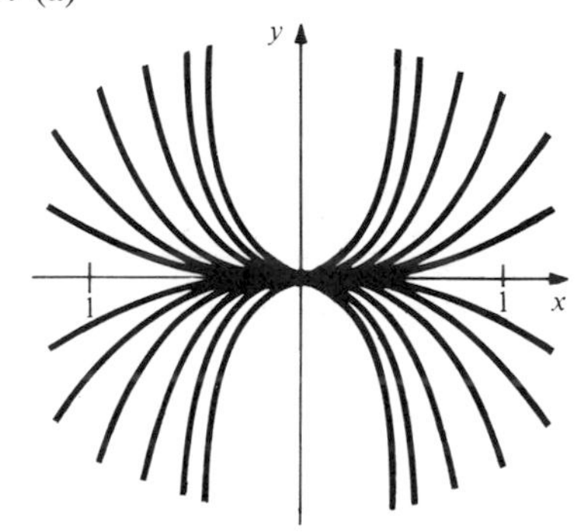

(b) $y' = 3cx^2$
(c) $y' = -1/3cx^2$; $y = 1/3cx + C$

27. (a)

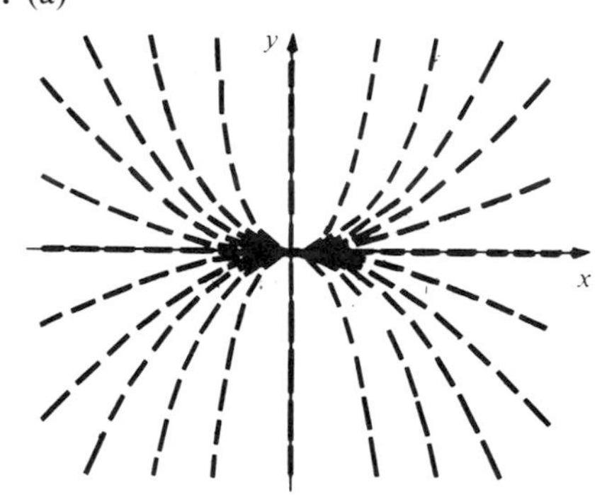

(b) $y = kx^2$

29. $y(1) \approx 2.2469$

31. $y(1) \approx 0.4683$

33. $\lim_{x\to\infty} y(x) = 3$

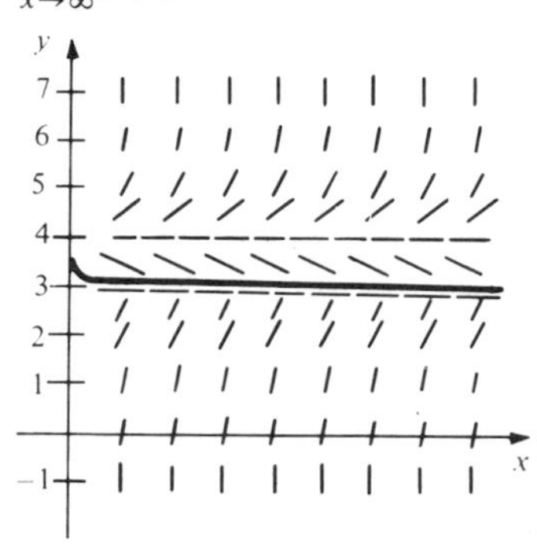

35. $\lim_{x\to\infty} y(x) = 1$

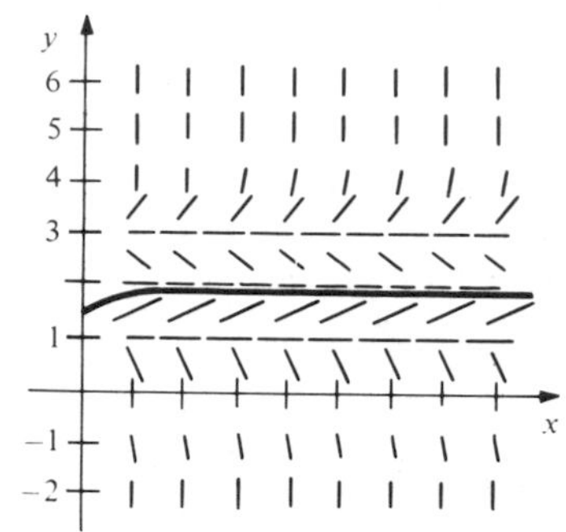

37. 61

39. $\int h(y)\,dy = -\int (1/g(x))\,dx$

8.6 Linear First-Order Equations

1. $y = 2 + (-3\ln|1 - x| + C)(1 - x)$

3. $y = 1 + C\exp(x^4/4)$

5. $y = -2 + 2\exp(\sin x)$

7. $y = (e^x - e)/x$

9. The equation is $L\dfrac{dI}{dt} + RI = E_0\cos\omega t + E_1$ and has solution

$$I = \frac{E_0}{L}\,\frac{1}{(R/L)^2 + \omega^2}\,\frac{R}{L}(\sin\omega t - \omega\cos\omega t)$$

$$+ Ce^{-tR/L} + \frac{E_1}{R}$$

11. $I = E_0C - E_0C\exp(-t/RC)$;
$I \to E_0C$ *as* $t \to +\infty$.

13. Set $y = .9 \times 2.51 \times 10^6$ and verify the value of t.

15. $6.28 \times 10^5 - (8.28 \times 10^5)\exp(-2.67 \times 10^{-7}t) - (2.01 \times 10^5)\exp(-1.07 \times 10^{-6}t)$

17. 15 seconds; 951 meters.

19. Use separation of variables to get
$v = \sqrt{mg/\gamma}\tanh(\sqrt{\gamma g/m}\,t)$

21. $\dfrac{FM_0}{M_1^2} - \dfrac{g}{2M_1^2}(M_0^2 + M_1^2)$

23. $y = -2(x + 1) + Ce^x$

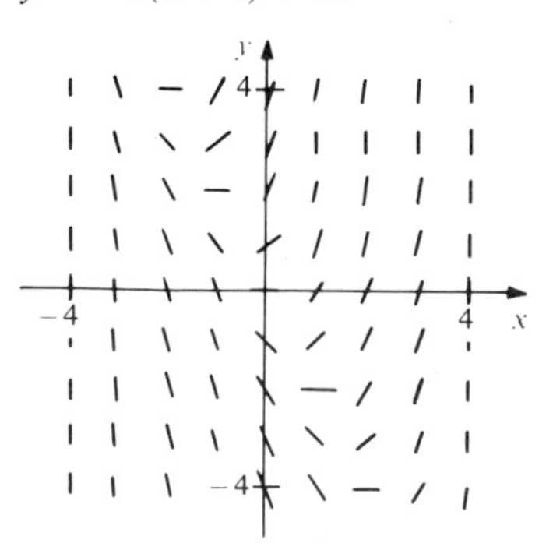

25. If y_1 and y_2 are solutions, prove, using methods of Section 8.2, uniqueness for $y' = P(x)y$ and apply it to $y = y_1 - y_2$. (This is one of several possible procedures.)

27. (a) $w' = (1-n)[Q + Pw]$

(b) $y = \pm 1/(x\sqrt{c - x^2})$

29. (a) $v = \dfrac{F}{\gamma - r} - \dfrac{g(M_0 - rt)}{\gamma - 2r} + C(M_0 - rt)^{\gamma/r - 1}$

where $C = M_0^{1-\gamma/r}\left(\dfrac{gM_0}{\gamma - 2r} - \dfrac{F}{\gamma - r}\right)$ and

where the air resistance force is γv.

(b) At burnout, $v = \dfrac{F}{\gamma - r} - \dfrac{gM_1}{\gamma - 2r} + CM_1^{\gamma/r-1}$.

Review Exercises for Chapter 8

1. $y = e^{3t}$

3. $y = (1/\sqrt{3})\sin\sqrt{3}\, t$

5. $y = (4e^{3t} - 1)/3$

7. $y = 4/(4 - t^4)$

9. $f(x) = e^{4x}$

11. $f(t) = \cosh 2t + \sinh 2t/2$

13. $x(t) = \cos t - \sin t$

15. $x(t) = (\sinh 3t)/3$

17. $y = -\ln(1/e + 1 - e^x)$

19. $x(t) = e^{-4t}$

21. $y = -t$

23. $g(t) = \cos(\sqrt{7/3}\, t - (2/\sqrt{7/3})\sin(\sqrt{7/3}\, t)$; amplitude is $\sqrt{19/7}$; phase is $-\sqrt{3/7}\tan^{-1}(2\sqrt{3/7})$

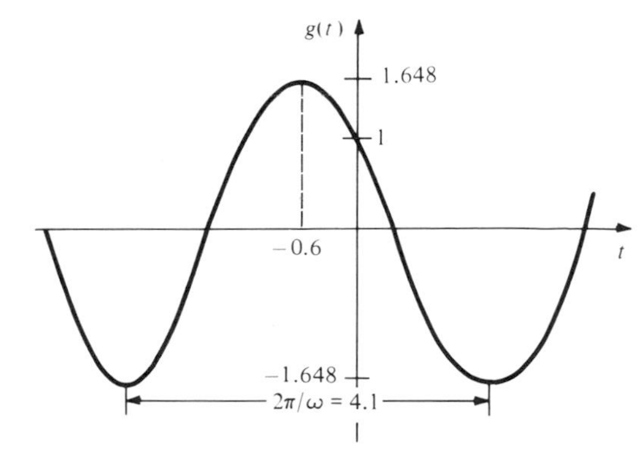

25.

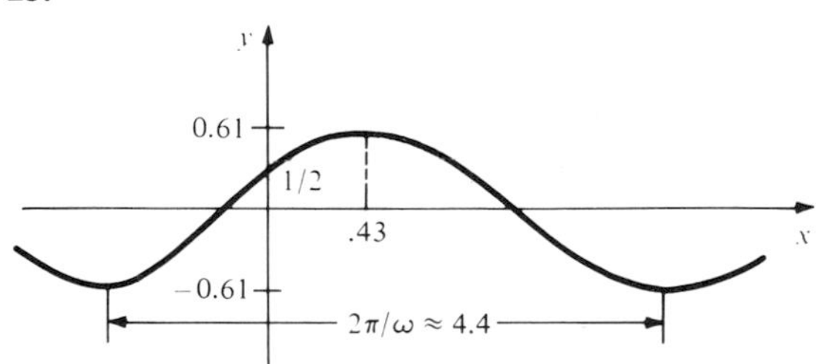

27.

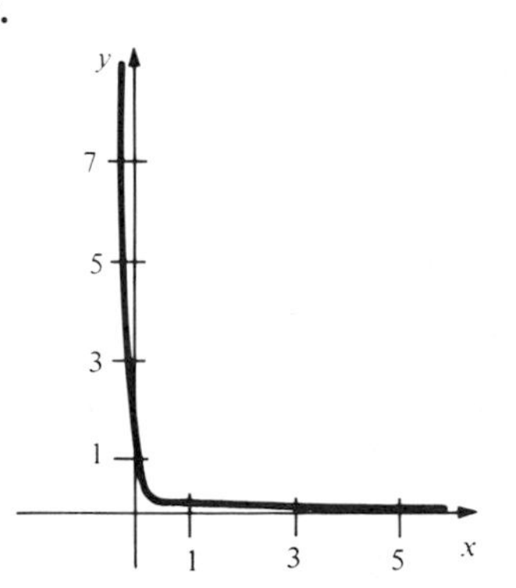

29. $\lim_{t\to\infty} x(t) = 3$

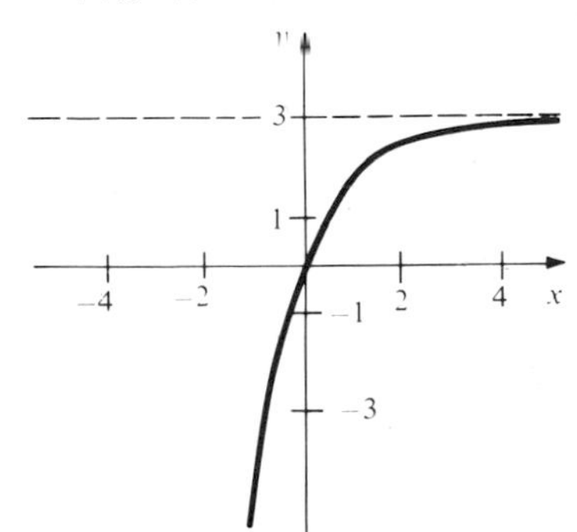

31. $x = e^t$

33. $y = x^2/2 - x - 2e^{-x} + 2$

35. $y(x) = \sinh 5x/\sinh 5$

37. $6x\cosh(3x^2)$

39. $2x/\sqrt{(x^2+1)^2 - 1}$

41. $\cosh 3x/\sqrt{x^2+1} + 3\sinh 3x \sinh^{-1}x$

43. $(-3/\sqrt{9x^2 - 1})\exp(1 - \cosh^{-1}(3x))$.

45. $\tan^{-1}(\sinh x) + C$

47. $(1/3)\tanh^{-1}(x/3) + C$ if $|x| < 3$
$(1/3)\coth^{-1}(x/3) + C$ if $|x| > 3$

49. $x\cosh x - \sinh x + C$

51. $x(t) = \sqrt{5/2.1}\,\sin\sqrt{2.1/5}\, t$

53. (a) $k = 640$
(b) -6400 newtons

55. (a) $y'' + (\omega^2 - \beta)y = 0$
(c) $x(t) = e^{-t}(\cos(\sqrt{3}\, t) + (1/\sqrt{3}\sin\sqrt{3}\, t))$

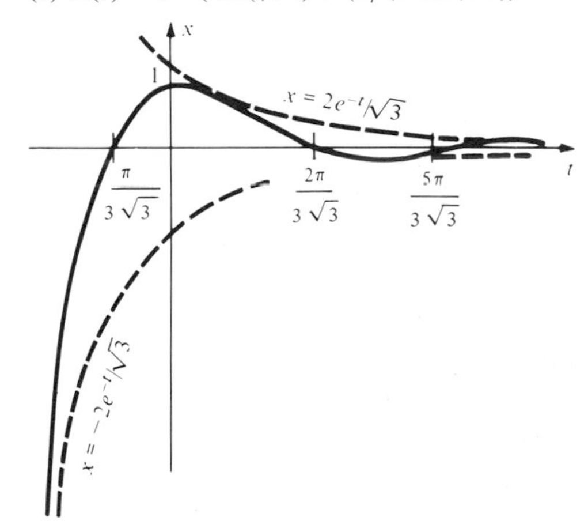

57. 66.4 years

59. 54,150 years

61. 27 minutes

63. (a) 73 years
(b) $S(t) = ke^{-\alpha t}$ where $k = S(0)$

65.

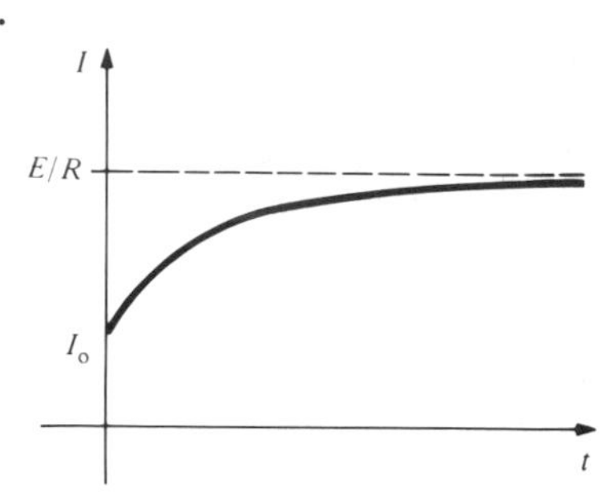

67. (a) $y^2/9 + x^2 = k$, $k = 2C/9$

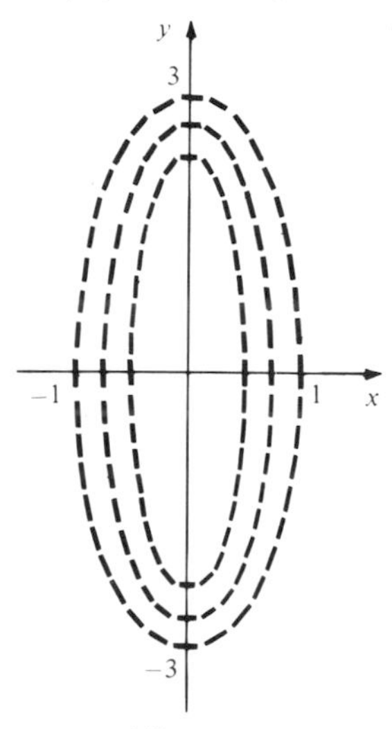

(b) $kx^{-1/9}$, $k = e^C$

69. 15.2 minutes, no. [The "no" could be "yes" if you allow a faster addition of fresh water after draining.]

71. $I = 2(3 \sin \pi t - \pi \cos \pi t)/(9 + \pi^2) + [1 + 2\pi/(9 + \pi^2)]e^{-3t}$

73. $y = -4/3 + Ce^{3x}$

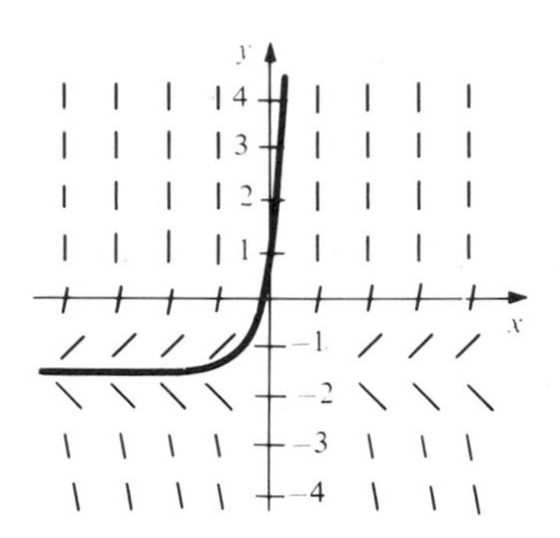

75. 1

77. $y = e^x$ is the exact solution; $y(1) = e \approx 2.71828$.

79. $y = -1/(x - 1)$ is the exact solution, it is not defined at $x = 1$.

81. $y = Ce^{at} - (b/a)$; the answers are all the same.

83. (a) Verify using the chain rule
(b) Integrate the relation in (a)
(c) Solve for $T = t$; the period is twice the time to go from $\theta = 0$ to $\theta = \theta_0$.

85. (a) $y = \cosh(x + a)$ or $y = 1$.
(b) Area under curve equals arc length.

Chapter 9 Answers

9.1 Volumes by the Slice Method

1. 3π

3. $Ah/3$

5. $2125/54$

7. $4\sqrt{3}/3$

9. $x_1 = (1 - \sqrt[3]{1/4})h$, $x_2 = (1 - \sqrt[3]{1/2})h$, $x_3 = (1 - \sqrt[3]{3/4})h$

11. 0.022 m^3

13. 1487.5 cm^3

15. 38π

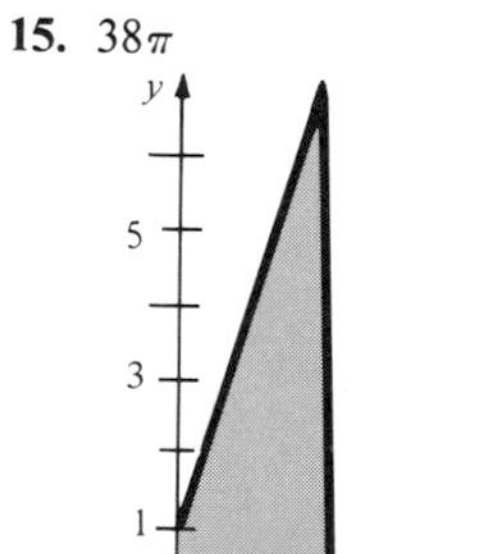

17. $3\pi^2$

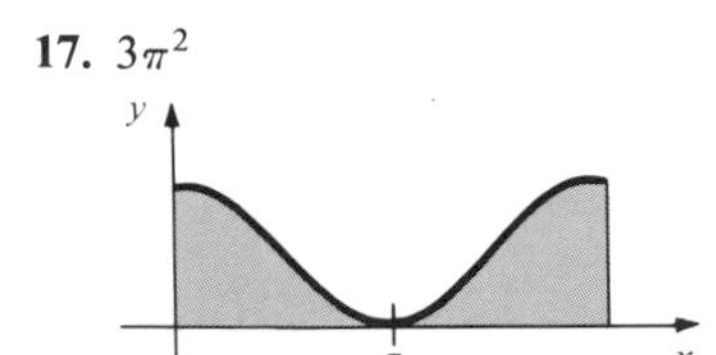

19. $71\pi/105$

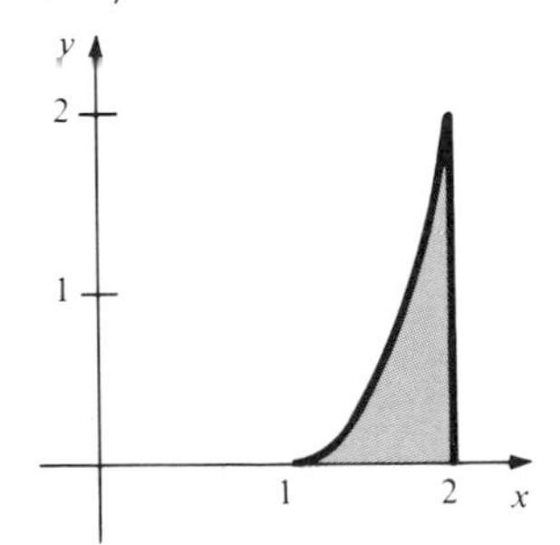

21. $\frac{4}{3}\pi r^3$

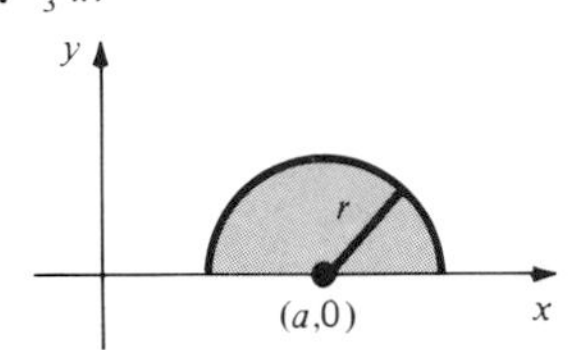

23. 13π

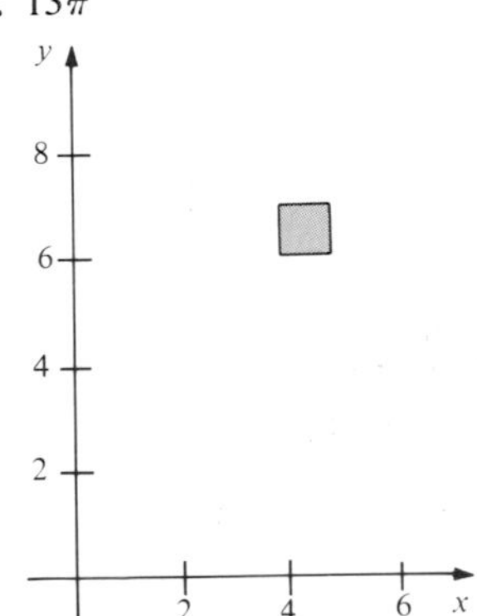

25. 13π (See Exercise 11, Section 9.2 for the figure.)

27. $5\ \text{cm}^3$

29. $V = \pi^2(R + r)(R - r)^2/4$

31. For the two solids, $A_1(x) = A_2(x)$. Now use the slice method.

9.2 Volumes by Shell Method

1. $2\pi^2$

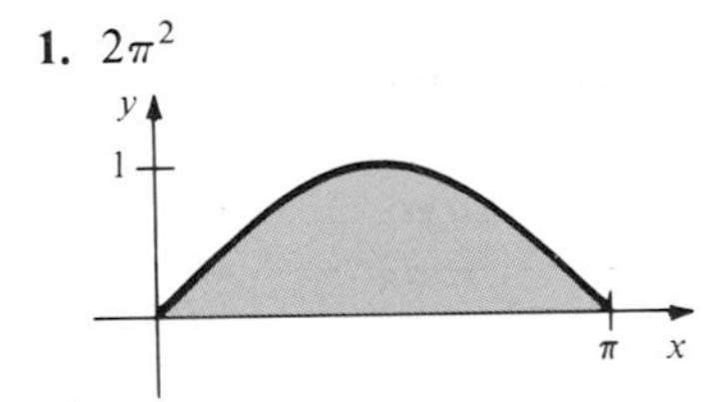

3. $20\pi/3$

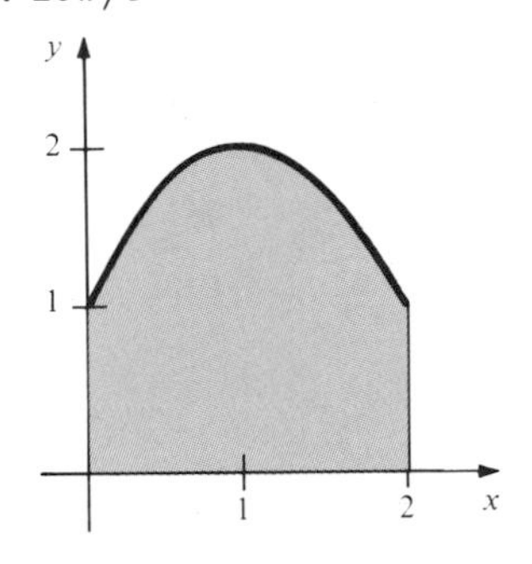

5. $\pi(17 + 4\sqrt{2} - 6\sqrt{3})/3$

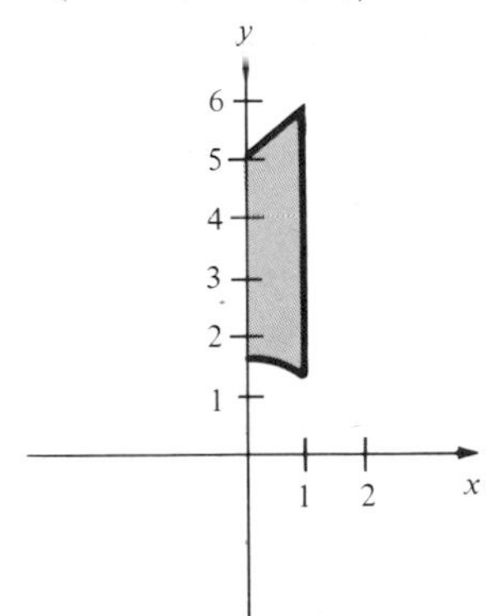

7. $2\pi^2 r^2 a$

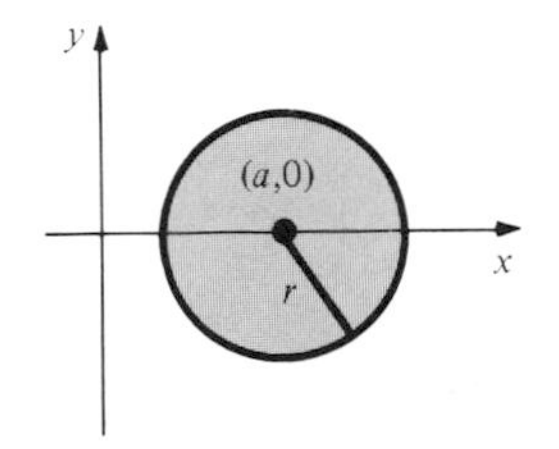

9. 9π (See the Figure for Exercise 23, in the left-hand column.)

11. 9π

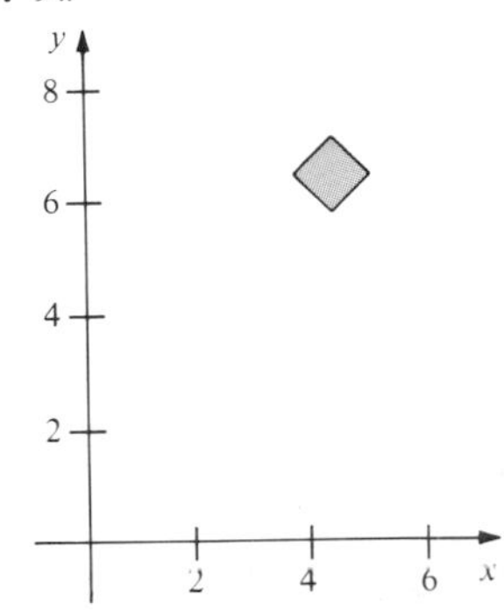

13. $4\pi/5$ (You get a cylinder when this volume is added to that of Example 5, Section 9.1.)

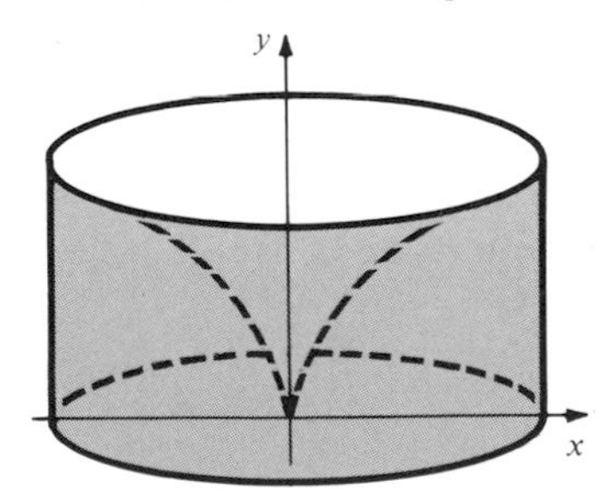

15. $\sqrt{3}\,\pi/2$

17. $24\pi^2$

19. (a) $V = 4\pi r^2 h + \pi h^3/3$

(b) $4\pi r^2$, it is the surface area of a sphere.

21. (a) $2\pi^2 a^2 b$

(b) $2\pi^2 b(2ah + h^2)$

(c) $4\pi^2 ab$

23. $\pi^3/4 - \pi^2 + 2\pi$

9.3 Average Values and the Mean Value Theorem for Integrals

1. $1/4$
3. $\ln\sqrt{5/2}$
5. 2
7. $\pi/4$
9. $\pi/2-1$
11. $-2/3\pi$
13. $9+\sqrt{3}$
15. $1/2$
17. 55° F
19. (a) $x^2/3+3x/2+2$
(b) The function approaches 2, which is the value of $f(x)$ at $x=0$.
21. Use the fundamental theorem of calculus and the definition of average value.
23. The average of $[f(x)+k]$ is $k+$ [the averge of $f(x)$].

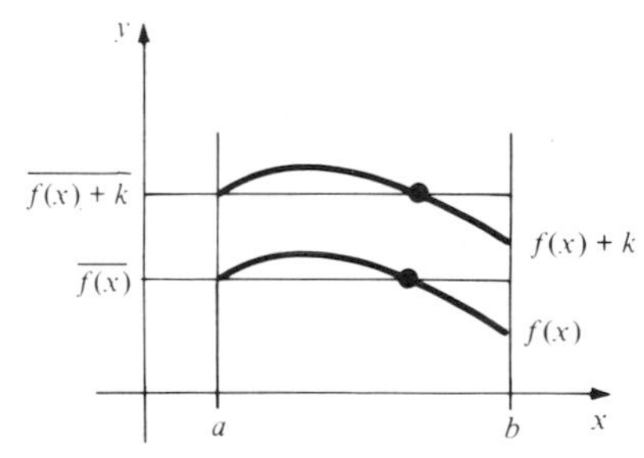

25. $f(b)-f(a)=\int_a^b f'(x)\,dx=f'(c)\cdot(b-a)$, for some c such that $a<c<b$.
27. $\exp\left[\int_a^b \ln f(x)\,dx/(b-a)\right]$
29. Write $F(x)-F(x_0)=\int_{x_0}^x f(s)\,ds$. If $|f(s)|\leqslant M$ on $[a,b]$ (extreme value theorem), $|F(x)-F(x_0)|\leqslant M|x-x_0|$, so given $\epsilon>0$, let $\delta=\epsilon/M$.

9.4 Center of Mass

1. $$\bar{x}=\frac{m_1x_1+(m_2+m_3)\left(\dfrac{m_2x_2+m_3x_3}{m_2+m_3}\right)}{m_1+(m_2+m_3)}=\frac{m_1x_1+m_2x_2+m_3x_3}{m_1+m_2+m_3}.$$
3. Let $M_1=m_1+m_2+m_3$ and $M_2=m_4$.
5. $\bar{x}=3$
7. $\bar{x}=67$
9. $\bar{x}=1,\ \bar{y}=4/3$
11. $\bar{x}=29/23,\ \bar{y}=21/23$
13. (a) $\bar{x}=1/2,\ \bar{y}=\sqrt{3}/6$
(b) $\bar{x}=3/8,\ \bar{y}=\sqrt{3}/8$
15. $$\frac{m_1x_1+(m_2+m_3+m_4)\left[\dfrac{m_2x_2+m_3x_3+m_4x_4}{m_2+m_3+m_4}\right]}{m_1+(m_2+m_3+m_4)}=\frac{m_1x_1+m_2x_2+m_3x_3+m_4x_4}{m_1+m_2+m_3+m_4}$$
17. $\bar{x}=3\ln(3/2),\ \bar{y}=26/27$
19. $\bar{x}=4/(3\pi),\ \bar{y}=4/(3\pi)$
21. $\bar{x}=4/3,\ \bar{y}=2/3$
23. Since $x_i\leqslant b$, $\bar{x}=\dfrac{m_1x_1+m_2x_2+m_3x_3}{m_1+m_2+m_3}\leqslant\dfrac{m_1b+m_2b+m_3b}{m_1+m_2+m_3}=b$.
Similarly $a\leqslant\bar{x}$. The center of mass does not lie outside the group of masses.
25. Differentiate $\bar{x}$ to get the velocity of the center of mass and use the definitions of P and M.
27. $\bar{x}=-4/21,\ \bar{y}=0$

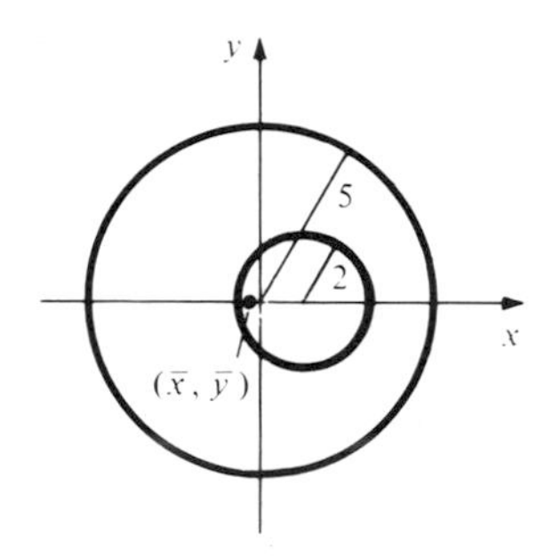

29. $\bar{x}=(\sqrt{2}\,\pi/4-1)/(\sqrt{2}-1),\ \bar{y}=\frac{1}{4}(\sqrt{2}-1)$
31. $\bar{x}=(x_1+x_2+x_3)/3,\ \bar{y}=(y_1+y_2+y_3)/3$

Supplement to 9.5: Integrating Sunshine

1. The arctic circle receives 1.25 times as much energy as the equator.
3. (a) $F=\sum_{T=0}^{364}{}'\left\{\sqrt{\cos^2 l-\sin^2 D}+\sin l\sin D\cos^{-1}(-\tan l\tan D)\right\}$
(b) Expressing $\sin D$ in terms of T, the sum in (a) yields
$$\int_0^{365}\left\{\sqrt{\cos^2 l-\sin^2\alpha\cos^2(2\pi T/365)}+\sin l\sin\alpha\cos(2\pi T/365)\times\cos^{-1}\left[\frac{-\tan l\sin\alpha\cos(2\pi T/365)}{\sqrt{1-\sin^2\alpha\cos^2(2\pi T/365)}}\right]\right\}dT.$$
This is an "elliptic integral" which you cannot evaluate.
5. $\sin l\sin D$
7. 0.294; it is consistent with the graph ($T=16.5$; about July 7).

9.5 Energy, Power, and Work

1. 1,890,000 joules
3. $360 + 96/\pi$ watt-hours
5. $3/2$
7. 0.232
9. 98 watts
11. (a) $18t^2$ joules
(b) 360 watts
13. 1.5 joules
15. (a) 45,000 joules
(b) 69.3 meters/second
17. 41,895,000 joules
19. 125,685,000 joules
21. 0.15 joules
23. 1.48×10^8 joules

33. $120/\pi$ joules

35. $\rho g\pi \int_0^a x^2[f(a) - f(x)]f'(x)\,dx$; the region is that under the graph $y = f(x)$, $0 \leqslant x \leqslant a$, revolved about the y-axis.

37. (a) The force on a slab of height $f(x)$ and width dx is $dx \int_0^{f(x)} \rho g y\,dy = \frac{1}{2}\rho g[f(x)]^2\,dx$. Now integrate.
(b) If the graph of f is revolved about the x axis, the total force is $\rho g/2\pi$ times the volume of the solid.
(c) $\frac{2}{3}\rho g \times 10^6$ Newtons.

Review Exercises for Chapter 9

1. (a) $2\pi^2$
(b) $\pi^2/2$
3. (a) $2\pi(2\ln 2 - 1)$
(b) $3\pi/2$
5. $64\sqrt{2}\,\pi/81$
7. $(57.6)\pi$
9. $5/4$
11. 1
13. 6
15. Apply the mean value theorem for integrals.
17. $1/3, 4/45, 2\sqrt{5}/15$
19. $1, (e^2 - 5)/4, \sqrt{e^2 - 5}/2$
21. $3/2, 1/4, 1/2$
23. (a) $\pi \int_a^b \rho(x)[f(x)]^2\,dx$
(b) $(14\pi/45)$ grams
25. $\bar{x} = 5/3$, $\bar{y} = 40/9$
27. $\bar{x} = 1/4(2\ln 2 - 1)$, $\bar{y} = 2(\ln 2 - 1)^2/(2\ln 2 - 1)$
29. $\bar{x} = 27/35$, $\bar{y} = -12/245$
31. (a) $7500 - 2100e^{-6}$ joules
(b) $\frac{1}{6}(125 - 35e^{-6})$ watts

39. (a) $\left\{ \frac{1}{b-a} \sum_{j=1}^{n} \left[k_j - \frac{1}{b-a} \sum_{i=1}^{n} k_i(t_i - t_{i-1}) \right]^2 (t_j - t_{j-1}) \right\}^{1/2}$
(b) $\left\{ \frac{1}{n} \sum_{j=1}^{n} \left[k_j - \frac{1}{n} \sum_{i=1}^{n} k_i \right]^2 \right\}^{1/2}$
(c) Show that if the standard deviation is 0, $k_i - \mu = 0$, which implies $k_i = \mu$.
(d) $\left\{ \frac{1}{n} \sum_{j=1}^{n} \left[a_i - \frac{1}{n} \sum_{i=1}^{n} a_i \right]^2 \right\}^{1/2}$
(e) All numbers in the list are equal.
41. Let $g(x) = f(\alpha x) - c$. Adjust α so g has zero integral. Apply the mean value theorem for integrals to g. (There may be other solutions as well.)
43. The average value of the logarithmic derivative is $\ln[f(b)/f(a)]/(b - a)$.

Chapter 10 Answers

10.1 Trigonometric Integrals

1. $(\cos^6 x)/6 - (\cos^4 x)/4 + C$
3. $3\pi/4$
5. $(\sin 2x)/4 - x/2 + C$
7. $1/4 - \pi/16$
9. $(\sin 2x)/4 - (\sin 6x)/12 + C$
11. 0
13. $-1/(3\cos^3 x) + 1/(5\cos^5 x) + C$
15. The answers are both $\tan^{-1}x + C$
17. $\sqrt{x^2 - 4} - 2\cos^{-1}(2/|x|) + C$
19. $(1/2)(\sin^{-1}u + u\sqrt{1 - u^2}) + C$
21. $\sqrt{4 + s^2} + C$
23. $(-1/3)\sqrt{4 - x^2}(x^2 + 8) + C$
25. $(1/2)\sinh^{-1}((8x + 1)/\sqrt{15}) + C$
27. $\sqrt{\left(x + \frac{1}{6}\right)^2 - \frac{13}{36}} - \frac{1}{6\sqrt{3}} \ln\left| \frac{6x+1}{\sqrt{13}} \sqrt{\frac{(6x+1)^2}{13} - 1} \right| + C$
29. $1, 0, 1/2, 0, 3/8, 0, 5/16$.
31. $\bar{x} = (\sqrt{5} - \sqrt{2})/\ln((\sqrt{5} + 2)/(\sqrt{2} + 1)) - 1$
$\bar{y} = (\tan^{-1}2 - \pi/4)/[2\ln((\sqrt{5} + 2)/(\sqrt{2} + 1))]$
33. 125
35. $\sqrt{3}, 9\sqrt{2}/4$
37. (a) Differentiate $[S(t)]^3$ and integrate the new expression.
(b) $[3(-t\cos t + \sin t + t/8 - (1/32)\sin 4t)]^{1/3}$
(c) Zeros at $t = n\pi$, n a positive integer. Maxima occur when n is odd.

10.2 Partial Fractions

1. $(1/125)\{4\ln[(x^2 + 1)/(x^2 - 4x + 4)] + (37/2)\tan^{-1}x + (15x - 20)/2(1 + x^2) - 5/(x - 2)\} + C$
3. $5/4 - 3\pi/8$
5. $(1/5)\{\ln(x - 2)^2 + (3/2)\ln(x^2 + 2x + 2) - \tan^{-1}(x + 1)\} + C$

7. $2 + (1/3)\ln 3 + (2/\sqrt{3})(\tan^{-1}(5/\sqrt{3}) - \tan^{-1}(3/\sqrt{3}))$
9. $(1/8)\ln((x^2 - 1)/(x^2 + 3)) + C$
11. $(1/2)\ln(5/2)$
13. $2\sqrt{x} - 2\tan^{-1}\sqrt{x} + C$
15. $(3/2)[(x^2 + 1)^{7/3}/7 - (x^2 + 1)^{4/3}/4] + C$
17. $-2/(1 + \tan(x/2)) + C$
19. $\pi/16 - (1/4)\ln|(1 + \tan(\pi/8))(1 + 2\tan(\pi/8) - \tan^2(\pi/8))| \approx -0.017$
21. $\pi \ln(225/176)$
23. $3(1 + x)^{2/3}/4 + (3/4\sqrt[3]{4})\ln|\sqrt[3]{4}(1 + x)^{2/3} + (2 + 2x)^{1/3} + 1| - (1/2\sqrt[6]{432})\tan^{-1}[(2(4 + 4x)^{1/3} + \sqrt[3]{2}/\sqrt[6]{108}] + C$
25. (a) $\dfrac{1}{20}\ln\left|\dfrac{x - 80}{x - 60}\right| = kt + \dfrac{1}{20}\ln\dfrac{4}{3}$

(b) $x = \dfrac{80(1 - e^{-20kt})}{\frac{4}{3} - e^{-20kt}}$

(c) 26.2 kg

27. (a) Using the substitution, we get

$$(q/m)\int u^{p+q-1}x^{r-m+1}\,du.$$

(b) If $r - m + 1 = mk$, the integral in (a) becomes

$$(q/m)\int u^{p+q-1}(u^q - b)^k\,du$$

which is an integral of a rational function of u.

10.3 Arc Length and Surface Area

1. $92/9$
3. $14/3$
5. $\int_a^b \sqrt{1 + n^2x^{2n-2}}\,dx$
7. $\int_0^1 \sqrt{1 + \cos^2 x - 2x\sin x\cos x + x^2\sin^2 x}\,dx$
9. $\sqrt{5} + \sqrt{2} + \sqrt{10}$
11. $\sqrt{5} + \sqrt{2} + \sqrt{17}$
13. $(\pi/6)(13^{3/2} - 5^{3/2})$
15. $2654\pi/9$
17. $2\pi(\sqrt{2} + \ln(1 + \sqrt{2}))$
19. $\pi[(3^{4/3} + 1/9)^{3/2} - (10/9)^{3/2}]$
21. $2\sqrt{2}\,\pi$
23. $\pi(6\sqrt{2} + 4\sqrt{5})$
25. $(1/27a^2)[(4 + 9a^2(1 + b))^{3/2} - (4 + 9a^2b)^{3/2}]$; the answer is independent of c.
27. $\int_{-1}^2 \sqrt{1 + 36x^4}\,dx \approx 19$
29. (a) $\int_0^{\pi/2} \sqrt{5 + \sec^4 x + 4\sec^2 x}\,dx$

(b) $2\pi\int_0^{\pi/2} (\tan x + 2x)\sqrt{5 + \sec^4 x + 4\sec^2 x}\,dx$

31. (a) $\int_1^2 \sqrt{1 + (1 - 1/x^2)^2}\,dx$

(b) $2\pi\int_1^2 (1/x + x)\sqrt{1 + (1 - 1/x^2)^2}\,dx$

33. Dividing the curve into 1 mm segments and revolving these, we get about 16 cm^2.
35. Use $|\sin\sqrt{3}\,x| \leqslant 1$ to get $L \leqslant \int_0^{2\pi} \sqrt{1 + 3}\,dx = 4\pi$.
37. Evaluate each integral numerically, using upper and lower sums.
39. $2\pi\int_a^b [1/(1 + x^2)]\left(\sqrt{1 + 4x^2/(1 + x^2)^4}\right)dx$; the integrand is $\leqslant \sqrt{5}/(1 + x^2)$.
41. (a) $\pi(a + b)\sqrt{1 + m^2}\,(b - a)$

(b) Use part (a).

10.4 Parametric Curves

1. $y = (1/4)(x + 9)$

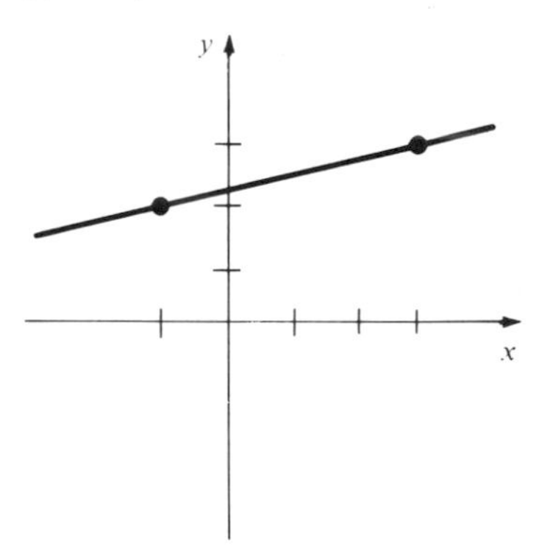

3. $1 = (x - 1)^2 + y^2$

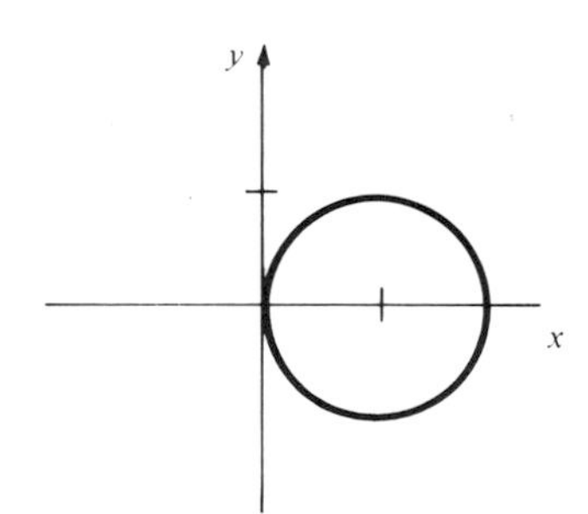

5. $x = t,\ y = \pm\sqrt{1 - 2t^2}$ or $x = \sin t/\sqrt{2},\ y = \cos t$
7. $x = t,\ y = 1/4t$.
9. $x = t,\ y = t^3 + 1$.
11. $x = t,\ y = \cos(2t)$.
13. $y = (1/3)(x + 3/2)$
15. $y = 1/2$
17. $(13, -7)$

19. $y = \cos\sqrt{x}$ $(x \geqslant 0)$, horizontal tangents at $t = n\pi$, n a nonzero integer. The slope is $-1/2$ at $t = 0$ although the curve ends.

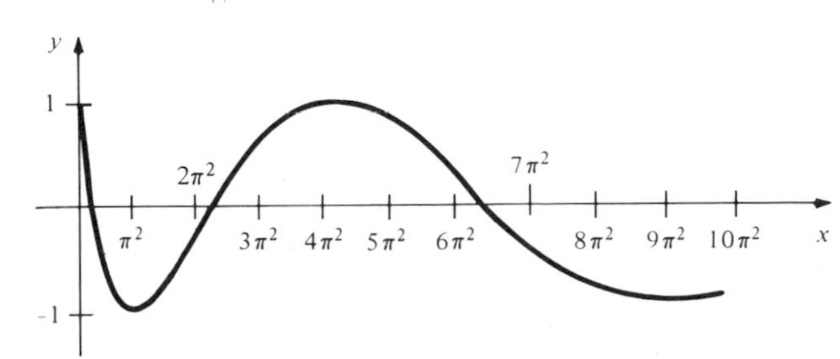

21. $y^2 = (1 - x)/2$, vertical tangents at $t = n\pi$, n an integer

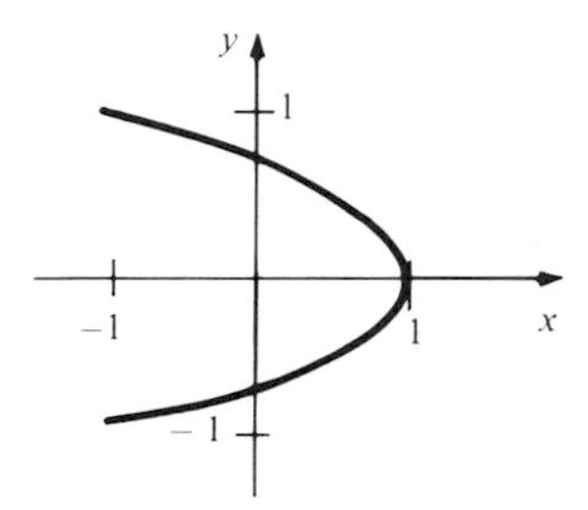

23. $(13^{3/2} - 8)/27$

25. $(1/2)[\sqrt{5} + (1/2)\ln(2 + \sqrt{5})]$

27. (a) Calculate the speed directly to show it equals $|a|$.
(b) Calculate directly to get $|a|(t_1 - t_0)$

29. (a) $y = -x/2 + \pi/2 - 1$

(b)

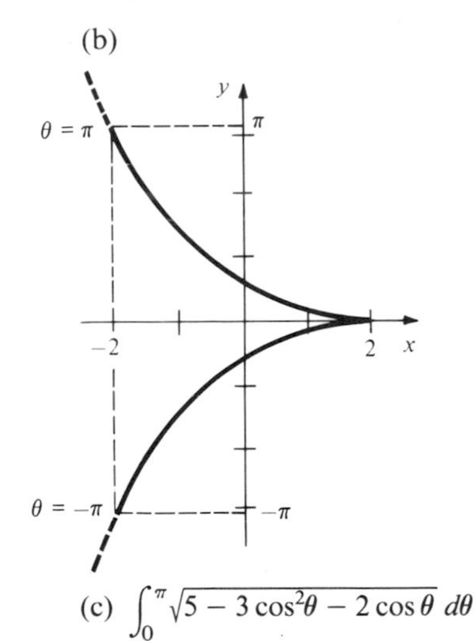

(c) $\int_0^{\pi} \sqrt{5 - 3\cos^2\theta - 2\cos\theta}\, d\theta$

31. 5

33. (a) $\dot{x} = k(\cos\omega t - \omega t \sin\omega t)$;
$\dot{y} = k(\sin\omega t + \omega t \cos\omega t)$.
(b) $k\sqrt{1 + \omega^2 t^2}$
(c) $2mk\omega$

35. (a) $x = t + (1 + 4t^2)^{-1/2}$,
$y = t^2 + 2t(1 + 4t^2)^{-1/2}$
(b) $x = \pm(1/2)\sqrt{1/(x^2 - y) - 1} + \sqrt{x^2 - y}$.

37. (a) We estimated about 338 miles.
(b) We estimated about 688 miles.
(c) It would probably be longer.
(d) The measurement would depend on the definition and scale of the map used.
(e) From the *World Almanac and Book of Facts* (1974), Newspaper Enterprise Assoc., New York, 1973, p. 744, we have coastline: 228 miles, shoreline: 3,478 miles.

10.5 Length and Area in Polar Coordinates

1. $12\sqrt{2}$

3. $(4/3)(13^{3/2} - 8)$

5. $9\pi/4$

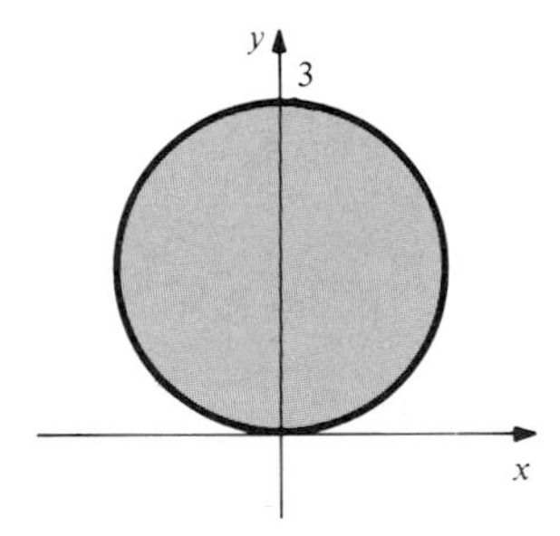

7. $9\pi^3/16$

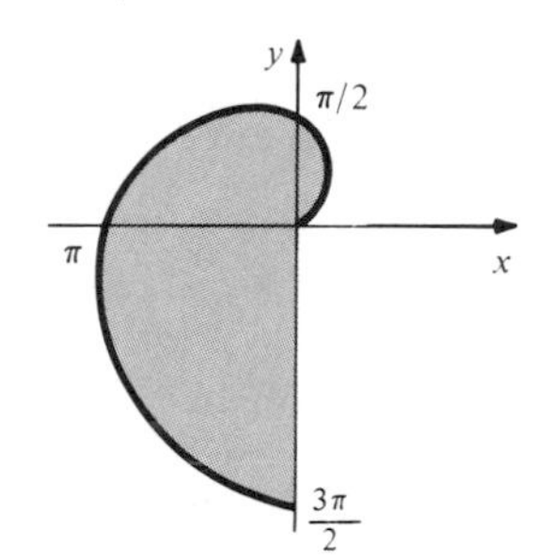

9. $33\pi/2$

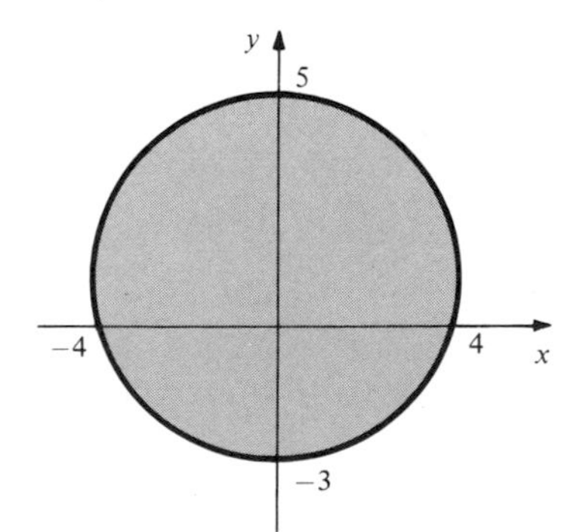

11. $2\pi r$

13. $s = \int_{-\pi/2}^{\pi/2} \sqrt{\sec^2(\theta/2)/4 + \tan^2(\theta/2)}\, d\theta$
$A = 2 - \pi/2$

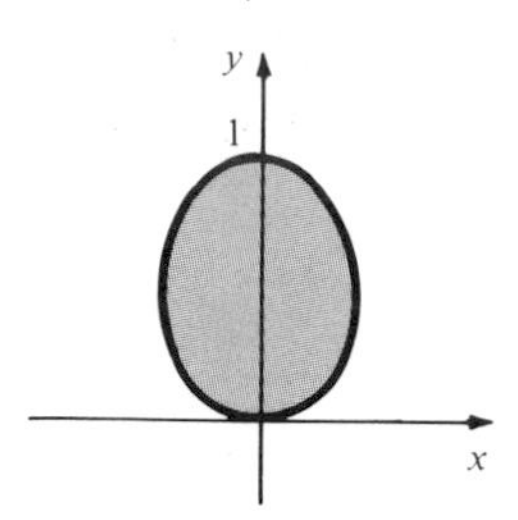

15. $s = \int_0^{\pi/4} \sqrt{\sec^2\theta \tan^2\theta + \sec^2\theta + 4\sec\theta + 4}\, d\theta$
$A = 1/2 + \pi/2 + \ln(3 + 2\sqrt{2})$

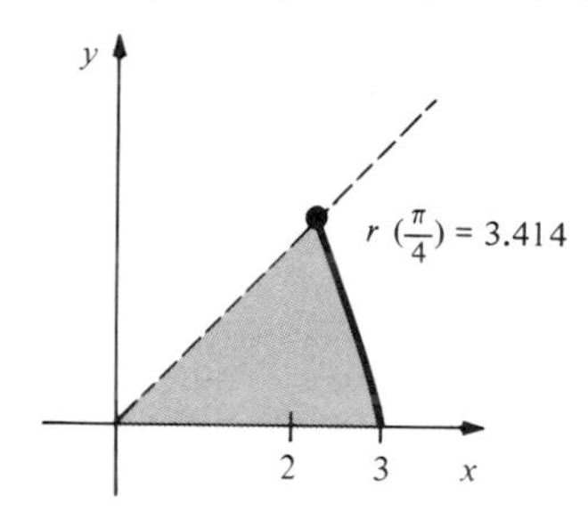

17. $s =$
$\int_0^{\pi/2} \sqrt{(1 + \cos\theta - \theta\sin\theta)^2 + \theta^2(1 + 2\cos\theta + \cos^2\theta)}\, d\theta$
$A = (1/2)[\pi^3/16 + \pi^2/2 - 4 - \pi/8]$

19. $s = \int_0^{\pi/2} \sqrt{(5 + 4\sin 2\theta)/(1 + 2\sin 2\theta)}\, d\theta$
$A = \pi/2$

21. $A = (1/4)(5\pi/6 - \sqrt{3})$
$L = (2 + \sqrt{3})\pi/6$

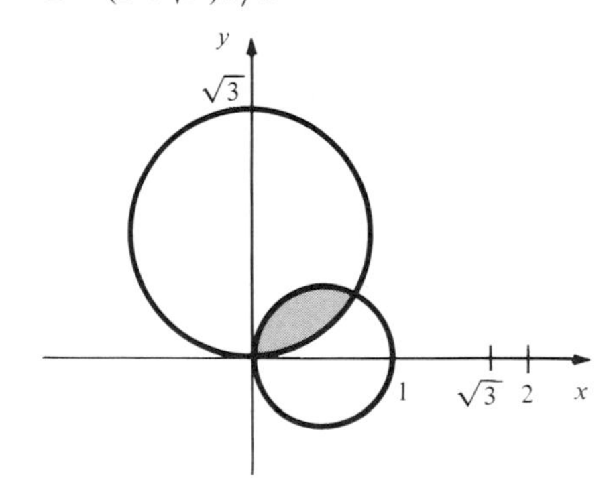

23. $A = \pi/2$
$L = 2\pi + 8$

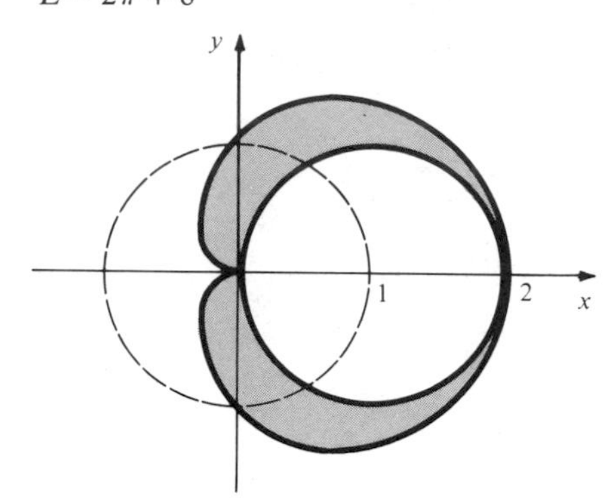

25. $\sqrt{2}\,(e^{2(n+1)\pi} - e^{2n\pi})$

27. (a) Use $x = a\cos t$, $y = b\sin t$, where $T = 2\pi$.
(b) Substitute into the given formula.

Review Exercises for Chapter 10

1. $\sin^3 x + C$
3. $(\cos 2x)/4 - (\cos 8x)/16 + C$
5. $(1 - x^2)^{3/2} - \sqrt{1 - x^2} + C$
7. $4(x/4 - \tan^{-1}(x/4)) + C$
9. $(2\sqrt{7}/7)\tan^{-1}[(2x + 1)/\sqrt{7}] + C$
11. $\ln|(x + 1)/x| - 1/x + C$
13. $(1/2)[\ln|x^2 + 1| + 1/(x^2 + 1)] + C$
15. $\tan^{-1}(x + 2) + C$
17. $-2\sqrt{x}\cos\sqrt{x} + 2\sin\sqrt{x} + C$
19. $-(1/2a)\cot(ax/2) - (1/6a)\cot^3(ax/2) + C$
21. $\ln|\sec x + \tan x| - \sin x + C$
23. $(\tan^{-1}x)^2/2 + C$
25. $(1/3\sqrt[3]{9})[\ln|x - \sqrt[3]{9}| - \ln\sqrt{x^2 + \sqrt[3]{9}\,x + 3\sqrt[3]{3}}$
$+ \sqrt{3}\tan^{-1}((2x/\sqrt[3]{9} + 1)/\sqrt{3})] + C.$
27. $2\sqrt{x}\,e^{\sqrt{x}} - e^{\sqrt{x}} + C$
29. $x - \ln(e^x + 1) + C$
31. $(-1/4)[(2x^2 - 1)/(x^2 - 1)^2] + C$
33. $-(1/10)\cos 5x - (1/2)\cos x + C$
35. $\ln\sqrt{x^2 + 1} + C$
37. $2e^{\sqrt{x}} + C$
39. $\frac{1}{2}\ln 2$
41. $\frac{1}{2}\ln(x^2 + 1) + C$
43. $x^4\ln x/4 - x^4/16 + C$
45. $\frac{1}{4}[(\ln 6 + 5)^4 - (\ln 3 + 5)^4] \approx 186.12$
47. $(1/4)\sinh 2 - 1/2$
49. 0
51. $(733^{3/2} - 4^{3/2})/243$
53. $59/24$
55. $\pi(5^{3/2} - 1)/6$
57. ≈ 31103

59. $x = (y+1)^2$

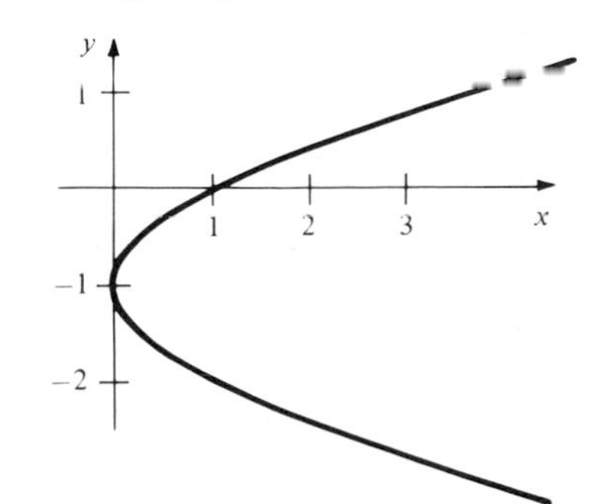

61. $y = 2x/3 + 1$

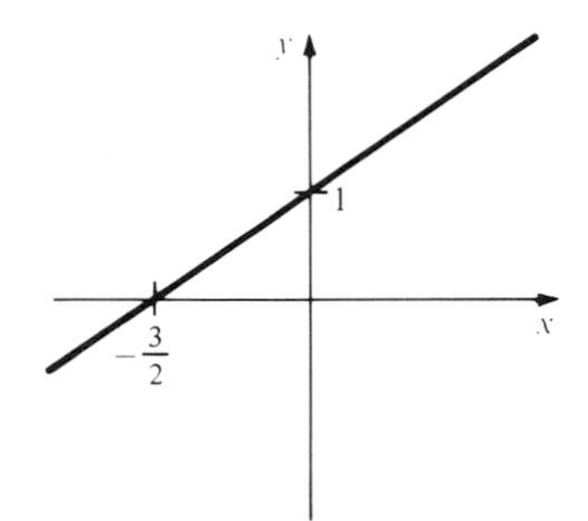

63. $x = 0, y \geqslant 0$

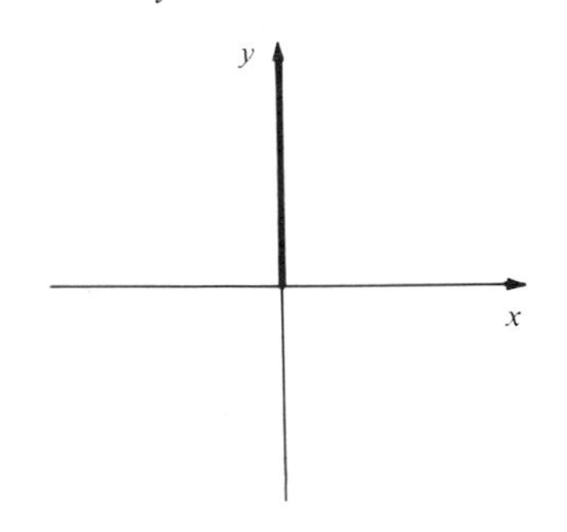

65. $y = 3x/4 + 5/4$

67. $(1/8)(\sqrt{257}\cdot 16 + \ln|\sqrt{257} + 16|)$

69. $L = (1/3)[(\pi^2/4 + 4)^{3/2} - 8]$
$A = \pi^5/320$

71. $L = \int_0^{\pi}\sqrt{(5/4) + \cos 2\theta + 3\sin^2 2\theta}\, d\theta$
$A = 3\pi/8$

73. $L = 5\sqrt{2}$
$A = 315\pi/256 + 9/4$

75. $b_2 = 1$, all others are zero.

77. $a_3 = 1$, all others are zero.

79. $a_4 = 3$, all others are zero.

81. $a_0 = 1$, $a_2 = -1/2$, all others are zero.

83. (a) $(1/k_2)\ln[N_0(k_1N(t) - k_2)/N(t)(k_1N_0 - k_2)]$
(b) $N(t) = k$, $N_0/[k_1N_0(1 - e^{k_2t}) + k_2e^{k_2t}]$
(c) The limit exists if $k_2 > 0$ and it equals k_2/k_1.

85. Use $(\cos\phi)\, d\phi = (\sin\phi_m)(\cos\beta)\, d\beta$ and substitute.

87. $a^{-1/2}\ln|2ax + b + 2\sqrt{a}\sqrt{ax^2 + bx + c}| + C$, $a > 0$
$(-a)^{-1/2}\sin^{-1}[(-2ax - b)/\sqrt{b^2 - 4ac}\,] + C$, $a < 0$

89. (a) $b - a + (b^{n+1} - a^{n+1})/(n+1)$ if $n \neq -1$. If $n = -1$, we have $b - a + \ln(b/a)$.
(b) $n = 0$: $L = b - a$; $n = 1$: $L = \sqrt{2}\,(b - a)$;
$n = 2$: see Example 3 of Section 10.3;
for $n = (2k+3)/(2k+2)$, $k = 0, 1, 2, 3, \ldots$

$$L = \left\{ \frac{n^{1/(1-n)}}{n-1}(1 + n^2x^{2n-2})^{3/2} \sum_{j=0}^{k}\binom{k}{j}\frac{(-1)^{k-j}}{2j+3}(1 + n^2x^{2n-2})^j \right\}\Bigg|_{x=a}^{x=b};$$

$n = \frac{3}{2}: L = \frac{1}{27}[(4 + 9b)^{3/2} - (4 + 9a)^{3/2}]$.

(c) Around the x-axis, we have

$$\pi\left[b - a + \frac{2(b^{n+1} - a^{n+1})}{n+1} + \frac{b^{2n+1} - a^{2n+1}}{2n+1}\right]$$

if $n \neq -1$ or $-1/2$. For $n = -1$ we have

$$\pi\left[b - a + 2\ln(b/a) - (a^{-1} - b^{-1})\right].$$

For $n = -1/2$ we have

$$\pi\left[b - a + 4\sqrt{b} - 4\sqrt{a} + \ln(b/a)\right].$$

Around the y-axis we have

$$\pi\left[b^2 - a^2 + \frac{2(b^{n+2} - a^{n+2})}{n+2}\right]$$

if $n \neq -2$. For $n = -2$, we have $\pi[b^2 - a^2 + 2\ln(b/a)]$.

(d) $A_x = 2\pi L$ (from 89(b)) + A_x (from 88(d))
$A_y = A_y$ (from 88(d))
Some answers from 88(d) needed here are:
88(d).

$$n = 0;\ A_x = 2\pi(b - a)$$

$$n = 1;\ A_x = \sqrt{2}\,\pi(b^2 - a^2)$$

$$n = 2;\ A_x = \frac{\pi}{32}\left[(1 + 8x^2)2x\sqrt{1 + 4x^2} - \ln\left(2x + \sqrt{1 + 4x^2}\right)\right]\Bigg|_{x=a}^{x=b}$$

$$n = 3;\ A_x = \frac{\pi}{27}(1 + 9x^4)^{3/2}\Big|_{x=a}^{x=b}$$

$$n = (2k+3)/(2k+1);\ k = 0, 1, 2, 3, \ldots$$

$$A_x = \frac{2\pi}{n-1}n^{(1+n)/(1-n)}(1 + n^2x^{2n-2})^{3/2}\sum_{j=0}^{k}\binom{k}{j}\frac{(-1)^{k-j}}{2j+3}(1 + n^2x^{2n-2})^j$$

$$n = 0;\ A_y = \pi(b^2 - a^2)$$

$$n = 1;\ A_y = \sqrt{2}\,\pi(b^2 - a^2)$$

$$n = 2;\ A_y = \frac{\pi}{6}\left[(1 + 4b^2)^{3/2} - (1 + 4a^2)^{3/2}\right]$$

$$n = (k+2)/(k+1);\ k = 0, 1, 2, 3, \ldots;$$

$$A_y = \frac{2\pi}{n-1}n^{2/(1-n)}(1 + n^2x^{2n-2})^{3/2} \times \sum_{j=0}^{k}\binom{k}{j}\frac{(-1)^{k-j}}{2j+3}(1 + n^2x^{2n-2})^j\Big|_a^b$$

91. (a) $2\pi\int_{\alpha}^{\beta} r\sin\theta\sqrt{r^2 + (r')^2}\, d\theta$
(b) $2\pi\int_{-\pi/4}^{\pi/4}\cos 2\theta\sin\theta\sqrt{1 + 3\sin^2 2\theta}\, d\theta$

93. (a)

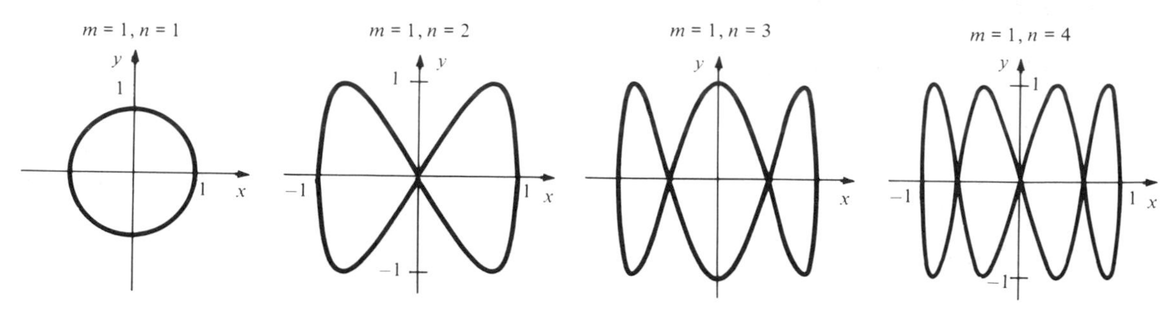

(b) Each curve will consist of n loops for n odd or even.

(c)

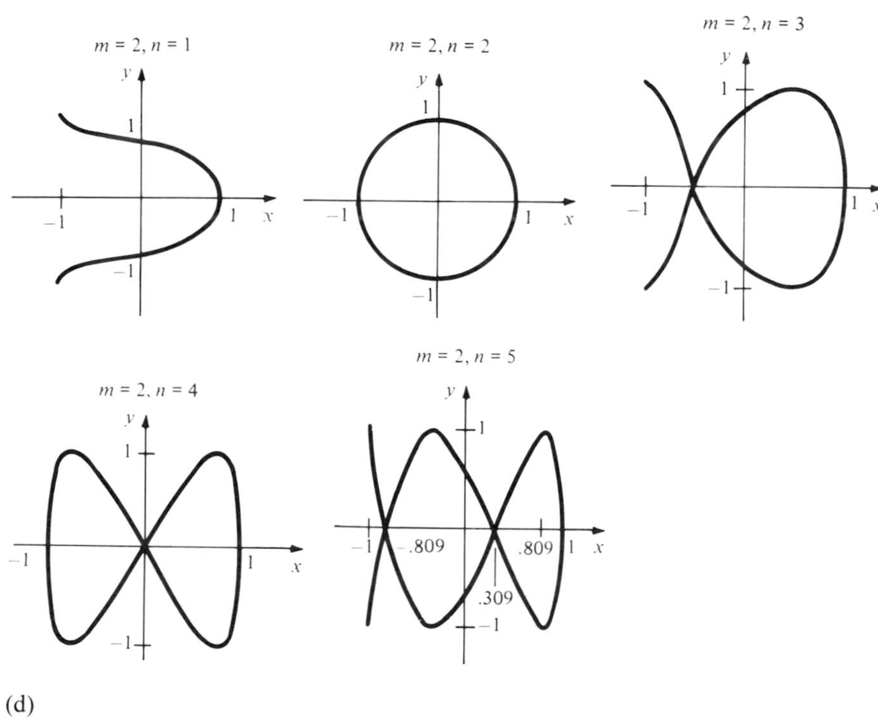

(d)

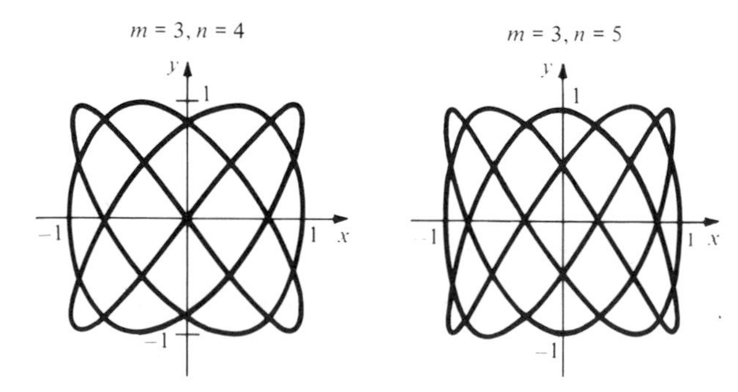

95. The last formula is the average of the first two.

Chapter 11 Answers

11.1 Limits of Functions

1. Choose δ less than 1 and $\varepsilon/(1 + 2a)$.

3. Write $x^3 + 2x^2 - 45 =$ $[(x - 3) + 3]^3 + 2[x - 3]^2 - 45$ and expand.

5. e^3

7. 5

9. -4

11. 6

13. $A = 1/\sqrt[3]{\varepsilon}$

15. $A = -\ln \varepsilon/3$

17. -2

19. $2/3$

21. $3/5$

23. $1/2$

25. 0. Consider $\sqrt{x^2 + a^2} - x$ as the difference between the hypotenuse and a leg of a right triangle. As x gets large, the difference becomes small.

27. $y = -1$ is a horizontal asymptote.

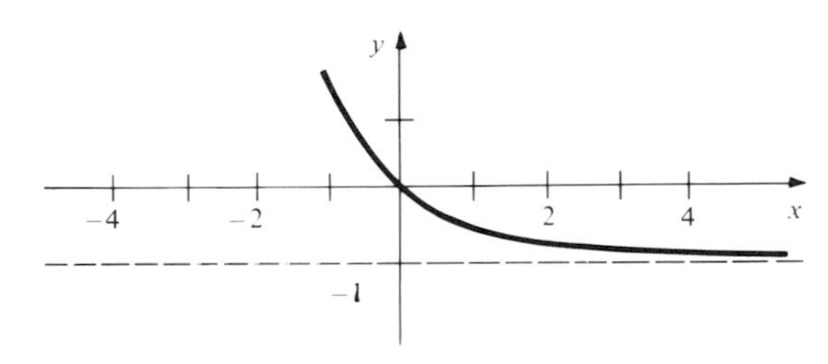

29. $+\infty$

31. $+\infty$

33. $+\infty$

35. $-\infty$

37. -1

39. -1

41. Vertical asymptotes at $x = 2, 3$.

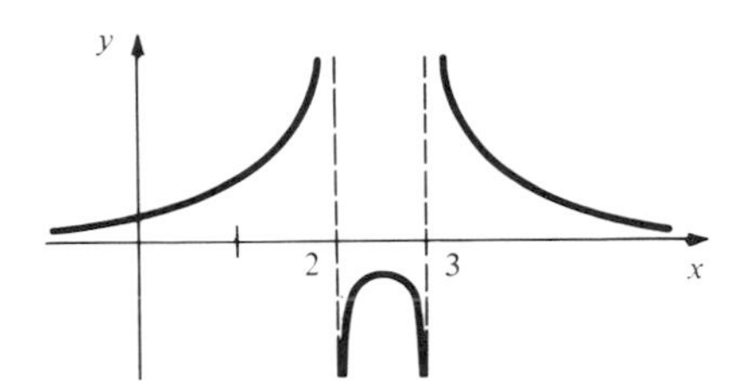

43. Vertical asymptotes at $x = \pm 1$.

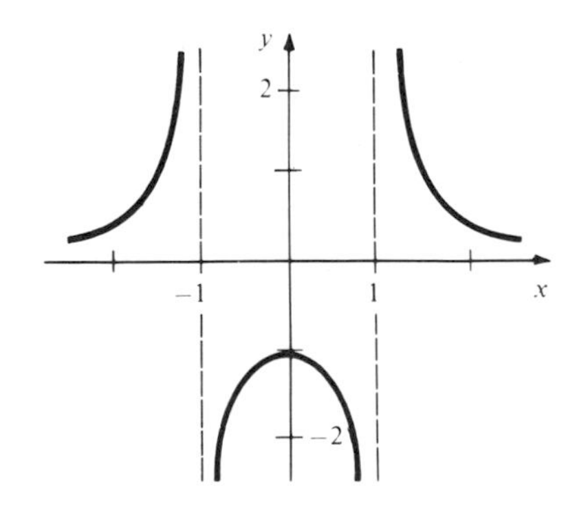

45. (a) Given ε, the A for g is the same as for f (as long as $|g(x)| \leqslant |f(x)|$ for $x \geqslant A$).

(b) 0

47. $7/9$

49. $3/2$

51. $4/5$

53. $2n + 1$

55. $16/17$

57. $+\infty$

59. $-\infty$

61. $y = 0$ is a horizontal asymptote; $x = -1$, $x = 1$ are vertical asymptotes.

63. $y = \pm 1$ are horizontal asymptotes.

65. If $f(x) = a_n x^n + \cdots$ and $g(x) = b_n x^n + \cdots$, show that $a_n/b_n = l$. If $l = \pm\infty$, then $\lim_{x\to-\infty} f(x)$ can be $\pm \lim_{x\to\infty} f(x)$.

67. (a) $f'(x) = -1$ for $x < 0$, $f'(x) = 1$ for $x > 0$, $f'(0)$ is not defined.

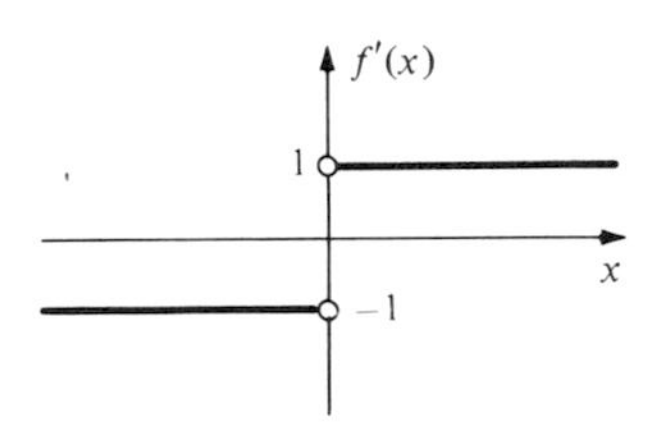

(b) As $x \to 0-$, the limit is -1, while as $x \to 0+$, we get 1.

(c) No.

69. (a)

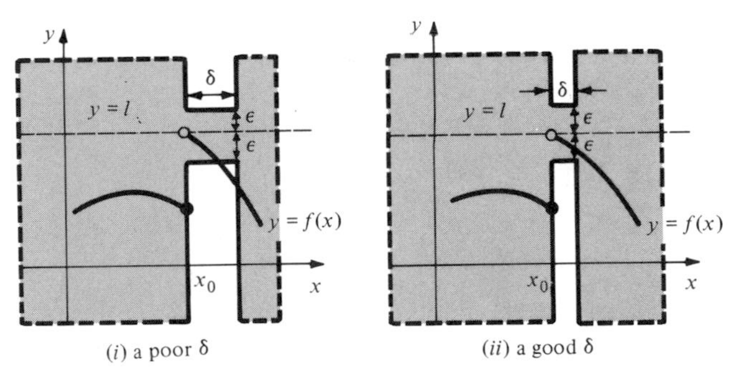

(i) a poor δ (ii) a good δ

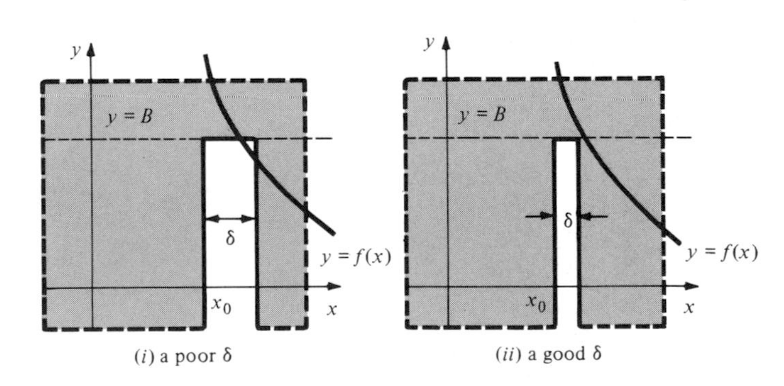

(i) a poor δ (ii) a good δ

71. N_0, which means that the population in the distant future will approach an equilibrium value N_0.

73. Use the laws of limits

75. Write $af(x) + bg(x) - aL - bM =$ $a[f(x) - L] + b[g(x) - M]$

77. Write

$$\frac{f(x)}{g(x)} - \frac{L}{M} = \frac{(f(x) - L)M - (g(x) - M)L}{Mg(x)}.$$

79. Given $B > 0$, let $\varepsilon = 1/B$. Choose δ so that $|1/f(x)| < \varepsilon$ when $|x - x_0| < \delta$; then $|f(x)| > B$ for $|x - x_0| < \delta$.

81. If $x \geqslant A$, $y \leqslant \delta$ where $\delta = 1/A$, $y = 1/x$.

11.2 L'Hôpital's Rule

1. 108 **3.** 2
5. $-9/10$ **7.** $-4/3$
9. ∞ **11.** 0
13. 0 **15.** 0
17. 1 **19.** 0
21. 0 **23.** 0
25. 0
27. does not exist (or is $+\infty$)
29. 0 **31.** $1/24$
33. 0 **35.** $1/120$
37. 0
39. The slope of the chord joining $(g(a), f(a))$ to $(g(b), f(b))$ equals the slope of the tangent line at some intermediate point.
41.

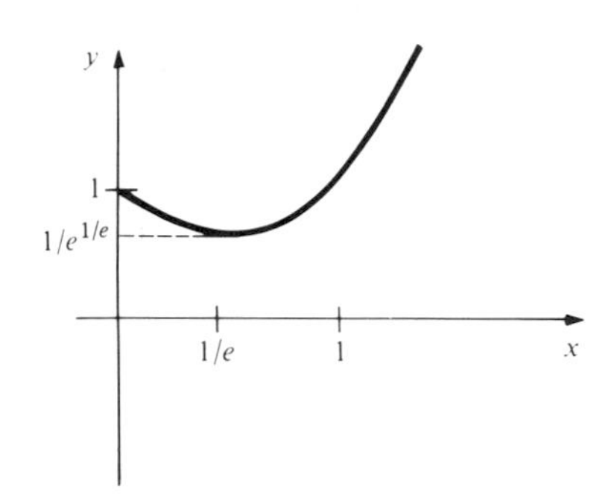

43. (a) $1/2$
(b) 1
(c) yes

11.3 Improper Integrals

1. 3 **3.** $e^{-5}/5$
5. $(\ln 3)/2$ **7.** $\pi/2$
9. Use $1/x^3$ **11.** Use e^{-x}
13. Use $1/x$ on $[1, \infty)$ **15.** Use $1/x$
17. $3\sqrt[3]{10}$ **19.** 2
21. Diverges **23.** Converges
25. Converges **27.** Converges
29. Converges **31.** Diverges
33. Converges **35.** Converges
37. Diverges **39.** Diverges
41. $k > 1$ or $k = 0$ **43.** ≈ 2.209
45. $6\sqrt{3}$ hours **47.** $\pi e^{-20}/2$
49. $\ln(2/3)$
51. Follow the method of Example 11.
53. (a) Change variables
(b) Use the comparison test. (Compare with $e^{x/2}$ for $x \leqslant -1$ and $e^{-x/2}$ for $x \geqslant 1$.)
55. (a) π
(b) $(p-1)(q-1) < 0$.
57. $f(x) = f(0) + \int_0^x f'(s)\,ds$; the integral converges.

11.4 Limits of Sequences and Newton's Method

1. n must be at least 6.
3. $\lim_{n\to\infty}(a_n) = 2$
5. $0, -1, 4 - 2\sqrt{2}, 9 - 2\sqrt{3}, 12$
7. $1/7, 1/14, 1/21, 1/28, 1/35, 1/42$
9. The sequence is $1/2$ for all n.
11. $N \geqslant 3/\varepsilon$ **13.** $n \geqslant 3/2\varepsilon$
15. 3 **17.** -3
19. 4 **21.** 0
23. 0 **25.** The limit is 1.
27. The limit is 1. **29.** 0
31. 0 **33.** 0
35. (a) $x = 0.523148$ is a root.
(b) $x = -0.2475, 7.7243$
37. $x = 1.118340$ is a root.
39. One root is $x = 4.493409$.
41.

	$\alpha = 2$	$\alpha = 3$	$\alpha = 5$
λ_1	1.1656	1.3242	1.4320
λ_2	4.6042	4.6407	4.6696
λ_3	7.7899	7.8113	7.8284

43. $1/e \approx 0.36788$
45. $a_n = 2^{2^n - 1}$
47. Use the definition of limit and let ε be a.
49. $1, 1/2, 1/4, 1/8, 1/16, \ldots, 1/(2^n), \ldots$; the limit is 0.
51. The limit does not exist.
53. $3/4$
55. (a) For any $A \geqslant 0$ there is an N such that $a_n \geqslant A$ if $n \geqslant N$, (b) let $N = 16A$.
57. (a) Assume $\lim_{n\to\infty} b_n < L$ and look at

$$\lim_{n\to\infty} b_n - \lim_{n\to\infty} a_n.$$

(b) Write $b_n - L = (b_n - a_n) + (a_n - L) \leqslant (c_n - a_n) + (a_n - L)$.
59. (a) Below about $a = 3.0$, iterates converge to a single point; at $a \approx 3.1$, they oscillate between two points; as a increases towards 4, the behavior gets more complicated.
(b), (c) See the references on p. 548.

11.5 Numerical Integration

1. 2.68; actual value is $8/3$ **3.** ≈ 0.13488
5. ≈ 0.3246 **7.** ≈ 1.464

9. ≈ 2.1824
11. Evaluation gives $a(x_2^3 - x_1^3)/3 + b(x_2^2 - x_1^2)/2 +$ $c(x_2 - x_2)$. Since $f''''(x) = 0$, Simpson's rule gives the exact answer. The error for the trapezoidal rule depends on $f''(x)$ and is nonzero.
13. 180, 9
15. 158 seconds
17. The first 2 digits are correct.

Review Exercises for Chapter 11

1. Choose δ to be $\min(1, \varepsilon/4)$.
3. Choose δ to be $\min(1, \varepsilon/5)$.
5. $\tan(-1)$
7. 1
9. 0
11. ∞
13. 0
15. 0
17. $y = \pm\pi/2$ are horizontal asymptotes.

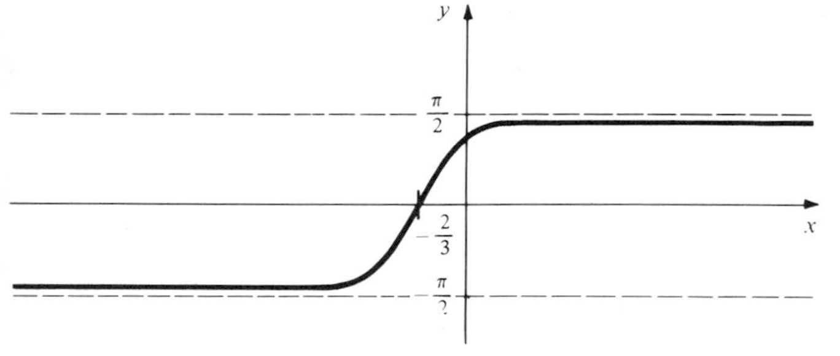

19. $y = 0$ is a horizontal asymptote.

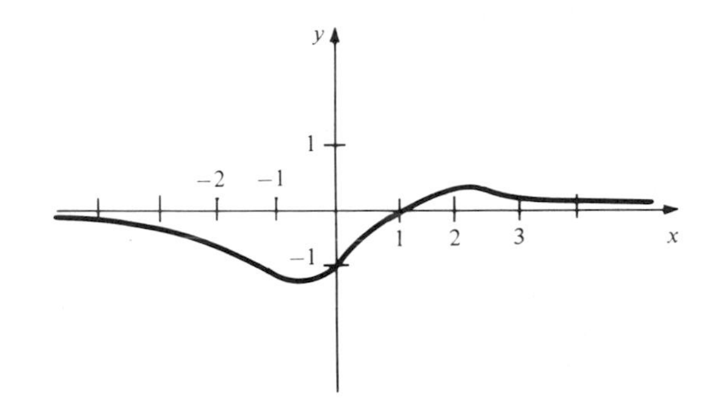

21. $1/4$
23. 0
25. 0
27. 5
29. $-1/6$
31. $\sec^2(3)$
33. 1
35. 0
37. 0
39. 1
41. e^2
43. 0
45. Converges to 1
47. Diverges
49. Converges to 2
51. Converges to $5/3$
53. Converges to $-1/4$
55. $2\pi/3\sqrt{3}$
57. $\pi/4$
59. 32,768
61. e^8
63. 0
65. 1
67. $\tan 3$
69. Does not exist
71. $-2/5$
73. 1
75. 0
77. -1.35530 (the only real root)
79. 1.14619
81. 2.31992
83. 50.154
85. Both
87. $1/\sqrt{x}$
89. (b)

$$\lim_{h\to 0}\{[f(x_0+2h) - 3f(x_0+h) + 3f(x_0) - f(x_0-h)]/h^3\}.$$

91. 1
93. S_n is the Riemann sum for $f(x) = x + x^2$.
95. The exact amount is

$$P(e^r + e^{364r/365} + \cdots + e^{r/365})$$

97. (a)

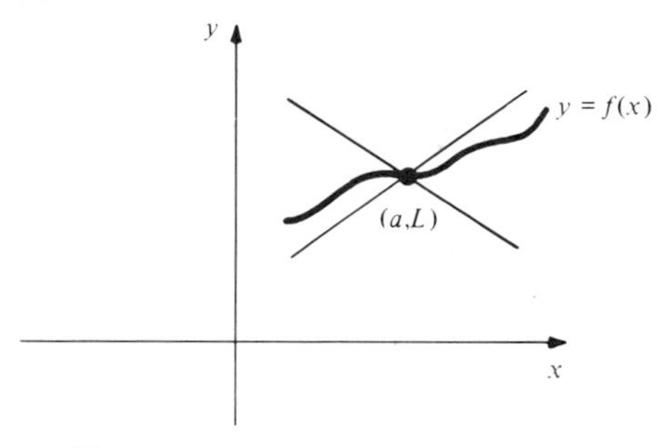

(c) Choose $\delta = \varepsilon/2m$, (or h, whichever is smallest).

101. (a) Use the definition of N
(b) Use the quotient rule
(c) $|N(x) - \bar{x}| \leqslant (Mq/p^2)|x - \bar{x}|^2$
(d) 5

Chapter 12 Answers

12.1 The Sum of an Infinite Series

1. $1/2, 5/6, 13/12, 77/60$
3. $2/3, 30/27, 38/27, 130/81$
5. $7/6$
7. 7
9. \$40,000
11. $1/12$
13. $16/27$
15. $81/2$
17. $3/2$
19. $64/9$
21. $\sum 1$ diverges and $\sum 1/2^i$ converges
23. 7
25. Diverges
27. Diverges
29. Diverges
31. Reduce to the sum of a convergent and a divergent series.
33. Let $a_i = 1$ and $b_i = -1$.
35. (a) $a_1 + a_2 + \cdots + a_n = (b_2 - b_1) + (b_3 - b_2) + \cdots + (b_{n+1} - b_n) = b_{n+1} - b_1$ (see Section 4.1).
(b) 1

37. (b) $\sum t_{2n+1} = \dfrac{12/27}{1-r}$ and $\sum t_{2n+2} = \dfrac{r \cdot 12/13}{1-r}$

The sum is 1.

12.2 The Comparison Test and Alternating Series

1. Use $8/3^i$
3. Use $1/3^i$
5. Use $1/3^i$
7. Use $1/2^i$
9. Use $1/i$
11. Use $4/3i$
13. Converges
15. Converges
17. Diverges
19. Converges
21. Converges
23. Diverges
25. Converges
27. Diverges
29. Converges
31. Diverges
33. Converges
35. 0.29
37. 0.37
39. Diverges
41. Diverges
43. Converges absolutely
45. Diverges
47. Converges conditionally
49. Converges conditionally
51. -0.18
53. -0.087
55. Converges
57. (a) $a_1 = 2,\ a_2 = \sqrt{6},\ a_3 = \sqrt{4+\sqrt{6}}$

(b) $\lim_{n\to\infty} a_n \approx 2.5616$

59. Increasing, bounded above. (Use induction.)
61. Increasing for $n \geqslant 2$, bounded above (Use induction.)
63. Show by induction that $a_2, a_3, \ldots$ is decreasing and bounded below, so converges. The limit l satisfies $l = \frac{1}{2}(l + \frac{B}{l})$.
65. $\lim_{n\to\infty} a_n = 4$
67. The limit exists by the decreasing sequence property.
69. Compare with $(3/4)^n$.

12.3 The Integral and Ratio Tests

1. Diverges
3. Converges
5. Converges
7. Converges
9. 3.00
11. Use Figure 12.3.2.
13. Converges
15. Converges
17. 11.54
19. (a) ≈ 1.708
(b) ≈ 1.7167
(c) 8 or more terms.
21. Converges
23. Diverges
25. Converges
27. Diverges
29. Converges
31. Converges
33. Converges
35. Converges
37. Show that if $|a_n|^{1/n} > 1$, then $|a_n| > 1$.
39. $p > 1$
41. $p > 1$

43. (a) $S - \frac{1}{2}f(n) = \sum_{i=1}^{n-1} f(i) + \frac{1}{2}f(n) + \frac{1}{2}\int_n^{n+1} f(x)\,dx + \int_{n+1}^{\infty} f(x)\,dx$
$\leqslant \sum_{i=1}^{\infty} f(i) + \frac{1}{2}f(n) + \frac{1}{2}f(n) + \int_{n+1}^{\infty} f(x)\,dx;$
now use the hint.
(b) Sum the first 9 terms to get 1.0819. The first method saves the work of adding 6 additional terms.

45. (b)

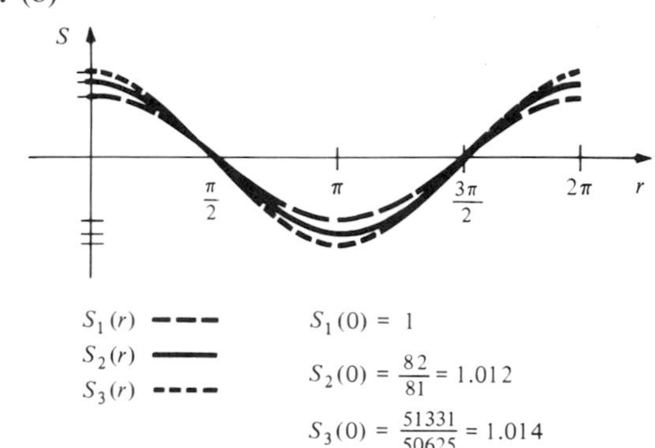

12.4 Power Series

1. Converges for $-1 \leqslant x < 1$.
3. Converges for $-1 \leqslant x \leqslant 1$.
5. Converges for $0 < x < 2$.
7. Converges for all x.
9. Converges for $-4 < x < 4$.
11. $R = \infty$
13. $R = 2$
15. $R = \infty$
17. $R = 1$, converges for $x = 1$ and -1.
19. $R = 3$
21. $R = 0$
23. Note that $f(0) = 0$ and $f'(0) = 1$.
25. (a) $R = 1$
(b) $\sum_{i=1}^{\infty} x^{i+1}$
(c) $f(x) = x(2-x)/(1-x)^2$ for $|x| < 1$
(d) 3
27. $\sum_{n=0}^{\infty}[(-1)^n x^{2n}/n!]$
29. $\tan^{-1}(x) = \sum_{n=0}^{\infty}[(-1)^n x^{2n+1}/(2n+1)]$, and $(d/dx)(\tan^{-1}x) = \sum_{n=0}^{\infty}(-1)^n/x^{2n}$.
31. $1/2 + 3x/4 + 7x^2/8 + 15x^3/16 + \cdots$
33. $x^2 + x^4/3 - 2x^6/45 + \cdots$
35. Set $f(x) = 1/(1-x)$ and $g(x) = -x^2/(1-x)$.
37. (a) $x + (1/3)x^3 + (2/15)x^5 + \cdots$
(b) $1 + x^2 + (2/13)x^4 + \cdots$
(c) $1 - x^2 + (1/3)x^4 - \cdots$
39. $\sum_{i=1}^{\infty}(-1)^{i+1}(1/i)x^i$
41. Use the fact that $\sqrt[i]{i} \to 1$ as $i \to \infty$.
43. Write $f(x) - f(x_0) = \left(f(x) - \sum_{i=0}^{N} a_i x^i\right) + \left(\sum_{i=0}^{N} a_i x^i - \sum_{i=0}^{N} a_i x_0^i\right) + \left(\sum_{i=0}^{N} a_i x_0^i - f(x_0)\right)$

45. Show that $f(x) = \int_0^x g(t)\,dt$

12.5 Taylor's Formula

1. $3x - 9x^3/2 + 81x^5/80 - 243x^7/1120 + \cdots$
3. $2 - 2x + 3x^2/2 - 4x^3/3 + 17x^4/24 - 4x^5/15 + 7x^6/80 - 8x^7/315 + \cdots$
5. $1/3 - 2(x-1)/3 + 5(x-1)^2/9 + 0 \cdot (x-1)^3$
7. $e + e(x-1) + e(x-1)^2/2 + e(x-1)^3/6$.
9. (a) $1 - x^2 + x^6 + \cdots$ (b) 720
11. Valid if $-1 < x \leqslant 1$ (Integrate $1/(1+x) = 1 - x + x^2 - x^3 + \cdots$.)
13. Let $x - 1 = u$ and use the bionomial series.
15. (a) $1 - (1/2)x^2 + (3/8)x^4 - (5/16)x^6 + (35/128)x^8 - \cdots$
(b) $(-1/2)(-1/2-1)\ldots(-1/2-10+1)\cdot(20!)/(10!)$
17. $f_0(x) = f_1(x) = 1,\ f_2(x) = f_3(x) = 1 - x^2/2,$ $f_4(x) = 1 - x^2/2 + x^4/24$.

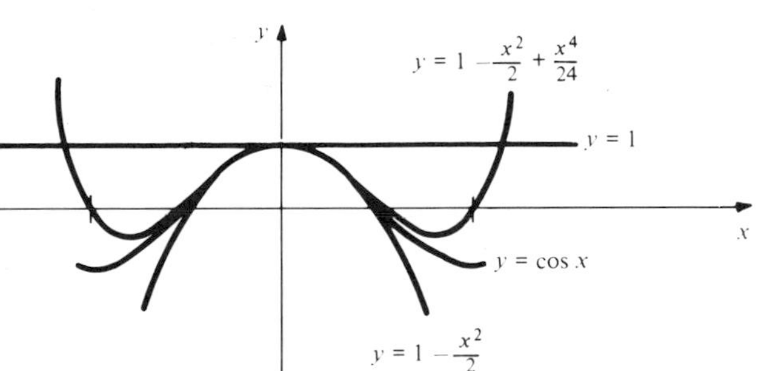

19. ≈ 0.095
21. ≈ 0.9
23. ≈ 0.401
25. (a) The remainder is less than $R^4M_3/12$ where M_3 is the maximum value of $|f'''(x)|$ on the interval $[x_0 - R, x_0 + R]$.
(b) 0.958
27. $-4/3$
29. $1/6$
31. $\sum_{n=0}^{\infty} x^n$ for $|x| < 1$
33. $\sum_{n=0}^{\infty} 2x^{2n+1}$ for $|x| < 1$
35. $\sum_{n=0}^{\infty} x^{2n}$ for $|x| < 1$
37. $1 + 2x^2 + x^4$
39. $\int_1^x \ln t\,dt = \sum_{n=2}^{\infty}\{(-1)^i(x-1)^i/[i(i-1)]\}$.
$x\ln x = (x-1) + \sum_{i=2}^{\infty}\{(-1)^i(x-1)^i/[i(i-1)]\}$.
Conclude $\int_1^x \ln t\,dt = x\ln x - x$.
41. $1, 0, 1/2, 0$
43. $0, -1, 0, -1/2$
45. $1/2 - x^2/4! + x^4/6! - \cdots$
47. $1 - 2x/3 + 2x^2/9 - 22x^3/81 + (38/243)x^4 - (134/729)x^5 + \cdots$
49. (a) $(x-1) - (x-1)^2/2 + (x-1)^3/3 - (x-1)^4/4$
(b) $1 + (x-e)/e - (x-e)^2/2e^2 + (x-e)^3/3e^3 - (x-e)^4/4e^4$
(c) $\ln 2 + (x-2)/2 - (x-2)^2/8 + (x-2)^3/24 - (x-2)^4/64$
51. $\ln 2 + x/2 + x^2/8 - x^4/192 + \cdots$
53. $\sin 1 + (\cos 1)x + [(\cos 1 - \sin 1)/2]x^2 - [(\sin 1)/2]x^3 + \cdots$
55. (a) 0.5869768
(b) It is within $1/1000$ of $\sin 36°$.
(c) $36° = \pi/5$ radians, and she used the first two terms of the Taylor expansion.
(d) Use the fact that $10° = \pi/18$ radians.
57. (a) $0, -1/3, 0$
(b) $1 - x^2/3! + x^4/5! - x^6/7! + \cdots$
59. Follow the method of Example 3(d).

12.6 Complex Numbers

1. $-i$
3. $-i$
5.

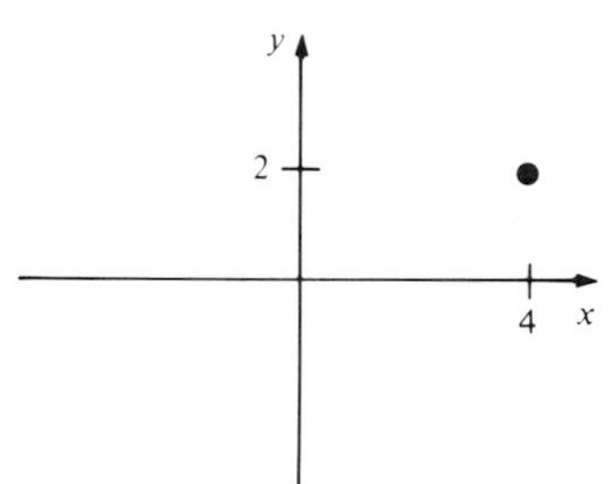

7.

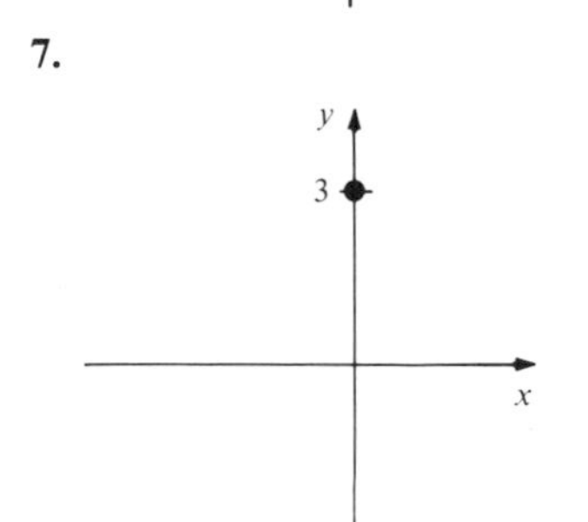

9.

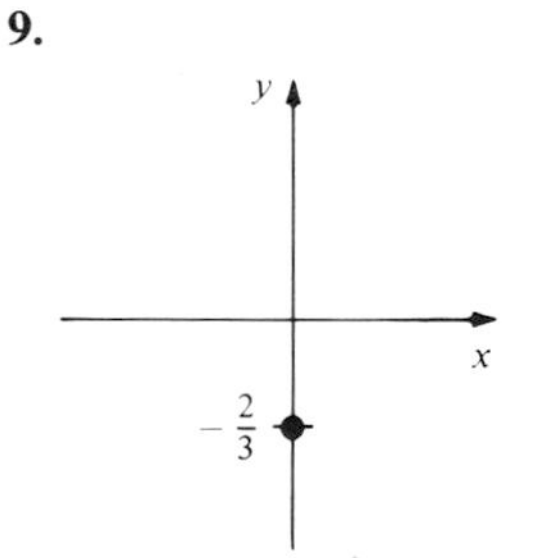

11.

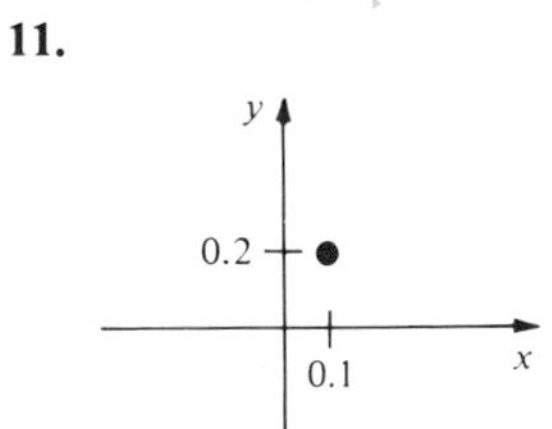

13. $-14 + 8i$
15. $3 + 4i$
17. $(5 + 3i)/34$
19. $(41 + 3i)/65$
21. $\pm\sqrt{3}$
23. $(1 \pm \sqrt{17}\,i)/6$
25. $(7 \pm \sqrt{53})/2$
27. $\pm 2(1 + i)$
29. $\pm 2\sqrt{2}(i - 1)$
31. -1

33. $-11/5$
35. $328/565$
37. $5-2i$
39. $\sqrt{3}-i/2$
41. $-1/3+2i/3$
43. $(-7+11i)/20$
45. 3
47. $|z|=\sqrt{2}$, $\theta=5\pi/4$

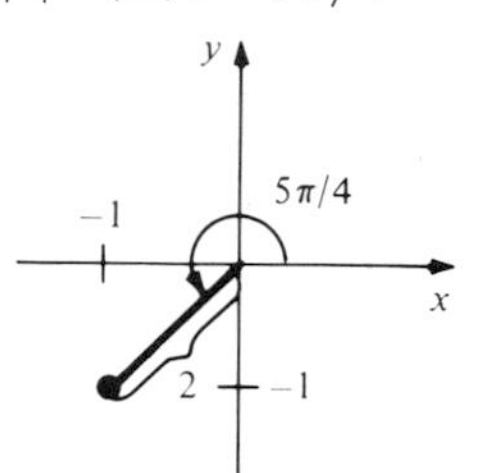

49. $|z|=2$, $\theta=0$

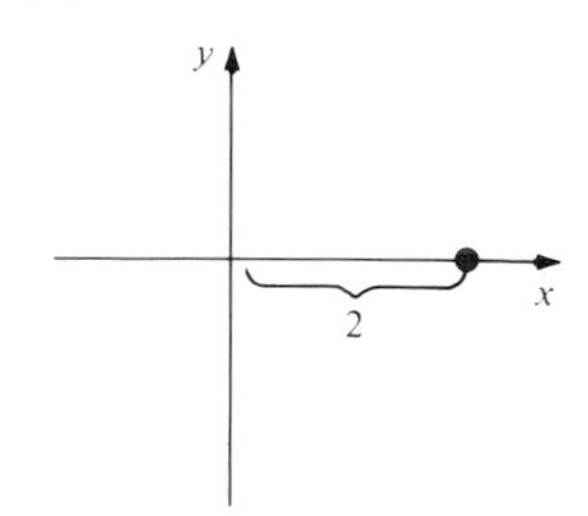

51. $|z|=5/6$, $\theta=-0.93$

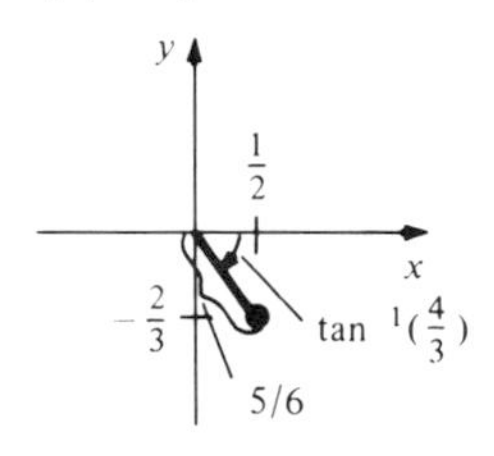

53. $|z|=\sqrt{74}$, $\theta=2.19$

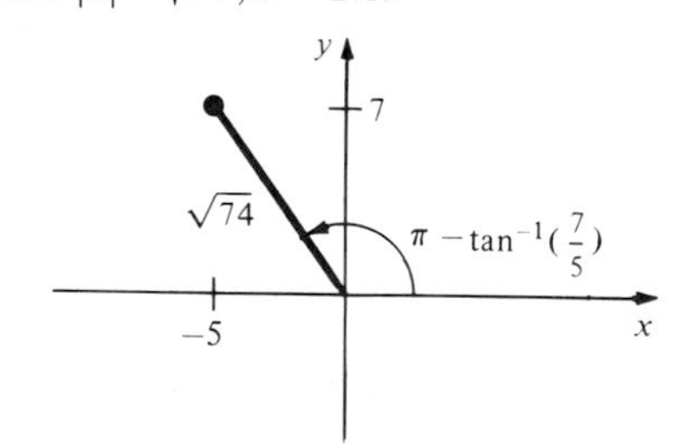

55. $|z|=\sqrt{68}$, $\theta=-2.9$ or 3.4

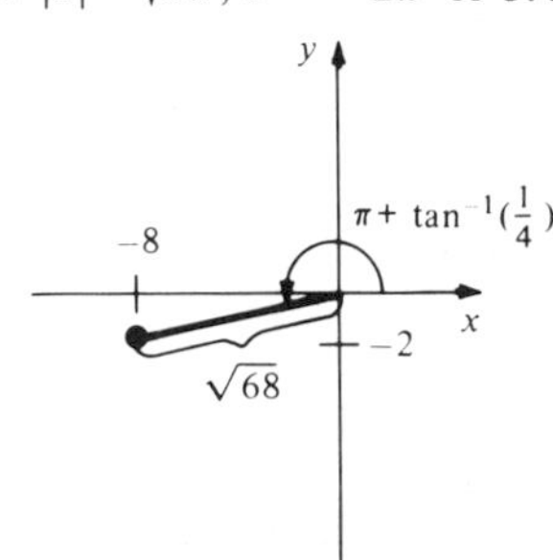

57. $|z|=\sqrt{1.93}$, $\theta=0.53$

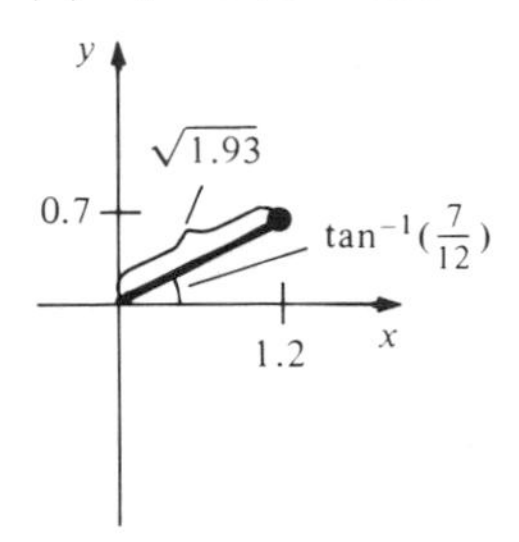

59. Let $z_1=a+ib$, $z_2=c+id$ and calculate $|z_1z_2|$ and $|z_1|\cdot|z_2|$.
61. $(8+3i)^4$
63.

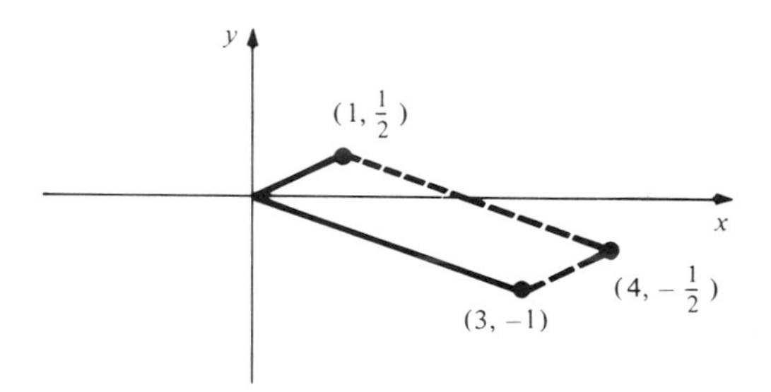

65.

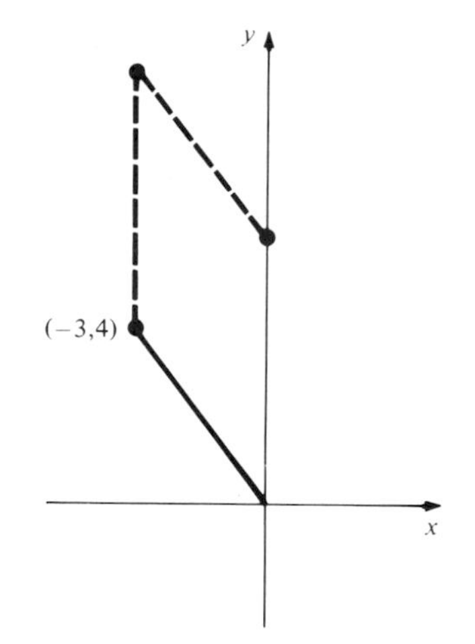

67. $2\sqrt{5}$
69. e^x, y
71. $1/2+\sqrt{3}\,i/2$
73. $(\sqrt{2}/2)(-1+i)$
75. ei
77. $(3-4i)/25$
79. (a) $e^{ix}\cdot e^{-ix}=(\cos x+i\sin x)(\cos x-i\sin x)$; multiply out
(b) Show $e^z\cdot e^{-z}=1$ using (a).
81. Show $e^{3\pi i/2}=-i$.
83. Use $(e^{i\theta})^n=e^{in\theta}$.
85. $\sqrt{2}\,e^{i\pi/4}$
87. $(\sqrt{5}/5)e^{i(0.46)}$
89. $\sqrt{58}\,e^{i(-0.4)}$

91. $(\sqrt{37}/2)e^{i(-1.74)}$
93. $25e^{i(1.83)}$
95. $e^{i(\pi/15+2\pi k/5)}$, $k = 0, 1, 2, 3, 4$; $(1.08)\, e^{i(0.22+2\pi k/5)}$, $k = 0, 1, 2, 3, 4$

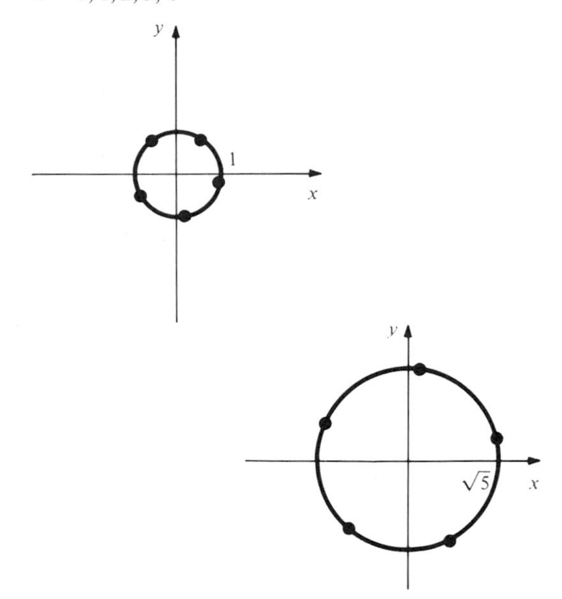

97. $\sqrt[6]{14}\, e^{i(0.155+2\pi k/6)}$, $k = 0, 1, 2, 3, 4, 5$; $\sqrt[6]{14}\, e^{i(0.11+2\pi k/6)}$, $k = 0, 1, 2, 3, 4, 5$

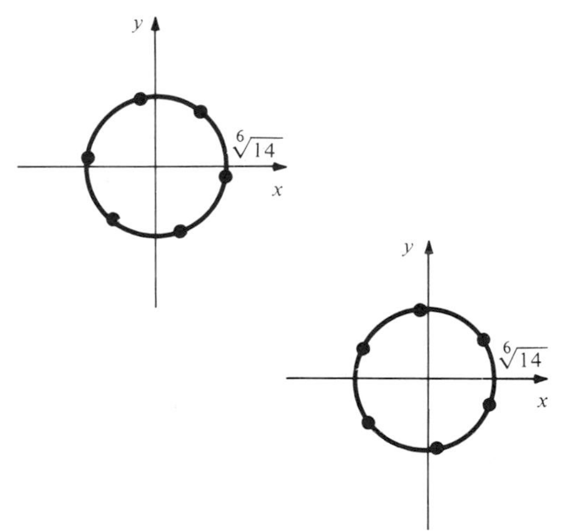

99. z is rotated by $\pi/4$ and its length multiplied by $1/\sqrt{2}$.
101. Show that $z^4 = 1$ and then that $z^2 = 1$.
103. Write $e^{i\theta} = \cos\theta + i\sin\theta$.
105.
$$\frac{1}{2}\left(\sqrt{2}\,z + \frac{1}{\sqrt{\sqrt{2}-1}} - i\sqrt{\sqrt{2}-1}\right) \times \left(\sqrt{2}\,z - \frac{1}{\sqrt{\sqrt{2}-1}} + i\sqrt{\sqrt{2}-1}\right)$$
107. $(z + 2i + 2)(z - 2)$
109. (a) $\tan i\theta = i\tanh\theta$ (b) $\tan i\theta = (\tanh\theta)e^{i\pi/2}$
111. $z_1 = aiz_2$, a a real number
113. (a) Factor $z^n - 1$
(b) Use your factorization in (a)
(c) $-1, i, -i$
115. The motion of the moon with the sun at the origin.
117. (a) $2\pi ni$ for any integer n.
(b) You could define $\ln(-1) = i\pi$, although there are other possibilities.

12.7 Second-Order Linear Differential Equations

1. $y = c_1\exp(3x) + c_2\exp(x)$
3. $y = c_1\exp(x/3) + c_2\exp(x)$
5. $y = \frac{1}{2}\exp(3x) - \frac{1}{2}\exp(x)$
7. $y = e^x$
9. $y = c_1\exp[(2+i)x] + c_2\exp[(2-i)x] = \exp(2x)[a_1\cos x + a_2\sin x]$
11. $y = c_1\exp[(3+2i)x] + c_2\exp[(3-2i)x] = \exp(3x)[a_1\cos 2x + a_2\sin 2x]$
13. $y = x\exp(3x)$
15. $y = (x-1)\exp(-\sqrt{2} + \sqrt{2}\,x)$
17. (a) Underdamped
(b) $x = (1/\bar{\omega})(\sin\bar{\omega}t)\exp(-\pi t/32)$, $\bar{\omega} = \pi\sqrt{255}/32 \approx \pi/2$.

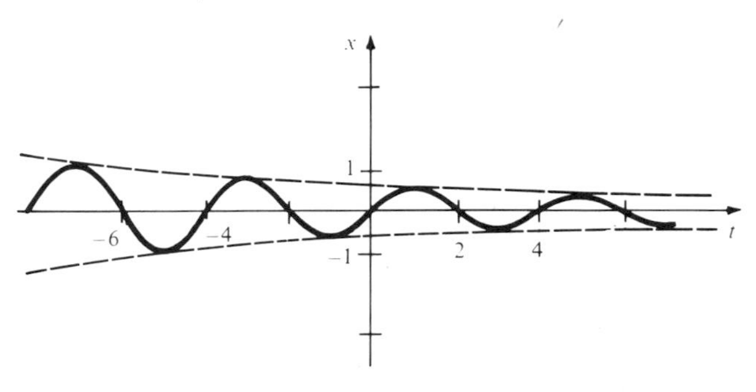

19. (a) Critically damped
(b) $x = t\exp(-\pi t/6)$

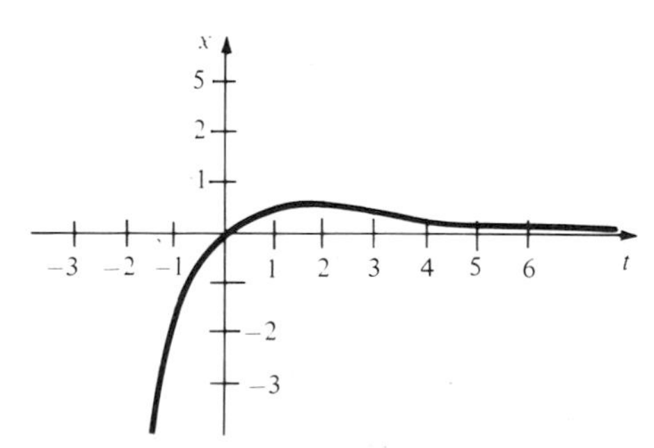

21. $y = c_1\exp(3x) + c_2\exp(x) + 2x + 6$.
23. $x = c_1\exp(t/3) + c_2\exp(t) + (2/5)\cos t + (-1/5)\sin t$
25. $y = e^{2x}(c_1\cos x + c_2\sin x) + x^2/5 + 13x/25 + 42/125$
27. $y = (c_1 + c_2 x)\exp(\sqrt{2}\,x) + [(1+2\sqrt{2})/9]\cos x + [(1-\sqrt{2})/9]\sin x$

29. $y = c_1\exp(3x) + c_2\exp(x) + 2x + 6$

31. $x = c_1\exp(t/3) + c_2\exp(t) - \sin t/5 - 2\cos t/5$

33. $y = c_1\exp(3x) + c_2\exp(x) + [\exp(3x/2)]\int(\tan x)\exp(-3x)\,dx - [\exp(x)/2]\int(\tan x)\exp(-x)\,dx$

35. $y = e^{2x}(C_1\cos 2x + C_2\sin 2x) + [e^{2x}\cos 2x/2]\cdot\int\{e^{2x}[(1-\cot 2x)(\cos 2x) - (1+\cot 2x)(\sin 2x)]\cdot(1+\cos^2 x)\}^{-1}dx + [e^{2z}\sin 2x/2]\int\{e^{2x}[(1-\tan 2x)\cdot(\cos 2x) + (1+\tan 2x)(\sin 2x)](1+\cos^2 x)\}^{-1}dx$

37. $x = -\cos 2t + \cos t = 2\sin(3t/2)\sin(t/2)$

39. $x = (-1/24)\cos 5t + (1/5)\sin 5t + (1/24)\cos t$

41. (a)
$$x(t) = e^{-4t}\left[\frac{-40}{101\sqrt{21}}\sin(\sqrt{21}\,t) - \frac{42}{505}\cos(\sqrt{21}\,t)\right] + \frac{2}{\sqrt{505}}\cos\left[2t - \tan^{-1}\left(\frac{8}{21}\right)\right]$$
(b) Looks like $(2/\sqrt{505})\cos(2t - \tan^{-1}(8/21))$

43. (a) $x(t) = \exp(-t/2)[(7/10)\cos(\sqrt{15}\,t/2) + (-1/2\sqrt{15})\cdot\sin(\sqrt{15}\,t/2)] + (1/\sqrt{10})\cos(t - \tan^{-1}(1/3))$
(b) Looks like $(1/\sqrt{10})\cos(t - \tan^{-1}(1/3))$.

45. Show that the Wronskian of y_1 and y_2 does not vanish.

47. (a) Subtract two solutions with the same initial conditions.
(b) Show that they are zero when $x = 0$.
(c) Solve algebraically for $y(x)$.

49. (a) Compute the derivative of the Wronskian
(b) If $(\alpha - 1)^2 \neq 4\beta$ and r_1, r_2 are roots, then $y = c_1x^{r_1} + c_2x^{r_2}$; if $(\alpha-1)^2 = 4\beta$ and r is the root, then $y = c_1x^r + c_2x^{(1-\alpha)/2}\ln x$. (Assume $x > 0$ in each case).

51. (a) Add all three forces
(b) Substitute and differentiate.

53. $c_1e^{\lambda} + c_2e^{i\lambda} + c_3e^{-\lambda} + c_4e^{-i\lambda}$ where $\lambda = (1+i)/\sqrt{2}$ or
$$e^{x/\sqrt{2}}\left[b_1\cos\left(\frac{x}{\sqrt{2}}\right) + b_2\sin\left(\frac{x}{\sqrt{2}}\right)\right] + e^{-x/\sqrt{2}}\left[b_3\cos\left(\frac{x}{\sqrt{2}}\right) + b_4\sin\left(\frac{x}{\sqrt{2}}\right)\right]$$

55. $\frac{1}{2}e^x + f(x)$, where $f(x)$ is the solution to Exercise 53.

12.8 Series Solutions of Differential Equations

1. $$y = a_0\left[\sum_{1=0}^{\infty}\frac{x^{2n}}{2^n n!}\right] + a_1\left[\sum_{n=0}^{\infty}\frac{2^n n!}{(2n+1)!}x^{2n+1}\right]$$

3. $$y = a_0 + a_1x + a_1\left[\sum_{n=1}^{\infty}\frac{(-1)^n x^{2n+1}}{(2n+1)(n+1)!}\right]$$

5. $y = x - x^3/3 + x^5/10 - \cdots$

7. $y = 2x - x^3 + 7x^5/20 - \cdots$

9. $$y = a_0\left(1 + \frac{x^3}{2} + \frac{x^6}{60} + \cdots\right) + a_1\left(x + \frac{x^4}{12} + \frac{x^7}{504} + \cdots\right)$$
The recursion relation is
$$a_{n+3} = a_n/[(n+3)(n+2)]$$

11. $$y = c_1\left(1 + x - \frac{x^2}{4} + \frac{x^3}{60} - \frac{x^4}{1920} + \cdots\right) + c_2\left(x^{4/3} - \frac{x^{7/3}}{7} + \frac{x^{10/3}}{140} - \frac{x^{13/3}}{5460} + \cdots\right)$$

13. $c_1x^{(1+11i)/6} + c_2x^{(1-11i)/6}$,
or $x^{1/6}\left[b_1\cos\left(\frac{11\ln x}{6}\right) + b_2\sin\left(\frac{11\ln x}{6}\right)\right]$ (no further terms).

15. (a) $$x^k + \frac{x^{k+2}}{4k+4} + \frac{x^{k+4}}{(4k+4)(8k+16)} + \cdots + \frac{x^{k+2j}}{4^j(k+1)(2k+4)\cdots(jk+j^2)} + \cdots$$
(b) $$x^{-k} + \frac{x^{-k+2}}{-4k+4} + \frac{x^{-k+4}}{(-4k+4)(-8k+16)} + \cdots + \frac{x^{-k+2j}}{4^j(-k+1)(-2k+4)\cdots(-jk+j^2)} + \cdots$$

17. Solve recursively for coefficients, then recognize the series for sine and cosine.

19. (a) Use the ratio test
(b) $x,\ -\frac{1}{2} + \frac{3}{2}x^2,\ -\frac{3}{2}x + \frac{5}{2}x^3$

21. Show that the Wronskian is non-zero

23. (a) Solve recursively
(b) Substitute the given function in the equation. (To discover the solution, use the methods for solving first order linear equations given in Section 8.6).

Review Exercises for Chapter 12

1. Converges to $1/11$.
3. Converges to $45/2$
5. Converges to $7/2$
7. Diverges
9. Converges
11. Converges
13. Converges
15. Diverges
17. Diverges
19. Converges
21. Converges

23. Converges
25. 0.78
27. -0.12
29. -0.24
31. 0.25
33. False
35. False
37. False
39. False
41. True
43. True
45. True
47. Use the comparison test.
49. $1/8$
51. 1
53. $R = \infty$
55. $R = \infty$
57. $R = 2$
59. $f(x) = \sum_{n=0}^{\infty} a_n x^n$, where $a_{2i} = \dfrac{2^{2i}}{(2i)!}$, $a_{2i+1} = \dfrac{(-1)^i 3^{2i+1} + 2^{2i+1}}{(2i+i)!}$
61. $\sum_{i=1}^{\infty}[(-1)^{i+1}x^{4i}/i]$
63. $\sum_{i=1}^{\infty}[(-1)^i x^{2i}/(2i)!]$
65. $\sum_{i=1}^{\infty}[x^i/[i(i!)]]$
67. $\sum_{i=0}^{\infty}[e^2(x-2)^i/i!]$, $R = \infty$
69. $\displaystyle\sum_{i=0}^{\infty} \frac{\frac{3}{2}(\frac{3}{2}-1)\cdots(\frac{3}{2}-i+1)}{i!}(x-1)^i$
71. $\pi^2/2$
73. 3
75. $3, 7, 3-7i, \sqrt{58}$
77. $\pm(1+\frac{1}{2}\sqrt{5})^{1/2} \approx \pm 1.46$, $\mp(-1+\frac{1}{2}\sqrt{5})^{1/2} \approx \mp 0.344$, $\sqrt{2+i} \approx \pm 1.46 \pm 0.344i$, $\sqrt[4]{5} \approx 1.50$
79. $z = \sqrt{2}\exp(-\pi i/4)$

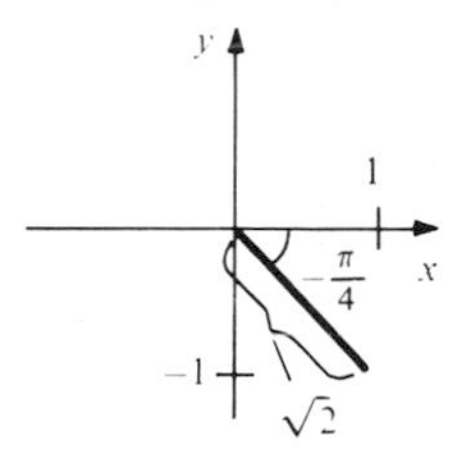

81. $z = \exp(\pi i)$

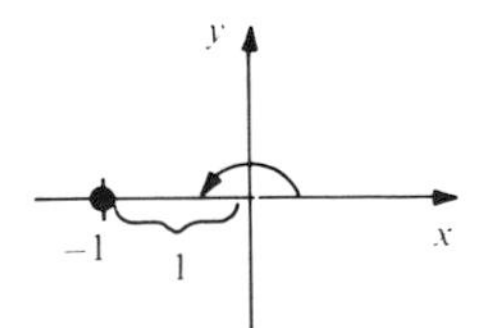

83. $1 \pm \sqrt{1-\pi i} \approx 2.4658 - 1.0717i$, $-0.4658 + 1.0717i$
85. $c_1\cos 2x + c_2\sin 2x$
87. $y = c_1\exp(-5x) + c_2\exp(-x)$
89. $y = -e^x/6 - 11\cos x/130 + 33\sin x/130 + c_1\exp(-5x) + c_2\exp(2x)$
91. $c_1 e^{3x} + c_2 x e^{3x} + \dfrac{140}{1369}\cos\left(\dfrac{x}{2}\right) - \dfrac{48}{1369}\sin\left(\dfrac{x}{2}\right)$
93. $-\cos(2x)\displaystyle\int \frac{x\sin(2x)}{\sqrt{x^2+1}}\,dx + \sin(2x)\int \frac{x\cos(2x)}{\sqrt{x^2+1}}\,dx$
95. $c_1 + e^{-x}(c_2\cos x + c_3\sin x)$
97. $m = 1$, $k = 9$, $\gamma = 1$, $F_0 = 1$, $\Omega = 2$, $\omega = 3$, $\delta = \tan^{-1}(\frac{1}{2})$. As $t \to \infty$, the solution approaches $\frac{1}{\sqrt{29}}\cos[2t - \tan^{-1}(\frac{2}{5})]$.
99. $m = 1$, $k = 25$, $\gamma = 6$, $F_{-} = 1$, $\Omega = \pi$, $\omega = 5$, $\delta = \tan^{-1}[6\pi/(25-\pi^2)]$. As $t \to \infty$, the solution approaches
$$\frac{1}{\sqrt{625 - 14\pi^2 - \pi^4}}\cos\left[\pi t - \tan^{-1}\left(\frac{6\pi}{25-\pi^2}\right)\right]$$
101. $a_0\left(1 - \dfrac{x^3}{3} + \cdots\right) + a_1\left(x - \dfrac{x^4}{6} + \cdots\right)$
103. $a_0\left(1 - x^2 + \dfrac{x^4}{6} - \dfrac{x^5}{5} + \cdots\right) + a_1\left(x - \dfrac{x^3}{3} + \dfrac{x^4}{6} - \dfrac{x^5}{20} + \cdots\right)$
105. $1 - x + \dfrac{x^2}{2} - \dfrac{11x^3}{6} + \cdots$
107. $x + \dfrac{x^3}{8} + \dfrac{x^5}{192} + \cdots$
109. (a) $m = L$, $k = 1/C$, $\gamma = R$
(b) $-e^{-10t}[.0000111\cos(7\sqrt{2}\,t) + 0.0000224\sin(7\sqrt{2}\,t) + 0.0000112\cos(60\pi t - 0.105711)]$
111. Factor out x^2.
113. (a) The partial sums converge to $y(x,t)$ for each (x,t).
(b) $\sum_{k=0}^{\infty}(-1)^k A_{2k+1}$
115. ≈ 0.659178
117. (a) ≈ 1.12
(b) ≈ 2.24. It is accurate to within 0.02.
119. $-1/2, 1/6, 0$
121. (a) ≈ 3.68
(b)

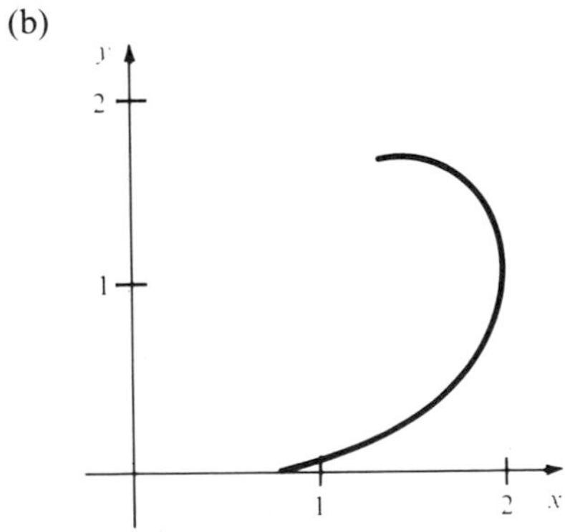

123. Show by induction that $g^{(n)}(x)$ is a polynomial times $g(x)$.
125. True
127. (a) Collect terms
(b) The radius of convergence is at most 1.
(c) e
129. Show that $\alpha < 1/k$ by using a Maclaurin series with remainder.

Index
Includes Volumes I and II

Note: Pages 1–336 refer to Volume I; pages 337–644 refer to Volume II.

Undergraduate Texts in Mathematics

Undergraduate Texts in Mathematics

Prenowitz/Jantosciak: Join Geometrics: A Theory of Convex Sets and Linear Geometry.

Priestly: Calculus: An Historical Approach.

Protter/Morrey: A First Course in Real Analysis.

Protter/Morrey: Intermediate Calculus.

Ross: Elementary Analysis: The Theory of Calculus.

Scharlau/Opolka: From Fermat to Minkowski: Lectures on the Theory of Numbers and Its Historical Development.

Sigler: Algebra.

Simmonds: A Brief on Tensor Analysis.

Singer/Thorpe: Lecture Notes on Elementary Topology and Geometry.

Smith: Linear Algebra. Second edition.

Smith: Primer of Modern Analysis.

Thorpe: Elementary Topics in Differential Geometry.

Troutman: Variational Calculus with Elementary Convexity.

Wilson: Much Ado About Calculus: A Modern Treatment with Applications Prepared for Use with the Computer.

29. $\int \operatorname{csch} x\,dx = \ln\left|\tanh\frac{x}{2}\right| = -\frac{1}{2}\ln\frac{\cosh x + 1}{\cosh x - 1}$

30. $\int \sinh^2 x\,dx = \frac{1}{4}\sinh 2x - \frac{1}{2}x$

31. $\int \cosh^2 x\,dx = \frac{1}{4}\sinh 2x + \frac{1}{2}x$

32. $\int \operatorname{sech}^2 x\,dx = \tanh x$

33. $\int \sinh^{-1}\frac{x}{a}\,dx = x\sinh^{-1}\frac{x}{a} - \sqrt{x^2 + a^2} \qquad (a > 0)$

34. $\int \cosh^{-1}\frac{x}{a}\,dx = \begin{cases} x\cosh^{-1}\frac{x}{a} - \sqrt{x^2 - a^2} & \left[\cosh^{-1}\left(\frac{x}{a}\right) > 0,\ a > 0\right] \\ x\cosh^{-1}\frac{x}{a} + \sqrt{x^2 - a^2} & \left[\cosh^{-1}\left(\frac{x}{a}\right) < 0,\ a > 0\right] \end{cases}$

35. $\int \tanh^{-1}\frac{x}{a}\,dx = x\tanh^{-1}\frac{x}{a} + \frac{a}{2}\ln|a^2 - x^2|$

36. $\int \frac{1}{\sqrt{a^2 + x^2}}\,dx = \ln(x + \sqrt{a^2 + x^2}) = \sinh^{-1}\frac{x}{a} \qquad (a > 0)$

37. $\int \frac{1}{a^2 + x^2}\,dx = \frac{1}{a}\tan^{-1}\frac{x}{a} \qquad (a > 0)$

38. $\int \sqrt{a^2 - x^2}\,dx = \frac{x}{2}\sqrt{a^2 - x^2} + \frac{a^2}{2}\sin^{-1}\frac{x}{a} \qquad (a > 0)$

39. $\int (a^2 - x^2)^{3/2}\,dx = \frac{x}{8}(5a^2 - 2x^2)\sqrt{a^2 - x^2} + \frac{3a^4}{8}\sin^{-1}\frac{x}{a} \qquad (a > 0)$

40. $\int \frac{1}{\sqrt{a^2 - x^2}}\,dx = \sin^{-1}\frac{x}{a} \qquad (a > 0)$

41. $\int \frac{1}{a^2 - x^2}\,dx = \frac{1}{2a}\ln\left|\frac{a + x}{a - x}\right|$

42. $\int \frac{1}{(a^2 - x^2)^{3/2}}\,dx = \frac{x}{a^2\sqrt{a^2 - x^2}}$

43. $\int \sqrt{x^2 \pm a^2}\,dx = \frac{x}{2}\sqrt{x^2 \pm a^2} \pm \frac{a^2}{2}\ln|x + \sqrt{x^2 \pm a^2}|$

44. $\int \frac{1}{\sqrt{x^2 - a^2}}\,dx = \ln|x + \sqrt{x^2 - a^2}| = \cosh^{-1}\frac{x}{a} \qquad (a > 0)$

45. $\int \frac{1}{x(a + bx)}\,dx = \frac{1}{a}\ln\left|\frac{x}{a + bx}\right|$

46. $\int x\sqrt{a + bx}\,dx = \frac{2(3bx - 2a)(a + bx)^{3/2}}{15b^2}$

47. $\int \frac{\sqrt{a + bx}}{x}\,dx = 2\sqrt{a + bx} + a\int \frac{1}{x\sqrt{a + bx}}\,dx$

48. $\int \frac{x}{\sqrt{a + bx}}\,dx = \frac{2(bx - 2a)\sqrt{a + bx}}{3b^2}$

49. $\int \frac{1}{x\sqrt{a + bx}}\,dx = \frac{1}{\sqrt{a}}\ln\left|\frac{\sqrt{a + bx} - \sqrt{a}}{\sqrt{a + bx} + \sqrt{a}}\right| \qquad (a > 0)$

$\qquad = \frac{2}{\sqrt{-a}}\tan^{-1}\sqrt{\frac{a + bx}{-a}} \qquad (a < 0)$

50. $\int \frac{\sqrt{a^2 - x^2}}{x}\,dx = \sqrt{a^2 - x^2} - a\ln\left|\frac{a + \sqrt{a^2 - x^2}}{x}\right|$

51. $\int x\sqrt{a^2 - x^2}\,dx = -\frac{1}{3}(a^2 - x^2)^{3/2}$

52. $\int x^2\sqrt{a^2 - x^2}\,dx = \frac{x}{8}(2x^2 - a^2)\sqrt{a^2 - x^2} + \frac{a^4}{8}\sin^{-1}\frac{x}{a} \qquad (a > 0)$

Continued on overleaf.

53. $\int \frac{1}{x\sqrt{a^2-x^2}}\,dx = -\frac{1}{a}\ln\left|\frac{a+\sqrt{a^2-x^2}}{x}\right|$

54. $\int \frac{x}{\sqrt{a^2-x^2}}\,dx = -\sqrt{a^2-x^2}$

55. $\int \frac{x^2}{\sqrt{a^2-x^2}}\,dx = -\frac{x}{2}\sqrt{a^2-x^2} + \frac{a^2}{2}\sin^{-1}\frac{x}{a} \qquad (a>0)$

56. $\int \frac{\sqrt{x^2+a^2}}{x}\,dx = \sqrt{x^2+a^2} - a\ln\left|\frac{a+\sqrt{x^2+a^2}}{x}\right|$

57. $\int \frac{\sqrt{x^2-a^2}}{x}\,dx = \sqrt{x^2-a^2} - a\cos^{-1}\frac{a}{|x|}$

$= \sqrt{x^2-a^2} - a\sec^{-1}\left(\frac{x}{a}\right) \qquad (a>0)$

58. $\int x\sqrt{x^2\pm a^2}\,dx = \frac{1}{3}(x^2\pm a^2)^{3/2}$

59. $\int \frac{1}{x\sqrt{x^2+a^2}}\,dx = \frac{1}{a}\ln\left|\frac{x}{a+\sqrt{x^2+a^2}}\right|$

60. $\int \frac{1}{x\sqrt{x^2-a^2}}\,dx = \frac{1}{a}\cos^{-1}\frac{a}{|x|} \qquad (a>0)$

61. $\int \frac{1}{x^2\sqrt{x^2\pm a^2}}\,dx = \mp\frac{\sqrt{x^2\pm a^2}}{a^2x}$

62. $\int \frac{x}{\sqrt{x^2\pm a^2}}\,dx = \sqrt{x^2\pm a^2}$

63. $\int \frac{1}{ax^2+bx+c}\,dx = \frac{1}{\sqrt{b^2-4ac}}\ln\left|\frac{2ax+b-\sqrt{b^2-4ac}}{2ax+b+\sqrt{b^2-4ac}}\right| \qquad (b^2>4ac)$

$= \frac{2}{\sqrt{4ac-b^2}}\tan^{-1}\frac{2ax+b}{\sqrt{4ac-b^2}} \qquad (b^2<4ac)$

64. $\int \frac{x}{ax^2+bx+c}\,dx = \frac{1}{2a}\ln|ax^2+bx+c| - \frac{b}{2a}\int\frac{1}{ax^2+bx+c}\,dx$

65. $\int \frac{1}{\sqrt{ax^2+bx+c}}\,dx = \frac{1}{\sqrt{a}}\ln|2ax+b+2\sqrt{a}\sqrt{ax^2+bx+c}| \qquad (a>0)$

$= \frac{1}{\sqrt{-a}}\sin^{-1}\frac{-2ax-b}{\sqrt{b^2-4ac}} \qquad (a<0)$

66. $\int \sqrt{ax^2+bx+c}\,dx = \frac{2ax+b}{4a}\sqrt{ax^2+bx+c} + \frac{4ac-b^2}{8a}\int\frac{1}{\sqrt{ax^2+b+c}}\,dx$

67. $\int \frac{x}{\sqrt{ax^2+bx+c}}\,dx = \frac{\sqrt{ax^2+bx+c}}{a} - \frac{b}{2a}\int\frac{1}{\sqrt{ax^2+bx+c}}\,dx$

68. $\int \frac{1}{x\sqrt{ax^2+bx+c}}\,dx = \frac{-1}{\sqrt{c}}\ln\left|\frac{2\sqrt{c}\sqrt{ax^2+bx+c}+bx+2c}{x}\right| \qquad (c>0)$

$= \frac{1}{\sqrt{-c}}\sin^{-1}\frac{bx+2c}{|x|\sqrt{b^2-4ac}} \qquad (c<0)$

69. $\int x^3\sqrt{x^2+a^2}\,dx = \left(\frac{1}{5}x^2-\frac{2}{15}a^2\right)\sqrt{(a^2+x^2)^3}$

70. $\int \frac{\sqrt{x^2\pm a^2}}{x^4}\,dx = \frac{\mp\sqrt{(x^2\pm a^2)^3}}{3a^2x^3}$

71. $\int \sin ax\sin bx\,dx = \frac{\sin(a-b)x}{2(a-b)} - \frac{\sin(a+b)x}{2(a+b)} \qquad (a^2\neq b^2)$

Continued on inside back cover.